3 0250 01111 6451
D0768615
WITHDRAWN

Landmarks in American Civil Engineering

This book was set in Helvetica and Gill Sans by Achorn Graphic Services and printed and bound by Halliday Lithograph in the United States of America.

Library of Congress Cataloging-in-Publication Data

Schodek, Daniel L., 1941–
Landmarks in American civil engineering.

Bibliography: p.
Includes index.
1. Civil engineering—United States. 2. United States—Public works. I. Title.
TA23.S36 1987 624′.0973 86-20162
ISBN 0-262-19256-X

Landmarks in American Civil Engineering

Daniel L. Schodek

MIT Press
Cambridge, Massachusetts
London, England

Contents

Chapter Five
Tunnels

Chapter Six
Water Supply and Control

Chapter Seven
Environmental Engineering

Chapter Eight
Dams

Chapter Nine
Buildings

Chapter Ten
Urban Planning

Chapter Eleven
Power Systems

Chapter Twelve
Surveying and Mapping

Chapter Thirteen
Coastal Facilities

Chapter Fourteen
Airports

Preface

The works described in this book are those that the civil engineering profession in America regards as its most remarkable and important achievements. They were drawn from a group of projects designated as landmarks by the American Society of Civil Engineers (ASCE), through a complex nominating and screening procedure that assures the significance of each. Although any process for specifying the most notable among the country's many civil engineering works must be judgmental, the judgments made in this instance represent the collective actions of people who not only know the field of civil engineering but also have a sense of the traditions of the profession. The book thus provides an insight into the values of the profession itself.

The projects were designated as landmarks for different reasons. Many are well known and their significance is fairly obvious; others seem, at first glance, humble at best. They frequently represent the first work of their kind, or the first successful trial of a new technology on a significant scale. At the time of their completion they may have set a standard for size, capacity, efficiency, or construction expediency. Projects were sometimes selected as significant because of the way they were built rather than the novelty of their design: many entailed new techniques of assembly or construction, new methods of financial or political cooperation. They may stand as records of perseverance over natural obstacles or community doubt. Most projects contributed significantly to the development of a region or even the nation. In most cases the works designated as landmarks fulfilled a combination of these criteria. No rule as to the date of construction was established; but since the importance of a project can generally be ascertained only through the perspective of time, most are from the eighteenth and nineteenth centuries, and the rest date from the early part of the twentieth century. Of each project at least some portion remains.

The ASCE's list of landmarks receives additions as important projects are identified and documented. It has not been possible to include the more newly nominated or designated landmarks in the main body of the text, and some other projects do not appear there because of a paucity of reliable information. Works in these two categories are briefly described in appendix 1. Finally, in a few cases a project has been defined here as including more than the work officially designated a landmark, because of contextual significance.

The projects described were built over a great span of time. In the early examples, persons from many different backgrounds play the role of civil engineer. In later examples, the civil engineer is a trained specialist with a clear professional identity. Yet common attributes are present in the builders of all of these projects. Rather than technical education or professional affiliation, it is their understanding of a need and their perseverance in its fulfillment through the building of a physical artifact that best characterize them as a group. They honed their expertise in the field or patiently sought better ways to analyze problems and create responses. Collectively, they represent an ideal to which civil engineers have traditionally aspired, even before all practitioners called themselves by the same name.

The accounts that follow thus also describe the changing definition of the civil engineer's role and responsibilities in our society through a look at specific actions in specific situations. Together they sketch the emergence of a professional identity in the context of solving a succession of problems. The landmark projects themselves in turn show how a profession's development fits within a broader history of national growth.

Acknowledgments

This book has been made possible through the active cooperation of the Committee on the History and Heritage of American Civil Engineering, a group established by the American Society of Civil Engineers. The committee is responsible not only for the designation of civil engineering landmarks but for maintaining archival material and general informational records on them as well. Individuals from across the country have contributed to these archives during the process of the designation of landmarks. Without this material it would have been virtually impossible to find documentation on many of the projects described. Special thanks are extended to Herbert R. Hands of the ASCE for his unflagging support in the development of the manuscript, and as general keeper of the archives. Neal FitzSimons, chairman of the committee, also lent not only his personal support but his time as well in reviewing the manuscript. He has consistently been in the forefront of endeavors to increase both the public's and the profession's awareness of the civil engineering heritage, specifically spearheading the effort to have important projects designated as landmarks.

Many individuals from local history and heritage committees in regional ASCE chapters participated in the preparation of the descriptive documents required as part of the landmark nomination and designation process. Much of this material made its way into the manuscript in one form or another, and I would like to acknowledge the often anonymous authors of these documents. Many other people from these same regional groups participated in reviewing various parts of the manuscript, as did engineers in charge of still-functioning projects. Since it is these local individuals who often know their projects the best, their cooperation was invaluable. Over forty such people participated in the review process, and their help is greatly appreciated.

Another indispensable source of information and illustrations was the Historic American Engineering Record (HAER), under the direction of Robert Kapsch. Established in 1969 by a tripartite agreement among the National Park Service, the American Society of Civil Engineers, and the Library of Congress, HAER reflects the federal government's recognition of our technological heritage. Its archives ensure that if projects are lost at least a permanent record remains. The Library of Congress maintains these records and makes them available to the general public as research material.

The collections of the Smithsonian Institution also proved invaluable, as did specific monographs prepared by Smithsonian members. The assistance of Robert Vogel is appreciated.

A number of other people, particularly students at the Harvard Graduate School of Design, were also involved in the research and writing phases; they include Joanne Aitken, Art Vogt, Kevin Ames, David McKenna, Hugh Morrow, Phil Rizzo, and many others. Kay, Ned, and Ben Schodek also gave their special form of support.

Introduction

Civil Engineering in America

The Brooklyn Bridge and the Erie Canal are among the handful of civil engineering feats that are well known and of generally recognized importance. Others, equally remarkable and of comparable significance, hold a less conspicuous place in the public consciousness. The Transcontinental Railroad, for example, played a vital role in the development of the United States. Many lesser known projects were of great regional or local importance. The immediate effect of a new canal or road in a region, for instance, was normally quite dramatic. The new artery would bring a sharp increase in trade between settlements along its route by cutting the time and cost of shipment. The population of the settlements themselves would often increase, or the distribution of population would be shifted from tributary areas to towns along the pathway. The migration of laborers for the project sometimes spawned new communities, which sustained themselves by other industries served by the new artery after the work itself had stopped. The economic focus of the region might shift to especially important points along the artery or to its endpoints, with some locations assuming specific roles as producers, processors, or distributors.

The engineers of projects such as these were frequently trustees of a public aspiration, and their work often articulated a community's hopes. Many projects became symbolic objects in the landscape. The physical artifact—a canal, tunnel, dam, or bridge—came to represent a point of order in nature, a proof of man's dominion. Such a structure often served as the focus of communal awe and celebration. These projects thus testify both to their makers' skills and to the role of those skills in the development of a society.

Who exactly were the engineers of these works? Where did they come from, and what was their training? The development of civil engineering in America has a special character that is marked by the inventiveness and exuberance of its practitioners on the one hand, and by a willingness to learn from precedents (primarily European) on the other. Traits such as innovativeness, brilliance, and perseverance also spring to mind, as does the phrase "occasional foolhardiness." Over time, practitioners in the field evolved a clear professional identity, with its own standards and mores. At the beginning of its history, however, America had few engineers with any sort of professional training, nor did these few possess any distinct sense of professional identity.

Early projects were often the work of technically minded individuals who were at once entrepreneurs, designers, and builders. These individuals were more or less familiar with European engineering precedents, depending on their own specific backgrounds. Some were of direct European origin and had a clear image of the way the built environment ought to be from this point of view. Others, born in this country, were isolated from these precedents and largely self-taught. Other than in isolated instances in prerevolutionary America, however—such as the carpenters' companies that evolved in Philadelphia and that reflected the prevalent English attitude toward apprenticeship as the proper mode of entry into the professional trades—there was initially little in the way of mechanisms for systematically training new practitioners in the field.

With time, there emerged two distinct thrusts in civil engineering training and education, existing more or less in parallel with one another. The first of these, based on field experience and apprenticeship, reflected the spirit of engineering in early America. This approach reached a high point during the building of the Erie Canal under Benjamin Wright; many individuals, such as John Jervis, went on to become leading civil engineers largely on the strength of what they had learned while working on the canal. Although some who received their technical training on the job had had only limited formal education—one such was the great surveyor Andrew Ellicott—others were university or college graduates. The latter included men like Loammi Baldwin, Jr., who built the dry docks at Boston and Norfolk. This apprenticeship form of education dominated the profession until the mid–nineteenth century. The second major thrust was the development of a formal program of civil engineering instruction that found its roots in the European college and university system of education. French models were particularly influential.

The origins of institutions offering formal education in civil engineering can be traced to the Revolutionary War era. The war created a demand for trained engineers for mapping, the construction of fortifications, and other needs. George Washington agitated for a means of training engineers. In 1775 the Continental Congress appointed Richard Gridley as chief engineer of the army; an additional authorization in 1776 led to the formation of an engineering corps. Lacking trained engineers, Washington and others turned to France for aid. This resulted in Gridley's replacement by Louis Du Portail as chief engineer, but it was not until 1778 that Du Portail was given any significant means to engage his responsibilities. After the war Washington, together with such men as Hamilton, Knox, Adams, Jefferson, and Monroe, argued for the establishment of a school for the practical application of military science. Hamilton in particular became a strong supporter and advocated a military academy with four schools. Finally, in 1794, Congress authorized President Washington to raise fortifications for the protection of harbors and coasts and to create a Corps of Artillerists and Engineers, to be educated and stationed at West Point. The school was destroyed by fire in 1796 and not reopened until 1801. Many political leaders at that time argued for a broader role for the school; and when Congress formally established the United States Military Academy at West Point in March 1802, a provision was included that the engineers of the academy were to be available for such duty and service as the president should direct. This laid the groundwork for the performance of public as well as military tasks.

Alden Partridge was appointed professor of engineering at West Point in 1813—the first position of its kind in the United States. A turning point at the academy came with the appointment of Sylvanus Thayer in 1817. Thayer, recently returned from a study of the educational system in France, reorganized the academy along strong pedagogical lines. Under his direction the curriculum was expanded to include courses in such topics as mechanics, chemistry, hydraulics, and optics. The French-trained engineer Claude Crozet succeeded Partridge, who had left after disagreements with Thayer. Crozet, later builder of a famous tunnel in the Blue Ridge Mountains and influential civil engineer in Virginia, introduced descriptive geometry into his courses in 1817. In 1823 a course entitled "civil engineering" was offered in which students were taught about the construction of bridges, roads, and canals. By 1832 the curriculum included both civil engineering and architecture, with classes in such topics as masonry and technical drawing. Even such subjects as the orders of architecture and perspective shades and shadows were included.

The development of a strong pedagogical approach to the applied sciences at this time was significant. Most of the colleges and universities then existent in America had been founded with the objective of instructing students in the conventional religious and moral beliefs of the day, although classical studies were soon added to most curricula. The real roots of Thayer's program at West Point lay elsewhere—notably in France. The Ecole Nationale des Ponts et Chaussées, founded in 1747, is generally regarded as the first significant engineering school in the world. There had been influential developments in engineering education in Europe, however, prior to the establishment of this institution: Christian Willenberg had instituted a formal course at the university in Prague around 1716, and the French Académie Royale d'Architecture had begun a systematic study of building materials in 1678. But the latter's role in technical education was largely taken over by the Ecole des Ponts et Chaussées. Under the vigorous direction of Jean Rodolphe Perronet, this school made France the leader in engineering education in the eighteenth century. Toward the end of the century, however, two new institutions, the Ecole des Beaux Arts and the Ecole Polytechnique, largely took the place of the Académie Royale d'Architecture and the Ecole des Ponts et Chaussées. By this time the fields of architecture and engineering were diverging in philosophy and intent, beginning a split between the professions that has yet to heal today.

The rise of the Ecole des Ponts et Chaussées and the later Ecole Polytechnique paralleled the emergence of a theoretical understanding of mechanics and other sciences important to engineering. The field of mechanics, for example, had undergone a period of rapid development, beginning with the substantial contributions of such giants as Galileo, Roberval, Mariotte, and Hooke, and furthered by such specific advances as Euler's solution of the buckling problem in 1744 and the development of the equation describing the shape of a catenary curve by Bernoulli (as well as Leibnitz and Huygens). With respect to applications in building, the analysis of St. Peter's dome conducted in 1742–1743 at the request of Pope Benedict XIV was of critical significance. This analysis, made by Le Seur, Jacquier, and Boscowitch, was one of the first major attempts to understand the forces present in building elements through the use of scientific methods. The techniques employed in the study created great controversy and further stimulated interest in the theory of structures. Subsequent work by many individuals—from Coulomb's solution of the bending problem in beams to Louis Marie Henri Navier's brilliant series of lectures on applied mechanics at the Ecole Polytechnique—added to the rapidly growing body of knowledge. Navier was a remarkable man who made numerous contributions to the field through an application of mathematics and scientific principles.

Throughout this period of great theoretical developments, French institutions maintained the leadership in formalizing an approach to engineering education based on scientific principles. Thus the curriculum that Thayer introduced at West Point was influenced by an evolved and tested model acknowledged by many at the time as the best in the world. It should also be noted that not only was the French system indirectly influential, but several notable American engineers—Charles Ellet, for example, builder of the longest suspension bridge in the world at the time of its completion in 1849 in Wheeling, West Virginia—trained at French schools.

The literature in the field was also beginning to develop. In 1823 Crozet translated Sganzin's *Elementary Course in Civil Engineering* (Crozet was a former student of Sganzin and had trained at the Ecole Polytechnique). The work of French theoreticians was also spread by a number of simplified derivative works in English, such as Jacob Bigelow's *Elements of Technology* (Boston, 1831) and James Renwick's *Elements of Mechanics* (Philadelphia, 1832), many of which became familiar to American engineers of the time. Important texts written by American engineers soon appeared. Charles S. Storrow produced *A Treatise on Water-Works* (Boston, 1835), the first

American work on the subject. Dennis H. Mahan, an instructor at West Point, produced a textbook entitled *An Elementary Course of Civil Engineering* (New York, 1838) that he had developed largely from his own lectures to cadets at the academy. In 1847 Squire Whipple published his classic work on bridge engineering, *A Work on Bridge Building,* which provided the first important treatment of truss analysis. Four years later Herman Haupt independently produced his *General Theory of Bridge Construction,* which paralleled Whipple's treatments.

While the French approach was reaching its full flower, the great builders of England pursued a different path in developing new technologies. They did not, by and large, rely on theoretical work; rather, a strong tradition of experimentalism and a more pragmatic approach prevailed. The influential Thomas Telford, for example, builder of the Menai Straits bridge and other significant structures, was a careful workman who learned from his own experiences and by observing the behavior of structures built by others. He also conducted many strength tests of materials before using them. The English tradition undoubtedly had an impact on many American engineers as well.

Still, it was the French system that primarily influenced formal education in civil engineering in America. After Thayer's innovations at West Point, the academy became a significant source of engineering talent in the formative period of the 1820s to the 1840s that saw the introduction of the railroad (with its demands for surveyors, bridge engineers, and other trained personnel) and the expansion of the country's canal and turnpike systems. Other engineering schools arose during the same period. After leaving West Point, Alden Partridge established the American Literary, Scientific, and Military Academy at Norwich, Vermont, where, according to the 1821 catalog, instruction was offered in civil engineering, including the construction of roads, canals, and bridges. In 1834 the academy became Norwich University and awarded two civil engineering degrees. Civil engineering was offered as an elective at the Gardiner Lyceum in Maine, started in 1822 by Robert Gardiner; the school closed in 1832. St. John's College in Maryland offered course work in the subject in 1826.

Stephen Van Rensselaer established the Rensselaer School in 1824 at Troy, New York. Amos Eaton became the school's first director and soon introduced a one-year technical course. Civil engineering was offered in the 1828 catalog. In 1835 the New York legislature authorized the school to establish a department of "Engineering and Technology," which led to the offering of a degree in civil engineering; four such degrees were granted that year. In 1849 the Rensselaer Institute was reorganized by B. Franklin Greene along the lines of the leading French technical schools; subsequently the school became known as the Rensselaer Polytechnic Institute. A curriculum was evolved that featured parallel sequences of studies in mathematics, the physical sciences, technical subjects, and the humanities. This innovative model became a prototype for engineering curricula throughout the country. By the Civil War, a number of schools offered civil engineering courses, including the University of the City of New York (from 1834) and the Virginia Military Institute (from 1839). Union College organized a formal school of engineering in 1845.

Nonetheless, the apprenticeship form of education continued to dominate the profession well into the 1850s. By this point, however, the application of design techniques founded in rational analyses or scientific principles, advocated in the works of engineers such as Squire Whipple and Herman Haupt, had become widely accepted, so that engineers felt an increased need for formal education.

Following the Civil War many schools began offering engineering studies in which civil engineering was a component. The establishment of the Thayer School of Civil Engineering at Dartmouth was notable because of its "professional school" program—which, in contrast to the Rensselaer program, adopted the model of medical and legal education in offering three years of study on general subjects followed by two years of strictly professional studies. The Morrill Land Grant Act of 1862, which laid the foundation for an extensive system of higher education, led to the formation of many civil engineering schools during this period. The growth in engineering education was fueled by the increasing demands of the country's expanding railroad network and by a rapidly industrializing society. By the end of the century formal civil engineering programs were entrenched in many schools across the country. The best of these produced students with superb levels of general education and technical training. Other programs, however, produced students with narrow visions and almost boorishly anticultural attitudes.

While a system of formal education was evolving, the profession itself reached toward a stronger identity and a clearer status in society. The rise of a professional identity among civil engineers was most obvious in the period after 1850. As early as 1839 a group of civil engineers had met in Baltimore and attempted to form a national organization. Failing to agree on specific terms for such a body, the group eventually recommended the establishment of four regional societies. In 1848 engineers in New York and Boston established societies; the New York society soon folded, but the Boston Society of Civil Engineers became—and remains today—a vital organization. One of its early directors went on to become the first president of the first national organization, the American Society of Civil Engineers, which was founded in 1852. For a time the national society was only marginally active, but it gained rapidly in strength in the period following the Civil War.

The relationship between the emergence of the profession, as reflected by the development of professional societies, and the rise of formal education in civil engineering in this country is particularly interesting in that from its beginnings the educational system has been largely directed by individuals committed primarily to education and has thus been marked by a collegiate rather than an apprenticeship character. In contrast, the formal educational systems in architecture, law, and medicine were more distinctly influenced by their respective practicing professions and hence more marked by an apprenticeship philosophy within formal educational curricula. In civil engineering, the schools and the professional societies emerged more or less in parallel, and the professional societies did not assume responsibility for guiding education in ways analogous to the development of some of the other professions in America.

The turn of the century saw not only a formal system of education in place but a clearly defined profession as well. By this time the tradition of the self-taught or exclusively apprenticeship-trained civil engineer had largely disappeared. The application of scientific methods, normally requiring some level of formal education, had become an integral part of the profession.

The 1920s in America were a period of societal infatuation with the sciences, which affected civil engineering as well, particularly in the colleges and universities, where emphasis on the application of scientific methods grew even stronger than it had been. Faculty involvement with research increased. Sophisticated analytical tools were developed, permitting engineers to undertake with confidence increasingly complex projects, such as the great high-rise buildings of the thirties; on the other hand, many engineers became preoccupied with these techniques at the expense of lessons to be gained from observation and practice in the field. The same period saw the rise of the civil engineer as administrator and public works official, as many engineers became involved with local, state, and federal government agencies.

Recent decades have seen further evolutions in the theory and practice of civil engineering—including the development of computer-assisted analysis and design techniques, which have enabled engineers to undertake even more complex analyses and potentially to design even more confidently than before, and have made the design process far more efficient and cost-effective. But the time and energy previously absorbed by number crunching seem not yet to be utilized in synthesizing really better civil designs in any widespread way. Only in the past few years, however, have classes of civil engineering students trained extensively on computer techniques been graduated, and the field is yet a young one.

While speculations about the future impact of new developments, such as computer technology, are fascinating, it is difficult to imagine changes as dramatic and rapid as those that occurred during the nineteenth century, when the American teaching and practice of civil engineering underwent their most spectacular transformations. The nineteenth century was the nascent period for the profession. We can but hope that the future holds improvements as remarkable as those that took place then.

Chapter One

Canals

Introduction

In 1674 the French explorer Louis Joliet wrote of the desirability of a water route between Quebec and the Mississippi River. This idea, surely one of the first suggestions for a canal in America, was made a reality in 1827 with the building of the Illinois and Michigan Canal. Close in time to Joliet's suggestion was that of some unknown New Englanders who discussed cutting a canal across the neck of Cape Cod in Massachusetts. Samuel Sewell of Sandwich mentions in his diary of October 26, 1676, that "Mr. Smith of Sandwich rode with me and showed me the place where some had thought to cut, for to make a passage from the south sea to the north." Such a canal was eventually built, but not until 1914. William Penn, the founder of Philadelphia, in 1690 proposed a canal to connect the Schuylkill and Delaware rivers with the Susquehanna. In time Pennsylvania became the location of many of the foremost canals in the country.

These early ideas reflect a vision, already present in some minds in the seventeenth century, of communication and commercial links between regions of the American continent. The advantages and potential of canals were well known, with precedents dating from ancient Egypt and the Middle East to more recent and extremely sophisticated European examples, such as the Braire and Canal du Midi canals in France, and the Sankey and the Trent and Mersey canals in England. A number of extremely competent canal builders, among them James Brindley and Thomas Telford, achieved fame in eighteenth-century Britain. Few among the European settlers in America, however, had direct experience with canals or canal construction, and it was not until the eighteenth century that any significant transportation canals were actually built in this country.

In 1750 a short canal was constructed in Orange County, New York; it is considered one of the first significant canals built in America (minor waterways associated with mills had been used for some time). The short Seneca Falls Canal in Virginia was started in 1774. The original Dismal Swamp Canal was opened for small craft in 1784 and enlarged in 1807 and 1812. The Carondelet Canal running eastward from New Orleans was opened in 1785, about the same time as the opening of a small canal in Maine. The Little Falls Canal in New York (1783–1785), later replaced by the Erie Canal, was a significant early canal. The country's first president, George Washington, was an active promoter of canal construction and helped create the Potowmack Canal—one of the first truly major canals in the United States (1785–1802). The Schuylkill and Susquehanna Canal (known later as the Union Canal) opened its first 15-mile section in 1792–1794. The innovative South Hadley Canal (1792–1795) in Massachusetts used an inclined plane to accommodate changes in level—the first of its kind in America. The James River Company completed a short canal near Richmond in 1789. The Middlesex Canal (1794–1803) in Massachusetts was a significant early canal, as was the Santee and Cooper Canal in South Carolina (1792–1800). Both of the latter were notable for their boldness in cutting across large territories rather than being simple bypasses or links between existing waterways.

American lack of experience with canal construction thus did not prevent burgeoning experimentation. By 1790 there were approximately 35 privately owned canal companies, subsidized in part by local and state funds. These companies were created as a means to raise money to cover construction costs. Stock was sold to the public, with the understanding that there would be profits from tolls collected along the canals' routes.

Many fine canals were built throughout the country during the early and middle parts of the nineteenth century. Most were successful; others were failures—either technically or financially—built in the euphoric days when the canal-building fever spread through the country. By 1827, with the completion of the 364-mile-long Erie Canal in New York State, enthusiasm for canal building had reached a peak—a response to the impressive growth of commerce and regional development that the project had brought about. Neighboring states watched the building and operation of the Erie Canal with interest. Some states in addition to New York, such as Ohio, benefited from the Erie; others watched as the Erie siphoned trade away. Many states reacted by developing their own extensive systems. Ohio, already greatly benefiting from the Erie, developed its own system, chiefly to feed into the Erie and enable Ohio merchants and farmers to reach eastern markets. Pennsylvanians, perceiving that the New York and Ohio canal systems might drain trade away from their own system, laid plans to push a complete canal route across the state to tap the western trade. By 1830 some 1,400 miles of canals in Pennsylvania were completed, planned, or under construction. Entrepreneurs in New Jersey built the Morris Canal, with its famous inclined planes to carry boats over a rise and fall of over 1,600 feet. The Delaware and Raritan was built to provide an inland water link between Philadelphia and New York and proved very successful.

But the years of canal fever also saw the rise of railroad transportation. As railways began their competition with waterways, enthusiasm for canals diminished significantly. The railroads proved to be efficient, fast, and inexpensive freight and passenger carriers. Competition for capital for construction became intense. Many canals faded out of existence almost as soon as the railroads came through, while plans for new ones were scrapped.

A few of the fine old canals built in the heyday of canal construction continued collecting tolls into the beginning of the twentieth century. Most, however, fell into disuse and were filled in so that the land could be used for other purposes. Several still function in one way or another. Some were eventually transformed or replaced with other types of waterways more suited to twentieth-century needs, such as the New York State Barge System. These new waterways carry on one of the country's finest transportation traditions.

The Potowmack Canal

The Potowmack Canal, started in 1785 and completed in 1802, was one of the earliest attempts to link the western territories with eastern markets. Although it did not endure as a profit-making venture, it initiated the era of canal building in this country that helped to settle the West and to establish a thriving trade between the cities and the territories.

By the end of the American Revolution, one half of the country's inhabitants lived beyond the foothills of the Appalachian mountain range, yet their only connection with the eastern coastal region was by an arduous journey over narrow and difficult roads. The isolation of the western areas was troublesome not only because it hindered the growth of political and cultural ties with the East but also because the riches of the West—wheat, whiskey, iron, and timber—were traded not through American ports but through foreign ones. New Orleans, which was then Spanish, and Quebec, then English, proved more accessible to the backwoods pioneers than Boston, New York, or Philadelphia.

George Washington had long been interested in connecting these two parts of the country. Washington was very familiar with the Appalachians, for he had been surveying estates and roads in that region from the age of sixteen. In 1748 he had been one among several Virginians who had organized the Ohio Company, which was formed to settle the northern reaches of the Potomac (originally spelled Potowmack) River valley, an area then known as "Ohio country." Headed by Thomas Cresap, this was the first incorporated effort to reach the great western areas and the first to make use of the Potomac River above tidewater. Washington became surveyor for the company and in 1749–1750 was a member of the party that blazed a trail up the Potomac River and down the Monongahela to the new acquisition. In 1750–1751 he helped survey the road leading to the area. Early on he began to develop plans for an improved and navigable Potomac that would carry settlers and trade all the way to the Ohio River. This was the beginning of a project that would occupy Washington periodically for the rest of his life. Even before

One of five locks at Great Falls, Virginia, constructed between 1785 and 1802 as part of the Potowmack Canal, the first extensive system of river navigation works undertaken in the United States. The system was operated by the Potowmack Company from about 1799 to 1828. The locks at Great Falls overcome a 77-foot difference in elevation. (Jack Boucher, Photographer, HAER Collection, Library of Congress)

the Revolution he attempted to obtain a charter to form the Potomac Company to improve navigation on the river. This early attempt failed, and during the war no further action could be taken concerning the enterprise. In September 1784 Washington set out from Mount Vernon for a 680-mile trip on horseback through the Allegheny Mountains. His goal was to obtain information about "the nearest and best communications between the eastern and western waters."

On his return in October Washington sent a letter to Virginia's governor, General Harrison, with a bill for an act incorporating a company to open the Potomac to navigation. This time, no doubt owing in part to the great respect and fame Washington had earned during the Revolution, the bill passed within the month. Representatives from the southern part of the state, who did not want to subsidize development in the northern part at their expense, included the James River in the plan.

Because Virginia had granted rights to the waters of the Potomac to Maryland, it was necessary for both states to cooperate in the enterprise. This was not a simple matter; the loose structure of the Confederation hampered attempts at interstate activities. Meanwhile, Washington wrote scores of letters to gain support for his plan. The necessary approval was won from both states when they were granted the provision that Washington would become the company's president.

The Potowmack Company was jointly chartered by Maryland and Virginia in 1785. Washington was installed as its president on May 17 and remained in that capacity until February 4, 1789, when he became president of the United States. For his services he was given fifty shares of company stock; but since it was his policy never to receive payment for public service, Washington accepted the shares only on condition that he could donate them for the construction of a college, later to become Washington and Lee. Washington did invest heavily in the Potowmack Company with his own money.

The first meeting of the Potowmack Company brought enthusiastic support. More than four hundred shares of stock with a value of $400 each were sold, comprising an amazingly large sum considering the uncertain economic situation that the penniless new nation faced. Washington's diary makes frequent reference to the early planning of the project. The company's goal was to build canals around the worst obstacles in the river and then, after the canals had proved profitable enough to justify the expense, connect them from Cumberland, Maryland, to Point Look Out, just below the District of Columbia. The total distance of the proposed route was 185 miles.

On October 18, 1785, Washington and other directors of the company visited one of the significant obstacles to free navigation on the Potomac River—Great Falls, located just above Georgetown. Here the river dropped 76 feet amid rocky outcroppings and dangerous rapids. In addition to the canal that was to be built around Great Falls, four others were planned, varying in length from 150 feet to two miles and placed on either side of the river, depending on which was more suitable. The longest canal was to be built around Little Falls, on the Maryland side, only a few miles above the tidewater head-of-navigation at Georgetown. The Great Falls Canal, the next one up the river, was to be located on the Virginia side and would be three quarters of a mile long. These two canals, because they required locks, were the most difficult to construct. The other three—the Seneca Falls Canal, on the Virginia side; the Shenandoah Falls Canal, on the Maryland side; and the House Falls Canal, on the Virginia side—circumvented dangerous rapids or falls. The latter two were located close to Harper's Ferry. All of the canals were approximately 6 feet deep, 25 feet wide on top, and 20 feet wide on the bottom.

An earlier attempt to provide passage around the Seneca Falls had been made by one John Ballendine, who had traveled to England to study canal building. He had begun his canal on the Maryland bank of the Potomac in 1774. Work was stopped by the Revolutionary War and never resumed. By the time the Potowmack Company was chartered, Ballendine, perhaps the only man in America to have studied a canal and locks, was at work on a canal to circumvent the falls on the James River.

Work by the Potowmack Company began in August 1785, when four flat boats and two "setts of hands"—about fifty men each—set to work removing loose rock and sand bars from the river bottom in order to smooth, straighten, and improve the channel. All of this was merely preliminary work, however. The most difficult part was yet to come, and Washington recognized the need for an engineer to design the five canals. He had met a man named James Rumsey in Shepherdstown, now in West Virginia, on his six-week journey through the Alleghenies in 1784. Rumsey, an educated man who was regarded as an inventor and engineer, had been experimenting with methods of propelling river boats with poles. He was also convinced that steam power was a viable means of water transport. Although Rumsey had never seen a canal, he was felt to be the best person available at the time for the job of designing one. Washington wrote to him on July 2, 1785, "As I have imbibed a very favorable opinion of your mechanical abilities, and have no reason to distrust your fitness in other respects, I took the liberty of mentioning your name to the Directors [of the Potowmack Company] and I dare say that if you are disposed to offer your services, they would be attended to under favorable circumstances." Rumsey, perhaps spurred on by the thought of having his river boats monopolize the Potomac River, accepted the position. On July 14, 1785, he was appointed. Richardson Stewart was made his assistant.

Rumsey put to rest any ideas of using slack-water navigation to skirt the falls and said that blasting and stone locks would be necessary in order to make the river navigable. He ordered two work boats built and divided the laborers into three groups to work simultaneously on the Shenandoah Falls, Seneca Falls, and Great Falls canals.

The shaky economy of the new republic created financial difficulties that plagued the Potowmack Company from the outset. An even more serious problem was the lack of skilled workers. Because of the shortage of free labor, the company originally began work with slaves, but eventually it settled on using indentured servants. The laborers' drunken brawls disturbed the citizenry and led more than once to an official quelling of their unruly behavior. The workers themselves were not too happy with the situation. Although they had agreed to the job initially, many found it more strenuous than they had reckoned on and ran away. Rewards for their return were advertised in the local papers, and they were usually found and brought back to the construction site.

The year 1785–1786 brought unusually heavy rains, which slowed the work, washed away uncompleted dams, destroyed chutes, and deposited new sand bars. In addition, Rumsey began quarreling with Stewart and made charges against him. In 1787, at a time when financial uncertainty had nearly brought construction to a halt, Rumsey asked for a raise, was denied it, and quit. Various superintendents followed in his place. By 1789 the company was forced into condemnation proceedings because it could not meet the terms of Henry Lee, owner and lessor of the Great Falls land. Despite the difficulties, work continued. By 1795 little more than half a mile had been completed, but this short distance included the important Little Falls canal and locks. The Little Falls locks numbered three, each 100 feet long and 15 feet wide, with a lift of 11 feet. They were built of wood, and no trace of them remains today.

By 1795 the Great Falls guard gate, leading downstream, was in place. Above this a dam had been constructed that forced excess water away from the canal. A mooring basin that would be used to accommodate boats waiting to pass through the locks had also been completed, and workmen were busy locating the lock seats. These were later abandoned in favor of a new location when the canal plans were altered.

The vertical distance between navigable waters above and below the Great Falls was normally 77 feet, though this varied. Five locks were planned. Owing to the topography of the river bank, it was determined that the 3,600 feet of horizontal distance could best be covered by a canal stretching in one piece along the higher level with the locks concentrated at the lower end of the waterway. The upper lock was 14 feet wide and 100 feet long, and the others 12 feet by 100 feet. Plans originally called for all the locks to be 14 feet wide, but when it was discovered that the 12-foot locks served the purpose and required less locking water, the others were narrowed. Each lock consisted of two long parallel walls whose ends were closed by miter gates: that is, two vertically hinged leaves (similar to barn doors), which, instead of closing into a horizontal plane, formed an angle that pointed upstream when closed. The pressure of the water against the gates, greater on the upstream side than on the downstream, tended to keep the gates closed. Each of the gates contained valves that were used to empty the chamber. Most of these were butterfly-type valves—flat plates that pivot horizontally at mid-height. The valves built into the first gate may have been the flat, sliding-plate type, an older variation of the butterfly type.

The walls of the first Great Falls lock were put in place between 1795 and 1797. The upper locks were constructed of huge, hand-hewn blocks of red Triassic sandstone, brought down the river by boat and painstakingly placed. The downstream locks were blasted out of the solid rock of the Potomac Gorge wall, the greatest obstacle along the Potomac. Blasting powder had not been developed at the time, so common black gunpowder, though dangerous and undependable, was used. Because the lower locks overcame more of the vertical distance than the upper ones, they required more water to fill them. Therefore, a small basin was constructed adjacent to the middle lock, and water was fed into it from the mooring basin. From this middle basin, in turn, water could be fed into the lower locks to supplement the water they received from the topmost locks. At this third lock the canal bent toward the southeast.

Even while work was in progress on the Great Falls locks, the Potowmack Company opened the rest of the canal, completed in 1797, to navigation. A bypass system was devised at Great Falls: Barrels of goods were hoisted from the boats at the mooring basin, loaded into carts, and taken to a wooden chute, while the empty boats were carried by wagon to the foot of the chute. The barrels were then rolled down the chute and back onto the boats. This laborious system continued until 1802, when the Great Falls locks were completed.

Traffic was heavy on the canal from the very beginning, yet the project was also in constant financial difficulty. Workmen had found it necessary to blast some 4,300 cubic feet of stone from the Potomac Gorge, and this had consumed an inordinate amount of time and money, exhausting company funds. In addition, the Potomac, long known for its variable water flow, was found to be navigable only three months of the year. In the spring the current was so swift that boats could not pole upstream, and during dry months the water was too shallow to navigate at all. The wood used in the Little Falls locks began to decay immediately, requiring constant attention and repair. Yearly freshets deposited sediment and disrupted the channel. Every two years floods silted the bed and eroded the banks until the sandstone walls began to give way. Because its charter stated that the canal was to be navigable during the dry season, the company went to great trouble to make it so; but attempts to deepen the canal only lowered the water level along with the canal bed. The legislatures of Maryland and Virginia were asked for help, but to no avail; the company tried to sell a new issue of stock but was not successful; and it finally conducted a lottery—also without success—in an attempt to raise funds.

Nevertheless, the locking process went smoothly and traffic on the canal increased enough to cause a shortage of boats. During the first year 45,000 barrels of flour, in addition to other goods, were transported; the cost of freight transport on the canal was half that of wagon transport. The Potowmack Company took in $10,000 in tolls the first year, distributing its first and only dividend of $5.50 per share. Although receipts would increase to $15,000 in 1807, no further dividends were issued.

Although Washington did not live to see his dream of inland navigation on the Potomac realized, the river was cluttered with freight-carrying boats not long after his death. There were narrow, double-ended vessels called sharpers that were about 75 feet long and 5 feet wide and drew only about 18 inches when fully loaded. They could transport 16 to 20 tons apiece and made the passage from Cumberland to Georgetown in

significant. The ascent from the Medford River to the Concord, originally put at 68.5 feet, was afterward found by a practicing engineer to be 104 feet. The original survey estimated the elevation of the Merrimack at Chelmsford to be 16.5 feet above that at Billerica Bridge, whereas in fact the water at Billerica Bridge is about 25 feet higher than at Chelmsford. This gives some idea of the difficulties Baldwin and the directors faced for want of engineering knowledge and adequate instruments.

Baldwin and Sullivan managed to convince the other directors that work could not proceed until the company obtained the services of a qualified engineer. It was decided that every effort must be made to get Weston for the job, and this was a considerable undertaking. Baldwin made numerous trips from Boston to Philadelphia, where Weston held a lucrative position that he was reluctant to leave. His employers too were anxious for him to stay. But Baldwin learned that Mrs. Weston, having heard that English visitors always enjoyed Boston society, had a passion to visit the city. Baldwin offered Weston a salary that equaled his Philadelphia earnings, plus a $1,000 bonus. In addition, he volunteered to draw up the plans for Robert Morris's proposed bridge across the Delaware at Trenton—a task Weston had been hired to do. Baldwin's visits and generous offers, together with Mrs. Weston's "passion," finally won the day, and Baldwin returned to Boston in the spring of 1794 with Weston's promise that he would follow shortly. Baldwin also brought with him one of Weston's measuring instruments, a definite improvement over Thompson's compass. An early form of the Wye level, this was one of the first accurate telescopic leveling instruments to be used in America. While awaiting Weston's arrival, Baldwin ordered two more such instruments from London. Finally, Baldwin carried with him to Boston his first glimpse of canal locks. During his stay in Philadelphia he had traveled to Georgetown to inspect the Great Falls locks, then under construction along the Potomac River.

Weston arrived in July, and while Boston society did its best to entertain Mrs. Weston, he and Baldwin made a nine-day trip over the proposed site. On August 2, 1794, he submitted his survey, which outlined two possible routes—one of which was taken by the canal and the other later became the path of a railroad—with a chart of the estimated placement of locks, bridges, culverts, and aqueducts. After only a three-week stay, during which time he offered chiefly support and some technical insights, Weston returned to Philadelphia, leaving the work essentially in the hands of Loammi Baldwin. Weston sent advice by letter on several occasions thereafter, but he never returned to aid the progress of construction. His levels were accurate, however, and for this rare service the promotors had good reason to be grateful.

Not much work could be done from the time Weston left in August until the following spring, except the buying of materials and the negotiation of contracts. The purchase of land along the proposed route also began. One hundred forty-two separate pieces had to be acquired. Some parcels were donated as outright gifts by enthusiastic sponsors of the project, but there were also negotiations that dragged on for months because owners feared their property might be damaged by floods from the proposed canal. In at least sixteen cases the company had to exercise the new power of eminent domain.

Construction was to begin at the Billerica mills on the Concord River, the high point of the canal. Water from the Concord would be stored in the millpond at Billerica and from there would flow in two directions through the canal—some filling the northwestern portion before dropping 27 feet through three locks at the Merrimack River in Chelmsford, and the rest flowing to the southeast over a distance of 21.5 miles to the Charles River at Charlestown, descending over this distance more than 100 feet through thirteen locks. These sixteen locks divided the 27 miles into eight different levels, from 1 to 6 miles in length. Four additional locks provided entrances into the Merrimack, Concord, Charles, and Mystic rivers.

On September 10, 1794, in a small ceremony on the western shore of the proprietors' millpond at the Concord River in Billerica, Loammi Baldwin turned over the first spadeful of earth. Construction was carried on in several places at once, since most of the work was being done by small landowners whose property lay along the canal. Each was paid to excavate a certain length of canal at a set price per length that was determined by the condition of the land—whether stony, cleared, or root-filled. The property owners generally enlisted the help of their neighbors, but work was done only intermittently when time could be spared from farm work.

Other work on the canal was contracted out to eight primary companies. At some points Baldwin had as many as five hundred men working for him. Common laborers received $8 per month in addition to room and board, while skilled craftsmen—masons, carpenters, stoneworkers, blacksmiths, and stone blowers—got $15 per month. Most could live at home and walk to work daily, but those who could not were lodged with villagers or housed in barracks under the direction of a steward. The men worked every day except Sunday, from sunup to sundown. In the summer they were permitted an hour-and-a-half break at noon, for which they worked an extra half hour after sunset. Construction ceased each winter.

In addition to organizing this army of men, Baldwin had to provide fodder and shelter for several hundred mules and horses. He also oversaw the various subsidiary enterprises necessary to complete the project. The company manufactured the tools used by the landowners and contractors—a variety of wheelbarrows, shovels, spades, hoes, pickaxes, crowbars, scythes, pitchforks, beetles, and wedges. Metal parts were made in the canal company's blacksmith shop in Billerica.

Baldwin supervised every detail throughout the project. At one point the directors wrote to him explaining that although they expected him to instruct the workers in their duties, anything more—such as taking part in the labor itself—was uncalled for and not really desirable in a supervisor. The directors themselves shared Baldwin's enthusiasm and often became similarly involved. The canal progressed slowly, and there was occasional friction between Baldwin and the directors. At one point the latter contemplated hiring an engineer from England, but they rejected the notion, concluding that they were unlikely to find anyone who was as thrifty or as effective with the workers as Baldwin.

One of Baldwin's most interesting innovations was the floating towpath at Billerica Mill Pond. The pond had been built in the seventeenth century to feed a grain mill. When the canal was built, the Middlesex Company took over the mill, and its operation fell to Baldwin, who grumbled to the directors that he hoped he would eventually find time to learn something about operating the thing. The millpond dam was 8 feet high and 150 feet long and crossed the Concord River beyond the canal. It was here that water would be stored to fill both northwesterly and southeasterly stretches of the canal—as well as to turn the mill wheels. Two waste gates regulated the flow to some extent. In 1798 Baldwin designed a floating towpath so that canal boats could be drawn across the pond from one section of the canal to the other. The towpath could also be opened to allow passage of timber rafts bound for sawmills on the Concord River.

The stretch from the Concord to the Merrimack was a relatively easy one to build, because the land was flat for most of the distance to the Merrimack, where all three locks were concentrated. The stretch from the Concord to the Charles proved much more difficult, for it involved many level changes, rocks to be cut through, streams to cross, and major cut-and-fill operations. The need for continual removal of rocks and boulders slowed work considerably. Wooden stone boats or drags (flat, runnerless sleds) were drawn in by oxen to be loaded with stone and carried away. As work progressed, a given section could be flooded with water, a boat floated in, the section drained, rocks loaded into the boat, the section refilled with water, and the boat towed away. In the low, swampy areas, earth fill was hauled in by ox-drawn carts, which maneuvered on a makeshift railroad of parallel planks tied together with cross-bracing.

The canal bed was 20 feet wide at the bottom and 30.5 feet at the water line, with banks sloping 33 degrees. The depth was 3.5 feet to begin with, but heavy silting reduced it to about 3 feet. Between the locks the canal

An interesting feature of the Middlesex Canal was the floating towpath at Billerica Mill Pond. Designed to enable canal boats to be drawn across the pond from one section of the canal to another, the towpath could be opened so that rafts and boats bound for sawmills on the Concord River could pass through. This watercolor, by William Barton, dates from 1825. (Courtesy of the Billerica Historical Society)

descended to ensure a steady flow. The banks extended 1 foot above the water line, with a 10-foot-wide towpath on one side and a 3-foot-wide bank, called the berm, on the other. Embankments had to be constructed beside long stretches of the canal. Those at Wilmington were 1,320 feet long and 25 feet high. The foundations for the embankments, known as "puddle gutters," were troughs 3 feet wide and deep enough that all stones and roots could be removed. They were filled with screened earth; this was well packed, and then the core of the embankment was constructed on top of it, also of screened and packed earth, giving a strong center to the whole.

Claims of damages because of floods, wells that had been drained, and water courses that had been diverted plagued the company continually. One of the greatest difficulties was how to achieve watertight construction. The fears of land owners that their grounds would be flooded were not unfounded. Most early canals always leaked—or "wept," as it was called. The Middlesex was no exception.

One solution to the weeping problem that had been successfully used elsewhere, particularly in England, was "puddling." A "puddle" was a mixture of well-tempered clay and sand, reduced to a semifluid state and rendered impervious to water by manual labor, as by working and chopping it about with spades. It was usually applied in three or more strata to a thickness of about three feet. Care was taken at each operation to work the new layer of puddling stuff so as to unite it with the stratum immediately beneath. Over the top course a layer of common soil was usually laid.

Early in the construction, attempts were made to puddle sections of the canal. The process proved, however, to be a long and tedious one that the Middlesex Company decided it could not afford. Baldwin attempted alternative solutions. Some banks were rammed; at other places the water was let in and out of the canal several times in order "to season and consolidate the banks"; the ox-drawn carts loaded with earth did their part also to compact the canal banks. Nevertheless, farmers continued to complain of flooded meadows and gardens.

It was during this work that Sullivan and Baldwin invented the dump cart, the forerunner of the dump truck. Sullivan had noticed the workers' practice of digging out the area beneath a hillside that was to be cleared away and then using crowbars on the top of the bank to collapse the whole downward. The loosened dirt was then shoveled into carts and drawn away. Sullivan reasoned that it would be simpler to collapse the hillside directly into a low cart, which could be made to fit into the undercut area. Baldwin built a few models to match Sullivan's descriptions and found that they worked well. They were used continuously thereafter.

A basic principle in canal building was never to let the waters of any stream mix with the canal water. This was to avoid floods on the canal, which would cause navigational problems and might endanger the banks. Baldwin carefully followed this rule. Smaller brooks were carried successfully under the canal bed in brick culverts. Larger ones had to be crossed over by means of "timber trunk" aqueducts.

The most impressive of the eight aqueducts Baldwin built was that spanning the Shawsheen River. This structure, which drew visitors from all over the country, consisted of raised embankments on both sides of the river, culminating in a water-filled trough across it, through which barges floated. The aqueduct proper was about 30 to 35 feet above the river and 188 feet long. Its two abutments and central pier of dressed granite were laid without mortar. The aqueduct was braced with timbers between the piers, and the bracing pieces rested in turn on horizontal timbers embedded in the piers. The granite blocks were lifted into place during construction by a guyed gin pole of Baldwin's devising, fitted with block and tackle. The wood construction of this and other aqueducts revealed Baldwin's early training as a cabinetmaker. The timber was framed together in the way used in carpentry, with tenons and mortises, and was strengthened by braces. The 30- to 35-foot-high embankments extended several hundred feet from each end of the aqueduct.

The Erie Canal Commission, which was sent to inspect the work in 1816, later criticized the practice of using mortise-and-tenon joints on the grounds that these wood joints would soon decay. Nevertheless, the Middlesex structures served their function well. The Shawsheen, for example, contained 300 tons of water but leaked only one barrel in 24 hours.

Because the aqueducts were narrow, only one boat could pass over at a time, so large waiting basins were formed at both ends. Later the Shawsheen aqueduct was shortened when the two end spans were filled with earth.

In addition to the aqueducts, Baldwin constructed over fifty bridges. Although a few served to carry post roads over the canal, most of them were "accommodation bridges" built to give passage to farmers whose land had been divided by the canal.

The other major task was lock construction. The triple locks on the Merrimack were the first to be built, and they took over three years. Weston sent to Baldwin his suggestions for their construction, which were altered by Baldwin for economy and, in some cases, in light of better design principles. The first locks were built of granite, for which there was desperate need of hydraulic cement (cement that can harden under water) in order to make the locks watertight. At that time hydraulic cement was made of granite dust and trass, a pumicelike volcanic substance. The latter was unavailable in the United States and very expensive to import. Baldwin searched in vain for a substitute, when finally a sea captain told him of a similar volcanic ash that was used in the Dutch West Indies. The Middlesex Canal Company immediately hired a sloop to send to the West Indies. It returned on August 23, 1796, with 40 tons of trass and instructions concerning its use. Baldwin altered the standard mixing procedure in a way that would influence later commercial cement manufacture in this country. The common practice was to beat together the trass, lime, and water for a long time, since the longer they were beaten, the harder the cement set. It was a day's work for one laborer to mix and beat two bushels of

The Shawsheen aqueduct was one of the more impressive structures on the Middlesex Canal. The granite blocks of its abutments and central pier remain in place today, as the photograph (*bottom*) shows. (Painting by Joseph Payro, courtesy of the Middlesex Canal Association Collection; photograph from the Library of Congress)

lime with one of trass. Baldwin greatly shortened the process by having the trass finely ground before mixing. The results worked well.

Work on the locks was constantly hampered by flooding. Baldwin devised a series of pumps that were used 24 hours a day during some periods. Acquiring iron for the hinges and valves on the lock gates was another difficulty. The first set was produced by a New York foundry from a mold made by Weston. Later ones, from Pawtucket, Rhode Island, failed even to fit together properly. At length, however, a satisfactory product was obtained.

Not only was the building of locks slow and difficult, it was also expensive. The foundations of one lock and the wing walls that buttressed the set of three locks alone required 1,000 tons of expensive, and difficult to procure, granite, as well as $2,000 worth of trass. Knowing that there were seventeen more locks to be built, the directors ordered Baldwin to make the remainder of wood. Baldwin argued against this, and Weston even wrote a letter to the directors enumerating the disadvantages of wood construction. The directors conceded it was an unwise economy but insisted it was a necessary one. After that, only those locks near the river were made of stone. All locks were 12 feet wide and long enough to accommodate 75-foot boats.

Water flowed through the completed canal for the first time on December 31, 1803. The project had taken ten years. The waterway immediately became the cheapest and most popular transit in Massachusetts. The boats that traveled the canal were flat-bottomed vessels, between 40 and 75 feet long and 9 to 9.5 feet wide. Each had a draft of about 1 foot and a maximum capacity of about 20 to 25 tons. Their speed was restricted to 3 mph. Each was drawn by one horse on a tow line connected to a short mast just forward of the boat's center. The crew consisted of a driver and a helmsman. Rafts, limited to 1.5 mph, were about 75 feet by 9.5 feet but could be linked together to form a train.

The advantages of canal travel over wagon transport were obvious at once. One horse, for example, could easily draw 25 tons of coal on the canal. On land the same horse could pull only 1 ton. One team of oxen could pull 100 tons, an amount that would take eighty teams on land. In the first eight months of the canal's operation, 9,405 tons were carried at a cost of $13,371. The cost for such a shipment by land would have been $53,484.

The brightly painted and decorated passenger packets had the right of way, cruising along at 4 mph. The passenger terminal was located in Charlestown at the corner of the present Rutherford Avenue and Essex Street. Writers of the day extolled the beauty, comfort, and cleanliness of canal travel, and outings became a popular pastime. Boston residents flocked to the country in order to stroll along the green towpath and breathe the fresh air. Canal cruises soon became acceptable therapy for those in delicate health. Since night passage was forbidden for fear of damaging the canal banks, taverns sprang up at all the locks, many becoming well known for their hospitality and jolly company. The taverns were all owned by the corporation.

The canal company directors, ever innovative, fitted a heavy Philadelphia-made steam engine onto one of the canal boats in 1812. This was the first use of a steam tow boat in the United States. It was used for three summers with minor modifications, until the vibrations of the engine threatened to shake the boat to pieces. The company then purchased a patent from Samuel Morey for his rotary steam engine, which Sullivan described as "a very small engine, and as yet without its condenser. The stroke of the cylinder is but 1 foot, and the diameter 9 inches. There is only one boiler 14 inches long and 3 feet in diameter. One part of the invention is an internal covering of the furnace, by which steam is generated with uncommon rapidity." Smoke was also generated with uncommon rapidity, since the fuel used was wood and tar. The steamboat worked well but could not be used efficiently on the canal. Its maximum speed of 7 mph threatened the banks with its wake, and the speed had to be limited to 3 mph. For this reason no others were developed for use on the Middlesex.

Despite the great popularity of the canal, the Middlesex Company's financial condition was never stable. The total cost of the canal had been $528,000, a staggering sum at the time. In addition, repairs and maintenance entailed considerable expense. Repairs had to be handled quickly before further damage was done to the banks, so a fleet of repair boats was constantly busy. Stop gates were positioned along the canal to permit operation in those parts not being repaired. In winter the water had to be drawn off the aqueducts and locks to prevent damage from expanding ice. "Bank watch" patrols were constantly treading the shoulders, watching for holes. Most of their work consisted of stuffing straw into these holes, which were made by minks and muskrats. The problem became so serious that in 1809 the canal company offered a bounty for the skins of both animals. Crews were also hired to scythe the long grass that grew in the canal bed and limited navigation. Once cut, it was allowed to float downstream to the next lock, where the lock tender would fork it out and spread it on the bank to dry. Proceeds from the hay were split by the reaper and the tender. Again and again the stockholders were asked to pay assessments so that the company could continue operation. They always complied and by 1817 had made one hundred payments. It was not until 1819 that the first dividend was returned to them.

The reasons for the Middlesex Canal's financial difficulties were numerous and present from the outset. The enormous enthusiasm for canals at the turn of the century, plus the backing of some of the most respected men of Boston, convinced the populace that the canal was a good idea. Unfortunately, the time was not ripe. Boston was a town of only 20,000 with a total assessed worth of $15,000,000. The territory through which the canal passed was too young and too sparsely settled to give it the support it needed. Although the first eight months of operation saw 9,405 tons of freight come into Boston via the canal, only 308 tons left Boston for the country; the exchange was one-sided and inefficient. The Embargo and then the War of 1812 severely limited shipment from the port of Boston. As a result the use of the canal for transshipment was also hampered.

Just when it looked as if the canal was at last about to enter a period of prosperity, the railroads developed. Ironically, the granite for the railroad sleepers and the engines themselves were transported into the area through the Middlesex Canal. The year the Lowell Railroad was put into operation, the Middlesex receipts were cut by one third. When the Nashua and Lowell opened, they were reduced by another third. In 1843 Caleb Eddy, seeing the inevitable end, petitioned the Commonwealth for the use of the canal to supply the city of Boston with drinking water. His plan, reasonable and well worked out, was rejected. On April 4, 1860, the property and assets of the Middlesex Canal Company were seized and forfeited. The forced sale of the property brought only $130,000; the Boston and Lowell Railroad was the principal purchaser. With the final dividend paid from this sale, the stockholders managed to come out even on their investment.

Nevertheless, the Middlesex was hardly a failure. It provided an invaluable service to the territory and contributed substantially to the business prosperity of Boston. One of the earliest American canals, it was also one of the last to succumb to competition from the railroads. Moreover, it served as a prototype for other canals, including the Erie.

Traces of the canal can be found today at various points along the route. Loammi Baldwin's house and a portion of the canal have been restored in Woburn. In Winchester, Middlesex Avenue marks the old canal bed, with Canal Street joining it where a basin used to be. The stonework of the Shawsheen aqueduct is still present. Parts of other aqueducts and towpaths can be seen in Wilmington and Billerica.

The Erie Canal

As the United States expanded westward, a reliable transportation route between the western territories and the Atlantic became a necessity. A route through upper New York State to the port of New York City seemed logical, especially since a wealth of waterways crisscrossed the state and a major natural break in the formidable wall of the Appalachian Mountains occurred in New York's Mohawk Valley. The Erie Canal, a 365-mile waterway from Albany to Buffalo, was eventually constructed through this valley. Begun in 1817, it took eight years to build and has often been called the first American school of civil engineering. Many of the fledgling surveyors and assistants who planned and completed the canal "graduated" from the project as highly skilled engineers.

In 1788 the innovative Robert Fulton wrote, after looking over the upper Mohawk River, that great benefit would result if Little Falls were to be "locked." Christopher Colles, an Irish-born engineer, made a survey and noted the Mohawk's obstructions to travel. Colles believed that transport on the Mohawk between Albany and Oswego was feasible and proposed the organization of a company to build seven bypass canals around unnavigable parts of the river and to dredge shallow areas. In 1791 Governor George Clinton recommended to the New York legislature that a road and river improvement policy be formulated. A legislative committee voted to examine internal improvements and recommended a survey by Land Office commissioners. In September Major Abraham Hardenburgh with Benjamin Wright conducted an initial survey, which was reported on favorably by the Land Office in January 1792. Legislation was enacted on March 30, 1792, incorporating two companies: the Western Inland Lock and Navigation Company, which would undertake to straighten the route on the Mohawk from the Hudson west to Lakes Seneca and Ontario, and the Northern Inland Lock and Navigation Company, which would build a navigable connection between the Hudson and Lake Champlain. Both companies were privately owned but were assured some state support, a rarity at the time.

In April 1793 the Western Inland began on the first short canal at Little Falls and started straightening Wood Creek from Rome to Oneida Lake. The Northern Inland started work at Stillwater but soon stopped for lack of funds. The Northern Inland venture was a failure; $100,000 was invested, but the canal was never completed. The Western Inland, however, succeeded in constructing a bypass canal around Little Falls. Just under a mile long, it was cut through solid rock and contained five locks. At one point during the construction the company nearly folded, but with the state's financial backing the canal was completed in November 1795. Large barges—which it had been impossible to carry around the falls—could now be used on that portion of the river. As a result, prices for transport fell. The public was not convinced that the outcome was worth the investment, however, and temporarily lost interest in canal construction.

Nonetheless, some activity did continue. In 1798 the state incorporated the Niagara Canal Company to survey and construct a canal around Niagara Falls from Lake Erie to Lake Ontario. This project was not completed. In 1802 the Western Inland made plans to improve navigation from Wood Creek to Little Canada Creek. Benjamin Wright made another survey in 1803. In the meantime economic conditions began to coalesce that would eventually result in the Erie Canal.

Jesse Hawley, a flour merchant from Geneva, New York, is often credited with the initial idea for the Erie Canal. In 1806 Hawley fled to Pittsburgh to escape his creditors. His financial troubles had been caused in part by the transportation difficulties across the state. While in Pittsburgh he published his thoughts in the December 1806 edition of *Commonwealth,* proposing the construction of a canal to connect Lake Erie with the Mohawk and Hudson rivers. Meanwhile, friends in New York urged him to return and give himself up to the authorities, pledging financial assistance. Hawley returned to Geneva, but his friends failed to keep their promise and he was sentenced to debtor's prison for twenty

Erie Canal, New York. Begun in Rome on July 4, 1817, this 365-mile-long canal cut through New York State from the Hudson River at Albany to the Great Lakes at Buffalo. It was the principal route for emigrants from the East and agricultural products from the West. In its day it was the world's longest canal and one of America's greatest engineering feats. The construction of the canal, largely under the direction of Benjamin Wright, provided training for many of the country's foremost civil engineers. Shown here is an early view of Lockport, where five pairs of double locks were cut through the solid rock of the Niagara Escarpment. (Library of Congress)

months. While in prison he wrote fourteen essays, published in the *Genesee Messenger* in 1807 and 1808, in which he laid out his plan for a canal in detail, suggesting a route very close to that ultimately selected for the Erie Canal. Hawley even estimated the construction cost at $6 million, not far from the final figure of just under $8 million. The result of the canal, Hawley predicted, would be great wealth for the state and a rapid growth in population. This plan, coming from an uneducated man, was considered a wild idea, but it nevertheless caught the public's attention.

Another early advocate of a canal was Gouverneur Morris, an important statesman and financier. His notion was to avoid the use of locks entirely and to cover the vertical distance between Lake Erie and the Hudson River by constructing the canal along a continuous incline with a uniform slope over the 363-mile route. This plan ignored the geographical fact that the New York countryside goes up and down significantly along the route: a canal built to Morris's specification would at places be 150 feet above the local grade level. Nonetheless, the plan received widespread support—though many wondered just what the momentum of the boats would be when they reached the end of their journey and whether, under those circumstances, the boats would be pulling the mules rather than the mules the boats. Although Morris seems not to have been the originator of the Erie Canal idea, he does appear responsible for persuading De Witt Clinton to back a canal project.

In 1808 Joshua Forman called for the legislature to look at the idea of a canal from the Hudson River to Lake Erie. The legislature appropriated $600 for a survey of possible routes and hired Judge James Geddes, an amateur surveyor, to do the job. Geddes spent most of 1808 walking the territory between the Hudson and Lake Erie, examining two possible routes. One plan would make extensive use of Lake Ontario; the other would cut more directly overland. The second plan was generally believed to be impracticable, because of the obstacle presented by the Irondequoit Valley in the critical section between Rochester and Palmyra; but when Geddes examined the valley, he reckoned that the tops of the high ridges of the valley could carry the canal across. His suggestion was to fill in the spaces between the ridges with an artificial embankment 68 feet high that would serve as a base for the canal. Geddes's report of 1809 presented his findings for both routes, but it was clear that he favored the inland one. Political and economic factors also supported this route. The inclusion of Lake Ontario would necessitate the transfer of goods from canal boats to larger lake vessels and then back to canal boats, a process that would require time and labor and would thus both raise the price of the goods and subject them to a good deal of wear and tear. A more persuasive argument at that time was that the British still controlled the area north of Lake Ontario, which made the route vulnerable should war again break out between America and England. The threat of the British cutting the lifeline to the western territories could not be ignored, and the inland route won out.

The question of how to finance the canal construction was of paramount importance. Secretary of the Treasury Gallatin's 1808 report to Congress on internal improvements had specifically mentioned a New York canal connection between the Great Lakes and the Atlantic, so New Yorkers had reason to believe they might obtain federal assistance for their project. They were disappointed, however. President Thomas Jefferson said it was "a splendid project and may be executed a century hence," and refused the request.

The New York delegation did not accept defeat. One of the problems facing canal proponents was finding someone influential enough to garner support from both political parties in New York, support that would be necessary for the passage of any canal bill. De Witt Clinton was perhaps the only man in the state who could gather such support. Clinton was a former mayor of New York City and a U.S. senator, and would soon be governor of the state of New York. He was greatly respected. In July 1810 Clinton joined Thomas Eddy, former treasurer of the Western Inland Lock and Navigation Company and an expert on canals, and other members of the newly formed Canal Commission in making a trip over the proposed route. The journey gave the men a first-hand taste of the terrible travel conditions that existed inland from Albany to Utica and that made western migration so difficult. Five days after they started, the commission members arrived at Rome, where they decided to divide and study both routes. The parties rejoined at Chippewa in early August. Clinton, who had followed the inland route, voiced strong preference for it, pointing out that in addition to being safe from British intervention and avoiding the necessity of transfer of goods, the inland route would encourage settlement along its entire length, something that a canal making extensive use of the lake could not do.

The commission made an enthusiastic report to the legislature on March 2, 1811, recommending that the canal be a state-owned enterprise, not a private concern. Public discussion followed. When at last the legislature had agreed to seek the money for construction, the outbreak of the War of 1812 put a stop to all plans. Clinton and the canal supporters continued to meet despite the war, and when it came to an end in 1815 they actively resumed their campaign. In December of that year they drew up a document pleading the canal cause. Clinton's language was effusive; he described the canal as one that would pass "through the most fertile country in the universe and would perhaps convey more riches on its waters than any other canal in the world." The memorandum also gave specifics: the canal was to contain 62 locks and would cost $20,000 per mile or approximately $6 million. Construction time was estimated at ten to fifteen years. Clinton concluded, "It remains for a free state to create a new era in history, and to erect a work more stupendous, more magnificent, and more beneficial than has hitherto been achieved by the human race."

Whether because of the details Clinton gave or because the idea had been before the public for many years, the canal now began to seem a reasonable venture and the appellation "Clinton's Ditch," previously used in derision, was now used with pride. But though the War of 1812 had demonstrated the desperate need for an east-west connection, political infighting continued to play a major role in the debate. Clinton was a powerful politician who had powerful enemies. While those whose land would appreciate with the canal's construction supported it loudly and 100,000 citizens signed a petition supporting the project, organized opposition formed to the south in New York City, on Long Island, and in the old patroon country along the Hudson—all areas whose residents believed that their commercial monopoly would be destroyed by the canal.

Debate over the plan preoccupied the legislature in 1816 and 1817. Clinton supporters won a victory when the legislature passed "An Act for Improving the Internal Navigation of This State" in April 1817. Apparently acting in the larger public interest, Martin Van Buren, a longtime political foe of Clinton and canal interests, suddenly supported the legislation and was influential in its passage. Although it did not yet fully authorize construction and required additional surveys, the act broke the bottleneck and the canal was on its way.

The canal commissioners divided the route into three sections and assigned one person to be in charge of each. James Geddes was given the western section, from Lake Erie east to the Seneca River; Benjamin Wright, the middle, from the Seneca River to Rome; and Charles Broadhead, the eastern portion, from Rome to Albany. Although some commissioners initially sought to obtain the services of the respected British engineer William Weston, they turned to Benjamin Wright to provide leadership for the whole undertaking. Wright soon hired Canvass White, later to become an accomplished civil engineer in his own right, to work on the canal. White was sent to England in the fall of 1817 to collect information on British canals and returned in 1818 with a collection of careful drawings and insights into canal construction techniques. Another assistant was Nathan Roberts, who eventually constructed the famous set of five pairs of locks at Lockport.

None of these men had any significant formal training in engineering or knew much about canal building before working on the Erie. In the end, engineers for the canal were not found—they were created. The civil engineering schools at both the Rensselaer Polytechnic Institute in Troy and Union College in Schenectady were outgrowths of the canal project, as were the careers of many individual builders, who, after the canal was finished, spread throughout the United States, constructing canals, railroads, bridges, and other structures. In time the Erie Canal came to be known by many as the nation's first school of civil engineering.

The overall plan devised by the engineers generally called for a canal 363 miles long, and about 40 feet wide at the top, 28 feet wide at the bottom, and 4 feet deep. Eighty-three locks would cover a vertical distance of 627 feet, and eighteen major aqueducts would carry the canal over streams and rivers. The shallow depth had been carefully calculated: it was the minimum depth that would float loaded boats but the maximum depth that the state could finance over that distance. When completed, the Erie would be the longest canal in the world.

Construction work at Lockport. (Library of Congress)

Groundbreaking took place in Rome on July 4, 1817. Rome was on one of the two long level stretches of the proposed canal that did not require locks. This choice of a starting point was excellent, both psychologically and strategically. The easy terrain offered a good place for training engineers and workers about canal construction and promised rapid progress that would encourage both builders and the public. The nearby Seneca and Mohawk rivers also afforded a good supply of water for the canal. Finally, this was an area Wright knew well. Although he had been born in Connecticut, his family had moved to Rome during his youth, and between 1792 and 1796 he had laid out 500,000 acres of land surrounding Rome and had surveyed for the Western Inland Lock and Navigation Company. His knowledge proved invaluable to the project.

The summer of 1817 was spent staking the 60-foot-wide swath that was to be cleared and marking a 40-foot-wide track within it for the canal itself. Following the stake crews came the boring crews, who took borings to a depth of 12 feet all along the canal's path, attempting to anticipate possible troubles. Third came the heavy workers, who cleared the land of the giant trees, pulled up roots, topped hillocks, filled depressions, dug ditches, and built embankments. These tasks gave rise to several ingenious devices. To pull down trees quickly, a machine was invented that made use of a screw, roller, cable, and crank. Next, engineers developed a stump puller, which rested on 16-foot-diameter wheels and had a smaller inner drum wrapped with a cable that was attached to the stump. The machine was so efficient that seven men and four horses could remove forty stumps each day. The workers even developed a new-style wheelbarrow and a special plow that grubbed out the underbrush.

In the beginning the canal builders used a procedure—unusual for its time—of letting out contracts based on sealed bids for work on sections as short as one quarter of a mile. The bidders furnished their own tools and hired their own workers. Since labor was not plentiful, this procedure looked as if it might solve the shortage problem, as well as involve local people in the project. Local farmers could do the work, and the money paid out in salaries would be fed back into the area. When the first 58 miles had been finished by fifty different contractors, however, the disadvantages became apparent. The quality of the work was inconsistent, and since new farmers were hired as the route lengthened, no gain could accrue from the experience of the earlier workers. In addition, farm chores interfered with the canal work, slowing the pace.

The decision was soon made to establish more permanent work crews, and soon workers from many parts of the country and abroad were laboring on the canal. With the development of experienced crews, the work sped up, particularly when workers were given the option of getting paid by the cubic yard of excavated earth. Some teams of three managed to dig three rods of canal (250 cubic yards of dirt) in five and a half days. Tales abound of the men who comprised these teams, and of the means used to increase productivity. Folklore holds that a major incentive came from the "jigger boss," who walked up and down the line doling out whiskey to the workers, a half gill at a time up to sixteen times a day (one quart per man each day) for work done.

Late in 1818 work on the canal had progressed to the Montezuma or Cayuga Marsh, which Canvass White described as neither lake nor land but "a streaky and unpleasant mixture of both." Men were forced to dig in water up to their chests, and "swamp sickness"—malaria and pneumonia—took its toll. A thousand or more men are said to have fallen ill, and many of these died. There was no cure except "Jesuit's bark," a Peruvian form of quinine that helped only a little. Many workers fled in fear, and a rumor spread across the state that an "effluvia miasmatica" had been released by the digging. Nevertheless, work continued, and the labor force grew as cold weather eliminated the mosquitos and hardened the ground underfoot. By October 23, 1819, the middle section was complete, and the water was let in with great ceremony.

The most difficult work lay ahead, and the anti-Clinton forces, well aware that the completed section represented the easiest part of the construction, forecast physical and financial disaster should the canal continue. Anti-canal leadership came from Tammany Hall in New York City and reflected the interests of a large number of New York merchants nervous about the possible loss of business due to the canal.

One of the greatest engineering challenges was the section that passed through the narrow and mountainous gap in the Mohawk Valley—in particular, the deep gorge through the Catskills. Canvass White engineered the canal's path, which shifted from the south to the north side of the river via an aqueduct and then ran along shelves cut into the rock wall of the gorge before shifting back to the south side of the river four miles from the Hudson by means of another aqueduct. This second aqueduct was 1,188 feet long. From there the canal bypassed the Cohoes Falls, stepped down the remaining 184 feet to the Hudson, turned south, and paralleled the Hudson River for five miles, widening out into a boat basin at Albany, its destination. It was also Canvass White who, in a sense, saved the entire project by locating near Chittenango deposits of limestone needed to make hydraulic cement, which was used to make waterproof locks. A half-million bushels of the special cement were used on the canal, an amount that the state could never have afforded to import from Europe had a local supply not been found.

By 1823 the canal had reached the eastern edge of the Irondequoit Valley. Geddes's original plan had been to fill in the valleys between the ridges to provide a continuous path for the canal, but this was deemed impractical. A subsequent proposal advocated the construction of a high wooden trestle that would rest on the ridges. Concern that the wind coming through the valley would destroy such a structure forced the builders to return to Geddes's suggestion. The ridges were eventually connected with two stone and earth embankments, one 1,320 feet long, the second 231 feet long, resulting in an enormous landfill skyway 76 feet above the valley floor. Beneath, the Irondequoit Brook flowed through a stone culvert.

The next important crossing was at the Genesee River at Rochester. There the experience-schooled engineers constructed a masterpiece—a magnificent 802-foot wooden aqueduct, bolted together into a single piece and placed on top of eleven 50-foot stone arches that had been constructed on piers sunk securely into the bed of the rapid-flowing stream. The structure aroused great public excitement and became a symbol of the canal's successful completion, which now seemed certain.

But this general public enthusiasm had not affected Clinton's standing with his political enemies, who had recently defeated him in his run for governor. While Clinton was in Buffalo, having come out in favor of that city as the canal's western terminus, he was ousted from his position as head of the canal commission. The public was stunned. Clinton's name was synonymous with the canal and he was by now recognized as a national hero, responsible for a project of national importance. The press took up his cause. New Yorkers expressed their anger in demonstrations and rallies all over the State; one in Albany drew ten thousand persons. There were rallies and praise even in New York City, a stronghold of anti-Clinton forces. When he ran again for governor in 1824, Clinton won by a landslide.

The Ditch crept westward until it reached the Niagara Escarpment, a large outcropping of solid rock that forms Niagara Falls. The area, filled with rattlesnakes and hardwood trees, was to become the location of the new town of Lockport, a town named for and created by the canal. The problem faced at Lockport was how to raise and lower boats a distance of 70 feet over a short stretch of land. The daring solution was planned by Nathan Roberts, a self-taught engineer who had done his first engineering while locating the Erie right of way. Accepting advice from no one, he designed two parallel sets of five consecutive locks that would transport boats in both directions at once. The New York City papers ran advertisements for 1,200 laborers to build the locks, and many who answered were Irish. Many stayed in Lockport permanently. The workers cut the locks through solid rock; the average cut was 25 to 30 feet over a distance of three miles. Lockport proved the most difficult and costly section to build. Only after a special highly tempered drill was developed could holes be made that would hold the explosive—usually common gunpowder. In some cases water was poured into the holes and allowed to freeze, expand, and crack the rocks. In a few instances the newly developed Du Pont blasting powder was used. According to local tales, stones several inches in diameter rained down upon the town following each blast, and the townspeople ran for shelter in fear of their lives. Householders took to cutting down small trees, removing the branches, and propping the trees against the eaves of their houses with the lower parts 8 or 10 feet from the house and the tops extending above the roof. This arrangement helped prevent further damage to the buildings, and the lean-tos also provided living quarters for the workmen.

All of the excavated rubble did not fly out of the canal, however. The problem of rubble removal was nearly insurmountable until Orange H. Dibble invented a horse-operated boom or crane. The horse could stand above the canal cut, safely out of the way of the blast, while lowering a wooden bucket into the ditch. The bucket, when filled, was raised and emptied. This ingenious device eliminated much hard labor, and when horse cranes were placed every 70 feet along the canal, work speeded up considerably. When construction was finished, the area was littered with rubble piles totaling millions of cubic yards of rock and earth.

By mid-1823 the decision had been made definitely to terminate the canal at Buffalo, and October 1825 was set as the finish date. The laborers were pressed to work quickly in order to finish on time. Legend has it that a new incentive system was adopted—barrels of whiskey were stationed all along the path, each within view of the next; when construction reached a barrel, the men were free to stop and polish off the contents, and heaven help the laggard who slowed the progress of his fellow workers when they were within easy reach of the barrel! The enthusiasm and spirit of the workers were enormous; they kept up a remarkable pace and finished the canal on schedule.

When the Erie Canal crossed the finish line in October 1825, all of New York exploded with joy. The opening ceremonies lasted for days, accompanying the progress down the canal of Clinton and other dignitaries, aboard the *Seneca Chief.* The boat left Buffalo on October 26, amid speech making, cannon fire, and innumerable toasts. Each town thereafter tried to outdo the others with its garlands, flags, bunting, and drinking, but the biggest celebration of all occurred in New York City, where the attitude toward the canal had turned favorable. Trade in New York had already increased greatly because of the canal. Cannons and bonfires lined the Hudson as Clinton floated down to New York on November 4 to preside over the "wedding of the waters," when water from Lake Erie, brought down the canal in casks, was poured into the Atlantic. In keeping with the extravagance of the occasion, the committee members then added vials of water from the Rhine, Ganges, Nile, Mississippi, Columbia, Thames, Danube, Seine, Amazon, La Plata, Orinoco, Indus, and Gambia rivers!

Only about 160 freight boats and a few packets plied the waters in 1826, but by 1836 there were 3,000 boats on the Erie, often forming an unbroken line in either direction. Worries about debts due to the high cost of the canal—$7.7 million—were unfounded. Revenues exceeded all expectations, and money poured into the state treasury, bolstering visions of a tax-free future. Almost from the beginning there were too many boats, and the traffic jams were serious enough to lead to an authorization in 1835 to widen the canal to 70 feet and deepen it to 7 feet. A feeder system was begun in 1828 with the Oswego Canal, between Lake Ontario and the Erie, and the Cayuga and Seneca Canal, which connected with the Finger Lakes. The Chemung Canal to Elmira was dug in 1833; the Oneida Lake Canal in 1835; the Chenango Canal, joining the Erie with the Susquehanna at Utica, in 1837; and the Genesee Canal, from Rochester to Olean, also in 1837. As the canal fever spread to other areas, the competition for ditch diggers grew fierce.

Although some of these other canals were successful, none was ever as instrumental in shaping the political and economic character of the nation as the Erie, for it was the opening of the Erie that gave the decisive impetus for commerce to move across the country east and west rather than north and south. The Erie and subsequent canals eventually made possible the commercial success of Chicago in the Midwest and the eastern centers of Boston, New York, and Baltimore. Meanwhile, the cultural connection with the Atlantic states was reinforced by the settlement of large numbers of New Yorkers and New Englanders in the new West. It was the Erie Canal that carried thousands of European immigrants into the heartland of America. Cities like Rochester became boom towns: Rochester experienced a 70-percent turnover in population during the period 1827–1834, and by 1855, 44 percent of its population was foreign born. Buffalo grew from 2,095 persons in 1820 to 42,261 in 1850, as the population pushed ever further inland.

Today, the New York State Barge Canal has largely replaced the great old canal, which was filled in in many places and fell into disuse in others. Nonetheless, vestiges of the canal can still be found all across the state. Communities such as Fort Hunter still take pride in their heritage as canal towns and contain locks and canal beds within their boundaries. Major parts of great structures—such as the Schoharie aqueduct—can also be found.

This early engraving shows the entrance to the Erie Canal from the Hudson River. (*Frank Leslie's Illustrated Newspaper,* November 22, 1856)

The Schoharie aqueduct was one of several structures built to carry the Erie Canal over existing rivers and streams. A wooden trough that has long since disappeared carried the canal and boats while the mules walked along the towing path on top of the stone arches on the right. John B. Jervis was responsible for part of the aqueduct design. Constructed between 1839 and 1841, the aqueduct was put into service in 1845. Nine of the original arches remain; the others were demolished around 1915 to reduce resistance to stream flow. (Jack Boucher, Photographer, HAER Collection, Library of Congress)

The Morris Canal

The Morris Canal, completed in 1831, was unique among the canals constructed in the era following the Revolutionary War. Running from Newark Bay to Phillipsburg on the Delaware River, the canal climbed an astounding 914 feet and then dropped 760 feet. A system of inclined planes was used to accomplish these grade changes. Canal boats were floated onto cars mounted on rails, which were in turn pulled up the inclines by hydraulic power. This system was soon recognized as one of the engineering wonders of America.

George McCulloch, a resident of Morristown, New Jersey, first conceived of the canal in 1820. Because of transportation difficulties the local iron industry was foundering for want of coal, which was also in great demand in New York City. It seemed to McCulloch that Lake Hopatcong could easily provide enough water to supply a trans-Jersey canal that would connect Pennsylvania's coal-rich Lehigh Valley with New York. Following some successful lobbying by McCulloch, the state of New Jersey chartered the Morris Canal and Banking Company in 1824. McCulloch was not pleased with the creation of a bank to finance the canal, and the eventual wild speculation conducted with company funds bore out his suspicions.

Preliminary surveys for the canal had been begun even before the company was chartered. The route was strategically placed to pass through the heart of the industrial district in the northern part of the state. This route also took the waterway through hill country, which required costly detours. While the rectilinear distance between the two terminal points is only 55 miles, the canal itself was 99 miles long.

The $2.5 million of stock that the company issued in 1824 was quickly sold, but investors were required to make only partial payment on their shares, and so the project suffered from inadequate funding from the beginning. In order to guarantee the completion of the entire canal, compromises were necessary. The width of the ditch was limited to 32 feet at the water line and 20 feet at the bottom, and the depth to 4 feet. The locks were

Hydraulic-powered inclined-plane system of the Morris Canal, northern New Jersey. This system was the key feature that permitted the successful completion of the project in 1831. The use of turbine-generated hydraulic power to supply the energy necessary to lift canal boats over the 914-foot topographic barrier was a bold technical feat. In this circa-1900 photograph a boat is at the foot of Plane 9 West, Port Warren, Stewartsville, New Jersey. (Courtesy of the American Society of Civil Engineers)

planned to handle boats of 25-ton capacity. This seriously limited traffic once the canal was completed, as the 50- to 75-ton craft in common use on the nearby Lehigh Canal could not pass through such a lock.

The great engineering challenge of the Morris Canal was the topography to be traversed. Of all the early canals it was the highest climber, rising 914 feet from the tidewater of Newark Bay to the summit at Lake Hopatcong before dropping 760 feet to the Delaware River at Phillipsburg. This gave the canal an average vertical movement of 18 feet per mile, which is quite dramatic when compared with the Erie Canal's relatively gentle slope of 1 foot per mile. If locks alone had been used to transport the boats over this vertical distance, nearly three hundred would have been needed. This would have made the entire project prohibitively expensive. The grade changes were handled instead by the inclined-plane system.

There were precedents for the use of inclined planes in both England and the United States. In England the system had first been successfully used in a canal built for the duke of Bridgewater. In the United States an inclined-plane system had been developed for use on the South Hadley Falls Canal, located a few miles downriver from the present city of Northampton, Massachusetts. This canal, built in 1793 and 1794, was constructed to improve navigation on the Connecticut River. Although only two miles long, it was a significant undertaking. It involved cutting a vertical cleft 40 feet deep and 300 feet long through solid rock. A 230-foot-long inclined plane, having a vertical lift of 53 feet, was also built. The face of the plane was of stone covered with heavy planks. The boat car itself consisted of a watertight box with gates at either end. Boats were floated into the car, the gates closed, and the car emptied of water through side ports. The car was then pulled up or let down the plane on three sets of wheels, graduated in size to hold the car level. Power was supplied by two 16-foot-diameter water wheels. The canal and its inclined-plane system were successful at first but soon became a target for complaints from both fishermen, who claimed that the canal dam prevented shad and salmon from making their annual spawning runs up the river, and farmers, who claimed that the dam caused flooding in nearby areas. The Massachusetts legislature was successfully petitioned in 1800 for removal of the dam and inclined planes. This marked the end of the first inclined-plane system in America. A subsequent use of an inclined-plane system occurred on the Delaware and Hudson Canal. In 1825 Benjamin Wright developed plans for a series of inclined planes—powered in some locations by hydraulic pressure and in others by stationary engines—to be built to serve coal mines. Dump cars would be loaded at the mines and lowered or raised along mountainous slopes to the canal head. Despite opposition by those seeking to build tunnels, Wright's ideas eventually prevailed.

The use of inclined planes on the Morris Canal was suggested by James Renwick, a British engineer and professor of natural and experimental philosophy at Columbia University in New York, whom the company had hired as a consultant on the project. Renwick, familiar with the English use of inclined planes in canal building, and perhaps familiar with the Massachusetts experiment, advocated a series of planes similar to those developed by his acquaintance Benjamin Wright for the Delaware and Hudson Canal. He adapted and improved on this method for changing grades. For the Morris Canal he designed 23 inclined planes, which overcame most of the vertical distance; the remaining distance was accomplished by 23 locks, which were used for elevation changes of 12 feet or less.

Digging began in 1827 at Upper Newark Bay; from there the canal followed the Passaic River, crossing it by a massive 80-foot stone arch aqueduct at Little Falls. Several miles west of Little Falls the waterway crossed the Pompton River on a 236-foot wooden aqueduct supported by stone piers, and continued to Boonton, Dover, and the southern tip of Lake Hopatcong. There it changed course to the southwest and ran down to Phillipsburg and the Delaware River. The digging was done by about a thousand men, many of whom were Irish, who worked with wheelbarrows, picks, and shovels, aided occasionally by horse-drawn scrapers. Skilled craftsmen were hired to make the locks, lock houses, aqueducts, and inclines. Water retention was a problem in the Morris Canal bed, as it had been on other canal projects, and puddling was used in places. Locks were constructed using hydraulic cement. Despite the difficult terrain through which the canal passed, work progressed rapidly.

By spring 1832 the canal was complete, and the first boat to go all the way through the canal—the *Walk-in-the-Water*—arrived at Newark on May 20. The effect of the waterway was immediate and pronounced: the prices of coal and wood fell and business was stimulated; the languishing New Jersey iron industry was revived; farmers began to import fertilizers and ship out produce on the barges; and towns grew.

The inclined-plane systems worked well. The inclines were, in effect, boat railroads, with cars—strong wooden cribs or cradles resting on iron wheels—that ran on rails bolted to longitudinal timbers. When a boat approached the incline from below, a car was lowered into the water, and the boat crew, having unhitched the mules, floated the boat onto the car and attached it securely. The car and boat were then drawn up the incline by a chain to the top, where a lock awaited them. After the car had entered the lock, the gates were closed and water was fed into the lock, floating the boat off the car; the boat was then drawn out the other end of the lock and continued its journey up the canal on the higher level. The inclines averaged a lift of 63 feet, with about a 10-foot rise for every 100 feet of track. Even the first trial, which was made on the eastern slope of the canal between Dover and Newark on November 1, 1830, was astonishingly efficient. Five boats left Dover loaded with iron ore, and the first passed over the incline in only 8 minutes, which was less time than would have been needed to pass through a lock. Soon a simplification of the incline brought even better results. Instead of a lock at the top, the new summit-type incline had a small mound or dike that held back the water of the canal. The boat and car were pulled up over the mound and dumped directly into the water at the higher level. Unique "hinged" canal boats

A boat at the foot of Plane 11 East, Morris Canal. Normally a boat was taken from the lower level to the upper level of the canal by a double car, each part of the car supporting half of the hinged canal boat. A track consisting of two rails guided and supported the car as it was drawn up by a wire rope up to 3 inches in diameter, winding on a drum up to 13 feet in diameter. The drum was revolved by a turbine water wheel, placed in the wheelhouse halfway down the incline (left of center in this photograph). Water from the upper level of the canal supplied the turbine and was then discharged into the lower level. A boat entered a car in the water, was hauled up the incline, and was discharged at the upper level by simply floating off the car. The process could be accomplished in just a few minutes. (HAER Collection, Library of Congress)

were devised, able to be easily mounted onto and dismounted from the cars; each boat consisted of two sections, fore and aft, joined in the center by a special connection that allowed the boat to bend at critical loading points.

In a typical inclined plane the power for the cables came from a Scotch turbine that was housed in a vaulted stone underground room. Water from the upper level of the canal was channeled through a wooden flume to a vertical cast-iron pipe, called a penstock, through which it dropped and traveled through a J-shaped curve at the bottom that forced the water in four jets back up and underneath a huge 24-foot iron wheel. The movement of water through the turbine caused it to revolve, moving the cable that hauled the boat-cars up the incline.

The simplicity of the system was a source of interest to engineers from all over the world, who came to see this unique, water-powered incline device; but it seems to have been regarded with suspicion by other American canal builders, and the design was not extensively copied (although James Geddes and Nathan Roberts had knowledge of the system when they set about building the inclines for the famous Portage Railroad, which also involved a canal system). Their doubts were not entirely unfounded, for the system had its drawbacks. One was that the incline cars did not provide the continuous support along the outside of the boat that the water in a lock did, and this, when the boat was fully loaded and being lifted, tended to weaken the sides of the vessel. The Morris Canal does not seem to have been affected by this problem, however, and the inclines functioned exceptionally well throughout the canal's life. Their cost compared favorably with the price of the Erie Canal's locks—$210 per foot of elevation on the Morris and $1,100 per foot on the Erie.

The canal was extensively utilized, and it soon became evident that it was too small. The locks could accommodate boats no more than 8.5 feet wide, 60 feet long, and with a 25-ton capacity. Other canals were serving 50- to 70-ton boats with the same number of operators. Because the final cost had been

A boat cradle on Plane 7 East. The powerhouse is on the right. (HAER Collection, Library of Congress)

This early photograph shows a typical canal boat at the summit of an incline—the hump of land at the top that held back the water at that level. (HAER Collection, Library of Congress)

$2,104,413, well over the original estimate of $1,000,000, and because many subscribers had defaulted in their payments on the stock, the company had no money to enlarge the canal. The Morris was also hampered because it touched only the upper end of Newark Bay and did not reach all the way to New York Harbor. A costly 12-mile extension was begun across the Bayonne neck to Jersey City, but by the time it was finished in 1836, the company's finances were in a dreadful condition. The company made use of its banking privileges, performing in every respect like a wildcat institution. With questionably obtained assets, it managed to finance an enlargement of the canal and locks in 1840 and 1841, and then collapsed in an ugly bankruptcy suit.

Nevertheless, the Morris Canal was leased to a new company, and with the enlarged capacity of the boats (60 to 65 tons) the annual tonnage increased. The peak years came between 1860 and 1870. After that the railroads, able to carry the same weight of goods a comparable distance in four hours instead of four days, began to take over the canal business. Attempts were made to obtain speedier boats by propelling them with gasoline, electricity, and steam, but speeds greater than 3 miles per hour created a swell that washed out the canal banks and caused the boats to settle on the bottom. Eventually the slower speeds, the longer route, and the fact that it froze over for four or five months of the year—during which it was used by skating commuters on their way to and from work—brought the canal to the end of its economic usefulness.

In 1924, in what seems a heartless attempt to remove all traces of this early American engineering marvel, the tracks were torn up, the ditch drained, and the aqueduct bridges dynamited. Nonetheless, parts of the great canal still remain. Enthusiasts have even restored parts of the inclined plane system. Traces of the original inclines can be seen at places such as Waterloo Village.

Morris Canal: a maintenance boat at the summit of Plane 9 West in 1898. Note the use of the unique hinge, which eased the troublesome problem of passing the boat over the lip that maintained the water level at the upper end of the plane. (Library of Congress)

This photograph of Plane 5 East, west of Dover, shows the canal bed. (HAER Collection, Library of Congress)

The Ohio State Canal System

Ohio's magnificent canal system, built between 1825 and 1848, included canals, slack-waters, locks, bridges, and dams, some of which continue in use today. Its engineers integrated natural and artificial waterways to produce a 1,015-mile transportation complex that opened up Ohio and linked it to the Atlantic states. This ambitious project was probably the single greatest impetus to Ohio's early development, making possible commercial agriculture and spurring industrial growth. It brought people, materials, products, technology, and general prosperity to what had been a relatively depressed frontier area. The record of the construction program speaks highly of the technical, managerial, financial, and political skills of Ohio's leaders at the time. Of the system's complex of waterways, some were publicly financed, others privately. The two major components were the Ohio and Erie Canal and the Miami and Erie Canal; tributary waterways are shown on the accompanying map. The early towpath canals have largely been abandoned since 1913 and are generally in a ruined condition, although significant remnants survive in scattered locations. Slack-water navigation is still pursued on the Muskingum River Improvement.

In many ways Ohio's early canals are offspring of the Erie. The possibility of connecting Ohio to the east-coast market via Lake Erie and the Erie Canal was an alluring one. An early proposal to construct a canal connecting Lake Erie with the Ohio River was widely supported, but it was unclear exactly where the canal should lie. Residents of the populous eastern part of the state insisted that the canal leave Lake Erie at Cleveland by way of the Cuyahoga River, proceed southward to Columbus along the Scioto River, and reach the Ohio at Portsmouth. Those in the western portion of the state demanded that the canal be located along the Miami River, connecting Cincinnati with Hamilton, Dayton, and—via the Maumee—with Toledo on Lake Erie. An impasse resulted. Finally a small sum of money was voted by the legislature for a survey of five proposed routes for a

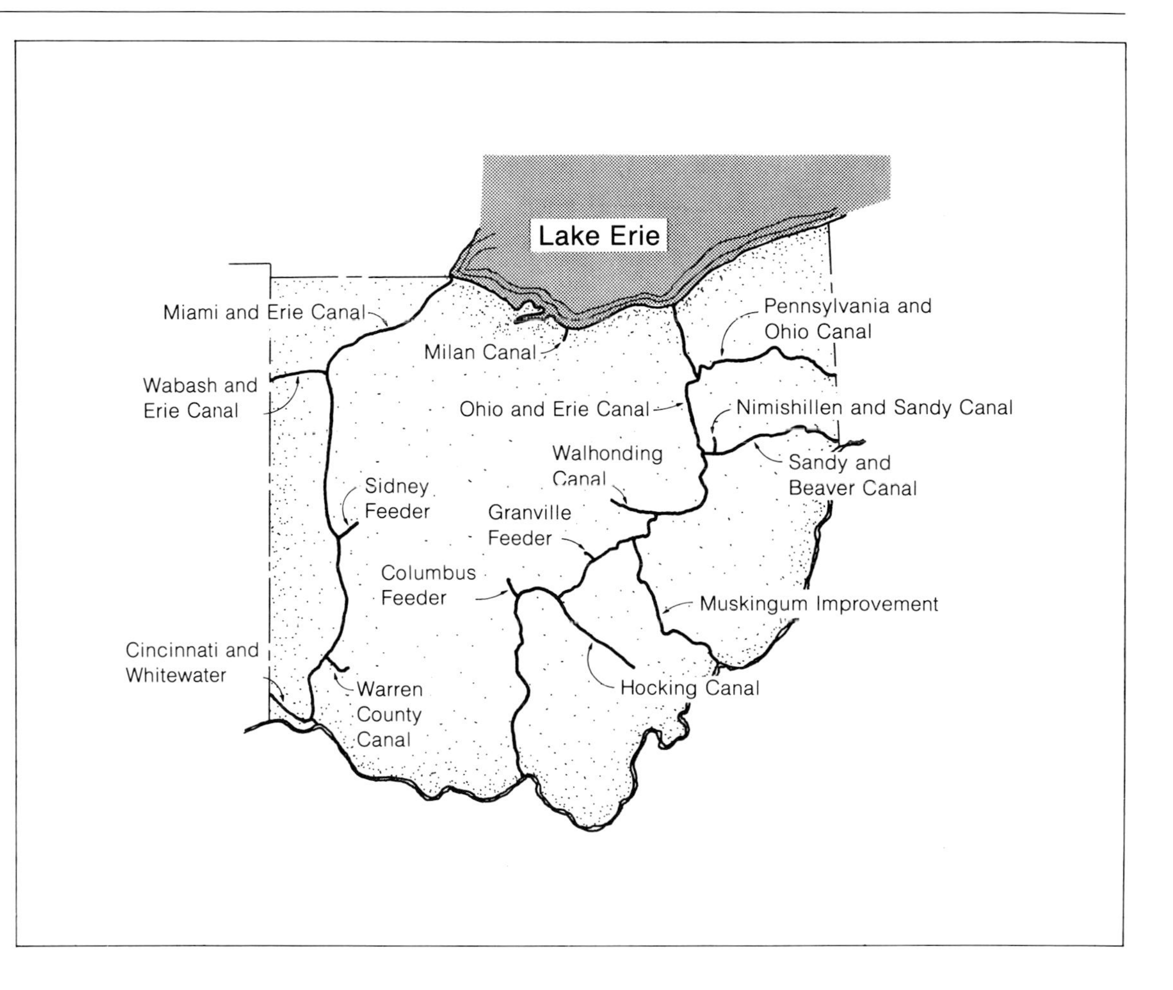

Ohio State Canal System. Under construction from 1825 to 1848, this complex of waterways included both locks and slack-waters, some of which continue in use today. By integrating natural and artificial waterways, its builders produced a 1,015-mile transportation system that opened the heartland of America to commerce with the Atlantic states. Among its engineering triumphs was the largest man-made lake in the world at the time.

north-south canal. James Geddes, who was then working on the Erie Canal, undertook the job. Two routes, eventually to become the locations of the Ohio and Erie Canal and the Miami and Erie Canal, were favored. But for some time no decision could be reached because of the intense sectionalism involved. At last in 1822 Governor Allen Brown succeeded in getting a canal commission appointed. Three years later, in a compromise measure, the Enabling Act of 1825 authorized the construction of the Ohio and Erie Canal and also provided for construction of the Miami and Erie Canal from Cincinnati to Dayton. The latter canal was to be extended to Lake Erie at a future time.

A final survey was made by David Bates, who had worked along with Benjamin Wright on the Erie Canal. He largely followed the route laid out by Geddes. Construction of the Ohio and Erie Canal began on July 4, 1825, with a ceremony in which New York's De Witt Clinton, of Erie Canal fame, was the principal speaker and turned over the first spadeful of earth. This took place at the Licking Summit level, three miles west of Newark; the official party then journeyed to Columbus and on to Middletown, where ground was broken for the Miami and Erie Canal.

Contracts had been let to professional companies whose crews had helped build New York's waterways, and to many local citizens as well. By late fall two thousand men were at work near Licking Summit. Their main camp was at Cuyahoga Falls (today's Akron). About a quarter of them were Ohio natives, and the rest were from New York State. Work also proceeded on the Miami canal. Construction was always hampered by lack of money, but the work never entirely stopped. This was due largely to the efforts of Alfred Kelley and Micajah T. Williams, two of the state's canal commissioners. Somehow they always managed to scrape together sufficient funding to keep the projects moving.

In 1827, at Cuyahoga Falls on the Ohio and Erie Canal, the *State of Ohio* was launched to make a journey along the completed portion of the canal to Cleveland. The short 37-mile section was declared open, and within a few months some $1,500 in tolls had been collected—a cheering sign. On December 1, 1832, the lock gates at Portsmouth were opened for the first time, and boats passed into the Ohio River.

The first through boat from Cincinnati reached Dayton in January 1829. Soon a clamor arose for the promised northern extension to the Maumee River and Lake Erie, but this was to be some time in coming. The Panic of 1837 caused severe financial problems, but somehow funds were found with which to continue. A cheering crowd greeted the first boat from Cincinnati to Piqua, the seat of Miami County, on July 4, 1837.

Of the many problems encountered on the Miami and Erie Canal, none could compare to the construction of the Grand Reservoir in Auglaize and Mercer counties. Formed by two dams, 25 feet high, one 2 miles and the other about 4 miles long, the reservoir covered about 27 square miles and was held to be the largest artificial lake in the world at the time. With its completion in 1845, water flowed into the channel, and the first boat passed from Toledo to Dayton in June.

While these canals were being built, other private canal projects were also launched—often with the intention of connecting to the Pennsylvania canal system. In 1836 the state legislature began subsidizing the highly publicized Ohio and Pennsylvania Canal. This canal opened its channel from the Pennsylvania border to Akron in 1840, but the Pennsylvania connection was not built and the project was not a success. Another private group built a short canal into the coal-rich Hocking Valley. The state took it over and extended it to Athens. It proved a difficult canal to keep in repair because of the destructive effects of the freshets on the Hocking River. The 73-mile Sandy and Beaver Canal, running from below Akron eastward to the Ohio River 40 miles downstream from Pittsburgh, was undertaken in 1835 and completed in 1846. It proved a financial disappointment. The Walhonding Branch of the Ohio and Erie, which originated at Coshocton, consistently lost money as well. The short Warren County Canal, chartered in 1830 to connect Lebanon with the Miami Canal, was also a doubtful proposition. The state stepped in to finish the half-completed canal in 1836. The Muskingum project, a combined canal and river connection that provided a short cut to Pittsburgh and the Pennsylvania canals, did prove profitable. The Whitewater Canal, which connected Cincinnati with Indiana's Whitewater River, never brought in the predicted revenues. It was abandoned in 1847, and the old channel was drained and converted for use as the right of way of the Cincinnati and Indianapolis Railroad.

If many of these minor canals proved less than profitable, nonetheless the main waterways were an enormous success. Business flourished for a long time, with towns such as Akron owing their phenomenal growth to the canals. The coming of the railroads caused the decline of the Ohio system, but many of its components hung on for a long time and parts remain in service today. The Miami and Erie was still collecting tolls in 1903. The Muskingum improvement still finds use. Remnants of locks and other traces of the canals can be found today at Canal Fulton and many other locations.

A mule team pulls a freight boat along the Ohio and Erie Canal, circa 1870. (HAER Collection, Library of Congress)

Chapter Two

Roads

Introduction

For the colonists living along the Atlantic seaboard in the seventeenth and eighteenth centuries, rivers and sheltered coastal waterways provided the chief transportation routes. Building a road through swamps and forests and across rivers was far from easy. Major roads connecting the larger settlements were few and with relatively unimproved surfaces that made them unusable for significant portions of the year. The purpose of many early roads was to provide direct routes for military movement. General Edward Braddock's 115-mile road from Fort Cumberland to present-day Pittsburgh, for example, was constructed in 1755 for military purposes. The King's Road, constructed through the territory that is now Florida and Georgia between 1766 and 1775, is another example of a road that was often used by the military but also contributed to the rapid growth and development of a large region.

By 1750 a regular stage provided service between Philadelphia and New York. Stages soon ran between other population centers as well. About the same time, a land postal service between principal cities was developed by colonial authorities. The roads themselves—their construction and their upkeep—were invariably the responsibility of local governments. Consequently, elected officials would often decide the course of a road, determine the requirements for its construction (for example, whether there should be side ditches or pavement with gravel or pounded stones), and raise funds for its upkeep. These governments were often authorized to require local inhabitants to work on the roads, although lotteries, donations, and private subscriptions occasionally supported roadwork as well. These means frequently proved ineffective, however, and by the end of the Revolutionary War there was considerable discussion of state assistance for primary roads. State governments responded by chartering private turnpike companies to build toll roads.

Virginia chartered a company that built a pike from Alexandria westward to near Berryville in 1785. A form of statewide transportation plan was developed in Pennsylvania in 1791 by the Society for Promoting the Improvement of Roads and Inland Navigation. The Philadelphia and Lancaster Turnpike Road Company was established by legislative acts in Pennsylvania in 1792. Its charter included requirements for road widths, gradients, and surface improvements (". . . an artificial road, which shall be bedded with wood, stone, gravel, or any other hard substance well compacted together, [of] a sufficient depth to secure a solid foundation to the same . . . shall be faced with gravel, or stone pounded . . . to secure a firm, and as near as the materials will admit, an even surface"). The resulting turnpike was financially successful and became a prototype for other comparable projects.

Toll roads became an accepted means for financing primary connecting roads. New York had chartered 67 companies by 1807, and Connecticut had chartered 50 by 1808. Toll roads were built by contractors or hired laborers who had been trained by road engineers. In general, the contracted work included the building of small-scale wood bridges over rivers and streams, the cutting of timber, and the laying of corduroy (logs placed side by side) in swampy areas.

The route from Philadelphia to New York was completely stone surfaced by 1812. To serve westward expansion and trade, four major transmountain roads were built: the Lancaster Pike was extended to Pittsburgh, and turnpikes were constructed in Maryland, New York, and Virginia. The road in Virginia, the Northwest Turnpike, was state owned. The practice of constructing turnpikes continued to spread until by mid-century there were hundreds of companies in operation.

did not come until 1956. By this time the technology of road and bridge construction had advanced considerably over the methods employed in the 1930s. Throughout these developments the American Association of State Highway Officials, the Bureau of Public Roads and its successor agency, the Federal Highway Administration, and other organizations provided the national perspective for the country's road-building efforts, while state and local bodies maintained their historic roles.

The King's Road

The King's Road, built between 1766 and 1775, extended from St. Augustine in the British colony of East Florida to Fort Barrington on the Altamaha River, near present-day Darien, in Georgia. It was a principal transportation route, facilitating settlement of the Florida interior as well as linking Florida with the colonies to the north.

The lands of North America were the subject of claims and disputes by European powers from the time they began their explorations into the new hemisphere. The Treaty of Paris of 1763, by which France relinquished all her claims east of the Mississippi (including Canada), also exacted terms from France's ally, Spain. Havana had fallen to the British, and in order to redeem the port, Spain agreed to cede Florida. Thus Spanish Florida, along with some French possessions, passed into British control. It was decided to divide the territory into two provinces, East and West Florida, with the border drawn along the Apalachicola River.

James Grant, the first English governor of East Florida, arrived in St. Augustine in 1764. The small population made it difficult for him to form a government; it took two months to assemble enough qualified men to form His Majesty's Council for East Florida. The population grew slowly: the expense of travel there was prohibitive to potential settlers, and living costs and land surveying fees were high. In addition, restrictions were placed on land agents, and there were problems with absentee proprietors, factors that tended to hamper settlement. Anxious to colonize the province, Grant addressed the essential question of transportation. On March 1, 1765, he wrote to the Lords Commissioners for Trade and Plantations in London, outlining the need for a road from St. Augustine to Fort Barrington, located on the route to Savannah at an important crossing of the Altamaha River. In the same report he recommended a site on the Ponce de Leon Inlet (then called The Mosquettos) as the location for a future settlement and proposed to build a road—a continuation of the one previously suggested—to this site (which was eventually selected by Dr. Andrew Turnbull for his New Smyrna colony).

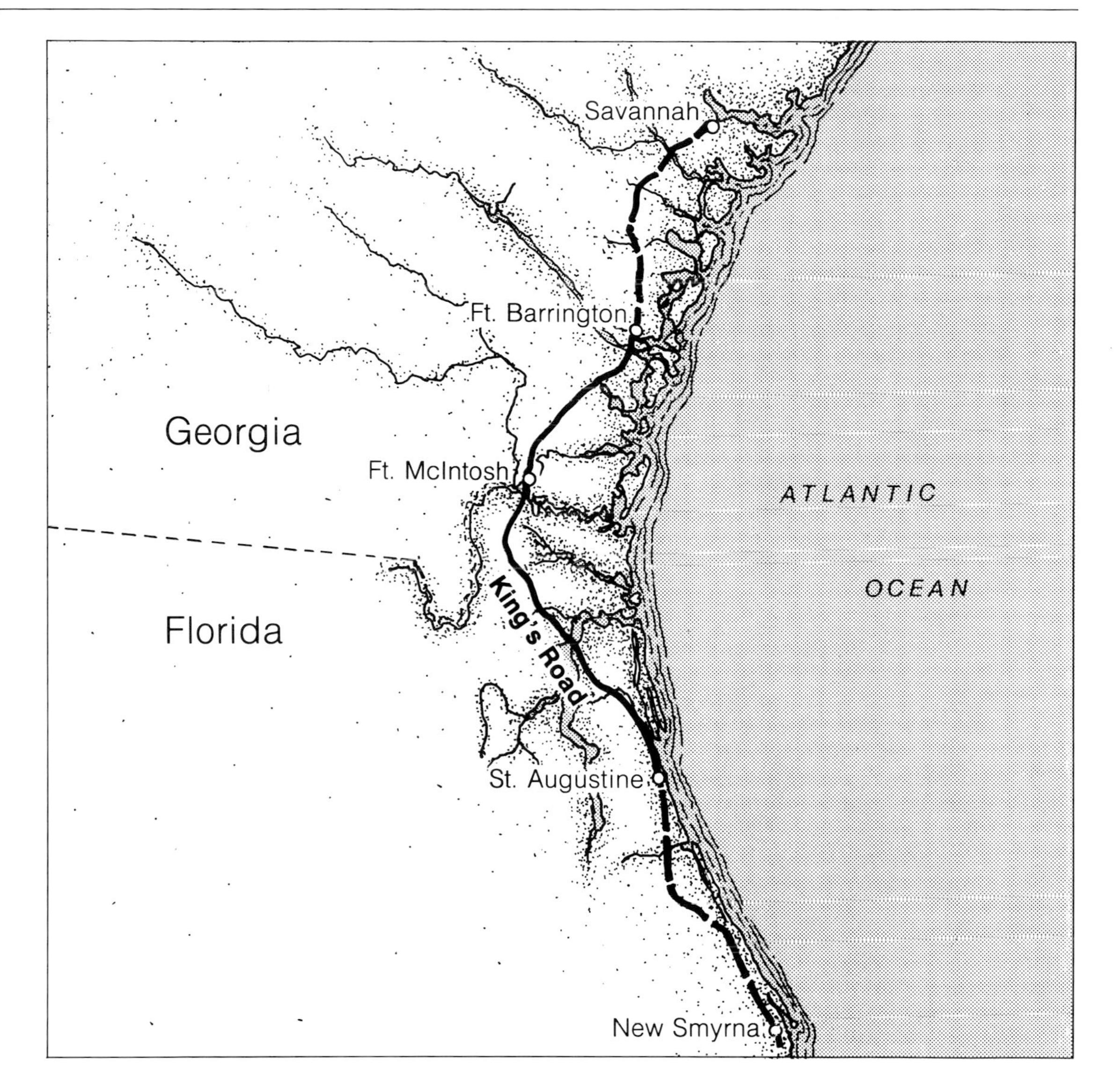

King's Road, Florida to Georgia. Built between 1766 and 1775, this was the principal overland transportation link between what was then British Florida and the colonies to the north. Construction of such a road was a remarkable feat for the time.

Grant was determined to raise funds for the road by subscription and himself signed first for 20 guineas to set a good example. Several hundred guineas were eventually collected locally.

In 1766 construction of the King's Road began. In all likelihood the route followed trails used by the Spanish settlers and the Indians before them; as early as 1565 friendly Indians had guided Menendez, the first Spanish military commander of Florida and the founder of St. Augustine, over trails northward to attack the French at Fort Caroline on the St. Johns River. A foot or horse path from Savannah to St. Augustine shows on a map drawn by J. Mitchell for the Board of Trade and Plantations in 1755. In June 1767 a Mr. Moncrief was paid £300 for building a bridge over the San Sebastian Creek (near St. Augustine) as well as a road over the adjacent marsh. Some form of the road north was known to exist by July 1770. Later accounts show payments to Joshua Yellowley of £500 between 1772 and 1776 for road construction from St. Augustine northward to near present-day Jacksonville, and £661 to Charles and Jermyrn Wright for a further connection to the St. Mary's River, the boundary of the colony. The Georgia assembly had previously committed to build a connecting road from the St. Mary's River to Fort Barrington. Not much is known about the actual road constructed, although a 1787 report describes it as an excellent road, 16 feet wide at the St. Mary's River, and an 1820 report notes that the road at St. Mary's had been widened to 25 feet. Most likely the road was simply a cleared strip making use of corduroy (logs laid in parallel on the ground) in low places.

In the meantime, Dr. Andrew Turnbull, a visionary wanting to establish a new colony, was preparing for his settlement in East Florida. He requested that the Board of Trade apply £400 to making roads, bridges, and ferries that would link his colony with others. In August 1767 Governor Grant informed the Board of Trade that he was "obliged to an Indian friend of mine, who took the trouble with his Gang about twenty in number to Blaze a Road . . . from this place [St. Augustine] to Mr. Oswald's plantation at Timouka and from thence to Doctor Turnbull's, South of the Mosquetto Inlett."

On June 26, 1768, Turnbull arrived in St. Augustine with about 700 Minorcans, Italians, and Greeks. The rest of the 1,403 original settlers arrived over the next several weeks and were on the plantation at New Smyrna by August 10. Serious building of the King's Road south of St. Augustine, however, did not get under way until 1770. Between 1770 and 1773 payments were made to Richard Payne for construction from the San Sebastian Creek south to Matanzas, and to Robert Bissett for a section from Matanzas to Indian River. Construction continued through 1775. The exact southern terminus of the road beyond New Smyrna remains unclear, but it was probably near the property of Elliot's Plantation, some 20 miles south of the Ponce de Leon Inlet. The road was difficult to construct, since it passed through many swamplands and crossed many rivers and streams.

Growth of the province proceeded slowly but predictably. The road was barely complete on the outbreak of the American Revolution. Colonists in the two Floridas were not sympathetic toward the independence movement. East Florida became a haven for fleeing Tories, many of whom traveled south over the King's Road. A stream of some 7,000 refugees came in 1778 alone. By then the road north of St. Augustine had seen increasingly warlike action, as border raids and plundering mounted. A group of Tories, organized as the "Florida Rangers" in 1775, conducted raids into Georgia. In their turn, the American rebels made numerous expeditions in an effort to annex East Florida. Charles Lee led one attempted invasion in 1776. Robert Howe's 1778 expedition penetrated as far as the St. Mary's River, generally following the King's Road.

The defeat of the British in 1781 precipitated a new and greater influx of Loyalists following the road, as Charleston and Savannah were evacuated. The colony was thrown into confusion, which was heightened in 1783 when it was announced that the second Treaty of Paris called for the cession of East Florida to Spain. A major evacuation was undertaken. Some refugees chose to return to the United States, leaving by way of the King's Road; of those who wished to remain British, some went to Nova Scotia, but most were resettled in the West Indies.

During the War of 1812 Americans once again made attempts to annex East Florida. Not only was it held that the United States should extend to the Gulf of Mexico; there was also an effort to stop Britain, who (with Spain largely under her domination) was using Florida as a provisioning place for her fleet. In their forays into Florida the American raiders traveled along the King's Road.

Eventually Florida became part of the United States, and the King's Road served the peaceable function of linking Florida to the rest of the country and opening up its interior for settlement. Today only scattered fragments of the road remain. Large stretches of it are buried under city streets, railroad tracks, and highways. Some lie beneath fields. Many place names and street names in the area contain the word "King" and serve to remind one that the King's Road once passed by.

The National Road

The National Road, built between 1811 and 1839. Eventually stretching from Cumberland, Maryland, to Vandalia, Illinois, this road was the precursor of today's federally funded interstate highway and represented the highest standards of design and construction at the time. This photograph, taken in 1910, shows a portion of the National Road near Washington, Pennsylvania. (Courtesy of the American Society of Civil Engineers)

The National Road, stretching from Cumberland, Maryland, to Vandalia, Illinois, was constructed between 1811 and 1839. It has been known by a variety of names, including the Cumberland Road, Ohio's Road, the United States Road, and Uncle Sam's Road. The original route is now followed by old U.S. 40 and Interstate 70. The first road to be built entirely with federal funds, it is significant not only for the magnitude of the project but also for its role in the westward expansion of the young United States. By facilitating trade and settlement, the National Road helped establish the economic and political unity of the growing country.

During the first two thirds of the eighteenth century, the lands west of the Ohio River were claimed by the French. The British, wishing to secure the allegiance of settlers who might otherwise find loyalty to the French more profitable, sought trade with the western settlements. The young Major George Washington proposed water routes to the Ohio River via the Potomac and James rivers. Land routes were also recognized as crucial to frontier growth. The only routes westward through the Appalachian and Allegheny mountains were Indian trails. It was one of these, Nemacolin's Trail, that Washington followed out of Cumberland, Maryland, in 1754. He had been dispatched by Governor Dinwiddie to meet the Indians and the French at Fort Duquesne, the site of present-day Pittsburgh. Forced to erect a quick shelter, Washington built Fort Necessity, near the present Uniontown, Pennsylvania. Unable to defend it, he surrendered the fort on July 4, 1754, and returned to Virginia.

In 1755 the British sent General Braddock to run the French out of the Ohio country. Setting out from Fort Cumberland, Braddock and his men followed Washington's route. Ahead of the army marched six hundred road builders, clearing the way for the four-mile column of men and equipment. The army was defeated by the French and Indian forces, and Braddock himself was killed. The 12-foot-wide road that the British had hewed from the forest, however, remained as the rudimentary beginning of the National Road.

The army was not the only group to make use of the Indian trails in opening the frontier. Ebenezer Zane had migrated westward to the mouth of Wheeling Creek in Virginia, where in 1770 he founded a settlement known as Zanesburg (later Wheeling). As he watched settlers arriving in Wheeling to begin their trips down the Ohio River, Zane realized that the usefulness of the river was limited by winter ice and spring flooding. Zane had already blazed a trail from Fort Pitt (Pittsburgh) to Wheeling; he proposed to extend the trail, following an old Indian trace. Zane saw the possibilities for enhancing settlement of the area and providing an overland mail route. One of the problems he anticipated was fording rivers, for the proposed road crossed the Muskingum, the Hocking, and the Scioto.

In 1796 Zane petitioned Congress for federal assistance. His modest request was for landing sites at each of the three river crossings, along with the money to "defray all expenses which may be incurred in surveying and laying off such lots of land." Congress gave him a square mile at each of the three sites; these are now the towns of Zanesville, Lancaster, and Chillicothe. Zane surveyed the land, widened the trace, and provided ferry service across the streams. Toll rates were determined by judges of the Northwest Territory. Ebenezer Zane became a wealthy man as he helped expedite the settlement of southern Ohio.

By 1800 the population of Ohio had reached 60,000, the minimum required for a territory to petition for statehood. With the active encouragement of President Thomas Jefferson, Ohio became a state in 1802. One of the provisions of the Ohio statehood act specified that 2 percent of the proceeds from the sale of public lands by the federal government would be used for building roads. The act called for "the laying-out and making of public roads leading from the navigable waters emptying into the Atlantic, to the Ohio, to the said state and through same, such roads to be laid out under the authority of Congress, with the consent of the several states through which the roads shall pass." This was the first provision for federal aid for road construction across state lines and provided a precedent for similar public aid to other territories on admission to statehood. It led to the passage by Congress of legislation establishing the first true national road to the West. A committee appointed by President Jefferson had recommended the right-of-way for the road, and in 1806 Congress appropriated $30,000 for the surveying and building of a road from Cumberland, Maryland, to the state of Ohio. Jefferson's secretary of the treasury, Albert Gallatin, once a surveyor himself, supervised the project.

Major advances in road construction technology were taking place about this time, some of which were incorporated into the National Road. These developments had been conceived abroad by Thomas Telford and John McAdam. Telford developed the notion of using heavy stone foundations with smaller stones for the upper layers of the road. McAdam extended Telford's ideas. He recognized the load-bearing capacity of dry, compact soil. An impervious surface, combined with crowning for drainage, he contended, would suffice for normal traffic. The impervious surface that McAdam devised consisted of broken stone of a small, uniform size. The compacting of the stone that occurred with use actually improved the resistance of the surface to water penetration.

The first section of the road, started in 1811, presented the most severe physical obstacles to the surveyors, engineers, and builders. Just west of Cumberland stood Negro Mountain, Big Savage Mountain, and Keysers Ridge, the highest points in the entire length of the road. To bridge the many small streams along the route, the most commonly employed structure was the semicircular stone arch. One that still stands is the bridge across Castlemans River in Grantsville, Maryland. Built in 1813, its 80-foot span was one of the longest in the country at the time. It can be seen today preserved in a park along Route 40. Another type of structure employed to cross streams was the more easily built S bridge, which curved like the letter S rather than extending straight across the stream.

The War of 1812 delayed work on the road; nevertheless, it was completed to Wheeling by 1818. Ebenezer Zane's town was by then a booming river port. Thousands of wagons rolled into Wheeling each year, with people and cargo bound for the western settlements. Among the vehicles were many Conestoga wagons, a type that had originated in the Lancaster region of Pennsylvania; the distinctively curved wagon beds made them much safer than straight-bed wagons. The cargo that came into Wheeling was sometimes loaded onto boats that traveled the Ohio River. Those wagons continuing west overland, however, had to cross the river by ferry. A real bottleneck existed in the National Road until 1849, when Charles Ellet built his famous suspension bridge across the Ohio.

Meanwhile, Ohioans waited for the road to reach their state. When construction began in Ohio in 1825, the route followed Zane's route between Wheeling and Zanesville. From there it continued west through Newark, Columbus, and Springfield. The most notable structure was the Y-shaped bridge at Zanesville. Originally constructed as a wooden covered bridge, it was rebuilt in 1900 as a concrete structure.

In 1829 the National Road was started in Indiana, with construction proceeding both ways from Indianapolis. Encountering none of the difficulties of mountains and streams that had challenged the builders of the first portion, the Indiana builders constructed a road that plowed straight through the flatlands. Pioneers swarmed into the area so rapidly that the unknown Indianapolis became the seat of government.

From its inception the National Road had been the subject of controversy. At issue were the rights of states versus the powers of the federal government. On campaign trails and in Congress the battle raged: was it proper for Congress to appropriate money for improvements that would benefit a state or an area rather than the whole country?

Each change of administration influenced the road. Jefferson had been a champion of the project; Monroe, on the other hand, in 1822 vetoed a bill to repair the road, arguing that Congress had no authority to pass such a bill and that, indeed, such power could be granted only by a constitutional amendment. Monroe's successor, John Quincy Adams, was a supporter of roads and canals. Con-

Many major bridges were built as part of the National Road, including this one across Castlemans River, constructed in 1813. (Courtesy of the Smithsonian Institution)

S-shaped bridges were built along the National Road when streams or rivers were encountered at an angle to the roadway. This shape allowed the bridge to be built perpendicular to the bank, thus minimizing spans and construction difficulties. (Courtesy of the Smithsonian Institution)

gressional opinion, however, was strongly divided on the matter of public works, and the partisans continued their battle into the Jackson administration.

During these years the condition of the road deteriorated as funds for repairs failed to be appropriated. In 1831 Ohio passed an act, which was approved by Congress, taking under its care the portion of the road within the state and providing for the erection of tollgates. Similar acts were passed in Pennsylvania and Maryland in 1832 and in Virginia in 1833. Pennsylvania and Maryland refused to accept the road until the federal government repaired it and erected tollgates; their final acceptances occurred in 1835.

The last regular appropriation for the National Road was made in 1838. Congress refused further help to the States, although a congressional committee reported in 1840 that the United States was bound in good faith, under compacts with Ohio, Indiana, Illinois, and Missouri, to complete the road. In 1848 the government transferred all rights to the appropriate portion of the road to the state of Indiana, and in 1856 it passed a similar act for Illinois. The road was national no longer.

The constitutional dispute and regional rivalries in Congress were not the sole reasons for differences over the National Road. Westward expansion of rail lines added to the legislators' hesitancy to allocate funds to a project that was in danger of becoming obsolete. With the advent of the automobile, however, a movement to repair the road took hold. The states through which it ran appropriated money and set to work to make amends for the deterioration it had undergone during the latter half of the nineteenth century. By 1923 the road was newly paved from Cumberland to Indianapolis. The Old Pike became part of the National Old Trails Road, extending westward to the Pacific coast; after World War II this was incorporated with U.S. Route 40 as an ocean-to-ocean highway. Today the National Road, partially superseded by Interstate 70, is still in service as one of America's most historic highways.

The Mullan Road

The Mullan Road, built between 1859 and 1862, stretched 624 miles from Fort Benton, Montana, to Walla Walla, Washington. It connected the head of navigation of the Missouri River to the terminus of the Columbia River transportation system.

Prior to the Mullan Road's construction there were three routes open to travelers to the Pacific Northwest: overland on the Oregon Trail, via ocean and portage across the Isthmus of Panama, and the long ocean voyage around Cape Horn. According to Captain John Mullan, the army officer who oversaw its construction, the road was an effort to "secure a route where the sea travel would be avoided in toto; and at the same time have the land transit the shortest minimum." The building of the road involved 120 miles of timber cutting 25 feet wide and 30 miles of excavation 15 to 20 feet wide; the remainder passed through open timbered country or rolling prairie and involved minor grading, with the laying of corduroy (logs laid parallel to one another) in swampy areas. Many bridges were constructed, including twenty over the Coeur d'Alene River alone. Unlike most other roads of the time, which followed existing trails, the route was selected on the basis of engineering reconnaissance surveys. (The information gathered was used in the building of later roads and railroads, including the Northern Pacific Railroad.) Although originally constructed for military reasons, the Mullan Road was used only once for such purposes. It proved valuable, however, in providing access to the Northwest for settlers, miners, and stockmen.

John Mullan was born in 1830 and graduated from West Point in 1852 as a second lieutenant. In 1853 he volunteered to participate in a project for which Congress had just appropriated $150,000: the exploration and survey of a railroad route from the Mississippi River to the Pacific Ocean. During 1853–1854 Lieutenant Mullan led a party that explored the wilderness between the Missouri and Columbia rivers and discovered a pass suitable for a wagon road. Construction of the road was delayed by the Indian Wars and by lags in the appropriation of funds. At last, on March 18, 1859, the War Department assigned $100,000 for the building of a military road from Fort Walla Walla, Washington, to Fort Benton, Montana, appointing Mullan as officer in charge of construction.

Between June 1859 and August 1860 Mullan and his party—100 enlisted men, 3 subalterns, and 100 workmen, including a few topographers and civil engineers—completed a road suitable for horses and wagons. There were changes to be made, however, and Mullan, after receiving the necessary endorsement from the War Department, resumed work in May 1861. The road was rebuilt from Walla Walla to the Coeur d'Alene mission to take advantage of routes developed by settlers and to incorporate ferries that had been established on the Snake and Spokane rivers. The work was completed by August 1862.

The Mullan Road contributed significantly to the growth of eastern Washington. During the decade from 1860 to 1870 it provided access from Walla Walla to newly discovered mining areas. From 1870 until 1883 it was a primary commerce route between Walla Walla merchants and settlers in the territory. Military use, a major goal of the original planners, was limited to one occurrence in the 1860s. Immigrant groups made use of the road, the last large caravan making the trip in 1867. However, the poor condition of the road—no federal funds had been made available for maintenance—hampered use by both soldiers and settlers. It was only the discovery of gold in the northern Rocky Mountains that kept the Mullan Road actively used.

The 1860s saw Walla Walla grow from a seven-house outpost to the largest city in the Washington Territory, with a population of 1,394. The city's location, at the junction of the Mullan Road and the Oregon Trail, was crucial. The Mullan Road became the supply route for mule trains going to the gold-producing areas of Montana, northern Idaho, northeastern Washington, and eastern British Columbia. Miners traveled to Walla Walla to buy goods, bringing gold dust as payment. In addition to serving as a supply depot for packers and miners, Walla Walla became the winter residence of many miners whose activities were curtailed by cold and snow.

Mullan Road, Walla Walla, Washington, to Fort Benton, Montana. This 624-mile road was the first major engineered highway in the Pacific Northwest. Surveyed between 1853 and 1854 and constructed between 1859 and 1862, it connected the Missouri and Columbia river basins, thereby accelerating the development of the northwest. (Courtesy of the American Society of Civil Engineers)

By 1867, however, mine productivity had decreased. The deteriorating condition of the Mullan Road made travel more difficult. Light-draft steamboats on the Missouri River began to replace the Walla Walla–based mule trains as suppliers to the Montana mines. As the mining business declined, farming in eastern Washington grew rapidly. Settlers flowed into the territory to stake land claims. Walla Walla became the center of trading and milling for the surrounding agricultural areas, which were reached via the Mullan Road. Sheepmen and cattlemen used the road to drive their stock to summer range and back to market.

The Mullan Road brought unforeseen results. Information compiled by its surveyors and builders was used in the construction of the Northern Pacific Railroad and in later highway building. The relative ease of rail transportation contributed to the decline in use of the road. On September 8, 1883, the Northern Pacific Railroad was completed with the customary "last-spike" ceremony. Among the guests invited by railroad president Henry Villard was Captain John Mullan. The records confirm that Mullan was appreciated and remembered as important to the development of the Northwest. Two towns, one in Idaho and the other in Montana, are named for him.

Chapter Three

Railroads

Introduction

The history of America's railroads is uniquely interwoven with that of the nation itself. The railroads had a profound social and economic significance, and a symbolic one as well, becoming inextricably associated with many of the ideals and aspirations of a young country fast on its way to becoming a world power. Whether on the relatively small scale of Boston and New York's subways and "interurbans," which facilitated commuting and thus an expansion of urban areas, or on the vast scale of the transcontinental railroad, which allowed people to reach places previously considered inaccessible and altered prevailing concepts of distance, the impact of rail transportation on the development of this country would be hard to overestimate.

The early history of railroad technology is predominantly British. During the first quarter of the nineteenth century, such men as George Stephenson, Richard Trevithick, and William Hedley developed the locomotives that ushered in the railroad era. The Stockton and Darlington Railroad opened in 1825, with Trevithick's steam engine pulling cars over a cast-iron tramway. The Liverpool and Manchester Railroad soon afterward opened as the first "modern" railway, operating on a double track of iron rails by steam locomotive, with such accessories as stations, bridges, tunnels, signals, cuttings, embankments, and cars of first-, second-, and third-class designation.

The progress of the British railways was watched by American industrialists and inventors. In 1818 Oliver Evans proposed a railway between New York and Philadelphia. Shortly afterward John Stevens received a charter from the New Jersey legislature for a line between the Delaware and Raritan rivers. Although the line was not constructed, Stevens did build a locomotive model in 1825 and ran it on a circular track near his home in Hoboken, New Jersey. The Quincy Railway in Massachusetts, built in 1826, employed horse-drawn rail cars to transport stone for use in building the Bunker Hill Monument, among other structures; its success demonstrated that rail transportation was both efficient and economically viable.

The Baltimore and Ohio Railroad Company was chartered in 1827 by the state of Maryland. The line was envisioned as a link between the east coast and the western frontier. Peter Cooper built his famous locomotive, the *Tom Thumb,* for the line, and it was run in the summer of 1830. But the small engine proved inadequate, and the Baltimore and Ohio used horse-drawn cars until the more reliable steam locomotive *York* had been built and put into service in the following year.

The Camden and Amboy Railroad, which ran between New York and Philadelphia in seven hours, carrying both passengers and freight, was opened in 1830. The Mohawk and Hudson line began laying track in the summer of the same year; its engine, the *De Witt Clinton,* made its first run in 1831. The Charleston-Hamburg Railroad, with its famous engine, *The Best Friend of Charleston,* was another early experiment. It provided America's first regularly scheduled train service, using steam locomotives.

Technological improvements occurred rapidly. John Lewis developed the swivel or "bogie" wheels that enabled trains to negotiate sharp curves at higher speeds than before. The equalizing beam devised by Joseph Harrison produced equal pressure on the driving wheels even where the tracks were uneven. In 1831 Matthias Baldwin built a steam engine with special joints that could contain higher steam pressures than before. The year before, Robert Stevens had devised the T rails that came into common use. All of these innovations helped make the railroads technically viable.

By 1837 over two hundred railroad companies were in existence. At first the huge amount of capital needed was mostly provided by private and state investors. The panic of 1837, however, caused enormous losses among investors and slowed development. The states-rights attitude prevalent at the time precluded extensive federal involvement in the railroads; an 1828 joint resolution of Congress actually made federal ownership of stock in railroads or canals illegal. Nonetheless, valuable federal support came through the General Survey Act of 1824, under which U.S. Army engineers completed a large number of surveys for private railroads. The first federal survey was made in 1826 for a line intended to connect the James and Roanoke rivers. By the time the program was terminated in 1838, sixty-one surveys had been completed. The system was stopped both because of abuses in its practice—some army officers held positions both in the army and with railroad companies; some companies had to pay for services, others did not—and because of opposition from congressional leaders from the South, who perceived that while northern states benefited from the program, southern states did not.

Federal aid came in other forms too: tariff revisions that reduced the duties on imported iron, for example, and land or right-of-way grants. In 1833 a statute was passed that allowed Illinois to use funds from public land sales to finance railroad construction. Although never carried through, the act set an important precedent. In 1835 Alabama and Florida received federal right-of-way grants that were forerunners of the more dramatic federal incentives to railroad building that came after the Civil War.

The states themselves also played a major role in encouraging railroad development. In 1828 Pennsylvania became the world's first governmental owner of a railroad when it undertook the construction and operation of the Philadelphia and Columbia Railroad. Frontier states such as Illinois were particularly involved in supporting railroads, viewing them as a way of developing their territories. Large federal land grants to states were frequently turned over to railroads. States also got involved in taking over the construction or operation of lines belonging to insolvent railroad companies within their boundaries. Unfortunately, many state subsidies were abused by profiteering individuals. States found themselves in debt because of their sponsorship, and a movement developed—particularly in the North—for divestiture of state funds from railroad construction. By 1857 Virginia and Georgia were the only states still involved in direct railroad support. Georgia operated the Western and Atlantic Railroad until 1871.

The railroads themselves for the most part continued to thrive and to benefit from technical developments that made their operations increasingly efficient. Railroad expansion demanded not only improvements to locomotives, cars, and roadbeds but also the ability to overcome geographical barriers, such as mountains and rivers. The Troy and Greenfield Railroad in Massachusetts, for example, had to build a major tunnel through the Hoosac Mountain in the Berkshires; in the course of that project the art of tunnel building advanced significantly. The need for improved bridge types capable of carrying the unprecedented loads presented by trains led to many innovations in the design and construction of bridges. Railroad engineers such as Wendel Bollman and Albert Fink devised prototypical bridges that could be rapidly reproduced. Many types of engineering skills were pooled together—surveying, equipment design, bridge and tunnel construction—in the effort to create a continuous and unified railway system connecting disparate parts of the country.

In urban areas rail cars pulled by horses were initially devised to meet transit needs. The first horse car, the *John Mason,* went into operation in New York City in 1832. Using rails greatly improved the efficiency of horse power, and soon such horse cars were in operation in many cities. The use of steam engines was generally restricted to outlying districts: their smoke and noise, while not problematic in the countryside, were intolerable in the cities. Inventors produced mechanical devices to combat these problems, but none proved particularly successful; and the horse car on rails remained dominant in urban areas until the late nineteenth century.

By the 1850s the interregional railroads had overtaken canals and turnpikes in importance. Many cities, such as Chicago, became major rail centers, and their populations grew rapidly. The increasingly complex and developed railroad systems in the northern and middle states tied these regions of the country together even more closely, while at the same time diverting traffic in goods and people from the Mississippi River and other networks important to the economy of the South. During the Civil War the developed railroad network of the North undoubtedly helped the Union cause, just as an inadequate system proved damaging to the South.

It was also during the Civil War that the first serious moves were made to implement the visions of earlier advocates of a rail connection between the east and west coasts. As early as 1819, the architect for the Bunker Hill and Washington monuments, Robert Mills, urged the building of a rail line between the Mississippi River and the Pacific. Asa Whitney's promotion of a Pacific railroad in the 1840s eventually sparked further interest. Finally, in 1862, Congress authorized the Union Pacific Railroad to lay track westward from Omaha, while the Central Pacific was chartered by California to begin building eastward from Sacramento. The "joining of the rails" occurred in 1869. In 1864 the Northern Pacific Railroad had been authorized to begin construction of a more northerly route. Such activities were encouraged by the federal land-grant policy in existence at the time. Some 130 million acres were in some way part of the land-grant programs that benefited the railroads so greatly. Direct rights-of-way were given, as well as land sections in proportion to the amount of track laid. Lands were also sold by the government to provide revenue so that long-term bonds could be issued to railroad companies.

The rapid expansion of the railroads and the tremendous amounts of capital involved in the booming construction led to fraud and corruption on a grand scale. Thomas Durant of the Union Pacific amassed huge profits from his enterprise the Credit Mobilier of America, a construction company that did poor work at exorbitant costs. Influential congressmen were given stock in the company—including James Brooks, who was the government-appointed director of the railroad's activities. Individuals connected with the Central Pacific were involved in several forms of misrepresentation for profit, having to do with amounts of track laid.

Riddled with corruption though they were, the railroads did tie the country together and open a remarkable era of trade and development. Railroad companies, especially in the 1880s, conducted large-scale advertising campaigns in which they propagated images of scenic beauty, wealth, and opportunity, aimed at widely varied groups of people, both to encourage travel by rail and to bring immigrant workers to the railroads and related industries. The comparative luxury the trains afforded the middle classes—allowing them to live, temporarily, like those whom they aspired to emulate—was symbolic of the opportunity awaiting them among the country's vast resources. The well-appointed Pullman cars were representative of civilization making headway through the untamed wilderness. A remarkable growth of population occurred in hitherto unoccupied regions of the West and Midwest.

Although urban mass-transit systems continued to rely mostly on means other than steam locomotives during this period, commuter lines and short lines connecting urban areas were developing rapidly. Eventually a scheme for an elevated railway won acceptance in New York City, a central selling point being the separation of the noise and pollution of the trains from other traffic. Four elevated lines began service in New York in 1879; they became popular and financially successful.

Other rail responses were tried in urban areas, notably the cablecar. Developed largely through the efforts of Andrew Hallidie in San Francisco around 1869, the cablecar offered many advantages. In 1881 Chicago opened a cable railway that attracted great attention. Other cities, such as Philadelphia, New York, St. Louis, Oakland, Cincinnati, Denver, Washington, and Baltimore, soon followed suit. High costs, however, restricted the cablecar's use to high-density urban areas, and such systems did not fully meet the needs of outwardly expanding cities.

The introduction of the trolley car was a perfect solution to many urban transportation needs. A small electric rail car had been built by Thomas Davenport of Vermont in 1834. Moses Farmer had designed and used small battery-powered cars in the 1840s in Dover, New Hampshire. Charles Page had demonstrated the utility of this mode of power in 1851 in Washington, D.C. But it was not until the time of Thomas Edison's work and further developments by John Henry in Kansas that the electric trolley became viable. From 1885 to 1887 Henry operated a system in Kansas City that was, however, beset with problems. The first really successful citywide trolley system was introduced by J. Van Depoele in Montgomery, Alabama, in 1886 (following attempts in Detroit, South Bend, and Minneapolis). Frank Sprague's Richmond line, which opened in 1888, was another pioneer. By the 1890s many cities had electrified trolley lines. These systems radically reformed transportation and proved one of the most significant factors in the shaping of American cities and suburbs.

For some of the country's larger and more densely populated cities, another rail solution—the subway—proved the key to better urban transportation. The success of the London subway, which opened in 1864, was well known in this country, and that system was the model for subway proposals in New York and Boston. Electrically powered engines provided the technical means for avoiding the obvious problems associated with the use of steam-powered engines in subways—although Beach's wonderful experimental pneumatic tube that ran below the streets of New York in 1870 will forever remain an intriguing solution for many. Boston opened its subway, the first in the nation, in 1897. New York followed in 1904.

While the cities struggled by various experimental means to solve their unique transit problems, the steam railroad remained supreme throughout the late 1800s in interregional transportation. The public image of the great railroads, however, steadily worsened. Corruption was widespread. The end of the nineteenth century saw increasing demands for reform in public sponsorship, rate structures, and many other areas. Organized resistance sprang up among shippers and businessmen faced with exorbitant rates and dubious practices. The Granger movement, which had originated as an effort to improve the social and intellectual lot of farmers, also took on the regulation of railroad practices as one of its issues. The establishment of the Interstate Commerce Commission in 1887 held hope for reformers, but this body at first proved ineffectual. Around the turn of the century, after the depression of the 1890s, populist demands for rate reductions were finally reflected in the development of rate regulations and in antimonopolistic court decisions against large railroad conglomerates. The filing of a suit under the 1880 Sherman Act against the giant Northern Securities Company and the subsequent 1904 Supreme Court decision to break up the company represented a major reform-movement victory. By the beginning of World War I the railroads were regulated by a system of controls that had been introduced in the name of the public interest. Other factors affected the bigger railroad networks as well, notably the growing automobile industry and the development of interurban or short-distance trains, both of which sharply cut into railroad passenger service between 1900 and 1916, causing many companies to abandon their passenger lines.

World War I saw further government involvement in railroads. A Railroad War Board was established to coordinate commerce, and in 1917 a presidential proclamation urged nationalization in order to meet wartime needs. Congress responded by creating the Railroad Administration. Under the aggressive leadership of William McAdoo, the new agency made great strides in unifying lines and augmenting cooperation among lines. Returned to their owners after the war, the nation's railroads entered a period of great prosperity and further development. They suffered, along with the rest of the nation, during the Great Depression of the thirties; but recovery came quickly through governmental intervention, particularly the National Recovery Act of 1933, which authorized the president to aid in financing railroad maintenance and equipment. The railroads were then able to help lead the whole country out of the depression by the efficient carrying of goods.

World War II found an efficient railroad system in place, and made tremendous demands on that system. The postwar period was marked by increased attention to passenger traffic and by the introduction of new technologies, notably diesel systems; but this era also saw a rise in the use of automobiles and airlines for carrying passengers and in the use of trucks for carrying goods. The development of the interstate highway system created direct competition for the railroads, as did the increasing popularity of air travel. Many railroads faltered, others collapsed. The creation in 1970 of the National Railroad Passenger Corporation, known as Amtrak, was intended to revitalize passenger service and to relieve the railroads of what had in a very short time become a burden. Further reorganization and help have since been directed toward the still-ailing industry—an industry whose development was identified with that of this country.

The Charleston-Hamburg Railroad

When completed in 1833, this line from Charleston to Hamburg in South Carolina ran a distance of 136 miles, which made it one of the longest railroads in the world at the time. The first railway in the United States to provide scheduled passenger service, it also became (in 1831, before it was completed) the first to carry mail under contract. This set an important precedent in an era when the government was seeking more frequent and rapid means of communication to link together the widely scattered population of the country.

Compared to its use in England, the steam-powered railroad engine was slow to gain acceptance in this country. In 1800 in England, Richard Trevithick, a Cornish mining engineer, had produced a high-pressure engine that could be built in relatively compact form. By the time the first railway charters were awarded in this country, the English had nearly finished the Liverpool and Manchester Railway, which in 1830—its first year in operation—carried 70,000 passengers. The earliest successful rail company organized in America was the Baltimore and Ohio, incorporated in 1827. Other companies were quickly organized in Philadelphia, New York, and Charleston. Boston already had a railroad of a sort—a horse-powered three-mile line constructed in 1826 to move granite from Quincy, Massachusetts, to the banks of the Neponset River.

In December 1827 a charter was granted to the South Carolina Canal and Railroad Company to serve Charleston, Hamburg, Columbia, and Camden. Charleston was experiencing economic difficulties, and many community leaders felt that it was important to establish additional linkages with the cotton-growing areas to the west of the city, as well as to connect Charleston with Hamburg, where the Savannah River became navigable. Subscriptions for stock were sold primarily to Charleston businessmen, and on May 12, 1828, the company was formally organized, with William Aiken as president. One of the prominent citizens who served as director of the new company was Ezra L. Miller, who would soon be designing equipment for the line.

Charleston-Hamburg Railroad, South Carolina. At 136 miles, this was one of the world's longest railroads when constructed in 1833. It was the first U.S. railway to operate passenger trains on an established schedule, the first to be completely locomotive powered, and the first to carry mail. The railroad was designed and built by Horatio Allen. A reproduction of the *Best Friend of Charleston,* which ran on the line, is shown. (Courtesy of the American Society of Civil Engineers)

The directors convened a "committee of inquiry" to gather information on the possible route of a new railroad, as well as on the materials and costs of construction. The committee submitted its report on November 11, 1828. Numerous difficulties and unknown factors were noted. While the committee was making its own investigations, the directors also applied to the War Department for assistance from the U.S. Army Corps of Engineers. Colonel William Howard, who had recently worked on the Baltimore and Ohio Railroad, was subsequently detached for work on the South Carolina line. A full report was made by the Corps of Engineers on August 27, 1829. This report, besides describing a possible route to be used and recommending construction techniques, pointed out the military advantages of the railway. The secretary of war and President Jackson were both interested in this aspect of the proposed line.

Funding for the railway was difficult to obtain. Although company records indicate the sale of nearly $350,000 in stock by March 1828, most of the company's assets existed only on paper. Because the purchase of stock required only a $10 down payment, ready cash for the implementation of the venture was very limited, and other sources for funding were explored. The company subsequently received grants from both the South Carolina legislature and the United States Congress. South Carolina also provided a $100,000 loan, and private investors soon helped to build the necessary capital.

Horatio Allen was hired as chief engineer early in the summer of 1829. His previous employer, John B. Jervis—then the chief engineer of the Delaware and Hudson Canal Company—was an early promotor of steam locomotives on rail lines. Allen, born in Schenectady, New York, in 1802, had distinguished himself at Columbia University and begun a career that would eventually involve some of the most daring and innovative engineering achievements of the century. After his work on the Charleston-Hamburg Railroad he became a leading engineer on the Croton waterworks project (see chapter 6), served on the board that reviewed John Roebling's Brooklyn Bridge proposal, and patented a rotary steam valve. At the time of his appointment on the Charleston-Hamburg line, he had lately been to England, where he is known to have met the important British railroad engineer George Stephenson and visited the Stockton and Darlington Railway (the first rail line in the world offering regularly scheduled service).

Allen's first task upon joining the project was to resurvey the proposed route for the railroad bed. He shaved several miles off the surveys made before his arrival. He also studied construction difficulties in detail. Finally, on January 9, 1830, construction was commenced at Line Street in Charleston, by Gifford, Holcomb, and Company.

In 1829 the railroad offered a prize for a horse-propelled car, which was won by Ezra Miller. The car was constructed by Thomas Dotterer, a mechanic, and Christian Detmold, a German engineer employed by the company as a surveyor. The design and building of the car proved an invaluable experience, although by this time Miller was an enthusiastic supporter of steam power (he even patented an improved steam boiler in 1830). He and Allen advocated steam locomotives for the Charleston-Hamburg in a report to the company in September 1829. The president and directors decided soon afterward to adopt mechanical propulsion for the line. But enthusiasm for steam power alone was not enough to convince all the board members to invest completely in such a new technology. On March 1, 1830, the company directors tentatively authorized Miller to build a locomotive meeting certain specifications, with the general understanding that the railroad would buy it. Miller, using $4,000 of his own funds, had a locomotive constructed by the West Point Foundry in New York. The locomotive finally built and attributed to Miller probably incorporated ideas from his own observations made during a visit to England as well as work by Detmold and others. At that time a number of locomotives were being built and tested, including the Baltimore and Ohio's *Tom Thumb.* It was an exciting period.

On October 23, 1830, the parts of the engine ordered by Miller arrived in Charleston aboard the brig *Niagara.* Miller engaged the firm of Eason and Dotterer to assemble the engine and help test it. About this time, the engine acquired its name, the *Best Friend of Charleston.* In initial tests the wheels were found inadequate for negotiating curves, but the design appeared basically sound; by mid-December the engine was pulling four loaded cars at speeds of over 20 miles an hour.

The board of directors agreed that their requirements had been met and bought the engine from Miller. Public announcement of the purchase of the *Best Friend* was made only one day before the line was officially opened on Christmas day, 1830. One hundred and forty persons were hauled on that day. The line at that time was only six miles long. Another demonstration was given on January 15, 1831.

This first engine was not in service for long. On June 17, 1831, as the *Best Friend* was being turned around on the revolving platform, the boiler blew up. The explosion was believed to have been caused by one of the firemen inadvertently holding down a safety valve. The fireman was killed. In the meantime another engine, The *West Point,* had been ordered in the fall of 1830 by Allen, had arrived on February 18, 1831, and had been given its trials the following March 5. Although not satisfactory at first, the *West Point* was put into regular service on July 15, 1831, and remained in service until June 4, 1833. Subsequently several other engines were put to work on the line. By November 1833 the line had 3 engines in operation, along with 8 passenger cars, 56 freight cars, 14 tender cars, and 11 lumber cars.

When the *Best Friend* had first been used in late 1830, the major part of the construction of the line had remained to be done. At the height of construction, the work force was 1,300 strong. Although the task was not a simple one, and grading by contemporary standards unheard of, work progressed quickly. The line was opened for service all the way to Hamburg on October 3, 1833.

Because the locomotives of the time were incapable of climbing steep grades, and the formation of embankments was a costly and time-consuming operation, large portions of the railroad were constructed on timber piles that served to level the roadbed. Almost 100 of the 136 miles of track were built on such pilings. At some places the track rested on ground level or ran through slight excavations, but for the most part it was 5 or 6 feet above the ground. When crossing gullies and small ravines, the track ran as high as 25 feet from the ground. Cross timbers for the trestle construction were supported by pairs of posts 10 to 15 inches in diameter and set 6 feet apart. Using a pile driver with a half-ton weight and a fall of 20 feet, workmen drove the pilings into the ground to a depth of 4 to 25 feet, depending on soil conditions. The tops were then leveled and cross timbers secured to them. There is evidence that some M-shaped pile structures were used, along with vertical piles with side braces. A timber rail was set in a groove fashioned in the cross timbers and wedged in place, and running rails were then spiked into place. Iron straps 2.5 inches wide and 0.5 inch thick were used. The rails were of yellow pine and rested upon sills of either lightwood (a pine abounding in pitch) or live oak. They were held in place by a locust or live oak key.

Despite the piling system used to level the roadbed, there were still inclines and grades that proved too much for the early engines. At a point 16 miles from the line's terminus at Hamburg, the railway was 510 feet above tidewater and faced a 360-foot descent to the town. An inclined plane was constructed over the first portion of the descent. The plane had three grades, the steepest of which dropped about 1 foot in every 13 feet of the run. It was laid with a double track, and at the top was placed a stationary steam engine that pulled up or let down cars on the incline by means of a cable and crank.

The remarkable construction of this railroad reflected a calculated belief that traffic would slowly develop on the line and improvements would be made later. There was no attempt to make it a permanent structure. The use of wood for both rails and sills and the laying of but a single track attest to this shrewd planning. The Baltimore and Ohio and several other early railroads, in contrast to the South Carolina model, started out with double tracks, granite sills, and ambitious systems of grades and bridges. These ventures encountered considerable financial embarrassment when their soaring construction costs appeared not to be justified by the anticipated traffic over the short term. In 1833 the Baltimore and Ohio was 14 miles of carefully graded, stone-silled track. The Charleston-Hamburg ran nearly ten times that distance. Other pragmatic reasons also influenced the nature of the construction, such as the fact that other than ballast from arriving ships, the Carolina low country lacked stone in any quantity.

From 1833 to 1837 fill was gradually placed around the pile construction up to the cross members under the rail. In 1835 the problematic grade near Hamburg was made uniform. In 1841 the stationary engine atop the hill near Hamburg was displaced, and from this date until 1852, when an entirely new grading was made, trains were helped over the grade and down the other side by another locomotive.

The Louisville, Cincinnati, and Charleston Railroad bought out the South Carolina Canal and Railroad Company in 1837. The entire system became officially known as the South Carolina Railroad Company in December 1842. The line was constantly improved as railroading promised to grow into a major industry. In 1837–1838 new heavier rail was laid, and from 1847 to 1852 this was replaced with the even heavier T rail. A branch line from Branchville to Columbia was completed in 1842, another from Camden to Kingville was completed in 1848, and a bridge across the Savannah River to Augusta, Georgia, was completed in 1853. With the opening of the bridge the town of Hamburg lost its importance as the line's terminus, while Augusta became a veritable boom town.

The Charleston-Hamburg Railroad was almost completely destroyed by General Sherman during his historic march in 1865. Following the Civil War it remained an entity on paper at least, until gradually rebuilt in the 1870s and 1880s after once again being sold. The line remained part of the South Carolina Railway Company until 1894, when it was sold to the South Carolina and Georgia Railroad Company. In 1899 the Southern Railway Company acquired the properties of the South Carolina and Georgia, and the line was incorporated into a system stretching from Charleston to Los Angeles.

Most of the original right-of-way of the Charleston-Hamburg Railroad is now a part of the Southern Railway System. Other portions of the roadbed became highways or streets and bear lingering testimony to the surveying skill of Horatio Allen. Little of the original construction remains. Commemorating the birth of regular steam-locomotive passenger service, however, is a collection of early railroad buildings in Charleston making up a National Historic Landmark District. These include the William Aiken House on King Street, home of the railway company's first president; the Camden Depot on Anne Street, immediately behind the house; and four other antebellum railroad buildings on Anne and John streets. A portion of the railroad runs through the collection of buildings. A modern structure houses a recent replica of the *Best Friend*.

The Transcontinental Railroad

Joining of the rails of the Transcontinental Railroad at Promontory Point, Utah, May 10, 1869. The linking of the continent by 1,776 miles of trunk line railroad over mountains and deserts was a turning point in American history, signaling the opening of the West and the emergence of a unified nation. (Library of Congress)

On May 10, 1869, the Union Pacific's Engine Number 119, coming west from Council Bluffs, Iowa, and the Central Pacific's *Jupiter,* coming east from Sacramento, California, met at Promontory Point, Utah, in celebration of the completion of a rail link between the Atlantic and Pacific coasts. Bottles of champagne were broken over both engines, a golden spike was driven, and a message telegraphed instantaneously around the country. Every participant in this jubilant ceremony felt the political and commercial significance of the event. The increased trade and settlement that followed brought such a demand for further service that within two decades the United States had four other transcontinental railroad lines.

Hailed as a link not only between the eastern and western states but between the Western world and the Orient as well, the first transcontinental railroad was built with amazing speed through some of America's most rugged terrain. It was constructed in two parts by separate companies, the Union Pacific and the Central Pacific—the former chartered to build from Omaha westward, the latter from Sacramento eastward, until they met. Both railroads are more notable for their feats in overcoming high, rocky mountains and spanning deep ravines than for any special mechanical features or innovative technology.

As early as 1819, Robert Mills, the architect of the Washington Monument, had suggested the construction of a railroad from the Mississippi River to the Pacific. Asa Whitney presented his trans-American proposal in 1836 and continued to advocate the cause in subsequent years. His efforts resulted in a railroad convention in 1849 whose purpose was to study the merits of a transcontinental railroad. Stephen A. Douglas was elected chairman. A resolution favoring a San Francisco–St. Louis route was passed. The gold rush of 1849, the prospect of expanding trade in the Far East, and the federal government's newly developed railroad land-grant policy contributed to agitation for a transcontinental rail link. During subsequent years Congress continually discussed the project, with sectional disputes between northern and southern congressmen hindering action. Funds were finally allocated in 1853 for surveys and explorations for routes between the thirty-second and forty-ninth north latitudes. The United States Corps of Topographical Engineers completed the surveys in December 1856. During the 1850s most planning focused on a southerly route along the thirty-second parallel, connecting Charleston, South Carolina, with San Diego, California. The famous Gadsden Purchase of 1853 ensured that the route would lie entirely within United States territory.

The Civil War caused the abandonment of this route in favor of the central or forty-second-parallel route, which would link San Francisco to Omaha. Legislation was passed by Congress in 1862 authorizing the Union Pacific Railroad to progress westward from the Missouri River. The eastern end of the route to be followed by the Union Pacific was fixed at Omaha by an act of Congress signed by Abraham Lincoln in 1863. The state of California had previously chartered the Central Pacific Railroad to build eastward from Sacramento.

Generous land grants were being made to railroads as a way of encouraging investment. The Union Pacific and the Central Pacific each obtained 200-foot-wide right-of-ways and ten alternate sections of public land for each mile of track laid. In addition, the secretary of the treasury was empowered to sell public lands and issue the companies long-term bonds payable in treasury notes. Over 130 million acres of public lands were given to the companies—a little less than 10 percent of the land in the United States, and worth at that time around $2 billion or more. The incentives for investment were not insubstantial.

Construction of the Central Pacific Railroad began at Sacramento, California, in 1863. In this photograph taken at Secrettown, California, in 1867, wagons are bringing fill to the ends of the wooden trestle. (Library of Congress)

From the gold rush of 1849 to the late 1860s several railroads were built out of Sacramento. Among them was the Sacramento Valley Railroad, built in 1853–1856, which ran 22 miles east to the mining town of Folsom. Theodore P. Judah was chief engineer.

In 1860 Judah surveyed possible routes eastward over the Sierra Nevada. He returned to Sacramento to find backers, and along with Collis P. Huntington and Mark Hopkins (of the firm of Huntington and Hopkins, Miners' Supplies and Hardware), Leland Stanford (soon to be governor of California), and Charles Crocker (a dry-goods dealer), he formed the Central Pacific Railroad of California with the object of building a road to connect with the East. After much persistence, government aid was obtained and ground was broken on January 16, 1863. Judah died only a few days before the ceremony. Samuel S. Montague assumed the role of chief engineer.

The Sierra Nevada presented the greatest challenge to the builders of the Central Pacific. There was little soil over the "granite mountains," and there were numerous ravines, which had to be bridged by trestles (the one over Deep Gulch was 500 feet long and 100 feet high). Fifteen tunnels had to be blasted through solid granite—an extremely arduous task, during which progess was made at the rate of about 2 feet per day. Unlike the Union Pacific workers, those of the Central Pacific were not troubled by Indian attacks. But severe winters offered another substantial hardship. One winter brought no less than 44 blizzards, one of which left drifts 60 to 70 feet high. During one phase of tunnel construction workers dug shelters out of the snow and then dug connecting tunnels to their work site, enabling them to remain protected at all times.

The first major task on the route was crossing the American River. Forty-foot redwood piles were driven into its bed to serve as bridge foundations. The bridge itself was a 2,200-foot-long trestle, only 6 feet above highwater level. Cape Horn, a cliff rising straight up from the American River, presented the chief obstacle. Its sides had to be cut away to make a ledge for the railroad. Workers were let down from the top in baskets and picked and drilled at the cliff, making impressions large enough to contain charges of black powder. Workers were then hauled up and the charge set off. It was no small feat.

The construction of the 1,659-foot Summit Tunnel, 202 miles from Sacramento, was representative of the hardships encountered. Begun at both east and west portals, it also required a shaft to be sunk so that work could proceed from the center as well. This was driven through granite at the rate of 7 inches per day. In order to hoist the excavated rubble out of the shaft, an engine was brought from Sacramento, hauled up the mountain—bolted to a flat-bottomed vehicle drawn by ten oxen—and let down slope with logging chains and chain tackle fastened to trees. The whole trip took six weeks.

The speed with which the Central Pacific was built was in good part due to the massive labor force. By the end of construction, 20,000 Chinese had been recruited from their native Canton Province; 300 mechanics, 900 horses, 100 oxen, and 800 wagons and carts were employed.

Some crude technology, combined with the precipitous nature of much of the trackage, contributed to a high accident and casualty rate on the Central Pacific once it was in operation. The couplers that held cars together had to be operated manually, resulting in frequent crushings of the brakemen responsible. The weakness of these devices produced some runaway cars, which would sometimes become loose as a train ran downhill and fly out of control over the sides of mountains. The brakemen (before the acceptance and use of George Westinghouse's air brake, invented in 1869) had the thankless task of manually operating brakes from the roofs of the cars, jumping from car to car over the 30-inch gap between them, making sure the process went as smoothly as possible. On a downgrade, cars sometimes piled up, as brakes were slow to take hold; and if a brake was tightened too much, the wheels were "skidded and flattened" and the brakemen docked $45 for each flattened wheel.

During the final days of the Central Pacific's rush eastward, grading and other crews preceding the tracklayers actually pushed past similar crews from the Union Pacific who were working westward, until there were some 100 miles of parallel grade cuts near Promontory Point. This absurd situation was brought about by financial incentives based on mileage of track laid.

The history of the Union Pacific Railroad goes back to the explorations of Peter Dey, chief engineer for the Rock Island Railroad.

Heading of the east portal of Tunnel Number 8 of the Central Pacific. (Library of Congress)

Around 1851 Dey engaged Grenville Dodge, an industrial and military engineer recently out of school, to help survey the Rock Island's Peoria line. Whether or not the railroad intended to expand beyond Iowa, Dodge himself envisioned a transcontinental railroad and spent the rest of his life working toward one. In 1853 Dodge and Dey left the Rock Island Railroad to work for Henry Farnam and Thomas C. Durant, owners of the Mississippi and Missouri Railroad. Dodge undertook reconnaissances of Iowa, under their private patronage, in order to determine the best railroad routes. Over the years he acquired an intimate familiarity with western topography; from his first surveys to the time of his appointment as chief engineer of the Union Pacific in 1866, he was constantly exploring train route possibilities.

During the Civil War Dodge, with the support of both General Grant and President Lincoln, expanded and protected railroad, telegraph, and stagecoach lines. General Sherman, another champion of Dodge's efforts, recognized the railroad's role in unification and declared that it offered a solution to both the Mormon and the Indian "problems"—by undermining the former's intransigent independent stance and by cutting the latter's territory in two. Dodge, who was made a general in 1862, undertook a series of campaigns against the Indians in 1865–1866, noted for their ruthlessness in the extermination of Indian populations.

Dodge discovered the Great Platte Valley route, along the forty-second parallel of latitude, which a number of previous explorations had somehow overlooked. Used for years by immigrants, traders, hunters, Indians, and buffalo, it was the obvious route for a railroad. Once past the Great Platte Valley, the Union Pacific as it was eventually built diverged into a northern branch, with its terminus in Oregon, and the main line, connecting with the Central Pacific.

Congress had granted a charter to the Union Pacific Railroad in 1862. The groundbreaking took place at its eastern terminus on December 1, 1863. On Dodge's recommendation President Lincoln had designated Omaha as the line's approximate terminus; the actual location was Council Bluffs, Iowa, on the eastern side of the Missouri River.

About 550 miles west of Omaha, Grenville Dodge's Union Pacific builders encountered the Dale Creek valley. The 126-foot-tall, 700-foot-long trestle bridge shown in this early photograph was built to cross the valley. Biggest of the Union Pacific's trestles, it was finished on April 23, 1868. (Photograph by Andrew J. Russell, Library of Congress)

As Union Pacific tracklayers pushed ahead, many temporary bridges were built, such as that shown here at Green River. On the left are the beginnings of a permanent masonry bridge. (Photograph by Andrew J. Russell, Library of Congress)

Excavations in Weber Canyon, autumn 1868. Shelves were made by blasting and pick-axing as part of the struggle to get down to grade level. Temporary tracks to haul rubble away are seen in the foreground. (Library of Congress)

Union Pacific workers building the permanent bridge at Green River. (Photograph by Andrew J. Russell, Library of Congress)

Due to the manpower shortage during the war, some Mormons and around 10,000 Irishmen were recruited to constitute the bulk of the labor force. A nearly total lack of supplies along the route itself necessitated bringing everything in by way of the Missouri River, which was only available for passage three months of the year. In 1867 the Northwestern Railway was completed as far west as Council Bluffs, greatly facilitating construction and the provision of supplies.

The first year or so was spent grading and conducting further surveys, for which Congress in 1864 allotted $500,000. It took about a month to grade every 100 miles. Each mile of track required about forty carloads of materials and supplies. Construction of the Union Pacific proceeded at all times under military protection, as Indian attacks posed a perpetual threat. The total length of the railroad was over 900 miles.

While there were no particularly innovative mechanical features used in the Union Pacific Railroad, some extraordinarily rugged terrain was overcome in its construction by means of inventive structures, choice of route, and no end of perseverance. Especially remarkable was the crossing of the Continental Divide—Dodge discovered a 34-mile stretch of open prairie across the Divide at the relatively low elevation of 7,000 feet—and of two mountain ranges. A maximum grade of 2.5 percent and a maximum curvature of 10 degrees were maintained at all times. One of the mountain ranges was the Wasatch, where, according to Dodge, "in the winter of 1869–1870 we had to blast the earth the same as the rocks." For years the pass over the Black Hills at elevation 8,236 feet, which Dodge discovered, was the highest point reached by any railroad in the United States. Many ravines were bridged by high, not-quite-sound trestles, and a couple of near-fatal accidents occurred on the Dale Creek Bridge, which was eventually strengthened.

The celebrated meeting of the Central Pacific and Union Pacific locomotives at Promontory Point in May 1869 was, as has been mentioned, not the first contact between the two railroad companies. Because the Union

Engine Number 119 of the Union Pacific crosses the newly completed Promontory Trestle just before the ceremony surrounding the joining of the Central and Union Pacific lines at Promontory Point. Evidence of the overlapping parallel grading—over 100 miles of it in all—that was done by the two competing lines prior to the establishment of Promontory Point as the meeting place is visible in this photograph. At times grading crews were within 100 yards of each other, often setting off blasting charges without notice to their competitors. (Photograph by Andrew J. Russell, Library of Congress)

Pacific and the Central Pacific were competing for land grants, which were based on miles of track laid, they would often grade far ahead of actual track in order to speed up the process. The official meeting place for the lines was not established by Congress until after the grading crews had met and passed each other; when construction ended, there were over 100 miles of extra roadbed, graded and never used. Sometimes the two parallel roadbeds were within 50 yards of each other. Colorful stories have survived of skirmishes between the Union Pacific's Irish workers and the Central Pacific's Chinese workers, when they found themselves within a literal stone's throw of one another. The incidents were not always in good spirit. Blasting was often done by one of the crews without giving notice to the other, for example.

The huge sums of money involved in the construction of the two railroads made corruption inevitable. Among the most notorious instances was the conduct of the Credit Mobilier Company. Founded by Thomas Durant, vice-president and general manager of the Union Pacific until 1869, this construction company did slipshod work for large sums of money. Various congressmen became involved with ownership in the Union Pacific and attempted to influence congressional votes on matters pertaining to the railroads. Congressman Oakes Ames, for example, gave stock in both the Union Pacific and the Credit Mobilier to other congressmen. Among those who accepted shares were Schuyler Colfax, speaker of the House and vice-president under Grant, and James Garfield, the future president. James Brooks, the government director of Union Pacific operations, also owned shares. In 1872 the whole situation came to light, resulting in one of the greatest public scandals in United States history and bringing to an end the political careers of several of the participants. The scandals were not, however, linked solely to the Union Pacific. President Collis Huntington of the Central Pacific reputedly collected huge sums from the government by completely misrepresenting the location of the base of the Sierra Nevada and hence inflating the length of mountainous country traversed, so that $48,000 per mile for construction in mountainous terrain could be obtained instead of the $16,000 for flat terrain. These episodes created significant public distrust.

The first trip on the Central Pacific took place on June 16–17, 1869; the 154-mile stretch (about one fifth of the entire route) from Sacramento to Reno involved a formidable crossing of the Sierra Nevada. Reno "sprang up like magic" after the railroad was built, becoming a town of stores, hotels, saloons, gambling houses, and stables. The Central Pacific was in continuous use until 1903, when the Southern Pacific, which had absorbed it, built the Lucin Cutoff across the Great Salt Lake. Until 1940 the Southern Pacific ran a weekly train over the "Old Golden Spike Route" to service livestock and wheat shippers in northwestern Utah. In 1942 the company removed the rails between Corinno, California, and Lucin, and donated them as scrap for the war effort.

The Union Pacific continued to upgrade its service. It was the first railroad in the West to offer certain passenger amenities and safety features. It was the first in the Far West to provide sleeping accommodations and to equip its through trains with dining-car service. In 1897 the Utah legislature permitted incorporation of the Union Pacific Railroad, and all "properties and appurtenances" of the old company were purchased by the new. The railroad was also authorized to expand its trackage, and new branches were built. (By the same act other companies were authorized to expand their lines and create a broader western network of railroads.) By 1919, 974 miles of second track had been laid, and some third and fourth track as well. Grade and curvature reductions and realignments took place, such that almost nowhere does the track exceed a grade of 0.9 percent.

At railroad crossings semaphores were installed, and electric lights were used in the trains. In 1919, amid a general modernization of equipment, further safety elements were installed, including electric signals. These were particularly important, for large numbers of people were killed or injured by trains while crossing tracks. Campaigns were launched in the second decade of this century to bring down the number of railroad fatalities that occurred at crossings and other places along the tracks. Between 1922 and 1934 there was a significant reduction in casualties.

The Mount Washington Cog Railway

The Mount Washington Cog Railway is one of the steepest in the world. The engineering feats responsible for its success, admirable today, were truly astonishing at the time of its completion on July 3, 1869. The railroad runs 3.5 miles up the last 3,719 feet of Mount Washington in New Hampshire, on an average grade of 25 percent. The steepest stretch, called Jacob's Ladder, has a 37.4-percent slope. Three of the 3.5 miles of track were built on trestles on account of the rugged, rocky terrain of the mountain. The railroad passes far beyond the treeline to the 6,288-foot summit, the highest point in the northeastern United States, where winds above 200 miles per hour—the highest in the world—have been recorded.

The railroad bears witness to the inventive genius of two men, Sylvester Marsh and Herrick Aiken. Marsh (1803–1884), engineer and inventor, made his fortune by a series of ingenious inventions and business endeavors in Ashtabula, Ohio, and in Chicago, a town of few inhabitants when he arrived there in 1829. In 1855 he returned to his native New Hampshire and began plans for a cog railway on Mount Washington. Marsh invented the train's essential mechanisms and was, moreover, chiefly responsible both for convincing the New Hampshire legislature of the project's worth and for financing it.

Marsh was granted a charter in June 1859 by the legislature, with the provision that the railway be built in five years. The charter was later extended, for construction had to be postponed until 1866. During these seven years Marsh developed and patented the cog device that allowed a train safe passage on steep grades.

Herrick Aiken (1797–1866), an inventor from Franklin, New York, also had a hand in the project. Apparently he had not only suggested the idea of a cog railway but planned the route on Mount Washington on which Marsh eventually built; he had even built a cog rail model with roadbed. Aiken's son Walter became closely involved with the railway as well. He provided much of the

Mount Washington Cog Railway, New Hampshire. When completed in 1869, this was the first significant mountain-climbing railway in the world. Grades on its 3.5-mile route to the summit reach 37.4 percent. (Courtesy of the American Society of Civil Engineers)

This contemporary engraving shows Sylvester Marsh's earlier 1864 proposal for a locomotive capable of ascending the White Mountains by steam. (*Scientific American*, March 5, 1864)

financial backing, ran the business for Marsh from 1877 until Marsh's death in 1884, and served as president until his own death in 1893.

The route for the railway—from the Marshfield base station, about 3,700 feet from the mountain's summit, along the western spur between the Burt and Ammonoosuc ravines—was thoroughly surveyed by Colonel Orville Freemen, a civil engineer from Lancaster, New Hampshire. Since the nearest railroad was 25 miles away, the first construction supplies were hauled by Marsh, driving oxen, and workmen, "carrying ox-yokes and other equipment on their shoulders." Trees were soon logged on site, a headquarters cabin constructed, and a water-powered sawmill erected on the Ammonoosuc River to provide timbers for the trestles.

The first locomotive, the *Peppersass,* was designed by Marsh and constructed by Campbell, Whittier and Company of Roxbury, Massachusetts. The *Peppersass* was brought in pieces to the site and reassembled at a forge built there for the purpose. The new locomotive aided in construction by hauling materials up Mount Washington before it carried its first passengers. Forty passengers rode the first stretch of track on a platform car pulled smoothly by the *Peppersass* in August 1865. A second engine, the *George Stephenson,* was added at this time. It could carry fifty passengers.

The *Peppersass* had several especially innovative features, all of which were important in handling grades of unprecedented steepness. Marsh developed friction brakes and a ratchet to grip the tracks during ascension. Downgrade control was achieved with air "admitted to the driving cylinders and compressed by pistons"; if necessary, steam could also be admitted. In addition, individual brakes were placed in each car. This first locomotive was steam powered—Marsh had invented several steam appliances while in Chicago—and had one pair of cylinders, which turned the front axle only, through gears. A vertical boiler was suspended from trunnions in order to remain upright while the locomotive ran on steep grades. After 1878 the boilers were placed in front of the locomotive, horizontally, and kept level while on slopes.

The Cog still serves the summit of Mount Washington and still runs by coal. Each 3.5-mile trip requires a ton of coal and 100 gallons of water, and takes approximately 70 minutes. The railway helps make the summit—with its post office (boasting its own zip code), observatory, museum, lecture areas, and refreshment and rest facilities—the popular tourist attraction that it is today.

The Cumbres and Toltec Scenic Railroad

The Cumbres and Toltec Scenic Railroad was built between 1879 and 1880 as a branch line of the extensive Denver and Rio Grande Western Railroad, which spread through western Colorado and much of Utah, serving booming mining industries in the 1880s and 1890s. Running 64 miles—35 as the crow flies—from Antonito, Colorado, to Chama, New Mexico, the Cumbres (Spanish for "summits") and Toltec traverses difficult but beautiful mountain terrain. Unlike most railroads, which follow streams or canyons, the Cumbres and Toltec was "cut from watershed to watershed," defying impediments in its tortuous course over the Continental Divide and standing as a striking testament to engineering perseverance and adeptness. The narrow gauge of 3 feet was necessary in order to negotiate the narrow, rocky shelves and precipitous paths it takes.

Gold and silver were discovered in the San Juan Mountains in 1873. As a direct result of the mining fervor, Colorado's population increased fivefold to 194,327 between 1870 and 1880. It became a state in 1873. To take advantage of the mineral wealth, General William J. Palmer founded the Denver and Rio Grande Western Railroad. During its construction, workers were recruited mostly from Denver and Pueblo, Colorado, but also from Canada, St. Louis, Chicago, and Kansas. Once in the mining country, many of the newly recruited workers deserted to prospect for themselves.

The route of the railroad lay through a break in the chain of the San Juan Mountains formed by the Rio de los Pinos to the east and Wolf Creek to the west. The summit at Cumbres Pass forms a gap in a nearly unbroken line of mountains and ridges. The route was initially laid out to follow the water grade of the Rio de los Pinos, but the water dropped off too steeply, and broad curves and loops were introduced along ridges and in valleys. Fills and bridges, such as the towering Cascade Creek trestle, were introduced, as well as a few short tunnels, in order to maintain acceptable grades. The trestle at

Cumbres and Toltec Scenic Railroad, Chama, New Mexico. This 64-mile branch line of the Denver and Rio Grande Western Railroad was built in 1879–1880 to serve the mining industry. It is one of the last remaining narrow-gauge railroads. The Cascade Creek bridge, shown here, the tallest on the line, was constructed in 1889. (Courtesy of the American Society of Civil Engineers)

Cascade Creek was originally a wooden one, built in 1881. It was replaced in 1889 by the current span, furnished by the famous Keystone Bridge Company. Composed of seven 54-foot spans, the structure looms 137 feet above Cascade Creek. No longitudinal cross bracing was used in this design, reputedly of German origin. The same design was later replicated in the Lobato trestle. Tunnel No. 1, the "Mud Tunnel," not far from Cascade Creek, is a 342-foot bore built through soft volcanic breccia, which flows when wet. Consequently, the tunnel is fully lined with timber. The eastern approach involved easier grades than the western end near Wolf Creek, where major bridges and other constructions were needed in order to maintain even a fairly steep 4-percent grade. The Cumbres line was placed whenever possible on the north slopes of valleys in order to get maximum sun exposure during the winter season and thus mitigate snow buildup.

The railroad winds up and down the San Juan Mountains, crosses the Continental Divide east of Cumbres, and here and there skirts mountains on ledges cut into cliff faces, including the 6-foot-wide granite shelf blasted from the rock nearly 1,000 feet above the Toltec Gorge, one of the sheerest drops in the United States. One passenger from the last century wrote of the Toltec Gorge, "How narrow apparently are these curved and smooth embankments that carry us across the ravines, and how spidery look the firmly braced bridges that span the torrents!" Nearly every stretch of the enormously varied route has been described with equal intensity. One section of track is called "the whiplash," because it doubles back on itself twice to wind its way up the Big Horn Mesa. The Cumbres Pass, at an elevation of 10,015 feet, is the highest railroad pass in the United States.

The mining industry began to decline during the first couple of decades of this century. Passenger service on the Cumbres and Toltec, which had always run regularly, ceased in 1951. Freight from local agricultural and ranching industries continued to be shipped by rail almost daily until 1968, when the Denver and Rio Grande Western petitioned the Interstate Commerce Commission for abandonment. Permission was granted; but in 1970, responding to popular protest, the Colorado and New Mexico legislatures established a joint railroad authority and purchased the line's trackage and facilities (including 9 locomotives and 130 pieces of rolling stock) for $547,210. In 1971 the two states appropriated funds for restoration, and in June of that year the newly restored train had its first run. Today the Cumbres and Toltec Scenic Railroad operates in the summer months, taking tourists through spectacular mountain scenery that remains basically unchanged since the line was originally built.

The Durango-Silverton Railroad

The Durango-Silverton branch of the once-thriving Denver and Rio Grande Western Railroad now takes tourists through 45 miles of the spectacular San Juan Mountains of southwestern Colorado in the summer months. During the fifty years between 1882 and 1932, however, the railroad served the smelting center of Durango, where most of the area's annual two million dollars' worth of ore was processed, and other mining towns along its route.

The discovery of gold and silver in the San Juan Mountains in 1873 spurred growth in the area, but the Denver and Rio Grande Western Railroad did not reach Durango until 1881. Durango was a classic case of a town that came into being and prospered because of the presence of the railroad, then languished when the fortunes of the railroad declined. It was also the classic Wild-West, lawless town, a place of saloons, gambling houses, and so many extralegal hangings that when Durango's only legal execution took place, the event was much celebrated, according to popular tales. More important, however, was that Durango possessed good deposits of coking coals and proved to be a good location for smelting operations. The Durango smelter, constructed in 1880, drew its ores from the rich deposits in the San Juan Mountains, including those at the boom camp of Silverton. Supplies were freighted to Silverton by wagon over Cunningham Pass.

In 1882, in the remarkably short time of nine months and five days, a branch line of the Denver and Rio Grande was pushed through from Durango to Silverton. The 45-mile route included some formidable stretches, such as the passage in the Animas Canyon, where the train runs along a rock ledge just wide enough to accommodate its tracks, with the Animas River a sheer 700 feet below. All of the track was made of steel, the first steel rail to be used on the Denver and Rio Grande, made by the Colorado Coal and Iron Company at Pueblo. Before this, all tracks were made of iron. Most of the Durango-Silverton branch is on native earth and rock fill, with a small portion on cinder

Durango-Silverton branch of the Denver and Rio Grande Western Railroad. One of the last surviving narrow-gauge railroads, it was built in 1882 to link the Colorado mining towns of Durango and Silverton. (Courtesy of the American Society of Civil Engineers)

ballast. Its maximum grade is 2.5 percent—a remarkable surveying and construction feat, given the mountainous terrain.

After completion of the Durango-Silverton branch, three other independent feed lines were built at the Silverton end to connect with mining districts to the north. These branch lines—the Silverton Railway, the Silverton Northern Railway, and the Silverton, Gladstone, and Northern Railroad—were profitable and fed the Durango smelter. Volumes on the Durango-Silverton branch were increased even more by ores from the Uncompahgre slope, which were freighted by wagon over Sheridan Pass for shipment by way of Silverton to the Durango smelter.

Operation of the Durango-Silverton was plagued by enormous rock slides and snow slides. Some slides carried ice, rock, and trees, and were so compacted that they had to be dynamited to be removed. Some were hundreds of feet long and up to forty feet deep. Four miles south of Silverton, in the area of the worst snow slides, a snow tunnel, created by numerous slides covering a wooden snowshed, kept the tracks free, until the timbers rotted and the whole structure had to be dismantled. Despite such hazards the Durango-Silverton was remarkably free from fatalities during its many years of operation; by 1948, when the route had been in use for 66 years, only four transportation workers had died in service, an excellent record for perilous mountain railroading.

In 1946 the Durango-Silverton had one of the last two scheduled narrow-gauge passenger runs west of the Ohio River. A few lead and silver mines were still in operation, though most activity had subsided by the 1930s. During World War II the towns of Silverton, Telluride, and Ouray (the latter two are close to Silverton, though not directly connected by rail) revived their vanadium, molybdenum, and radioactive ore mining, but afterward they reverted to cattle and truck-farm service. Local pressures from farmers and cattlemen kept the freight trains in service, even after Continental Trailways bought the Durango-Silverton in 1948. The railroad is now a popular tourist excursion route.

The Boston Subway

On September 1, 1897, the first subway system in North America—and one of the first in the world—opened for service. In both its functional concepts and its engineering elements, the Boston subway was a distinguished achievement. New and challenging transportation problems were resolved by a sound engineering procedure that was determined after extensive studies and the consideration of many alternatives—a practice now considered virtually mandatory for all major engineering work. The success of the Boston system spurred the construction of subways in other American cities faced with similar transportation problems, most notably New York. The subway became an important factor in the subsequent development of these American cities.

The narrow, winding streets that had developed in the old colonial town of Boston, and that lend the city what we now consider a unique charm, caused a traffic nightmare in the bustling Boston of the 1890s. Horse cars on surface rail lines had been introduced to the city quite early, and electric streetcar lines were adopted in the latter part of the nineteenth century. These systems greatly facilitated travel into and out of the downtown area, but the rail lines on the already congested streets created an impossible traffic situation. At busy times streetcars, horse-drawn wagons, and carriages would be lined up for blocks on Tremont Street, unable to move.

In June 1891 the Rapid Transit Commission of the city of Boston was appointed to study traffic problems. The commission held fifty-one public hearings during its deliberations. In April 1892 it submitted a report to the state legislature that made several recommendations, including an elevated railroad along the east side of Washington Street in downtown Boston and a subway or tunnel along the line of Tremont Street. Although none existed in this country, a subway had been in operation in London for some time and others in Paris and Budapest were under construction. Alfred Beach had experimented with a pneumatically powered underground car in New York in 1870 (see New York Subway), and a number of other subway proposals had been advanced in that city over the past three decades.

A special committee was appointed by the legislature of 1893 to consider the report. After forty more hearings, two acts were passed, one for the creation of a Metropolitan Transit Commission and the other for a Board of Subway Commissioners. The Metropolitan Transit Commission was charged with developing a route from the outlying Franklin Park to the heart of the city. This act was rejected by the citizens of Boston during the next election. The second act gave authority—pending City Council approval—to the Board of Subway Commissioners to construct a subway running from near Tremont Street to the Scollay Square area (the area now occupied by the city hall and a federal building complex). The City Council of Boston gave its approval to the plan.

Due to defects in the plan, the appointed commissioners recommended an amended act that would grant authority to build a subway for four tracks from Pleasant Street to Causeway Street, along with branches and other outlets. Furthermore, permission was requested to locate stations under the malls on Tremont and Boylston streets. After legislative review a composite act was passed in 1894 providing for the incorporation of the Boston Elevated Railway and for the creation of the Boston Transit Commission. A strong fight preceded passage of the act, with business interests opposing public sentiment against any encroachment on the sacred ground of the historic Boston Common. The act gave significant powers to the transit commission, including the authorization (but not the requirement) to build a subway or subways in the vicinity of the Common, to build a tunnel under Beacon Hill, and to construct a bridge over the Charles River. Howard A. Carson was named chief engineer, and Henry H. Carter consulting engineer.

By this time a series of remedies for congestion had already been advanced, which the commission systematically explored. One of the more discussed—the "alley" plan—had been rejected by Boston's citizens in 1893. The downtown part of the plan involved laying out a new central avenue between Washington and Tremont streets, which would carry tracks on two levels in an elevated structure. This plan was perceived by many to threaten to contract rather than enlarge the central business district. Other plans investigated included the widening of Tremont Street and the laying of additional tracks; laying streetcar tracks in the Common—a major open space; the elimination of through cars on Tremont Street and the use of a shuttle service only; and the construction of a subway.

The commission sponsored several engineering studies to provide background information. These studies included careful surveys (made by borings and excavations in cellars of existing buildings) of subsurface conditions to determine not only the soil types present but also the character and dimensions of building foundations along proposed routes. Maps of existing subsurface utility networks were reviewed, and surveys made of the character of surface elements—poles, sidewalks, and so forth.

In the end, a subway plan was recommended. The reasons advanced by the commission included the following: a subway would not destroy existing property; dangerous grade crossings would be eliminated; total traffic capacity would be increased; surface posts and other networks for trolleys could be eliminated; cars could run quickly and safely; maintenance should be decreased (removal of snow and ice, for instance); construction costs were estimated to be smaller than for widening streets. Noted as militating against the subway idea were the costs for the installation and maintenance of pumping, ventilating, and lighting elements and the expense of building and maintaining the stations themselves. After some deliberation the Massachusetts legislature authorized the subway's construction.

The system was designed in several sections. Sections 1 and 3 consisted of a subway that started at the foot of an incline at the Public Garden west of Charles Street and ran to a point just below Temple Place where the

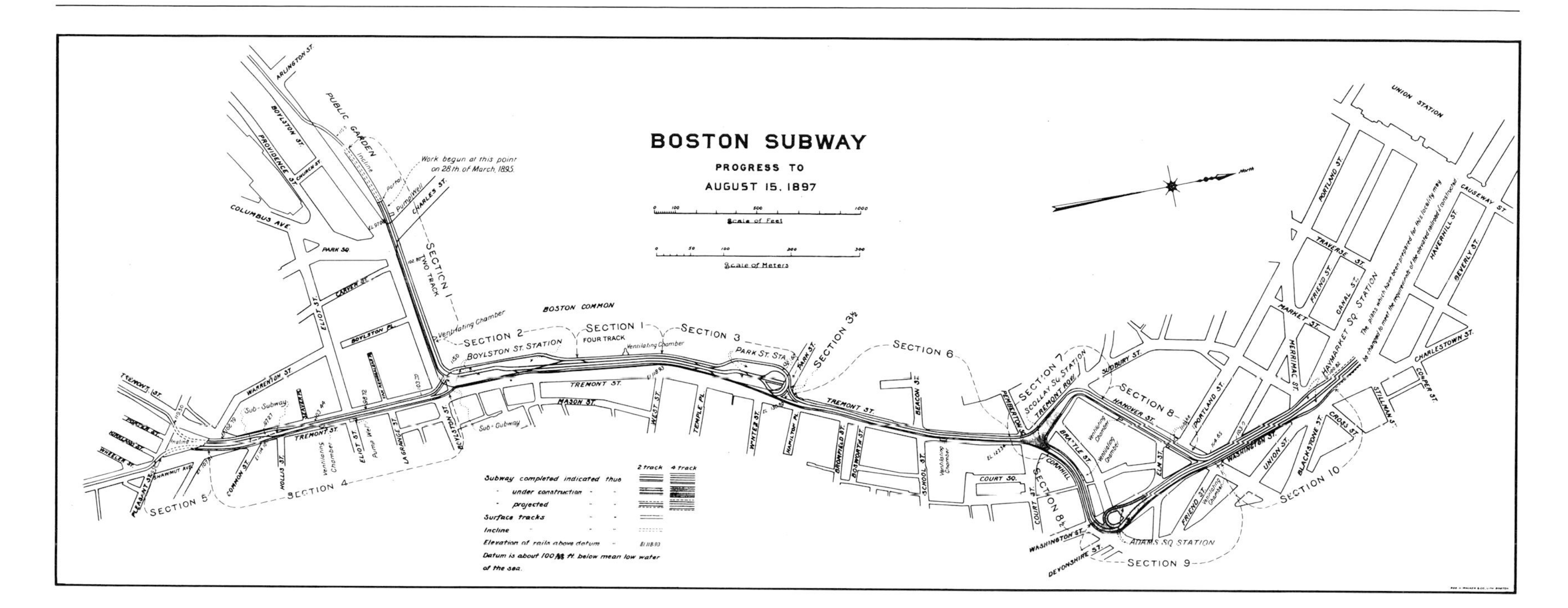

The Boston subway, Boston, Massachusetts. Following an extensive series of engineering studies made under the direction of Chief Engineer Howard A. Carson, ground was broken for the first subway in the United States on March 28, 1895. A cut-and-cover method of construction was primarily used to make the tunnels. (Courtesy of the Massachusetts Bay Transportation Authority)

tracks separated to form the Park Street station. Section 2 was essentially the Boylston Street station and its approaches. Sections 4 and 5 ran up to Warrenton Street and emerged from the subway by an incline. Other sections were to run between the Park Street station and the Haymarket and Scollay Square areas. Careful consideration was given to the number of tracks and to the design of the stations themselves for maximum efficiency and minimum confusion during heavy passenger rushes. The size of the platforms was partly determined on the basis of a count taken on the Saturday before Christmas, 1894, of the people taking regular trolley cars (a maximum of about 3,500 per hour). Twenty-three studies were conducted for the Boylston Street station alone.

Extensive studies were made in various areas. The design and construction methods of comparable works elsewhere—including the London, Glasgow, and Budapest underground railways, as well as a number of specific tunnels—were reviewed. Boston's underground utilities were further examined, in the anticipation that the route would be nearly filled with gas pipes, sewers, electrical conduits, and so on. The original construction drawings of these elements were carefully compiled and analyzed, when available; the fact that little documentation existed added to the difficulties of the study. Another aspect of the project, the question of ventilation within the tunnels, was taken up by Professor S. Homer Woodbridge. A special ventilation system was eventually designed.

The nature of the cross section of the tunnel had next to be determined, and a review was made of possible alternatives, such as a single wide arch, a double-barreled arch, or a beam-and-column system. Early in the planning stages a decision had been reached to place the subway close to the surface whenever possible. This, it was felt, would minimize excavation, cause the least possible damage to adjacent buildings during construction, and make the stations easy to enter and leave. For much of the subway, then, the engineers felt that the shallow depth necessitated a structure that did not rely on any restraint provided by the surrounding earth for stability. This requirement militated against the arched masonry profile in common use in other tunneling projects. Restraint against the arches' outward thrust could not easily be provided in a shallow tunnel without building costly buttressing elements. The steel beam-and-column system, which exerted no lateral thrusts, was generally felt to be more appropriate, particularly in view of the cut-and-cover construction method anticipated. Though concerned over possible rusting and electrolysis, the engineers felt that these problems were surmountable and that a properly built and maintained structure would provide many years of good service.

A structure was devised that used closely spaced, flat I-beams supported directly by I-sections placed vertically. Brick arches between the horizontally spanning I-beams formed the roof. Diagonal struts connected the horizontal and vertical steel members. This basic design was eventually used throughout much of the subway, such as in the important Tremont Street tunnel. In some sections, however, specific conditions suggested the use of a single wide arch or a double-barreled arch. Each of the primary sections of the subway was generally to be done by a different construction contractor. Different steel contractors were used also.

Associated with the decision to maintain a shallow tunnel depth was the decision to use what was then described as the "slice" method of construction, rather than an alternative method such as a shield. Trenches were to be excavated, the tunnel structure put in place, and the top backfilled to the original grade level. It was expected, and later proved to be the case, that this general process would have to be modified at some locations because of special conditions.

Construction began on March 28, 1895, on the incline in the Public Garden. Shrubs and trees were removed: decayed trees were cut down, while others were transplanted. Excavation proceeded through layers of ashes, sand, oyster shells, gravel, and fibrous peat. Laborers shoveled earth into carts on rails for removal until a depth of about 10 feet had been reached. After this the earth was shoveled into skips, lifted to the surface, and dumped into the carts. Fixed and movable

Construction workers who helped build the Boston subway, October 28, 1895. (Boston Transit Commission, *Second Annual Report,* 1895)

derricks, each operated by a steam hoisting engine, were used. Work on section 1 proceeded quickly, with the open incline largely completed by August 1895. Construction was, however, temporarily halted by an injunction initiated by groups still opposed to the subway. Eventually the Supreme Judicial Court allowed the work to proceed.

The Tremont Street tunnel, which runs under Tremont Street on the south side of the Boston Common, was also constructed by the slice or cut-and-cover method. The cut was 17 feet deep by 24 feet wide in the two-track section and 48 feet wide in the four-track section. A very difficult provision in the construction contract stipulated that on Tremont Street two thirds of the roadway and all the car tracks had to be open to traffic daily from 7:00 a.m. to 7:00 p.m. Between 11:30 p.m. and 6:00 a.m. all the tracks and two thirds of the street could be closed. This provision was complied with. A section 10 feet by 12 feet by 6 feet was typically excavated and covered before the next section was opened.

As subsequent sections were built, various construction and tunneling techniques were employed. Because of the particular conditions existing in section 6, for example, a steel shield had to be used in tunneling. A wide arch structure was consequently put into place. Section 4 consisted in part of two single-track subways under Boylston Street that connected to the Boylston Street station. The two single tracks converged to form a double-barreled arch in places and then diverged again. The slice method was again generally used for excavation.

As the work progressed, many buildings had to have their foundations underpinned. The contractors for the sections were generally responsible for any damage to buildings adjacent to their routes. At one point excavation was delayed when human graves were encountered. The line was then progressing through a portion of the old burying ground in the Common. The bones of some 910 persons had to be removed and reinterred in another part of the graveyard.

On March 4, 1897, a major explosion occurred at the corner of Boylston and Tremont streets, caused by a leakage of gas into an

Excavation operations on section 2 of the Tremont Street mall, July 9, 1895. (Boston Transit Commission, *Second Annual Report,* 1895)

The steel structural system used in the Tremont Street section of the subway is evident in this engraving from 1895. (*Scientific American,* September 18, 1895)

excavation between the top of the subway and a temporary bridge carrying streetcars and pedestrians. Six people were killed and many others injured.

Simultaneous with the construction of the subway was a massive grade-changing operation on parts of the Common and Public Garden, according to plans prepared by the famed landscape architect Frederick Law Olmsted. Sections of these open spaces had long been poorly drained, and odors from the underlying peat were often quite noticeable as well. Regrading of the southeastern part of the Public Garden occurred early in 1895. In some places the new grade was as much as 6 feet higher than the old. Fill came from the ongoing excavations for the subway. In order to preserve the large old trees, each was carefully surrounded by trenches and then undercut to form a great earth ball. Chains were passed under the tree, and it was then raised by four screw jacks to the new grade level. This operation was done in winter so that the earth balls were more or less solid during the lifting operation.

A series of ventilating chambers were built throughout the system. The use of electric fans for ventilation in the stations and tunnels was an innovation of the Boston project and was subsequently emulated elsewhere. The cars were run and lighted by one electrical system, the tunnels and stations lighted by another.

The eight buildings constructed on the Boston Common over the Boylston and Park Street stations were designed by the firm of Wheelwright and Haven and constructed by the Norcross brothers. All had walls of cut granite, linings of white enameled brick, and copper and glass roofs. The Scollay Square and Adams Square buildings were designed by Charles Brigham, who also designed the handsome Haymarket Square station, which was, as noted in the fourth annual report of the Boston Transit Commission, "devoid of ornament, but intended to be light, durable, economical in design, and appropriate for the work it is called upon to do."

Construction of the Park Street station, July 28, 1896. (Boston Transit Commission, *Third Annual Report,* 1896)

Cars running on temporary tracks laid over the roof structure of the Park Street station, November 27, 1896. (Boston Transit Commission, *Third Annual Report,* 1896)

Sections 1 and 2 were opened for limited service in August 1895. The major part of the system was opened on September 1, 1897. One hundred thousand passengers rode the line the first day, and over fifty million the first year; the reduction in surface traffic was dramatic. This subway system and its subsequent additions continue to serve a city whose metropolitan region has since more than tripled in size. The Tremont Street section is still a key element in Boston's transit system.

On March 4, 1897, a gas explosion ripped through part of the subway line on the corner of Boylston and Tremont streets. Several people were killed and considerable damage done. (Courtesy of the Prints Department, Boston Public Library)

The New York Subway

Built between 1900 and 1904, the New York subway was one of the country's first significant underground rapid transit systems and had a major impact on the growth and development of the city it served. In its construction and operation were demonstrated several technological innovations and advances. Of particular significance in the subway's construction and in its expansion during the following years were the cut-and-cover digging techniques, rock tunneling, underwater tunneling by shield and sunken tubes, steel bent construction, and elevated steel structures.

Demand for a transit system surfaced in New York as early as 1831, when a proposal was made that the city build a railroad down Broadway. Although nothing came of this idea, horse-drawn streetcars were soon operating down Fourth Avenue. This form of transportation proved immensely popular, and horse-drawn car lines spread throughout the city. Population increases soon led to overcrowding on the lines, however.

By midcentury New York was feeling a pressing need for a better public transportation system. The city's population explosion had led to a host of interrelated social, economic, and political problems. Overcrowding was common, since the city was largely confined to the island of Manhattan and could expand only northward. Population densities skyrocketed in lower Manhattan, where the poor were largely trapped. An underground railway, it was hoped, would increase mobility and open up the less densely populated upper Manhattan to greater numbers of people. Workers who were now living in congested tenements could acquire single-family houses in an area where property prices were low. The subway would also provide a separation between home and the workplace—a common social goal.

In 1864 a rapid transit bill was introduced in the New York state legislature that would allow private investors to construct an underground railway similar to the one opened in London in 1863. The bill was introduced by Hugh B. Willson and Henry Poor, who together had founded the Metropolitan Railway

The New York subway, opened on October 27, 1904. A typical cross section on Fourth Avenue, showing the method of supporting trolley tracks and gas and water mains during construction. (*Scientific American*, September 1902)

Company. Upstate politicians and the railroad companies, however, were powerful adversaries of the plan. Two technological obstacles presented themselves as well—coal-burning engines promised to be prohibitively noxious and dirty in an underground tunnel, and the danger of undermining buildings along the proposed route was great. A. P. Robinson, an engineer hired by Willson, was working at the time on developing compressed-air locomotion, which may have somewhat allayed concerns about cleanliness. The bill died in committee, however.

In 1865 a second transit bill was introduced and passed in the legislature, but the governor vetoed it. Overcrowding on the horse cars led to continued demands for improvement, and in 1867 the state senate's Committee on Rapid Transit recommended a system of underground railways. Two lines, one on the West Side and one on the East Side, were proposed. An experimental elevated railroad opened in 1868.

Meanwhile, Alfred Ely Beach, a brilliant inventor and editor of *Scientific American,* had been designing a pneumatic subway. First introduced at the 1867 fair of the American Institute, a small close-fitting car that carried twelve passengers was blown through a short tube by a fan. Aware of the plans of Boss Tweed (then at the zenith of his power as Tammany leader) to build a 40-foot-high stone viaduct for a railroad running the length of Manhattan—and of Tweed's consequent resistance to competing transportation plans—Beach sought and received a charter to build a pneumatic mail-carrying tunnel. In actuality he began construction of a passenger-carrying system, presumably believing that once it was built, Boss Tweed could do nothing about it. Beach began the sub-rosa excavation of his tunnel from the basement of a store near Broadway.

The tunnel was built in loose, sandy soil by means of a tubular movable shield that permitted workmen to buck up the wall at the rear while those in front were excavating. Beach's innovative shield used 18 hydraulic rams set around the circumference to force the shield forward—a great improvement over the shields used by Brunel and Greathead in England, which used propelling screws. A single hand-operated pump controlled the forward course of the shield. Eight horizontal shelves with sharpened edges extended across the front of the shield. The hydraulic rams forced the edges into the soil for the length of the stroke. The material that entered and was supported by the shelves was then removed and hauled off. Tunnel linings were installed as movement progressed (masonry in straight sections, cast iron in curved). This was the first example in this country of the use of this innovative and subsequently widely used tunnel-building technology.

In February 1870 Beach opened his wonderful 312-foot-long clandestinely built subway, which ran under Broadway. A waiting room was excavated at the Warren Street terminal, and a large blower installed to move the 22-passenger car. The blower created a pressure in the waiting room and tunnel area (of about 0.25 pounds per square inch) that created a thrust against the close-fitting passenger car and forced it down the tube. The car was returned by reversing the blower's direction and creating a suction.

Following what was probably the only surprise subway opening ever to occur in a major city, the system remained in use for about a year. During 1870 more than 400,000 people paid the 25-cent fare to ride the line. Beach then proposed the expansion of his successful project to the New York assembly, which was simultaneously considering Tweed's bill for his railroad viaduct. Tweed was, of course, furious over Beach's subway. Twice, Governor John T. Hoffman vetoed Beach's bill, although Tweed's bill was signed. Governor John A. Dix finally signed Beach's bill, but Beach could not obtain the financing and the project died.

After Beach's tube project collapsed, a number of other companies received charters from the legislature to construct underground railways. All of them failed to acquire the necessary private financing, however: the ventures were not attractive to investors because of high risks and high capital outlay. In the meantime the elevated rail system continued to be developed, but the trains were slow and did not at all solve the city's transit problems. During this same period, however, electrically powered trains were developed that held promise for eliminating the chief drawbacks of steam-powered underground railways—smoke and noxious gases.

The mayor of New York between 1887 and 1889, Abram S. Hewitt, recognized that the els and the street railway had failed to provide adequate mass transportation. He believed that a subway was the only practical solution and that some sort of balance between private and public interests was the best route to take. He announced such a plan in 1888 and continued to advocate it after his term of office. In 1891, his successor appointed a five-member Board of Rapid Transit Commissioners to develop specific plans, map out routes, secure rights-of-way, and sell the right of construction and operation to a private company. These ends were all met except the critical one of finding a responsible bidder to undertake the construction and operate the system. In the same year the state legislature allowed cities for the first time to use their own funds for transit construction. In 1894 New York voters approved public ownership by referendum, and preparations for a subway were under way.

William Barclay Parsons, a civil engineer educated at the Columbia College School of Mines, provided technical expertise to the various commissions dealing with the subway issue. He had previously worked with the railroads in several capacities and had served as the engineer for two companies that had sought to build a Manhattan tunnel. He had also visited transit systems in Britain and on the Continent. In 1894, as chief engineer for the Board of Rapid Transit Commissioners, he proposed the plan eventually adopted.

The first contract was prepared under Parsons's direction and adopted on November 15, 1899. One of the two bidders, John B. McDonald, who was experienced with railroad construction, was selected as the lowest bidder on January 15, 1900. There was some question as to whether McDonald had access to the necessary capital to carry out the contract. This issue was resolved by August Belmont, a financier, who incorporated a new firm, the Rapid Transit Subway Construction Company, to do the work. Belmont had been involved with Parsons in several previous subway schemes. McDonald assigned his lease to the firm. At the same time Belmont formed a second company, the Interborough Rapid Transit Company, to operate the subway once it was completed.

Ground was broken on March 25, 1900. Construction for the Brooklyn extension began on November 8, 1902. Five primary types of construction were used: flat concrete roofs with steel beams across and to the sides, flat reinforced-concrete roofs supported by steel bulb angle columns between tracks, concrete-lined tunnels, cast-iron tubes, and elevated steel structures. Most of the first subway was built by the cut-and-cover method. Deep excavations were made in which to construct the flat-roofed structures of the tunnel. The steel and concrete structures were made with columns spaced 5 feet apart, which in turn connected to steel roof beams; concrete arches spanned between the roof beams. The reinforced-concrete structures used spans from 18 to 30 inches thick.

Buildings encountered were underpinned by extending their foundations to sound rock. The network of water and gas mains, sewers, steam pipes, pneumatic tubes, electrical conduits, and other elements already existing underground proved a major difficulty in construction. These elements had all to be rerouted.

In places where rock was tunneled, a concrete arch system was used. At some places the tunnel is very wide—up to 37.5 feet. Cast-iron tubes consisting of cylinders 16 feet in diameter and encased and lined with concrete were used elsewhere. In the river areas a trench was dug (instead of tunneled) and a caisson built. The roof of the caisson was the bottom of the tubes, whose bottom halves were then sunk and the top halves floated over and lowered. Finally, concrete was poured over the whole. In other locations cast-iron tubes with cast-iron segment linings were used. The tubes were partly constructed by the use of air pressure through rock and with hydraulic shields through silt and sand. In some places, often because of grade requirements, elevated steel structures were

Construction in progress on a later phase of the New York subway. (Courtesy of the New York City Transit Authority)

Later subway extensions involved the sunken-tube approach to tunnel building. This 1913 photograph shows a tube being towed to the tunnel site. (Courtesy of the New York City Transit Authority)

used. Among the more notable was the Manhattan Valley viaduct on the present 125th Street, where a two-hinged arch spanning 168.5 feet was used.

The first part of the subway (Contract 1) cost $35 million to build. Though it was constructed before the introduction of the power shovel and the air hammer, it was the most quickly built section of the entire system. It opened on October 27, 1904—having been anticipated and planned for forty years. Mayor George B. McClelland drove the first train, packed with dignitaries sitting and hanging from the straps, from City Hall through most of the 9.1-mile route, which terminated in the Harlem area of New York's Upper West Side. The trip took 26 minutes, at the rapid rate of 22 miles per hour. On October 28, 150,000 of New York's 2.5 million people inaugurated the first phase of what was to become the nation's most extensive rapid transit system, one that soon became the established example for other cities across the nation in design and mechanical features. By the third day of operation 350,000 passengers were riding the subways, causing rush hour congestion and train delays. Over the next several decades the subway system, which had quickly become essential to the life of New York City, continued to expand.

Several innovative safety features were incorporated into the system, some of which are still used today. If a motorman removed his hand from the control handle, the train stopped automatically. In order to avert collisions, as the trains ran at frequent intervals, trippers automatically kept trains a certain distance apart if a motorman ran through a red light or stop signal.

The first subway was uncomfortably hot in the summer (5 to 10 degrees hotter than outside) and cold in the winter—much to the surprise of the builders, who had assumed the earth would insulate against both summer heat and winter cold. The friction of the train and the heat from the crowds of people, however, were sufficient to significantly increase summer temperatures. A system for cooling the stations was devised in 1905–1906 and built at the Brooklyn Bridge station, the first stop on Contract 2, the extension of the subway completed in 1908. Two artesian wells were installed at the station. Cold water from the wells passed through pipes along the station walls, cooling the air contained in a second set of pipes. The cooled air then passed through ducts into the station and brought the temperature to about 5 degrees below that of street level. The system proved too costly, however, and was discontinued. Not until sixty years later was another air conditioning system seriously considered.

The subway system profoundly influenced settlement patterns and the location of commercial establishments in New York. One hundred hotels were built along Broadway during the first several years after its completion, and Broadway and Times Square rapidly became the center of the city. Other expected retail centers, such as the area between 14th and 25th streets, never developed.

The subway did not, however, prove to be the panacea for urban problems that some people had hoped. The new express trains did indeed break down barriers to the outward expansion of the city and stimulated a large amount of construction beyond its traditional confines. This expansion, however, resulted in overcrowded express lines, which in turn decreased mobility. Moreover, the low land prices of northern Manhattan, which, it had been hoped, would foster the availability of housing for all, simply disappeared as land and building speculators stepped into the picture. Investors often had privy knowledge of coming transit improvements in an area, which allowed them to buy up large tracts of land cheaply beforehand. After the subway came, property values would rise and the investors would sell out. There was little incentive to build high-quality housing. Tenements were erected, not the one- and two-family structures envisioned by social planners.

The subway, then, has done much to determine the habitability, style, and development of one of the world's great cities. Its effects, both positive and negative, can be debated at length, but it is almost impossible to conceive what New York might have been like without it.

Chapter Four

Bridges

Introduction

Bridges are commonplace but richly symbolic elements in our landscape. In a unique way, they serve as an emerging society's statements of aspiration and intent. In no other type of structure do so many aspects of engineering, technology, civil design, economics, and social organization come together to produce a single form. The history of American bridge building, then, epitomizes the development of the nation as a whole; it also encapsulates the remarkable evolution of bridge-building theory from a long-established but empirically based design approach to a highly developed engineering art. With this evolution of structural theory came both a clarification of the forms of many existing bridge types and a host of new forms with undreamt-of spans and load-carrying capabilities. At the same time, however, it must be noted that the mature forms of several innovative bridge types (especially suspension and truss types) appeared in early examples built by individuals who had a larger amount of vision—coupled with the pragmatic skills of the craftsman—than they had of underlying scientific principles. It is, in truth, this dialogue among vision, craft, and theory that characterizes the evolution of bridge building in America.

The first bridges in the colonies were simple spans—groups of logs resting on opposite banks of a creek or supported on pilings driven into the earth. Occasionally surface planks were added, or wooden cribs filled with stone were used instead of pilings. Over time this bridge type was improved—the logs were hewn, the planks were dressed, and guard rails were added—but its principal aspect, its short, simple span, remained the same. A pile-and-beam bridge was built around 1630 across the Mystic River in Medford, Massachusetts, and was followed by others such as the Great Bridge across the Charles River at Cambridge (1662). The bridge built by Samuel Sewall across the York River in Maine in 1761, which was supported by pile bents, also exemplifies this type, as does Sewall's bridge over the Charles River at Boston. Completed in 1786, the latter boasted forty lamps for night lighting.

A second bridge type used by colonial builders was the floating span. A series of timbers were tied together and anchored in place floating over a watercourse. The bridge over Glenview Pond at Lynn, Massachusetts, was of this type.

Both pile-and-beam and floating bridges had inherent difficulties. Both types were vulnerable to ice floes and river flotsam. The pile-and-beam bridge, due to the limited height of its supporting pilings and the relatively short timber spans, formed a virtual barrier across a watercourse. Nor were these bridges easy to build in deep water. Floating spans, though they could be moved out of the way of river traffic, posed other problems. The timber that formed the platforms tended to sink over time. The Glenview Pond bridge, for example, simply received additional layers of timber as successive platforms subsided until 1916, when the bridge was a solid wall of timber resting on the pond's bottom.

To construct a span of sufficient height and length not to impede river traffic or be damaged by it, new bridge-building techniques were necessary. Very rapidly in the early nineteenth century, many new types were explored by innovative builders. The basic forms developed during these years—trusses, arches, suspension spans, cantilever forms, and combinations of these—still characterize much of today's construction. Even the more recently perfected types, such as tubular box bridges of steel or precast concrete, have their origins in this remarkable period.

Of all the bridge types brought to a high level of development in the nineteenth century, the arch is perhaps the most familiar. The stone arch descended to America through a continuous tradition reaching back to antiquity. The Frankford Avenue

Bridge in Philadelphia (1697) is a notable early example of this tradition. Yet while the arch's qualities of strength, durability, and beauty were obvious to eighteenth- and nineteenth-century Americans, its use presented difficulties. The greatest of these were the relatively high cost of stone construction and the necessity for great numbers of highly skilled craftsmen. The long construction time normally required was also problematic for a country characterized by rapid development and expansion. Nonetheless, where the demand for quality was high enough, stone arches were erected. The introduction of railroads, with their unique loadings, brought many such demands. Among the more beautiful examples are the Starrucca Viaduct (1848) in Pennsylvania, the Great Northern Railway Bridge (1883) in Minneapolis, and the Rockville Bridge (1902) across the Susquehanna in Pennsylvania.

The arch form was also explored in a new way through the use of timber, iron, and steel. Richard Delafield's cast-iron arch bridge over Dunlap's Creek in Brownsville, Pennsylvania (1838), was the first major bridge of this type in this country. The metal arch form of construction, however, is best epitomized by the Eads Bridge across the Mississippi at St. Louis (1874)—still one of the nation's most celebrated bridges. The introduction of reinforced concrete opened the way for further developments in the arch form. The Alvord Lake Bridge (1889) is generally credited with being the first American bridge of this type.

While the arch form provided a high-quality bridge, its costliness and difficulties of construction prevented it from becoming a common solution to the nation's bridge-building problems. Early-nineteenth-century innovators naturally turned to timber as a widely available and inexpensive building material, using it for other bridge types besides trestles. Enoch Hale's bridge across the Connecticut River at Bellows Falls, Vermont, built in 1785, was a transitional type of structure, particularly notable for its use of timber as a material for structural framing. But the true potential of the material became evident with the introduction and development of the truss form, wherein long spans could be achieved by a triangulated arrangement of short timber elements. While simple forms of the truss had long been in use in buildings, it was in the bridge-building field that the structural possibilities of the form were fully explored. Timothy Palmer used a form of arched truss in his 1792 bridge across the Merrimack River at Newburyport, Massachusetts. His larger and more well known Permanent Bridge in Philadelphia (1798–1806) introduced a new era in bridge building in this country and abroad. One of its innovative features was the use of an exterior wood covering to protect structural members from weathering. A combination arch and truss system, introduced by Theodore Burr, came into wide use; a major example is the bridge built in 1804 over the Hudson River at Waterford. Lewis Wernwag used a form of arched truss in his famous "Colossus" bridge over the Schuylkill at Philadelphia (1809–1812). This remarkable bridge spanned 340 feet with a rise of only 20 feet.

By the 1820s new forms were being introduced—including Ithiel Town's lattice truss, which was completely free of any arch action. Stephen Long advanced the design of trusses even further with his well-thought-out bridges—his first was built in 1829 for the Baltimore and Ohio Railroad—and subsequent patent designs. William Howe of Massachusetts developed and patented one of the most widely used timber trusses. The Howe truss evolved into a very efficient and elegant structural form. Iron, newly developed as a building material, was introduced into various elements that made good use of its special properties: wrought iron, for instance, was used in elements subjected to tension forces.

The coming of the railroad led to changes in the form of the covered truss bridge and eventually to a reduction in its use. For a while heavier train loads could be carried by increasing the size of timber members, but the emerging all-iron bridge soon proved more appropriate. The covered truss bridge continued to be used in many settings, however, particularly for highway bridges. The period from the turn of the century to the 1840s represented the golden age of the timber truss form. In the course of this period one can also see a change in the form of trusses, from earlier redundant configurations associated with intuitive design approaches to very sophisticated triangulation patterns that reflect a deep (albeit not necessarily mathematical) understanding of structural actions.

With the advent of various forms of iron and steel, the truss form was naturally adapted to these new materials—and the great long-span bridges we know today were born. The development of the all-iron truss is fundamentally an American accomplishment, although the truss form itself has European precedents and all-iron forms have their European counterparts. Although presaged by a design for an all-iron bridge by August Canfield of New Jersey, Earl Trumbull's 77-foot span of 1840 that carried a road over the Erie Canal in Frankford, New York, was the first of its kind.

The railroad imposed new demands on bridges—greater loadings, longer spans, and increased rigidity. Richard Osborne constructed an all-iron truss for a railroad in 1845 in Manayunk, Pennsylvania, using a form of Howe truss. The innovative Pratt truss, patented in 1844, was soon recognized as superior to the Howe form when used with all-iron members; it remains in common use today. The emergence of more formal methods of structural analysis, such as that developed by Squire Whipple and recorded in his important 1847 publication, *A Work on Bridge Building,* and that described by Herman Haupt in his 1851 *General Theory of Bridge Construction,* also contributed to the widespread adoption of trusses and to improved truss forms. Whipple's bridges, used largely in connection with the Erie Canal, were very dependable and the first scientifically designed iron bridges in the country.

In the 1850s Wendel Bollman and Albert Fink, engineers for the Baltimore and Ohio Railroad who worked under the direction of Benjamin Latrobe, Jr., devised innovative truss forms especially for railroad use. As structural forms these trusses had shortcomings, but they met the needs at hand. Both forms became prototypes, and many were built across the country. Fink's truss, the first example of which was completed at Fairmont, West Virginia, in 1852, soon superseded that of Bollman for most heavy line uses. Other engineers too devised (and often patented) new truss forms. The truss developed by S. Post, an engineer for the New York and Erie Railroad, came into widespread use following his 1865 patent. An especially interesting type was the lenticular, or Pauli, truss, which was initially developed in Europe but explored by several American engineers, including Gustav Lindenthal in his Smithfield Street Bridge of 1884 over the Monongahela River at Pittsburgh. This form enjoyed popularity until just after the turn of the century.

The first all-steel bridge was William Sooy Smith's truss bridge over the Missouri River at Glasgow, Missouri, built in 1878–1879. (Eads had used an alloy steel in his famous bridge at St. Louis, but only in parts of the structure.) The long-span railroad cantilever truss was first employed by Charles Smith and L. F. Bouscaren in 1876, with 375-foot spans over the Kentucky River at Dixville, Kentucky. The bridge was modeled as a continuous structure, with hinges inserted at points of inflection or zero bending moment.

In addition to the many specially constructed iron or steel bridges, the later nineteenth century also saw the standardization and industrial production of the truss form in iron and steel. The patented bridges by Bollman and Fink, for example, were important not only for their configurations but also for what they represented as reproducible design solutions for the country's rapidly expanding transportation network. Companies such as the Berlin Iron Works of Berlin, Connecticut, began turning out industrially made bridges by the score. The Phoenix Bridge Company was another big builder. This development met a significant need, since outside the railroads and certain centers there were still relatively few engineers specifically knowledgeable about long-span bridge construction.

The new medium of all-iron construction was also explored through the development of other structural forms, including various forms of beams or plate girders. In 1846–1847 James Millholland built a double-web form of box girder for a 50-foot span for the Baltimore and Susquehanna Railroad at Bolton Station, Maryland. Many different girders were used during the Civil War, but they were invariably limited by the characteristics of wrought-iron fabrication and construction to short and moderate spans. As steel was introduced, however, spans grew in length. A major plate-girder bridge was completed over the Harlem River in New York City in 1888.

While the truss type became the workhorse of bridges, it was the suspension span that caught the imagination of the times. The composition formed by the very high supporting towers and slender horizontal deck, united by the lyrically graceful sweeping curve of the suspension cable, possessed a dramatic beauty that neither the staccato rhythm and rococo filigree of the truss nor the sheer power of the hulking cantilevers could approach.

The first suspension bridges in this country were designed by James Finley and built on his patent. Finley's bridges contain most of the key elements of mature forms, especially the stiffened bridge deck. The suspension "cable" patented by Finley was in reality an eyebar chain whose link system was flexible enough to form the curve of the channel span. Finley's first bridge was built over Jacobs Creek in Pennsylvania in 1801. His largest undertaking was the Schuylkill River bridge at Philadelphia in 1809. The introduction of wire for use in cables marked a turning point in the evolution of the suspension form. Later notable designers include Charles Ellet, whose greatest success was the Wheeling Suspension Bridge across the Ohio River, completed in 1849. This bridge, with its 1,000-foot span, was the longest of its kind in the world at the time of its construction. Drawn wrought-iron wire was used for the main cables.

The giant of the age, however, was John Roebling. The German-born Roebling began his career as a designer of cable suspension structures with an application that now seems unlikely. In the period 1843–1850 Roebling erected a series of aqueducts over several rivers of northeastern Pennsylvania, including the Delaware and the Lackawaxen. Later he built the Niagara Bridge (1851)—the first suspension bridge designed for railroad use in America—and the bridges at Cincinnati (1857–1867) and Brooklyn (1870–1883). The Brooklyn Bridge, Roebling's final triumph, was completed after his death by his son. Built between Manhattan and Brooklyn over the deep and very busy East River, the bridge succeeded from all points of view—engineering technique, logistics, process, utility, and beauty. It became a symbol of progress for an entire age. As an example of the suspension form, the Brooklyn Bridge would remain preeminent for fifty years, until challenged by the monumental works of O. H.

Ammann and others in New York and San Francisco. Works such as the George Washington Bridge and the Golden Gate Bridge finally redefined an evolution toward thinness, lightness, and extraordinary span.

Less exciting but, like the truss bridges, eminently serviceable were the movable bridges developed most extensively toward the end of the nineteenth century. From a crude drawbridge in Chicago at Dearborn Street in 1834, the movable bridge has evolved in its own way to become a very sophisticated type. Pontoon bridges, swing or pivot bridges, vertical lifts, bascules, and other forms were explored at one time or another in many American cities, but especially in Chicago. The development of the Chicago type of trunnion-bascule (or rolling-lift) bridge represents an important advance in the evolution of movable bridges. The first bridge of this type was opened in 1894.

The development of reinforced concrete marked another turning point in the history of American bridges. Ernest Ransome built the first reinforced-concrete bridge in America, the Alvord Lake Bridge, in San Francisco in 1889. Fritz von Emperger helped popularize reinforced-concrete construction with a number of successful bridges, including the elegant structure across the Housatonic River at Stockbridge, Massachusetts, of 1895 and an even longer-span structure in Eden Park, Cincinnati, also built in 1895. A trussed-bar system was utilized in an arch bridge at Lake Park in Milwaukee in 1905. A three-span girder bridge was also completed in 1905 in Marion, Iowa. By 1910 many designers were using reinforced concrete in essentially its modern form. A notable innovation occurred in 1937 with the introduction of the hollow box form of girder by the Washington State Department of Highways for a bridge over Henderson Bay at Purdy. Another significant innovation was the development of prestressing and posttensioning techniques for reinforced-concrete structures. Presaged by P. H. Jackson's patent in 1886, and following several European examples, prestressing was adopted for the Walnut Lane Bridge in Philadelphia in 1949–1950. This important structure ushered in the extensive use of prestressing techniques in bridge construction.

The evolution of the bridge types briefly described above reflects the almost complete transformation of the theory of bridge design from the tentative, experimentally based approach employed with early timber forms into a well-understood, quantitively based discipline—one that nonetheless requires vision to produce truly exceptional bridges. The tradition of bridge building in America is indeed a proud one.

The Frankford Avenue Bridge

The old Frankford Avenue Bridge spans Pennypack Creek in Philadelphia. Built in 1697, it is one of the oldest extant bridges in the United States and also holds the distinction of being the oldest stone arch bridge still in service on a modern highway. The structure, often called the Pennypack Bridge, is said to have been "erected by the community, each adult male inhabitant contributing his share, either in work or in money." The three-span stone arch bridge served as an important link in what was by then known as the King's Road to New York.

The King's Road was originally a forest trail, used frequently by travelers in wagons and stagecoaches. It became the post road from Philadelphia to New York and the preferred route to the north. The first portion of what was to become the King's Road to New York was in use in 1677. The need for a bridge over Pennypack Creek was recognized by William Penn at this time, and subsequently he urged local authorities to construct such a bridge so that he might visit New York. The Second Court of the County of Philadelphia reported in 1683 "the want of a bridge or a ferry over . . . the Pennebecca and, in general, bridges or ferries over the creeks in the King's Road." In 1686 the Provincial Court in Philadelphia ordered the King's Road to be laid out, making it the first public road surveyed in Bucks County. There seems, however, to have been no regular transportation along the road until 1725. The first stagecoach to go from Philadelphia to New York took three days, this service beginning in 1756. The road became known as the Frankford and Bristol Turnpike in 1803.

When constructed in 1697, the bridge was so narrow that two teams of horses could barely pass each other on it. In 1773 a Miss Sarah Eve, while passing through the area, noted, "The creek is not very wide, so that trees on each side might almost shake hands, and what adds much to the beauty of the whole are the shrubs and bushes that blossom along the banks." In 1775 it is documented that an express rider galloped across the "Pennypack Bridge" bringing news of the first battle of the American Revolution, which

Frankford Avenue Bridge, Philadelphia. This three-span stone arch bridge over Pennypack Creek was built in 1697 and has served as an important roadway ever since. It is the first known major stone arch bridge to be built in this country and one of the oldest extant American bridges of any kind. (Courtesy of the American Society of Civil Engineers)

had occurred in Lexington. The bridge became part of the Frankford and Bristol Turnpike in 1803.

The bridge was widened on the downstream side in 1893, when the trolley route was extended northward on Frankford Avenue. A sidewalk was also provided at this time, and grade changes were made, the old stone arches remaining intact throughout the alterations.

The Carrollton Viaduct

The Carrollton Viaduct, over Gwynns Falls in the southwestern part of the city of Baltimore, was the first major engineered structure on an American railroad. Completed in 1829 as part of the first line of the Baltimore and Ohio Railroad, this two-span masonry arch bridge—the first masonry viaduct to be constructed in the United States—remains in full service today.

In 1827 several investors in the Baltimore area applied for a charter to build a railroad—the country's newest form of transportation. An act of incorporation was secured and the Baltimore and Ohio Railroad came into being. On April 17, 1827, Peter Little petitioned the government "to procure the cooperation of the United States in the execution of said project, and to ascertain to what extent such assistance may be contributed." The secretary of war replied with the offer of three surveying brigades to assist in laying out the railroad. The company thus became the first American railroad to ask for and receive government engineering aid.

The three engineering parties were headed by Colonel Stephen Long, Dr. William Howard, and Major William McNeill. Since Secretary of War Barbour considered the undertaking to be of national importance, the direct costs of the survey were to be "charged to the public appropriation for Internal Improvement." President P. E. Thomas of the railroad company also asked for the use of Jonathan Knight, who was then employed by the War Department in the laying out of the National (or Cumberland) Road (see chapter 2), to help in determining the best method of locating and constructing railroads. Knight resigned his position with the National Road and came to work on the Baltimore and Ohio line.

Preliminary surveying was begun in 1827 by the army engineers. The company sent Major McNeill, Knight, and Lieutenant George Whistler to England in 1828 to examine railroad-building practices there. Later, McNeill and Knight served with Colonel Long as the company's board of engineers, and Whistler became superintending engineer. Limited appropriations made it necessary to withdraw some of the surveying assistance in 1828, and William Howard and his party were subsequently ordered to South Carolina to survey the proposed railroad route from Charleston to Hamburg.

As the surveying progressed, Long and Knight (in charge of engineering operations) advertised for bids for constructing the first few miles of the road. The first track was laid in October 1829, under Whistler's superintendence. Construction was completed from Baltimore to Ellicott's Mills in 1830.

Recognizing that many bridges would have to be constructed for the railroad, the engineers took up the important policy question of what type to use. Those who had examined the British railways preferred heavy, solid masonry bridges with stone foundations under the rails. Caspar Wever, superintendent of construction, also urged that these structures be masonry. Colonel Long, however, strongly favored wooden bridges, pointing out that they were economically very attractive. After an intense internal struggle, a decision was reached to use stone bridges everywhere on the first division of the road, with one exception: Colonel Long was to erect a wooden truss bridge where the Washington and Baltimore Turnpike was carried over the new railroad near Baltimore. Long's design for this bridge was eventually patented.

At one time or another, some fourteen army engineers took part in the survey, location, and initial construction of the railroad. In 1830 a dispute over construction and disbursement procedures arose within the company's board of engineers, resulting in the group's dissolution and the end of the involvement of the army engineers in the project. Colonel Long continued his engineering career elsewhere and made extremely important contributions to the development of truss structures; the "Long truss" came into wide use (see the Cornish-Windsor, Blenheim, and Bridgeport Covered Bridges).

The Carrollton Viaduct was built for the line in 1829, with McNeill, Whistler, Knight, and Wever all associated with its construction. Also involved was James Lloyd, architect. This was a true viaduct, in that it crossed a

Carrollton Viaduct, Baltimore. The first major engineered structure on an American railroad, this two-span masonry arch viaduct was completed in 1829 and remains in service today. (Courtesy of the American Society of Civil Engineers)

relatively wide valley and not just a stream or river. Such a structure, quite uncommon at the time, was demanded by the need to maintain limited grades for the rail lines.

The construction process was unique for its day. It involved the handling of great masses of huge granite blocks. Some 12,000 perches of stone went into the structure, which was divided into a main arch, having an 80-foot span, and a side one. The total length of the viaduct was 312 feet and its height above the stream was 51 feet 9 inches. Gwynns Falls passed through the larger opening, while the side arch originally spanned a wagon road. Building the structure involved the construction of a lot of temporary wooden centering, which was kept in place, supporting some 1,500 tons of granite, until the arches were finally keyed. The stone came mainly from the vicinity of Ellicott's Mills, with some from Port Deposit on the lower Susquehanna. It was carefully dressed and measured before being swung into place. Construction started in the middle of May, 1829, and was largely complete by November. This successful project was followed by the construction of the Patterson Viaduct, over the Patapsco River, and the Thomas Viaduct. The Carrollton Viaduct is still in service, a fact that attests both to its careful design and to the inherent strength and stiffness of masonry arches.

Dunlap's Creek Bridge

Dunlap's Creek Bridge in Brownsville, Pennsylvania, has the distinction of being the first all-metal bridge built in the United States and has been in continuous use since its construction. The "Old Iron Bridge" was conceived by Captain Richard Delafield, of the Army Corps of Engineers, and completed in 1838. The bridge is recognized as having a unique type of construction for the period in which it was built.

Several previous bridges had occupied the site upon which the all-metal bridge was built. The first bridge across Dunlap's Creek was constructed of wood some time prior to June 1794. This wooden structure was repaired in 1801 but is said to have "floated off" in the spring of 1808. In 1809 a chain-link suspension bridge, described as being "30 feet above the water and very long," was built on the site. This bridge collapsed in March 1820, the event triggered by the combined weight of deep snow and a heavy wagon.

The crossing of Dunlap's Creek between Brownsville and Bridgeport was essential to the completion of the new National Road. In 1821 another wooden bridge was built on the site, at a cost of $2,050. Its fate is undocumented, but records indicate that it was no longer in existence in 1825.

Plans for a new bridge were drawn up by Captain Richard Delafield, who had come to the region in 1832 as a captain in the Corps of Engineers to take charge of repairs on the National Road east of Ohio (this was before the various states took over this responsibility). Delafield suggested a new bridge in 1833 and proposed a new site as well, some distance away from the original location. Political disputes arose, however, in connection with the siting: Bridgeport merchants objected to the suggested location because they felt their businesses, established along the old route, would suffer, while Brownsville residents saw advantages in the new location. The dispute reached President Jackson, who issued an executive order stipulating the siting of the bridge in its old location. Difficulties with that site, however, led to a

Dunlap's Creek Bridge, Fayette County, Pennsylvania. This structure is the oldest extant all-metal arch bridge in the United States. It was completed in 1838 and demonstrated the feasibility of using iron in bridge construction. (Library of Congress)

compromise location some distance away from both Delafield's site and the old location.

The design of the structure had been pretty well defined by Delafield by 1835. As early as 1833 he had argued for a cast-iron structure in a letter to his commanding officer, noting his belief that a wooden structure was vulnerable to rot and fire. At this time there were no other all-iron bridges in America, although some bridges made use of some iron elements and all-iron bridges had been in use abroad for some time (beginning with the completion in 1779 of the famous iron bridge at Coalbrookdale spanning the River Severn in England). Cast-iron was also making its way into use in building structures.

By 1835 the idea of an iron structure for the Dunlap's Creek location was firmly established through Delafield's efforts. Lieutenant George Cass was assigned to purchase pig iron from furnaces in Portsmouth, Ohio. The casting was to be done in Brownsville by John Snowden, who rented the facilities of the Herbertson Foundry for the task. The abutments for the bridge were started in 1836 by Messrs. Keys and Searight. Sandstone was used in the construction of these 14-foot-thick, 25-foot-long, and 45-foot-high structures. Wing walls were added to further protect the abutments from the wash of the creek. The arch, as finally constructed, spanned 80 feet and had a rise of 8 feet. It was composed of five identical arch ribs that each in turn comprised nine identical segments. The design economized on the casting process by reusing the same patterns, since Delafield recognized the benefits of using "standardized, interchangeable, manufactured parts." In this respect the design was unique, departing from contemporary French and English bridge-building practices. Delafield and Cass were in charge of all construction specifications and pursued their duties with great skill and attention to detail.

The erection of the structure was not without its problems—one instance being the settling and cracking of a masonry wing wall in November 1837. Cass wrote, "Everything seems to have gone wrong since the commencement of this work and I do hope that I may never have such another job in my life

A recent close-up of Dunlap's Creek Bridge, showing both the old structure and the more recent steel modifications. (Jet Lowe, Photographer, HAER Collection, Library of Congress)

again. It has from the beginning to this time given me more trouble and uneasiness than a work of ten times the magnitude ought to have done."

The final bridge consisted of a plank deck supported by stringers and beams, which were in turn supported through a frame by five hollow cast-iron segmental tubes built up in short lengths and bolted together at cir cumferential flanges. The bridge was first used in July 1838. It had been erected at a total cost of $39,811.63, which was quite high for its day.

In deciding to use cast iron for the structure, Delafield seems to have known quite well that he was indeed designing the first all-metal bridge in the United States. Stone was available locally, as were the skills for constructing durable timber structures. The size of Dunlap's Creek did not seem to merit the honor of being spanned by the first cast-iron bridge in America. Certainly Delafield's ambitious character had something to do with it; before the bridge was quite completed, he moved on to his next assignment, as superintendent of the U.S. Military Academy at West Point.

In 1920 sidewalks were added to the structure, necessitating the addition of modern steel members. Samples of the original iron were rated by metallurgists at the time, revealing a purity comparable to that of iron parts manufactured in 1920. The bridge is now, unfortunately, almost obscured from view by surrounding buildings and the later alterations designed to strengthen the structure.

The Starrucca Viaduct

One of the handsomest railroad bridges ever constructed is the Starrucca Viaduct, built in 1848 to cross Starrucca Creek in Pennsylvania. This extraordinarily proportioned bridge still carries trains of the Erie-Lackawanna Railroad, its row of slender piers rising over 100 feet above water level.

In 1848 the New York and Erie Railroad became the longest rail line in the world. The route began at Piermont-on-Hudson, a small town not far from New York City, and ended 484 miles away at Dunkirk on the shores of Lake Erie. Its optimum path would normally have been along the course of a natural waterway, since changes in grade would be slight. There was no natural waterway for the New York and Erie to follow, however—only an occasional stream; and formidable mountains were strewn in its way. Compounding the problem were the conditions of the railroad's charter. The New York state legislature had approved the railway solely on condition that it stay within the boundaries of New York, and at the same time had decreed that it be built far from the route of the Erie Canal. The former restriction was reluctantly waived when the railroad reached two major obstacles, the Delaware valley at Port Jervis and the Susquehanna River.

In 1834 when the engineer Benjamin Wright surveyed the route of the New York and Erie, it was with the intention of giving the railroad a maximum grade of less than 100 feet per mile. This was a very gentle slope and quite an ambitious goal, given the nature of the terrain. Political considerations modified the route that Wright had charted, however, and the plan was soon altered. Some communities paid to be included on the route, while others, with less foresight or dominated by other interests, wanted no part of it.

A major problem was the crossing of the deep gorge of Starrucca Creek at Lanesboro, Pennsylvania, four miles west of the Susquehanna. The railroad reached the valley in 1844. Company engineers entertained many possible solutions, including building an enormous embankment across the end of it, but finally concluded that a viaduct was the

Starrucca Viaduct, Lanesboro, Pennsylvania, built in 1848. This key structure of the New York and Erie Railroad was among the earliest major links between the Eastern Seaboard and the Midwest. It was constructed in record time and was among the first important engineering works to utilize structural concrete. (Courtesy of the American Society of Civil Engineers)

only sensible choice. The approach of using a massive fill was discarded because of the lack of good borrow pits in the area. Estimates indicated that the fill would cost more than a bridge. The viaduct eventually proposed had an enormous length of over 1,000 feet and needed to be about 110 feet high to maintain grade.

Two men, Julius Adams and James P. Kirkwood, were responsible for the final design and construction of the great viaduct. Adams, at the time superintending engineer of the New York and Erie's Central Division, was closely related to several of the most prominent railroad builders of the day. George Whistler and William Swift, who had helped build the Western Railroad in Massachusetts between Worcester and Albany from 1836 to 1842, were his uncles. A third uncle was the engineer Joseph Swift, who served as superintendent of West Point; and one of his cousins was an engineer for the New York and Erie. (Another cousin was the artist James Whistler.) Adams spent some time at West Point but soon joined other family members in the civil engineering profession. His uncle George Whistler was engaged at the time, along with his colleague William McNeill, in building the Carrollton Viaduct, one of the country's first major railroad viaducts. Later McNeill and Whistler completed the great granite Canton Viaduct over the Neponset River near Boston (which is still in existence). The Canton Viaduct, 580 feet long and 58 feet high, had no rival for size among American masonry structures at the time except for the Baltimore and Ohio's Thomas Viaduct. Undoubtedly, Adams followed these activities closely. He himself served with the Paterson and Hudson Railroad and then with the Providence and Stonington as assistant engineer, did a survey for the eastward extension of the Lockport and Niagara Falls Railroad in 1841, and afterward spent some time with the Mohawk and Hudson. Eventually he joined the New York and Erie. While there he designed a great wooden arch bridge, 275 feet long, called the Cascade Bridge. Adams by now was an accomplished designer.

While there is some apparent confusion in recent discussions of the Starrucca Viaduct as to exactly who designed it, all the evidence certainly points to Julius Adams. Adams did, however, enlist his brother-in-law, James Pugh Kirkwood, as superintendent in charge of building the structure, and Kirkwood may have helped shape the final design. Kirkwood, a recent immigrant from Scotland, had begun by working for McNeill and Whistler on the Norwich and Worcester line and had worked on the Boston and Providence road and the Stonington road as well. He and Adams had worked together on the Western Railroad in Massachusetts, Kirkwood serving as resident engineer in the difficult mountain division where Whistler had designed a number of remarkable stone-arch culverts. Recovering from the death of his wife and a bout with a "malaise," Kirkwood was at loose ends when approached by Adams to become superintendent. By this time, it seems, work on the great viaduct had already begun.

Construction started in 1847. Adams's letter book shows that a contract with Messrs. Baird and Collins to build the entire viaduct for $8.50 a cubic yard was signed on August 7, 1847. The job was eventually taken from them because of their slow progress and completed by a series of small operators under Kirkwood's direction. But at the start Baird and Collins opened quarries and laid a 4-mile track to the main quarry up the Starrucca Creek. This line's cars were animal powered. Apparently the line crossed the creek several times and had a branch ascending the north hillside near the site. Here derricks lifted the stone from the cars. Other stone was hauled by wagon.

Work on the main piers, with one of the first significant uses of concrete in this country, progressed the first year. The relatively shallow depth of the foundations, some 6 to 9 feet in places, was later a cause for concern. But Adams had discovered that a broad shelf of hardpan existed in the area, rather than alluvial material, which would have required deeper foundations. The foundations for the two short piers on the north hillside and the single short pier on the south hillside began with stone laid directly on the hardpan. The footings for the thirteen full-height piers, however, were concrete slabs made of Rosendale cement from the works at Rosendale in Ulster County, New York. The Rosendale works had been in operation since 1828 and had furnished cement for the Delaware and Hudson Canal. The footings were 3 feet thick and 18 feet 6 inches by 40 feet 3 inches in plan. Blocks of stone were then stair-stepped to ground level, where they measured 13 feet by 33 feet 6 inches.

Under Kirkwood's direction the falsework for the great arches was erected. It consisted of a combination of trestles, scaffolding, and arch centers. Lumber for the falsework was provided largely by Simon H. Barnes, a local businessman. In order to meet the tight schedule, both manpower and equipment were worked to excess. Screws instead of wedges were used by Kirkwood to strike the centering.

The last arch was closed on October 10, 1848. None of the stonework was fully dimensioned except the coping. Occasionally rounded stones protruded from the work. The quoins (the exterior angles of the piers and arches) were chiseled to a neat edge. All of the arch stones went throughout the arches, but only the stones of the outer rings (the voussoirs) were made the same depth. Each pier was topped by a solid backing or haunch that extended upward from the springing line of the arches. From the haunch a tie or head wall some 3 feet thick continued to the base of the floor. Behind the flanks or side walls of the arches the viaduct was made to be largely hollow. Relieving or spandrel arches of brick paralleled the side walls. The bluestone slabs of the deck were covered with concrete. The main structure was built of random ashlar bluestone; it has a stark, unornamented simplicity.

The viaduct was ready for track on November 10, 1848, and construction was completed on November 23. The first locomotive crossed on December 9. This was the New York and Erie's Norris 4-4-0 "Orange," which had been brought to Binghamton via the Chenango Canal a month before.

The Starrucca Viaduct is one of the finest pieces of masonry construction anywhere. Its eighteen slender semicircular stone arches, each spanning 50 feet, hover 110 feet above the stream bed. It is 1,040 feet in length and carries two sets of tracks. Built to carry the 50-ton locomotives of its day, it now bears the weight of 400-ton locomotives and heavy freight cars. The final cost of the viaduct (which was built with the lavish backing of British capital) was high for its day—about $335,000—making it one of the most expensive railroad bridges in the world, as well as one of its most beautiful.

Excursionists aboard the steamboat *Ermini* enjoy a splendid view of the Starrucca Viaduct, over which a new 4-6-0 engine pulls a passenger train. The photograph dates from the latter part of the nineteenth century.

The Delaware Aqueduct

The Delaware Aqueduct, built for the Delaware and Hudson Canal by John Augustus Roebling, is the earliest still-standing suspension bridge of this famed engineer-inventor, who is considered the father of the modern suspension bridge. It is further distinguished as the oldest existing cable suspension bridge in America that retains its original principal elements. The sole survivor and largest of several suspension aqueducts erected by Roebling between 1847 and 1850 to carry the Delaware and Hudson Canal over river impasses, this structure, which spans the Delaware River between Lackawaxen, Pennsylvania, and Minisink Ford, New York, was a key link in increasing the capacity and efficiency of the canal system.

The Delaware and Hudson Canal Company was formed in the spring of 1823 by Maurice and William Wurts. At the time, a canal was the only feasible means of carrying coal from the rich fields of northeast Pennsylvania to New York City, a very profitable seaboard market. Charters were granted by the New York and Pennsylvania legislatures to construct a system of navigable waterways from the coal fields along the Lackawaxen River to the Delaware River, and from the Delaware to the Hudson River by way of the Neversink River and Rondout Creek. From the port at Rondout, New York, coal could be easily shipped by schooner along the Hudson to New York City. After four years of construction, the canal opened in 1829. Covering a route of 108 miles with a lockage of 1,086 feet, it shipped 7,000 tons of coal in the first year of operation.

From the start the canal was a pioneer venture. Modest initial capital and the uncertainty of future needs dictated compromise and cost-efficiency along the first route. In the 1840s, as competition from the Erie Railroad became a threat, and as profits rose, the directors of the waterway launched an almost continuous program to increase the canal's capacity. It was in this phase of development that the Delaware Aqueduct was to play its critical role. From an initial 4-foot depth the canal was eventually enlarged to a 6-foot

Delaware Aqueduct, Lackawaxen, Pennsylvania, to Minisink Ford, New York. This is the earliest extant suspension bridge by John Augustus Roebling, rightfully acknowledged as the father of the modern suspension bridge. It is also perhaps the oldest existing wire-cable suspension bridge in the world that retains its original principal elements. Built in 1847–1849 to carry the Delaware and Hudson Canal across the Delaware River, it is one of the nation's most significant engineering relics. This photograph was taken in the 1970s prior to restoration. (Courtesy of the American Society of Civil Engineers)

depth; the path was widened; and larger-capacity boats were used. By 1846 enlargements required the rebuilding of all locks and aqueducts. Increased speed was also a priority, and this prompted a study of how to remove the worst bottleneck in the system—the slack-water crossing of the Delaware River between Lackawaxen, Pennsylvania, and Minisink Ford, New York, just above the mouth of the Lackawaxen River.

Lack of capital had prevented the erection of an aqueduct up to this point. Instead, a dam had been created, forming a still pool across which the canal boats were towed. Not only were delays a problem, but tension developed between canal officials and those involved in the considerable traffic of timber rafts on the Delaware. Under new pressures and prosperity, the canal company directed its chief engineer, R. F. Lord, to plan enlargements of the canal at this strategic point. His proposal to relocate the path below the mouth of the Lackawaxen, south of the old crossing and the dam, required the construction of a second aqueduct over the Lackawaxen. Problems with ice floes from the Lackawaxen and with the particular conditions of the earlier site were enough to offset the cost of the new aqueduct.

In February 1846 canal officials authorized an aqueduct over the Delaware River and another over the Lackawaxen. By December of that year two proposals had been submitted. One was for a conventional trussed timber structure of six spans. The other, submitted by John A. Roebling, was for a wire-cable suspension aqueduct of four spans. By using a suspension system, Roebling was able to achieve greater span lengths and therefore fewer river piers. Not only was this favorable in reducing both costs and the obstruction of ice, flood waters, and river traffic, it could also be erected without falsework. The cables could be put in place without support; and after suspenders were attached, timber cross frames for the trunk could be hoisted into position from barges anchored below. The rest of the aqueduct structure could then be easily assembled.

Roebling's plan was tentatively accepted on January 6, 1847. On the 29th of January, R. F. Lord traveled to Pittsburgh, Pennsylvania, to inspect a similar structure erected by Roebling in 1844–1845. This aqueduct, built to carry the Pennsylvania Canal over the Allegheny River, was Roebling's first major bridge of any kind, and it was here that he had introduced his practical method of cable construction. The bridge proved a great success and became a prototype for subsequent Roebling suspension bridges. Lord was impressed by Roebling's work at Pittsburgh, concluding that his abilities were far ahead of his time. Shortly after Lord's return, Roebling was awarded the contracts for both final design and construction of the Delaware and Lackawaxen aqueducts.

By March 1847 Lord had received Roebling's initial working drawings and had begun constructing the foundations and laying the pier and abutment masonry: Roebling's construction contract covered only the superstructure of suspended spans, "including all iron, timber, and wire work, the company to do all masonry and cement." Roebling's chief role at this time was in the coordination and careful setting of the great iron anchor plates embedded in the masonry of the abutments. These huge castings resisted the pull of the anchor bars that rose up through the abutment masonry, ultimately to restrain the main cables. Roebling began his main effort in the summer or fall of 1847. Working on the Delaware and Lackawaxen aqueducts simultaneously throughout 1848, he was able to complete both in time for the opening of the canal season in April 1849.

Following closely the design used at Pittsburgh, these aqueducts were built to pass only a single boat, although towpaths flanked both sides. The main body had a heavy wooden trunk, or flume, to carry canal boats. The trunk held approximately 6 feet of water and was 19 feet wide at the waterline. The trunk sides were made of two thicknesses of 2.5-inch untreated white pine plank, placed tightly in opposite diagonals and caulked up to the waterline. This formed a rigid, solid lattice structure, whose stiffness was sufficient to carry its own dead weight, leaving the cables to carry the water load. The floor was also of double beams, hung from the suspenders as in a conventional suspension bridge. The 8-foot towpaths were bracketed out from the sides and level with the trunk top.

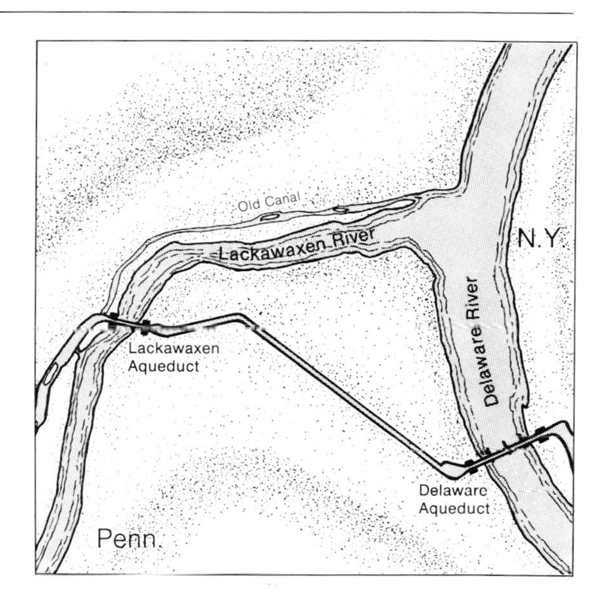

The Delaware Aqueduct spanned the Delaware River. The shorter Lackawaxen Aqueduct was nearby. The old canal had run parallel to the Lackawaxen River and then into the Delaware.

This photograph, circa 1885, shows the Delaware Aqueduct in use. The water-carrying wooden trough is clearly visible. The view is toward Pennsylvania from New York. (Courtesy of the Smithsonian Institution)

One of the last boats through the Lackawaxen Aqueduct, a companion to the Delaware Aqueduct, is shown coming through in this circa-1898 photograph. (Courtesy of the Smithsonian Institution)

The trunk system's points of suspension, moved in from the extreme ends of the transverse floor beams to points just outside the trunk sides, effectively reduced the clear span length of the beams and increased their supporting capacity. The weight of towpaths and bracing, and the pressure of the water against the trunk sides acting through the inside diagonal struts, bore down on the cantilevered ends of the floor beams, thus providing a counteracting force to that exerted by the water load at the centers of the beams and consequently reducing the total bending moments—hence stresses—in the beams.

The original trunk has not survived, but the rest of the structure is largely as originally built. The two continuous main cables carry the main load. At the bottoms of their sags they are slightly above floor level, and at the piers and abutments they rise to be carried over cast-iron saddles mounted on squat stone towers that stand about 4 feet above what would be the trunk top. The suspenders are plain 1.25-inch wrought-iron rods, doubled over the cables into stirrup form, their bottom ends threaded for the floor beam nuts. They bear upon the cables on small cast-iron saddles. Those nearest the towers, where the slope of the cable is greatest, are prevented from sliding by wrought-iron restraining links or stays. Each cable is composed of 2,150 wires, making up seven strands. Each strand was formed by carrying the wires across from anchorage to anchorage. At each anchorage a loop was made that passed over a cast-iron strand shoe pinned to the anchor bars, anchoring the strand. The strands were thus formed of a single, continuous wire, spliced at the ends. Between the towers these strands were compacted into cylindrical form, then varnished and coated with a continuous wrapping of iron wire for protection.

Each cable is 8.5 inches in diameter. Extending for a total length of 576 feet, the cables drape into four spans. From the western abutment the first span is 142 feet; the next three are 131 feet each. As originally designed, the ultimate strength of the pair of cables was 3,870 tons and the force on them from the loaded trunk 770 tons. The cables are restrained by the anchor bars, which pass down through the abutments, terminating in 6-foot-square cast-iron anchor plates upon which the masonry bears. Its dead weight resists the pull of the cables.

The Lackawaxen Aqueduct, a half mile west of the Delaware Aqueduct, was almost identical in construction but had only two spans, approximately 115 feet each, with a single river pier. Both aqueducts were such unqualified successes, structurally and operationally, that Roebling was contracted to build two other aqueducts for the canal. In 1849–1850 he erected the structures at Cuddebackville, over the Neversink River, and at High Falls, over Rondout Creek. Both were operating by the 1851 canal season.

The 1847–1850 improvements were an important asset to the canal system. In particular, the Delaware and Lackawaxen aqueducts allowed the seasonal nine-day stoppage of transport due to high water to be avoided, and cut a full day from passage time under normal conditions. Rates could thus be reduced significantly, permitting lively competition with the railroad for bulk coal haulage into the 1870s.

Peak operations were in 1872, when 3 million tons of coal were carried. After this, railroad technology and capacity increased rapidly, outstripping the canal. Canal service continued under declining conditions until 1898, when the last boat passed through the system. In 1899 the company was liquidated and operations were transferred to rail networks.

Of the suspension aqueducts built by Roebling, only the Delaware had any adaptive reuse. The others were abandoned and eventually fell prey to salvaging operations and natural deterioration. The Delaware Aqueduct, miles from any other crossings of the river, was purchased privately and converted into a highway bridge. The towpaths were removed, and a low railing was run along the downstream side of the trunk floor to provide a separate pedestrian walk. A toll house was built at the New York end, and grading was done to accommodate existing roads.

After a fire in 1932, all the original timber was removed, including the trunk and floor beams, and a simplified floor system substituted. Today the bridge consists of transverse floor beams, hung from the original suspenders and cables; longitudinal stringers; and a plain, transverse plank decking. An accident closed the bridge to traffic in 1977. It has since been acquired by the National Park Service, which is currently carrying out a restoration. A decision was made to allow some automobile traffic on the bridge, because of its useful role in the immediate transportation network, but to limit vehicle use and make the bridge very accessible for pedestrians as well. Studies of the current state of the structure have revealed the principal elements to be still sound. The wooden towpaths and water-carrying trough will be recreated, with some modifications. The sidewalls will be opened, exposing the original cables so that Roebling's construction can be viewed and understood. A concrete slab will be installed to weigh the bridge down as canal water once did, thus mitigating the slipping and consequent fraying of cable prone to occur as a result of moving traffic on the bridge. This slab will also provide the roadway.

A view of the Delaware Aqueduct after its conversion to a highway bridge. The floor of the wooden water-carrying trough became a roadway, and a railing was run along one side to provide a pedestrian walk. (Courtesy of the Smithsonian Institution)

A recent view of the structure. (HAER Collection, Library of Congress)

The aqueduct is undergoing extensive restoration by the National Park Service. (HAER Collection, Library of Congress)

The Wheeling Suspension Bridge

Wheeling Suspension Bridge, Wheeling, West Virginia. The original bridge, built in 1847–1849 by the charismatic Charles Ellet, Jr., was one of the first major long-span wire-cable suspension bridges in the world. It was the first suspension bridge to surpass the 1,000-foot-span mark and was the longest in the world at the time of its construction. The main span was destroyed in a gale in 1854 but was rebuilt soon afterward. (HAER Collection, Library of Congress)

The Wheeling Suspension Bridge, at Wheeling, West Virginia, was built by Charles Ellet, Jr., in 1847–1849 to serve as a link in the National Road from Cumberland, Maryland, to the West. The longest suspension bridge in the world at the time of its construction and the first to span over 1,000 feet, it was among the first major wire-cable bridges and is now one of the world's oldest surviving examples of the type.

Charles Ellet, Jr., who became one of America's leading practitioners of suspension bridge construction, was born to a rural Pennsylvania family and received little formal education. On the recommendation of the pioneer civil engineer Benjamin Wright, he left home at the age of seventeen to join a surveying crew, and later worked on the Chesapeake and Ohio Canal project. In a momentous decision, Ellet determined to study in France, where he entered the Ecole des Ponts et Chaussées in November 1830.

The school, then under the direction of the famous mathematician-engineers Vicat and Navier, was well known as a center for applied mechanics and structural analysis. Navier had published an important work on suspension bridges. Ellet's interest in such structures was stimulated in this environment. He was quite possibly already familiar with the 1816 footbridge by White and Hazard over the Schuylkill—probably the world's first major wire suspension bridge (a smaller one was built at Galashiels in Britain also in 1816)—and with some of James Finley's early iron-chain bridges.

Ellet came back to the United States in 1832 a confirmed advocate of suspension bridges. That year, in response to a government advertisement, he submitted a proposal for a suspension bridge across the Potomac; the proposal was rejected. He then accepted a position under Benjamin Wright as surveyor of part of the New York and Erie Railroad. His next job was as assistant engineer of the James River and Kanawha Canal; he was appointed chief engineer in 1836 but left the company in 1839. Never a small thinker, in 1840 he submitted a proposal for a suspension bridge across the Mississippi at St. Louis that would be over 3,000 feet long, with a clear span of 1,200 feet. Again his proposal was rejected.

Interest in suspension bridges kept rising among both engineers and the general population, however. Along with some failures, there had been successes as well, in this country and abroad. The suspension type appeared economically advantageous and held the promise of long-span construction. The possibility of using a superior material such as drawn wire was extremely attractive in an era when steel was not available in quantity, nor were large wrought-iron sections. The fact that suspension bridges could be quantitatively analyzed was another attraction to engineers considering long-span structures: Navier had published some straightforward techniques for the static analysis of certain types of suspension structures. It was only later that analytical techniques for trusses were readily available—primarily through the work of Whipple and Haupt in America, although Navier had earlier published some important analytical work on this type too. Ellet continued to advocate suspension structures and did a great deal of writing about them. In view of the eventual collapse of his own Wheeling bridge, it should be noted that he was not unaware of the problem of instability and the danger of "vertical oscillations" of suspension structures.

Ellet again advanced a proposal for a suspension bridge in a bid to replace Lewis Wernwag's famous "Colossus" across the Schuylkill, which had burned in 1838. This time the proposal was accepted by the bridge commissioners. John Roebling contacted Ellet to offer his services, but Ellet would not have them. This contract had been the object of intense competition, and the rivalry that grew up around it may have been the origin of the rift that eventually developed between the two gifted engineers. Ellet completed the 358-foot span—the Fairmount Park Suspension Bridge—in 1842; it was America's first important wire suspension bridge. It was replaced in 1874.

Ellet next proposed a suspension bridge across the Niagara Gorge, in a competition that involved John Roebling, Samuel Keefer, and Edward Serrell. Ellet's plan was chosen by the bridge company. Between 1847 and 1848 he built a light suspension span to serve as a pedestrian bridge. He intended to use this as a service bridge to aid in the construction of a larger structure. Ellet solved the problem of how to get an initial wire across the gorge with its towering sides by a method that has become almost an American legend—he offered five dollars to any boy who could fly a kite across the gorge. A boy named Homan Walsh succeeded. The kite was landed and its string used to pull a series of increasingly heavy cords, and finally a wire rope, across the gorge. Ellet pulled himself across in a basket. Ever the showman, he later rode his horse across the almost completed bridge before the side railings were put on, to the amazement of crowds of onlookers. His service bridge was opened to the public and attracted many tolls. After quarreling with the company over money and other matters, however, Ellet eventually left the project. John Roebling secured the commission and went on to build his famous railroad suspension bridge there; it was completed in 1856.

It was while he was still involved with the Niagara Gorge project that Ellet received the contract for what was to become his greatest work, the Wheeling Suspension Bridge. The need for a bridge across the Ohio had long been discussed. In 1816 the legislatures of Ohio and Virginia by concurrent action had incorporated the Wheeling and Belmont Bridge Company to build such a bridge. Nothing immediately resulted. But the necessity of a bridge as part of the National Road had been repeatedly urged upon Congress as nationally important. Ellet himself considered the need for a bridge at Wheeling in 1836 and developed a sketch of a possible design. In 1838 government engineers developed a plan for suspending a bridge between two piers, but again nothing further happened. Virginia amended the charter in 1847 to permit the construction of a suspension bridge.

In 1847 the Wheeling and Belmont Bridge Company, acting under new directors, sought subscriptions to the capital stock. A wooden bridge built by Colonel Ebenezer Zane, an early road builder, across the back channel of the Ohio at Wheeling was absorbed into the new company, along with the road across Wheeling Island. In response to invitations from the directors, Charles Ellet and John Roebling both submitted plans for a bridge across the east channel of the river to the island. Roebling suggested a bridge with a pier in the middle. Ellet proposed a single span. Bearing in mind that a river pier might interfere with navigation, the directors approved Ellet's structure. The proposed bridge was some 1,010 feet in length, with a light, narrow roadway. It was a bold bridge, its clear span longer than any other in the world at the time. He accepted 200 shares of stock in the bridge in lieu of a $5,000 payment for his services.

Construction began in 1847. The stonework on the towers was done under the direction of William Otterson. The iron wire for the cables was supplied by a local firm, Richards and Bodley. The flooring was supported by twelve iron wire cables suspended from the towers, ten large ones and two small ones. The cables were anchored by eyebars into masonry walls built under Main Street at Wheeling and into abutments on the island. The structure was completed in 1849. It spanned 1,008.5 feet from tower to tower and 1,307.5 feet overall.

The bridge served well for several years. Then, on May 17, 1854, disaster struck. A great storm arose and the suspended portion of the bridge collapsed into the river. A newspaper man who saw it happen gave a full account of the catastrophe in the *Wheeling Intelligencer:*

> We saw persons running toward the river bank; we followed just in time to see the whole structure heaving and dashing with tremendous force. For a few moments we watched it with breathless anxiety, lunging like a ship in a storm; at one time it rose to nearly the height of the tower, then fell, and twisted and writhed, and was dashed almost bottom upward. At last there seemed to be a determined twist along the entire span, about one half of the flooring being nearly reversed, and down went the immense structure from its dizzy height to the stream below, with an appalling crash and roar. . . . The great body of the flooring and the suspenders, forming something like a basket swung between the towers, was swayed to and fro like the motion of a pendulum, each vibration giving it increased momentum."

An early view of Wheeling, West Virginia. (Library of Congress)

Clearly the reporter was describing a now classic mode of failure due to a form of aerodynamic instability—the same fate that overtook the ill-starred Tacoma Narrows Bridge (often now referred to as Galloping Gertie) in this century. Even the eyewitness descriptions are remarkably similar.

The collapse was complete. Except for some of the cables, the entire suspended structure had fallen into the river, whence it was eventually dragged to the shore by a steamboat. Only the towers remained undamaged.

Ellet was summoned to rebuild the bridge temporarily and to draw up a long-range plan for its complete reconstruction. He had a 14-foot-wide version of the bridge open again within three months. The temporary span continued to serve until the summer of 1859, when Captain William K. McComas, engineer and superintendent, rebuilt the bridge. McComas had spent $50 on a trip to Niagara Falls to study Roebling's bridge there, and some Roebling principles may have been introduced during the rebuilding; but the rebuilding was not done by Roebling himself, despite a myth to that effect that is still widespread today. Roebling's company apparently did become involved with the bridge in 1871–1872 during additional improvements. John Roebling's son, Washington, was involved in the planning. Wire stays were added at this time, along with improvements in the flooring system. The bridge has continued in constant use, with further overhauls occurring in 1886, 1956, and 1983.

This great bridge, still considered one of the most significant structures in America, serves as a monument to the dashing Charles Ellet, who was killed during the Civil War. Ellet must be regarded as one of this country's most visionary and innovative engineers.

The Cornish-Windsor, Blenheim, and Bridgeport Covered Bridges

The Covered-Truss Bridge

The great era of covered timber truss bridges in the early nineteenth century in the United States was the start of a revolution in the way bridges were conceived, designed, and built. The evolution of the timber truss bridge reflects a transition from an innovative but highly empirical approach to bridge design to one having its basis in the emerging theoretical developments in the young field of structural engineering. While many notable early bridges still survive, three of them—the Cornish-Windsor Bridge spanning the Connecticut River between New Hampshire and Vermont (1866), the Blenheim Bridge across the Schoharie Creek in New York (1855), and the Bridgeport Bridge in Nevada County, California (1862)—serve as particularly fine examples of this remarkable period of bridge building.

Certainly the types of bridges most familiar to early American builders were those found in Europe, where various forms of masonry construction prevailed (although timber was widely used in some places). The circumstances surrounding bridge building in America were very different. Distances were much greater. Rivers were an obstacle to every sort of north-south communication. Sophisticated European-style dressed block bridges were prohibitively expensive and required labor skills that were almost nonexistent in the New World at the time. Moreover, there was a pressing need to build quickly, a need difficult to satisfy when building with masonry. Therefore it was no small blessing that the land held a plentiful supply of giant timbers—which had, in prerevolutionary times, gone into the royal ships. Even though labor was costly in the United States, carpentry skills were ubiquitous. All the factors that might encourage wooden bridge construction were present.

Until the 1790s Americans carried on as best they could with simple forms of bridges. Most were timber trestles, although some stone arch bridges were built. The pile-and-beam system primarily employed was a little-developed descendant of an approach used by Roman military engineers that had come down to the colonists through their European predecessors. Some developments, notably Enoch Hale's framed timber bridge at Bellows Falls, Vermont, built in 1786, presaged more innovative uses of timber in bridges.

The seminal development in wooden bridge construction occurred in the 1790s when Timothy Palmer revived the timber truss after its years of near oblivion. Simple truss forms now known as king and queen posts were, of course, used in many buildings both in Europe and in America. Yet they were never consciously developed as a uniquely viable structural form capable of spanning large distances. The specific form of the wooden truss bridge appeared early in the heavily forested areas of Europe. There are records of them from as early as the fourteenth century. Andrea Palladio was the first, however, to publish designs for such bridges that were widely distributed. In his *Four Books of Architecture* (1570) he provided four well-detailed truss bridge designs; he noted that a friend of his had seen truss bridges in use in Germany, but that Italy possessed nothing like them. Few paid these designs any heed, however. In 1742 Palladio's *Four Books* were translated into English by Giacomo Leoni, not for the wooden truss designs but for the architectural theories. Nevertheless, the chapter including bridges was also translated. William Gibbs about the same time published his Palladian-influenced work, and the inspired Lord Burlington went so far as to have an 87-foot Palladian arched truss constructed on his estate. The structure was merely fanciful, however. Some timber truss bridges were evolved in Switzerland, but again the truss seems never to have been consciously exploited as a structural device potentially capable of spanning long distances and carrying heavy loads.

Palmer's sudden development and application of the form was thus an extraordinary act, seemingly not an outgrowth of preceding events. But, in fact, his reintroduction of the truss seems to have caused no great stir. It was quietly introduced by a man who saw it as an expedient, not as a major breakthrough. It was picked up and developed by several other practically minded persons, and by the time it reached the attention of European theorists, it had evolved far beyond the imagination of Continental engineers.

Palmer was from Newburyport, Massachusetts. He apprenticed to ship- and millwrights and, though considered eccentric by his neighbors, was acknowledged to be an "ingenius housewright." He was over forty when he began designing his massive bridges, built of square timbers with lapped and mortised joints. He built at a time that can best be described as a free-for-all in bridge design. Such disparate men as Charles Willson Peale and Thomas Paine took great interest in building bridges, although their designs were structurally impossible. Palmer's designs were not merely visionary, however. Indeed, they were such sophisticated works that it seems necessary to assume that Palmer saw Leoni's translations of Palladio's work, yet there is no evidence that this was the case.

Palmer's first bridges functioned structurally as arches, with the interstitial trusses serving as stiffening elements, and consequently were not instances of the clear and unambiguous truss forms we know today. His first bridge was built in 1792 over the Merrimack river between the towns of Essex and Merrimac in Massachusetts. The bridge deck on these initial structures followed the bowed shape of the arches, and diagonal bracing underneath the deck served to strengthen against wind. By building these trussed arches out of massive timbers, Palmer obtained enormous spans. His most important bridge was the Permanent Bridge in Philadelphia (1798–1806), which carried Market Street over the Schuylkill River and provided the last link in the post road connecting Philadelphia and Lancaster, a connection of great advantage to Philadelphia merchants. The bridge contained three spans and was the first to be built with multiple spans of trusses. The longest arch was 194 feet 10 inches. A center truss divided the road into two 13-foot sections.

The Permanent Bridge was also possibly the first covered bridge in America. It was made so at the insistence of the president of the bridge commission, after Palmer told him that the structure would last approximately 10 years without weatherboarding and 30 to 40 years with. Palmer resisted the idea at first, since he wanted his structure to show, to advertise his abilities. He was apparently persuaded of the advantages of weatherboarding and covered all his succeeding structures.

Theodore Burr made the next major advance in bridge construction with his invention of the "Burr arch," which he patented in 1817. Burr was from a millwright family in Torringford, Connecticut, but he moved west to New York to operate a new branch of the family saw- and gristmill near the Chenango River. He built his first bridge so that customers from the other side of the river might do business with him.

Burr's design was basically a pair of timber arches connected to trusses in the same plane for stiffening. The first major bridge to utilize Burr's system was built between 1803 and 1804 over the Hudson River at Waterford, New York. Three parallel trusses formed the structure, one on each outer edge and one dividing the two traffic lanes. The trusses had parallel top and bottom chords, with two diagonals between each pair of posts. The two outer trusses were bolted to a timber arch. The loads were shared by arch and truss. The system was massive and redundant. Of significance, however, is the separation of truss and arch. Eventually the arch would be discarded by other builders and the timber truss further developed on its own. Another important feature of the Burr arch-truss was its level roadbed.

Burr became a prominent bridge architect, and his arch-truss became a prototype. Many covered bridges still standing in this country make use of the Burr arch-truss. Burr ended his career by simultaneously building five giant spans across the Susquehanna—the river that was the greatest challenge to any bridge architect at the time. The largest of these had a main span of 360 feet. After Burr's death the boss-carpenters from these projects took the designs throughout New York and back to New England. Dozens of such structures were built in the Northeast. The Burr arch-truss became the bridge of Pennsylvania, Maryland, and Virginia, and for awhile in the Far West any arched bridge was called a "burr."

A German-born American, Lewis Wernwag, was responsible for other wooden giants. One was the "Colossus" over the Schuylkill in Philadelphia, a single-span arched truss of 340 feet with a rise of 20 feet. Although it was a truss-strengthened arch and the load was probably carried almost entirely by the arches, it did make use of two diagonals in each panel, as did Burr's, as well as wrought-iron struts acting as supplementary tension members. The arch was formed of five parallel laminated ribs, 3 feet 6 inches deep in all. Robert Mill provided it with a Greek revival weatherboarding.

Wernwag's best designs consisted of a double arch that sandwiched a truss composed of flared king posts and an unusual system of diagonal braces that did not meet at the panel points. Connections were made by cast-iron boxes set into the upper and lower chords. In some, dry rot was prevented by holding the beams slightly apart with iron links and screw bolts that could be tightened if the timber shrank or needed to be replaced.

Until 1820 most timber bridges were variations of the Palmer, Burr, or Wernwag designs. Their structures were unlike anything ever seen before or since—enormous covered flying roadways, untouched by flood waters or ice floes. Popular as the new truss bridges were, they were not without problems. For one thing, the bridges all used a lavish amount of timber, which required felling, curing, and forming. Because the bridges were constructed of giant members, their building required enormous manpower, including specialized workers who could splice the massive chords and work the intricate joints. A bridge that could be built easily by ordinary local carpenters was needed. In 1820 the answer was given when Ithiel Town patented his lattice truss.

Town, born in 1784 in Thompson, Connecticut, was an architect in New Haven. He traveled all over the country designing large public buildings (large-span public halls provided another impetus in the development of timber truss design) and while in North Carolina devised his truss, which was free of any arch action. It had parallel top and bottom chords with vertical end posts but no intermediate posts. Its lattice web of closely intersecting diagonals formed a type of rigid plane. Vertical loads only were transferred to the supports, and abutment connections were simple. It was ideal for multispan bridges, as it required no individual fitting to the piers. The diagonals were fastened to one another at each intersection, and the web itself was sandwiched between double top and bottom chords. This further increased the load-carrying capability of the design. The members were lightweight, standard-sized pieces of ordinary spruce or pine, with holes drilled into them for connecting pins. Easily assembled by untrained workers, this structure proved to be the bridge-construction equivalent of the balloon frame in building. Customarily, panels were assembled flat on land and then placed in position. Word spread that the truss could "be built by the mile and cut off by the yard." This was almost literally the case: whereas other bridges had to be constructed as a complex and unique system in place, Town's trusses were undifferentiated throughout—they had no natural midpoint or end. Town strengthened his design for rail traffic simply by doubling the number of planks and pins, adding a second parallel truss wall to each side. Occasionally an arch was added for extra strength.

For a short period of time (1820–1840) the Town truss enjoyed great popularity, partly because of its ease of construction but partly also because of Town's astuteness as a promoter. Beyond a few test or exhibition models, he built few lattice bridges himself. He advertised widely, however, and set up agents for sale of the design in every town. The simple royalty procedure was a charge of $1 per foot in advance—or, for one of the pirated spans found by agents who scouted the country for them, $2 per foot (or the illicit builders would face court action).

The lattice truss had, along with its virtues, some serious flaws. It was quite unstable laterally, and its poor rigidity led to lateral buckling of the top chords under heavy loads or vibration. And it was still, despite the absence of an arch, a highly redundant structure whose behavior was difficult to understand, much less predict. Nevertheless, it cleared the way for a rapid progression of new forms in truss design. While it lasted it proved very practical, particularly for long multiple spans. The longest was constructed over the James River at Richmond, Virginia, in 1838. The 2,820-foot bridge consisted of nine spans of up to 152 feet 6 inches each. It functioned well until its destruction by Confederate forces in 1865.

Another major advance was made by a colonel in the Army Engineer Corps, Stephen H. Long, a native of Hopkinton, New Hampshire, who had previously worked on the Carrollton Viaduct (described earlier in this chapter). His bridge, patented in 1830, rivaled Town's in popularity during the decade 1830–1840. Long, another great promoter, experimented on bridge designs for thirty years and wrote leaflets with directions for construction that were distributed by his extensive network of agents and subagents. Each was equipped with a suitcase-sized model of the Long truss to present to town aldermen at a moment's notice.

Long's truss was a true panel truss. It consisted of rectangular panels with crossed diagonals and used no arch. Originally it featured an auxiliary king-post truss at the center, the point of maximum bending, but this redundant feature was later dropped. The relatively short posts were compression members, and the diagonals were designed to withstand tension, although they could function in compression if necessary. This was an important feature in a bridge, which, by definition, carried moving loads that often caused stress reversals in many members. Long's sway bracing was another important innovation. Sway was becoming an ever greater design consideration because of the heavier and faster-moving loads produced by trains. Long's floor bracing consisted of an arrangement of pieces in the shape of a K. The short diagonals were more resistant to buckling than were longer, single diagonals. The K-brace is still commonly used in bridges.

Most important of all, however, was the apparent introduction of analytical techniques into the design process. The careful proportions of Long's bridges indicate that he had an excellent understanding of the distribution of forces in truss members. Indeed, when the famous European engineer and theorist Karl Culmann made his 1849 visit to the United States to see what American bridge builders were doing, he was greatly impressed by Long's work and wanted to know where Long had got his knowledge of static analysis, since no treatises on the subject were available in the United States until years after Long had received his patent. Apparently Culmann never found out, nor is the question answerable now. Perhaps Long's excellent design was the result of intuition. If so, it was an intuition that failed him in his later work, which was regressive compared with that of the 1830s.

By 1840 both the Town and Long trusses had been eclipsed by the Howe truss. Howe, a millwright from Spencer, Massachusetts, developed a truss that looked much like Long's. It was a panel truss with two parallel chords and crossed diagonals. Howe's design, however, introduced iron rods that were devised to function well under tension in place of Long's wooden uprights, which also had to carry occasional tension forces. This solved one of the traditional problems of wooden truss design—the pulling apart of joints under tension. Iron rods can be fastened to timber members easily to prevent pullout. In addition, Howe's iron rods were fashioned with adjustable turnbuckles. Thus, the bridge could be adjusted after repeated loadings had caused the members to change shape; this also meant that unseasoned wood could be used for bridge construction, since shrinkage could be compensated for later.

By this time it had become the practice to promote one's design, and Howe followed suit. He divided the country into territories, and exclusive franchises went to his friends and relatives. They were besieged with orders from railroad companies that needed many bridges for their new western routes. Hundreds of spans were cut from standard plans. Precut boards were shipped west by flatbed car from timbered eastern areas. Iron rods and nuts, along with specially patented angle blocks, were manufactured in New England and shipped out as well. The frame could be erected in one or two days by unskilled labor; and as soon as it was finished, train traffic began. The weatherboarding was put on at leisure by local carpentry crews while the bridge was in service. Hundreds of these spans were built simultaneously across the country.

Howe's design was used for enormous railroad bridges. Occasionally one of the diagonals was omitted, permitting exact analysis of the way loads were carried. A few very large bridges were built with supplementary arches and double diagonals. In some locations, Howe timber trusses were constructed as a matter of course for automobile traffic well into the twentieth century.

Several of Howe's bridges collapsed when some of the members thought to be under one state of stress were actually under another, and failed accordingly. In fact, some localities outlawed the Howe truss because of this problem. A variation that combated it was introduced in Vermont. Here double compression members were combined with a single iron tension member in an easy-to-construct design that reduced the failures.

Another significant truss design was introduced by Thomas Pratt, a graduate of Rensselaer Polytechnic in Troy, New York, who had studied architecture, building construction, mathematics, and natural science. Pratt's truss consisted of timber parallel top and bottom chords with vertical timber posts. The double (crossed) diagonals, however, were of wrought iron. One or the other of the diagonals always functioned as a tension member, depending on the loading present, with the uprights always in compression. After several disasters with Howe trusses, Pratt's became a standard railroad feature. Unfortunately, his truss could not be adjusted as easily as could Howe's, a significant disadvantage. The Pratt truss proved structurally preferable, however, because it put the shorter members in compression and the longer in tension, unlike Howe's, which did exactly the opposite.

In 1847 Squire Whipple published his classic of structural engineering, *A Work on Bridge Building.* He was one of the first engineers to analyze correctly the stresses in a bridge truss. Whipple understood the concept of force resolution, and of statics in general, that underlies modern structural analysis and design procedures. His treatise also included tables on maximum allowable stresses in timber and iron. Whipple introduced America to its first all-metal truss and is, in fact, more famous for his work with metal bridges than for his work with timber. It was his book, however, coming at the end of a fifty-year era of unparalleled advance in timber truss construction, that brought the development of the wooden truss to its peak and drew the attention of European engineers, leading to a resurgence of wooden bridges both on the Continent and in England, despite the scarcity of wood in many regions.

Today, although the giants of the age are gone, over 1,500 wooden bridges remain in the United States and Canada. Their hidden structures, little noticed by most covered-bridge enthusiasts—who delight in the picturesque qualities that were far from most builders' minds—reveal the direct parentage of today's modern truss bridges.

Cornish-Windsor Covered Bridge, Cornish, New Hampshire, to Windsor, Vermont. This two-span, 460-foot bridge is the longest extant covered bridge in the United States. It is a Town lattice-truss design of a type widely used on many early timber bridges. It was built in 1866 and is still in use. (HAER Collection, Library of Congress)

The Cornish-Windsor Bridge

The two-span covered bridge that crosses the Connecticut River between Cornish, New Hampshire, and Windsor, Vermont, measures 460 feet between abutments and is thus the longest covered bridge remaining in the country—though not the longest single-span covered bridge. Built in 1866, it is in excellent repair and is still in use today.

In colonial times the only means of crossing the Connecticut at Windsor—unless the river was sufficiently frozen to drive across—was by private boat. In 1784 the New Hampshire legislature chartered a ferry to make the passage, protecting the proprietors by giving them exclusive rights for several miles up and down the river. During times of low water or partial freezing, however, little communication between the two sides took place. It became increasingly clear to the residents of Windsor that if they wanted to compete commercially with other towns along the river, they would have to invest in a bridge: those up- and downstream—one of which was Enoch Hale's innovative 1784–1785 structure at Bellows Falls—provided such easy crossing from one side to another that commerce was beginning to bypass the Cornish-Windsor area. Thus, in 1795 the town constructed a bridge. Designed by Moody Spofford, it comprised two wooden arches that met on a central pier. The entire structure measured 521 feet between abutments and was 34 feet wide. The central pier was enormous, measuring 41 feet by 45 feet. Since the roadbed followed the rise and fall of the arches, this pier was probably made so large in order to provide a level transitional space between the two adjoining spans. The bridge was not a covered bridge in the conventional sense; that is, its roadbed was not protected by a roof, but the arches and supporting members were enclosed for protection from the weather. The bridge was destroyed by high waters in 1824 and was replaced by a three-span structure that had level floors and rested on two piers. In 1849 this bridge too was destroyed by a flood. Its replacement was exactly like the present structure, except that a few of the timbers were smaller in cross section. The 1849 bridge was replaced by the now-standing one in 1866. Both of these bridges were of the Town lattice-truss design.

The bridge builders were two local men, Bela Fletcher and James Tasker. Although uneducated, they were experienced at erecting large bridges, and they successfully used rule-of-thumb design principles to produce a durable structure. The trusses were constructed in a meadow near the bridge site, with the use of only simple tools—the hammer, axe, saw, and auger. When finished, they were moved out over the water, and then the roadway, roof, and siding were added. Typical of many wooden bridges, this one for years had a sign reading "Walk Your Horses or Pay Two Dollars Fine." This precaution was a response to the perceived possibility that the rhythm of galloping horses could set up vibrations in the bridge, causing it literally to shake itself to pieces.

Until upstream flood control was effected, the bridge had to withstand several deluges. The design of the structure allowed it to escape major damage from the trees and houses that crashed against it during storms in 1927 and 1938. With the exception of minor alterations to accommodate automobile traffic, the bridge survives in its original state.

View of the Town-lattice structure of the Cornish-Windsor Bridge. (HAER Collection, Library of Congress)

The Blenheim Bridge

Completed in 1855, the covered wooden truss bridge at North Blenheim, New York, is one of the longest bridges of its kind in the world. The clear span of 232 feet was achieved by an ingenious interlocking of truss and arch elements.

The Blenheim Bridge Company was incorporated on April 19, 1828, by an act of the New York state legislature. This first charter expired some years later and was then extended by a petition of George W. Martin and William Fink, directors of the company. Around 1850, impetus for the construction of a bridge at North Blenheim was given by the arrival of Major Hezekiah Dickerman, who built a tannery in the village. The hemlock bark used in the process of tanning the hides was obtained from trees on the opposite side of Schoharie Creek. Fording the river proved hard on his horses and wagons. By 1854 Dickerman was president and shareholder of the bridge company and had decided, along with Martin and Fink, that the time had come to build a bridge. The company then began seeking a contractor.

It was soon found that Nicholas Montgomery Powers was due in Schoharie to repair a covered bridge there. Powers was already known as a bridge builder. Born in Pittsford, Vermont, in 1817, he had undertaken his first covered bridge at Pittsford Mills at the age of twenty-one. His father had had to go his bond and agree to make good any timbers that the "boy" might spoil. None were spoiled and the bridge was a success. Soon Powers had no shortage of bridge contracts. Although he had little formal education, his practical skills soon caused him to become one of Vermont's most well-known builders. But it was in New York at Blenheim that he undertook one of his greatest bridges.

The bridge at Blenheim was constructed between 1854 and 1855 at a cost of some $6,000. Between its two roadways Powers built a great single-span arch, largely of oak, consisting of three separate but interconnected ribs placed one on top of another. The ribs are separated by periodically placed spacing blocks. Long connecting bolts run

Blenheim Covered Bridge, North Blenheim, New York. Built by Nicholas Powers in 1854–1855, this covered wooden truss bridge was the longest bridge of its kind in the world at the time of its construction. It still spans the Schoharie Creek. (Courtesy of the American Society of Civil Engineers)

The great center arch of the Blenheim Covered Bridge. (HAER Collection, Library of Congress)

through the ribs and spacing blocks. This arch is sandwiched between two trusses, each having a repetitive pattern of wooden X's. On the outer sides of the roadways are similarly designed trusses. Thus the structure consists of three primary spanning lines—two trusses (one on each side) and a center arch (stiffened with trusses). The whole structure spans 232 feet. The rising arch seems to tower over the level of the roadways.

Powers was paid $7 a day for this job (about $2,000 total). He picked his own men for the construction work. He built the bridge piece by piece in the village and then assembled it on temporary scaffolding on the bridge site. To demonstrate his confidence in the quality of his work, Powers sat on the roof of the bridge, while the workmen removed the scaffolding that was temporarily supporting it. "If the bridge goes down," he declared, "I never want to see the sun rise again." He lived to see many more sunrises.

In the spring of 1869 a severe freshet washed out a channel near the western end of the bridge. A small wooden extension was erected across this channel. A herd of cattle crashed through the extension span in 1887, and it had to be rebuilt. In 1891 the bridge came under the jurisdiction of the state. Soon afterwards, the extension collapsed under the weight of a threshing machine. An iron structure was built over the channel in 1895.

In the 1920s many began to feel that the bridge had outlived its usefulness. The iron extension had cracked under a load of ice and snow. A new bridge was slated to be put up across the creek nearby. A community effort, however, was mounted to save the old wooden bridge. Petitions were circulated and poems written ("Snow quilts the ground! Sleepy-eyed children, in a horsedrawn sleigh, are warm and cosy! Homeward bound, their Papa halts at the bridge to pay the toll—his hands cold, cheeks rosy. Old Nellie's hoofs 'clop-clop' on the planks; sleigh bells jingle merrily through the night . . ."). On July 8, 1931, the Schoharie County Board of Supervisors voted unanimously to retain the bridge as a public historical relic. The county became its custodian. The iron extension was taken down. Eventually designated a national historic landmark, the bridge still spans Schoharie Creek.

The Bridgeport Bridge

At the other end of the country, the Bridgeport Bridge, in Nevada County, California, is one of the oldest bridges in the West and has a remarkably long clear span of approximately 233 feet. It crosses the south fork of the Yuba River about 10 miles northeast of Smartville. Built in 1862 to replace an earlier bridge washed out in a flood, it was part of a 14-mile toll road operated by the Virginia City Turnpike Company.

Since the counties had no funds to construct the roads needed for accommodating the influx of new settlers that had begun with the 1848 gold rush, the California legislature authorized the formation of private turnpike companies. The Virginia City Turnpike was a portion of the major roadway that connected Virginia City, Nevada, and the rest of the silver-producing Comstock Lode with commercial centers in California. The road tolls were set by the counties. In 1862 the price for the 14-mile section, including passage over the Bridgeport Bridge, was 25¢ for foot travelers, 50¢ for horsemen, and $6—the maximum toll—for teams of eight animals.

The bridge was built by David Ingerfield Wood, a native of the Midwest who had come to California with an interest in commerce and road building. He and some colleagues had formed the Virginia City Turnpike Company, with Wood serving as its president. The firm also owned a lumbering business in the High Sierra, which provided the Douglas fir for the bridge's construction.

The design is a double intersectional truss, sandwiched between two laminated arches that span the entire length of the bridge. The structure is a composite of iron and wood; the vertical members are constructed of wrought iron, the bearing blocks of cast iron, and the remainder of Douglas fir. The bridge is usually regarded as a form of Warren truss, although in its final form it could be described in several ways. The Warren truss was not a popular design in the United States until the late nineteenth century, when it began to be used extensively. The configuration was developed in this country in 1849 by Squire Whipple, who was apparently unaware of its English precedent, patented in 1848 by James Warren and Willoughby Monzani. At the time of its invention it was the simplest of all the designs, having no verticals but only diagonal members, alternately sloped in opposite directions to form a giant zigzag. The design was later modified by adding either posts or a second set of diagonals (which produced a pattern of continuous X's). This latter modification was the one used on the Bridgeport Bridge.

The two arches are formed of laminated 5-by-12-inch hand-hewn timbers and are held on either side of the truss by means of iron through bolts. The structure is protected from the weather by a gabled roof and siding, all of which is covered with about 27,000 split sugar-pine shakes. The arch is visible both inside and outside the bridge, and the entire structure rests on enormous granite blocks.

Bridgeport Covered Bridge, Nevada County, California. Built in 1862, this is the longest extant single-span covered bridge west of the Mississippi River. Still in service, it originally carried heavy freight as part of the turnpike between Maryville, California, and Virginia City, Nevada. (HAER Collection, Library of Congress)

An interior view of the Bridgeport Covered Bridge. (Courtesy of the American Society of Civil Engineers)

The Whipple Bowstring Truss

The Whipple bowstring truss now located on the Union College campus in Schenectady, New York, is one of the few surviving trusses built by Squire Whipple in New York in the mid–nineteenth century. The bridge is of a type patented by Whipple, who is generally considered to be the father of scientific bridge building in the United States.

Squire Whipple (Squire was his given name, not a title) was born in Hardwick, Massachusetts, but soon moved with his family to New York State, where he eventually graduated from Union College. Realizing that the expansion and widening of the Erie Canal would necessitate many new bridges, he saved up $1,000 and constructed his first iron bowstring bridge over the canal at Utica. He patented the design and details of his bridge in 1841. The bridge found immediate and widespread acceptance, and was quickly imitated—often with slight variations to avoid paying Whipple royalties. The state of New York adopted his patented bridge as a standard for its canals, but then evaded royalty payments by decreeing that the bridges were "erected for the public good."

During this period Whipple wrote a small book that was eventually to have great impact on the practice of structural engineering. Entitled *A Work on Bridge Building* and published at Whipple's expense in 1847, it was the first comprehensive American work on the scientific design of truss bridges and on the physical properties of wood and iron. In it Whipple described an accurate method for calculating the nature and magnitude of the forces acting in each of the members of a truss under a given loading condition. The basis for understanding the distribution of forces in a truss had been established abroad earlier. An essential work, Emiland Gauthey's *Traité de la construction des ponts* (posthumously published between 1809 and 1813 by his nephew, the famous mathematician Louis Navier of the Ecole des Ponts et Chaussées in Paris), had provided the foundation for several subsequent works in the area. Gauthey's treatise included a discussion of what he termed the principles of "equilibrium of position" and "equilibrium of resistance."

Whipple bowstring truss bridge (circa 1855), Schenectady, New York. This Whipple truss was of a type patented in 1841 by Squire Whipple and was one of the first scientifically designed truss bridges in the United States. The bridge originally spanned Cayadutta Creek in Johnstown, New York; in 1979 it was relocated on the Union College campus. (Courtesy of the American Society of Civil Engineers)

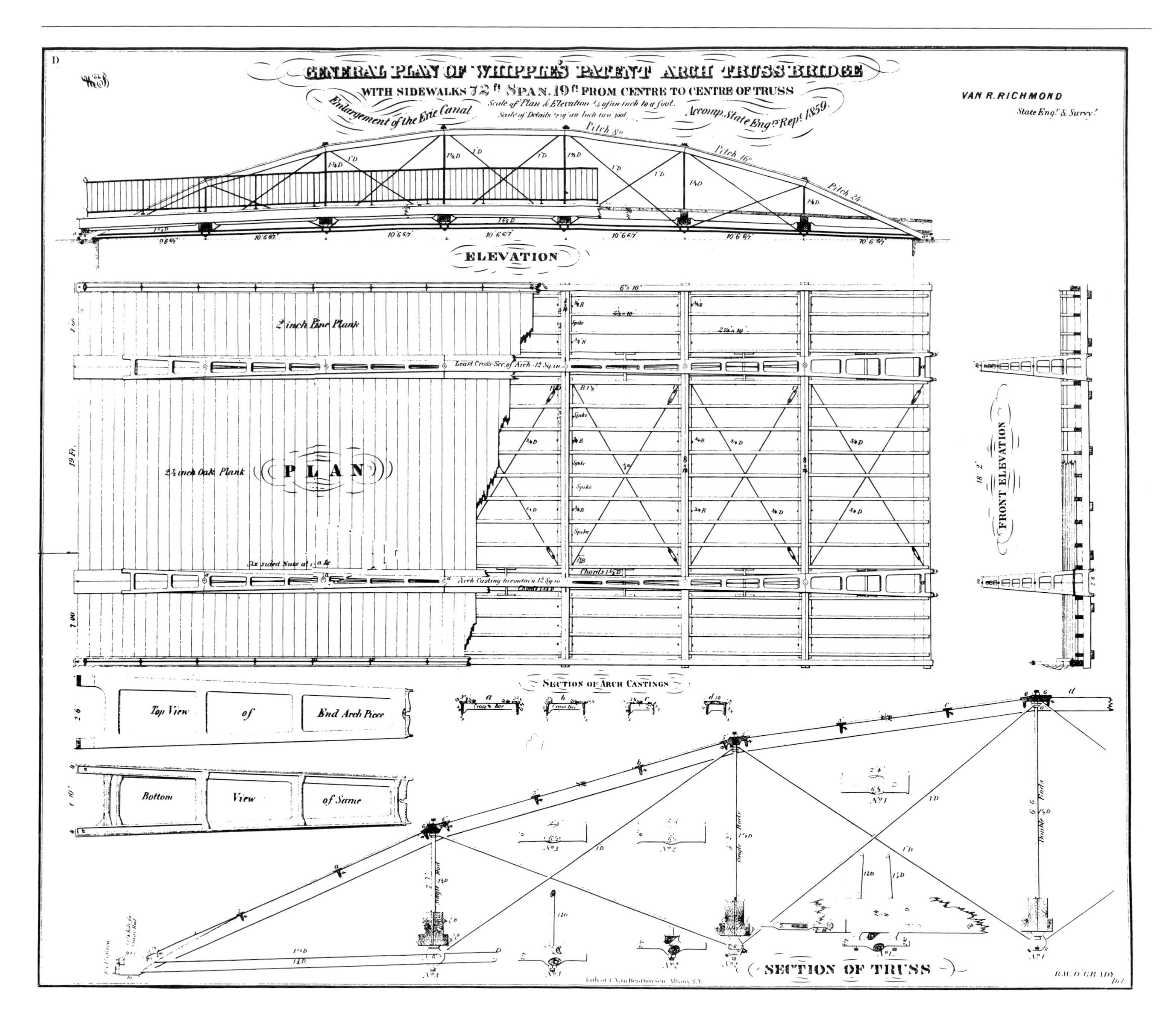

Squire Whipple's bowstring truss was widely used as part of the canal system of New York. Shown here is a "General Plan of Whipple's Patent Arch Truss Bridge" of seven panels and a 72-foot span, published by the state engineer's office in 1860. (*State Engineer and Surveyor of the Canals, 1860, Plate D*)

This bridge at Normansville, New York, was constructed according to Squire Whipple's patented design by Simon DeGraff, a builder from Syracuse. Many builders followed Whipple's basic pattern. (HAER Collection, Library of Congress)

The former was an initial attempt to resolve bridge loads into components in individual members. The latter dealt with material properties and the sizing of members. Most American engineers, however, were only marginally acquainted (if at all) with these European works.

Whipple's treatise was a masterful combination of theoretical and empirical knowledge of bridge design. The method he worked out for determining force distributions in trusses was logical and simple. His calculations took into account unfavorable load positions on a truss and the fact that varying the location of a load caused different force distributions to be developed in a truss. In an apparent attempt to address readers trained only in arithmetic, he substituted rhetorical expositions for the trigonometric and algebraic manipulations involved in the mathematical analysis of trusses. He also addressed the properties of materials and included tables delineating the tensile and compressive strengths of wood and iron. He recommended that owing to uncertainties in the qualities of the materials, the effects of fatigue, and other considerations, working stresses equal to more than one-sixth to one-fourth of the known strength of the material should not be used.

At first the book had little impact. In fact, when Herman Haupt published his *General Theory of Bridge Construction* in 1851, apparently developed independently of Whipple's work, he was under the impression that he had written a pioneer American work on the subject. It took nearly thirty years for the contents of Whipple's book to be fully appreciated. But during this era bridge building slowly transformed itself from an empirically based craft to a scientifically based profession.

In the meantime Whipple continued his productive career. A truss form that often bears Whipple's name—but is quite unlike his patented bowstring truss—was first built in 1846. A unique feature was the presence of diagonal members that crossed two panels rather than one. The diagonals were designed for tension, and the posts and top chords for compression. This truss was widely adopted for railroad bridges. In 1849 Whipple built the first Warren-type truss (consisting of oppositely sloped diagonals and no vertical members), apparently without knowledge of its English precedent (it had been developed by James Warren and Willoughby Monzani, who patented it in 1848). As late as 1869 he was still developing new bridge types, including lift and draw bridges. He died in 1888.

The Whipple bowstring truss now located on the Union College campus is one of a handful of surviving Whipple trusses, which were once built by the hundreds. Until April 1979 the bridge spanned Cayadutta Creek in Johnstown, New York. At that time it was a little-used pedestrian bridge, although apparently it was once part of a mill route. It is possibly the oldest of the known Whipple trusses in eastern New York. The city of Johnstown donated the bridge to Union College, and it was reerected over the Hans Groot Kill by civil engineering students under the direction of Professor Richard R. Pikul and Professor Francis E. Griggs.

Built of cast and wrought iron, the bridge has arched top chords carried down to join the horizontal bottom chords at the ends of the structure. The spaces between the top and bottom chords are subdivided into panels consisting of vertical members and crossed diagonal members. There is some evidence that Whipple thought of this form primarily as an iron arch with a stiffening truss system, but the structure is a valid truss form.

The bridge is 58 feet long. Its top chords are rectilinear open-web tangential castings (5 each). The bottom chords are round wrought open links (7 each). The diagonals are round rods. The top chord bears two inscriptions: "S. Whipple's Patent. Shipman & Son Builders, Springfield Centre, N.Y.," and "Patented 1841 by S. Whipple, Albany, N.Y.—Patent Renewed, 1855." In its placement on the Union College campus, some new wooden elements (notably side handrails unlike anything previously present) were added and the structure unusually painted.

The Bollman Truss Bridge

The Bollman truss bridge that spans the Little Patuxent River at Savage, Maryland, on an abandoned spur of the Baltimore and Ohio Railroad, is a surviving example of a unique type of bridge that helped to facilitate the rapid expansion of railroads in this country. The Bollman truss was among the first to use iron in all principal structural members.

Wendel Bollman (1814–1884), a self-taught engineer from Baltimore, achieved a reputation as one of the most successful iron-bridge builders in the world. He began his career as a carpenter's apprentice in 1828, working for the newly formed Baltimore and Ohio Railroad. He later became a rodman and helped lay the first lengths of track on the original stretch of the line out of Baltimore. He was supervised from 1829 to 1830 by George Whistler and Benjamin Latrobe, Jr., and undoubtedly learned a great deal from these accomplished engineers. For a period he returned to the carpenter's trade and went into business for himself. In 1837 he returned to the Baltimore and Ohio and worked on several timber bridges. Soon made foreman of bridges, he reconstructed the bridge at Harpers Ferry. Throughout this period Bollman was engaged in an intensive program of self-study. He rose within the company until he was appointed master of the road by the railroad's chief engineer, Benjamin Latrobe, Jr., in 1848. He was thoroughly familiar with Latrobe's wooden truss designs and soon began exploring the use of iron.

A number of all-iron bridges already existed. Earl Trumbull had built one in Frankfort, New York, in 1840. (Earlier August Canfield had proposed an all-iron design for a bridge in New Jersey, but it had not been built.) Richard Osborne had constructed an all-iron truss in 1845 in Manayunk, Pennsylvania, using a form of Howe truss. The first really extensive use of iron in bridge construction in this country, however, was by Squire Whipple, in a series of trussed highway arches over the Erie Canal in the early 1840s. These bridges promoted an interest in the material, and by 1849 Latrobe favored its use in spans to be produced by the Baltimore and Ohio. Latrobe himself had used both cast iron and wrought iron in his radiating truss

This Bollman truss bridge in Savage, Maryland, is a remaining example of a patented design by Wendel Bollman that facilitated the rapid expansion of American railroads. It was built in 1852 for the Baltimore and Ohio Railroad and moved to Savage in 1888. (Courtesy of the Smithsonian Institution)

over the Patapsco River at Elysville, Maryland, but it was limited to certain members; the principal loads were carried by heavy diagonal timbers from each panel, except the four at the center, directly to the bearing points at the piers. This system, in which the panel members acted more or less independently of one another, was to have an effect on Bollman's design.

Another apparent influence was a commonly used technique of the period known as "trussing," in which a simple timber beam was strengthened by the addition of an iron rod below. The rod was anchored at the ends and held a certain distance from the center of the beam by a vertical post. Tension forces were developed in the rod, reducing the bending present in the timber beam. By increasing the effective depth of the system, one could reduce stresses in the beam. Bollman found that by increasing the number of rods and posts, he could amplify the advantages of this technique to the point where considerable distances could be spanned.

Bollman went a step further and replaced the timber with cast-iron members. Cast iron was the likely choice because it was cheap and could be produced in cross sections large enough to resist buckling. The trussing rods, subjected only to tension, were of wrought iron. This material was limited in application at the time since thick sections could not be produced, but it was used effectively here in that the rods needed only sufficient thickness to resist primary tension stresses.

The system that Bollman finally developed was a composite one in which diagonal truss links or eyebars were used in combination with a cast-iron compression chord, which he called a "stretcher." The spacing between the chord and the junction of each pair of links was maintained by a vertical post or strut of cast iron. Most important in his scheme was that each floor beam or panel point was supported independently by two pairs of diagonals tied to cast-iron abutments. The apparent rationale was that if any section should fail, all others would continue to support their loads undisturbed, preventing total collapse. This was a significant feature at a time when a general mistrust of ironwork—due to its known high rate of failure and its spindly appearance—prevailed. The lower chord in the Bollman truss was apparently considered primarily an alignment member for the vertical struts, not a main load-carrying member.

The first all-iron Bollman truss bridge was constructed over the Little Patuxent River at Savage Factory, near Laurel, Maryland, in 1850. It was built at a cost of $23,825 and spanned 76 feet. It was soon removed and the present Savage Bridge built in 1852 at the same location. When that structure became inadequate due to heavier loads, it was moved in 1888 to its present site on the Savage spur. Its isolated location allowed it to survive while other Bollman trusses were being replaced. A double-span through truss, it extends 80 feet in each section and was one of the first to use iron towers.

The success of the pilot model at Savage and of another at Bladensburg, Indiana, gave Latrobe sufficient justification to adopt the use of iron in subsequent major bridge structures on the Baltimore and Ohio. Almost immediately after the completion of the Savage span, Bollman undertook the design of replacements for the Patapsco River bridge and the Winchester span over the Potomac at Harpers Ferry, which was originally built by Lewis Wernwag. The Winchester span, completed in 1851, possessed a high degree of refinement in detailing and was considered to be a prototype of the Bollman design. Bollman patented his system in 1852, calling it a "suspension and trussed bridge," as it was a hybrid of both. Later it became known to many as a suspension truss, a clear analogy since each pair of primary rods could be compared to a suspension cable.

With the establishment of Bollman's patent in 1852 and the publication of a booklet on the subject in the same year, his ideas became rapidly promoted both in the United States and abroad. Although there is no record that the type was ever reproduced in Europe, it was nonetheless studied with respectful interest there. Of general interest was the success of the structural use of iron, and its endorsement by an engineer of Latrobe's status. As the railroad launched a massive campaign to replace its timber spans with Bollman trusses, the bridge design became even more widely noted.

Bollman truss bridge, Savage, Maryland. Bollman's bridges were conceived of as "suspension and trussed" structures. He believed that train loads were picked up at the bottom of each of the panels of the structure by a series of independently acting suspension rods. Each suspension configuration picked up a load from one of the transverse floor beams carrying the railway and carried it diagonally back to the tops of the end posts of the structure, where it was transferred to the foundations. The top chord served as a common horizontal compression strut for absorbing the horizontal thrusts of the suspension configurations. In addition, each of the panels was triangulated, which stiffened the entire structure. The lower horizontal chord was conceived of as primarily an alignment piece. (HAER Collection, Library of Congress)

This photograph of a Bollman bridge at North Branch, Maryland, was taken in June 1858. The train is a Baltimore and Ohio excursion special, making a trip from Baltimore to Wheeling, with stops at points of interest for the benefit of the forty artists and photographers aboard. (Courtesy of the Smithsonian Institution)

Bollman designed and built bridge types other than his "suspension and trussed" form. The marvelous bridge shown in the circa-1871 photograph above was at Lake Roland, Baltimore. (Courtesy of the Smithsonian Institution)

In time most of the Bollman bridges were replaced, often because of increased train loadings. The photograph below, taken around 1923, shows the replacement of Bollman trusses with plate girders. (Courtesy of the Smithsonian Institution)

By the late 1850s iron was well established as a bridge construction material. As railroad loadings and performance demands increased, however, and as the engineering profession began to experiment with new spanning methods, certain weaknesses of the Bollman truss became apparent. First, although its components were easy to fabricate and its analysis and design straightforward, it was not as economical in its use of material as the more conventional panel truss. Second, there was an inherent lack of rigidity due to discontinuities in the system, with the use of essentially nonstructural lower chords particularly leading to reductions in rigidity. Third, difficulties arose because of the unequal lengths of the links in each group except the center one: the resulting unevenness in the thermal expansion and contraction of the framework made it difficult to keep this type of bridge in adjustment. Consequently, the system was typically limited to intermediate span lengths of about 150 feet. For longer spans the Baltimore and Ohio adopted the Fink truss, which was similar in concept except for a different suspension arrangement.

Despite its problems, the Bollman suspension truss system was widely used. By 1870, for example, all of the timber spans on the original Harpers Ferry bridge had been replaced with Bollman trusses. These remained in full main-line service until 1894, and some were used for highway traffic until destroyed in the flood of 1936. Use of the system began to decline in the 1870s. The increasing distrust of cast iron for major structural members (due to its brittleness), advances in structural theory, increasing train loads, and the greater availability of a variety of rolled sections, all contributed to its demise. None were built after 1875. Many of those already built remained in service for some time, until their replacement was demanded because of higher train loads.

In the meantime Bollman himself had resigned from the Baltimore and Ohio in 1858 and had formed with John Tegmeyer and John Clark, two former assistants, a bridge-building firm in Baltimore—W. Bollman and Company. Along with work in the United States, the firm constructed bridges in Chile and Mexico. It was dormant during the Civil

War period of 1861–1865 but resumed operation as the Patapsco Bridge and Iron Works (with Bollman as sole proprietor) immediately afterward. Bollman's firm built many other types of bridges in addition to his own "suspension and truss" system; eventually he produced hundreds of structures. He also helped develop an important column type (made of segmental wrought-iron sections with standing flanges for riveting into a circular shape) that was the forerunner of the famous column shape normally coupled with the later Phoenix Iron Works. The Patapsco Bridge and Iron Works ushered in what soon became a huge private bridge-building industry in the United States, for the design, fabrication, and erection of iron structures.

The Fink Through-Truss and Deck-Truss Bridges

The Fink through-truss bridge at Hamden, New Jersey, and the Fink deck-truss bridge at Lynchburg, Virginia, are two examples of a railroad bridge construction system in common use in the mid–nineteenth century in the United States. Invented and patented by Albert Fink, one of the foremost civil engineers of his day, the Fink truss system provided a unique structural answer to the need for long-span railroad bridges and contributed significantly to the expansion and growth of the American railroad. The Lynchburg bridge remains in place, but the Hamden bridge, badly damaged in 1978, has been dismantled and is awaiting reerection.

Albert Fink emigrated in 1849 from Germany to Baltimore, where he was trained as an architect and civil engineer. He was first employed by the Baltimore and Ohio Railroad as an assistant to Benjamin Latrobe, Jr., along with another notable bridge designer, Wendel Bollman. It was while he was working under Latrobe that Fink designed his deck and through trusses, for which he received a patent on May 9, 1854. The principle of his system was not unlike that of Bollman's: Fink described the truss in his patent application as a "peculiar combination of different systems of triangular bracing . . . so that a weight coming on one of the systems of the truss . . . is not carried over one or more of the other systems before it is carried back to the abutment." Thus, the system could be conceived of as a series of independently acting simple suspension structures.

Fink's truss design was a clever solution to a common design problem of his time. Steel had not yet been developed for extensive structural use, wrought iron was expensive, and wooden truss bridges tended to rot or wear out at the joints. Fink's truss used cast iron for members that were in compression only (where cast iron can work effectively) and, since cast iron is unreliable in tension, wrought iron for members that were in tension (wrought iron was more expensive than cast iron, and difficult to make in thick section at the time, but it did work quite well in tension). Its particular triangulation arrangement made it more suitable for long-span structures than was Bollman's truss.

Fink's system was soon adopted for long-span bridges on the Baltimore and Ohio Railroad and the Parkersburg branch. Major bridges using the system included one at Green River, Kentucky, and another at Louisville, spanning the Ohio River. Fink became resident engineer for construction on the Parkersburg branch, and later joined the Louisville and Nashville Railroad, which he ran with notable efficiency during the Civil War. He served as vice-president of the railroad from 1870 to 1875. An outstanding railroad administrator, Fink developed the science of railroad statistics and worked out a system of pricing based on mathematical techniques instead of rule-of-thumb estimates.

Although Fink trusses were being built into the 1870s, the combination of increased locomotive weight and advancing bridge technology helped to make Fink's system obsolete—as had happened with Bollman's. Between 1869 and 1900 the average locomotive weight rose from 20 tons to 150 tons, rendering many bridges inadequate. By the 1870s the scientific principles of bridge design were at last clearly understood, and steel was available for widespread use in both tension and compression members. Fink's trusses thus represent American bridge engineering practice at a time when this practice was undergoing a transition from empirical to theoretical design methods.

The Fink through-truss bridge near Hamden, New Jersey, is the only example of this particular truss type known to exist. In a through-truss bridge the roadway is suspended from the lower chords of the supporting trusses. The train thus passes between, or through, the two parallel trusses. The history of this bridge begins in September 1857 when, after some hesitation, the Hunterdon County Board of Freeholders made the decision to erect a metal truss bridge instead of a wooden one to replace a bridge that had recently been washed away. Authorization was given to the Trenton Locomotive and Machine Company

Fink through-truss bridge, Hamden, New Jersey. This bridge, constructed of cast and wrought iron, was completed in 1858. It is an example of a structural system that was patented by Albert Fink of the Baltimore and Ohio Railroad in 1854 and used extensively on American railways, as well as for roof trusses and road bridges, between 1854 and 1875. The bridge was severely damaged in 1978 and has been dismantled and stored, awaiting reconstruction. (HAER Collection, Library of Congress)

Fink deck-truss bridge, Lynchburg, Virginia. The only known survivor of the deck-truss type, it was constructed about 1870 and moved to its present site in 1893. (HAER Collection, Library of Congress)

to manufacture and put up the bridge following the patented Fink through-truss design. The work went quickly: on February 1, 1858, the bridge was tested by an estimated eleven tons and fifty men. Following the application of a coat of paint and a plaque, the bridge was opened to traffic.

The bridge is 99.9 feet long, 15.7 feet wide, and 19 feet high. It has a frame structure and is divided into eight bays. Between the cast-iron vertical posts, tie rods of wrought iron provide diagonal bracing. Two sets of diagonal braces tie the whole structure together: those in the first set span two bays each, and those in the second set each span half the bridge, joining at the center compression post just below the deck.

The Hamden bridge was one of the oldest metal truss bridges in continuous service without having had significant structural modifications, until a vehicle collision in 1978 severely damaged it and caused its collapse. The remaining pieces were salvaged and are now stored, awaiting recasting of damaged pieces and reerection. The damage to this significant structure was a great blow, removing as it did a remaining example of a most remarkable and important bridge type from public access. It is to be hoped that reconstruction on its original site will eventually become practicable.

The Fink deck-truss bridge at Lynchburg, Virginia, is the only known survivor of this type, in which the roadway lies on top of the upper chords of the supporting trusses. The history of this particular bridge is somewhat murky. It is known to have been placed at its present location—on a section of the old Norfolk and Western Railroad then known as the Halsey Spur—in 1893, but seems to have been originally constructed for and used at another site on one of the lines that preceded the Norfolk and Western. By 1893 Fink trusses had for some years been considered inadequate for the existing railroad loads, and it was probably for that reason that the bridge was adapted and reused at a site on a secondary line. The original construction is believed to have taken place in the 1870s, although it is possible that it dates from as early as the 1850s.

The bridge has a span of 52.5 feet and presently serves as a siding for a Lynchburg Ready Mix plant. The truss is rather simple, consisting of vertical and diagonal members of wrought iron, all below the deck. The top chords, which support the floor system, are made of wood, apparently untreated oak of about 15-inch thickness. This bridge is scheduled to be removed from service in the near future. Because of its uniqueness, however, efforts have been made to preserve it in another location, perhaps as part of a public museum or park.

This rare early photograph clearly shows the structure of Fink's deck-truss bridge design. Note the absence of a bottom chord. (Courtesy of the Smithsonian Institution)

This bridge across the Monongahela River at Fairmont, West Virginia, was designed by Albert Fink. It is no longer standing. (Courtesy of the Smithsonian Institution)

The Bidwell Bar Suspension Bridge

California's Bidwell Bar Bridge is typical of the suspension bridges that were so significant to the development of the northern part of the Sierra during the gold rush. Of the suspension spans constructed in northern California in the 1850s—at Whiskey Bar, Rattlesnake Bar, Condemned Bar, Comanche, and several other locations—none are left, however, except the Bidwell Bar Bridge. It spanned the Feather River about ten miles north of Oroville and has been reconstructed near its original site.

A previous bridge had briefly occupied the site. Licensed as a toll bridge to T. A. Sherwood and J. E. Lewis by the California Court of Sessions in May 1851 and immediately built, it was lost in April 1852. For a time a ferry operated at the spot. In October 1852 a new group was granted a license to construct another toll bridge, but nothing came of it. At last in December 1854 a license was granted to the Bidwell Bridge Company to build a suspension toll bridge. The company promptly advertised in the *Butte Record* for construction bids and in May 1855 awarded a contract to Jones and Murray, builders in Sacramento, for a sum of $26,500. The bridge was completed later in the same year by fifteen men working under Murray at a cost of about $35,000 and immediately opened to traffic. In spite of heavy use, it was never a money maker. Stock in the bridge company dropped from the original $100 per share to 12½¢ per share. The county of Butte purchased the bridge in 1883 for $5,000. Though not a financial success, the Bidwell Bar Bridge contributed greatly to the development of the region, just as its counterparts did to other areas in the northern Sierra. Further, the technology used was remarkable for its location.

The bridge had four wire cables, two on each side, each several inches in diameter and made up of a large number of parallel wire strands, spirally wrapped with a wrought-iron wire and then heavily painted. The completed cables were 407 feet long and anchored in solid rock and concrete. The anchorages were bent bars buried deep in the foundations. They were sealed in oil. Each of the bridge towers consisted of four cast-iron

Bidwell Bar Bridge, Oroville, California. Built in 1855 during California's gold rush, this suspension bridge spanned the Feather River about ten miles northeast of Oroville and has been reconstructed near its original site. It is typical of the suspension bridges constructed in the West during this period and is the only remaining example of its kind. (HAER Collection, Library of Congress)

posts tied together with a cross cast-iron plate at the base. These posts were capped with a specially fabricated cast-iron cap, topped with a cast-iron saddle that could move with the cables. The towers were made by the Starbuck Iron Works in Troy, New York. The roadway was 18 feet in the clear and the abutments 12 feet high.

Practically all of the material used in the bridge had to be shipped by sea via the dangerous "around-the-horn" route. Many of the materials employed had been developed only within the past ten years or so. The manufacture of wire in significant amounts had become possible in the middle 1840s. It is remarkable that builders during the gold rush brought these advances to California—that they had the capabilities of designing, ordering the materials for, and constructing structures of this magnitude under the extreme frontier conditions of northern California in the 1850s.

In 1964 the Bidwell Bar Bridge was dismantled because of the construction of the Oroville Dam and Reservoir. Owing to efforts by members of the American Society of Civil Engineers, the bridge was reconstructed rather than destroyed. It was erected at a site 600 feet above and about half a mile west of its original location, using the original cables, hangers, towers, and saddles, as well as some original stone for abutment facings. Ernest James was the project manager and Alfred Golye the chief engineer for this endeavor. The site is part of the Kelly Ridge recreational area near Oroville Dam.

The Cabin John Aqueduct and Bridge

At the time of its completion in 1864, the 220-foot stone arch of the Cabin John Aqueduct was the longest masonry arch in the world. The structure is still providing water for the District of Columbia, as well as carrying traffic across the Cabin John Valley in Montgomery County, Maryland.

When he was planning the city of Washington, Pierre Charles L'Enfant investigated many possible sources for future water supply. Although Washington was bordered by three rivers, serious water shortages were anticipated. Several plans were proposed, but it was not until 1851, when the population of Washington and Georgetown was nearing 50,000, that Congress became concerned about having adequate water supplies. An appropriation of $5,000 was made to determine the best manner "of affording to the cities of Washington and Georgetown an unfailing and abundant supply of good and wholesome water."

Surveys were made in the winter of 1852–1853 by Montgomery C. Meigs, a bridge and hydraulic engineer and a general in the U.S. Army Corps of Engineers, who in his report of February 12, 1853, proposed three plans for obtaining the necessary water supply and submitted cost estimates. The plan that was adopted recommended taking water from the Potomac above the Great Falls, some fourteen miles away. A dam, a masonry conduit, two reservoirs, and five bridges were the primary components of the water system.

Meigs was a farsighted man who had earned his reputation by superintending the replacement of the old wooden dome of the Capitol with the present cast-iron dome (it was Meigs who hired Constantino Brumidi to paint the famous frieze in the rotunda). He argued that waterworks in the United States had invariably been designed on an inadequate scale, and that rapid population growth and increased consumption justified the construction of a conduit capable of carrying nearly four times the amount of water then furnished to the city of Paris, one and one-half times that furnished to London, and two and one-half times that furnished to New York. The architect Robert Mills told Congress, "This would be, indeed, a work worthy of our Republic, and would place it on a footing with the proudest of the ancient governments—even Rome herself."

The first appropriation for construction of the aqueduct was made in March 1857, and ground was broken in November. In his report on the proposed line of the conduit, General Meigs stated that the only serious obstacle in its whole course was the valley of the Cabin John Branch, seven miles from the Great Falls. (The stream is named for a character called "Captain John" or sometimes "John of the Cabin," who is said to have lived in a small log house in the valley.) This valley, he said, might be crossed by pipes, but "they always occasion a loss of head or else exceed in cost the bridges they replace." He therefore first proposed to cross the valley by a bridge 482 feet long and 20 feet wide, supported upon six semicircular arches of 60-foot span, resting upon piers 7 feet thick at the top and of various heights, the highest being 52.5 feet. The estimated cost was $72,409. This plan was not put into effect, however, and instead a single, long-span, stone arch was built.

The structure was designed by Meigs and his assistant Alfred L. Rives, a Virginian who was a graduate of the University of Paris. Rives was undoubtedly familiar with the work of Perronet, the brilliant French engineer whose experimentation in the direction of longer and flatter arches anticipated design features used later extensively in concrete. It may well be this indirect influence that inspired the Cabin John Bridge's unusually flat span of 220 feet with a rise of only 57.25 feet. The total length of the bridge including abutments is 450 feet; its width is 20 feet. The arch has a 110-degree segment with a center crown radius of 134 feet. The roadway is 101 feet above the creek bed, and the brick conduit is 9 feet in diameter.

In appearance the Cabin John Bridge is bold and geometrically simple. A long, narrow arch is supported between plain battered abutments. The double arch rings are unique. The bottom ring is of radially layered, cut and dressed granite, and the inner ring is of radially layered sandstone, the material used

Cabin John Aqueduct, Cabin John, Maryland. Built between 1857 and 1864 under the direction of Montgomery C. Meigs, this was the longest masonry arch in the world until 1903. The structure is still serving the purpose for which it was built, supplying water to Washington, D.C., as well as carrying modern traffic loads on a major highway. (Courtesy of the American Society of Civil Engineers)

on the rest of the bridge. Arch sides and abutments are random ashlar, with some horizontality expressed. The spandrel is distinguished by a simple single band of dressed stone trim. A red Seneca sandstone parapet tops the ashlar structure. The flat top is accentuated by strong moldings. The top section appears so thin that one wonders how it could contain water conduits.

The bridge is not, as its outward appearance leads one to believe, of solid stone construction. Like those of many long-span masonry arch bridges, its spandrels, hidden behind sandstone side walls, are hollow. This device was used to reduce the weight of the structure upon the haunches and thus to obviate the danger of its springing up at the center and collapsing. The structure has five spandrel arches at the west end and four at the east end. The material for the arch ring is dressed Quincy granite voussoirs, 4 feet deep at the crown and 6 feet deep at the springs. The spandrel walls are a thin stratum of sandstone. The unusual radial layering of the sandstone slabs for some distance above the upper surface of the granite arch ring was presumably intended to increase the compressive strength of the arch, although this may have been unnecessary in view of the depth of the granite ring. The entire bridge contains 13,283 cubic yards of stone masonry, concrete, and brickwork.

When work began on the bridge, the creek was dammed to form a pond and connected to the Chesapeake and Ohio Canal by a lock. Materials were transported to the site by boat and an overhead timber crane. A heavy timber center was constructed and a quarry opened for sandstone a few hundred feet up the valley. Construction was delayed by a shortage of funds, and contractors were also slow in making deliveries. Of course, the advent of the Civil War brought further delays.

The Civil War contributed significantly to the bridge's history. Meigs was in charge of the project from the time of the first survey until July 1860, when he was relieved by Captain H. W. Benham of the Corps of Engineers. Benham, in turn, was succeeded by another engineer from the corps. Meigs had recognized ability in the field of aqueduct

This 1859 photograph shows the centering for the arch of the Cabin John Aqueduct. (Library of Congress)

construction, however, and Congress conditioned one appropriation on his supervision. This greatly offended President Buchanan, who saw in it an infringement on his power as commander-in-chief. Meigs was transferred to Fort Jefferson in the Tortugas for seven months. When he returned, he was again put in charge of the aqueduct.

Alfred Rives resigned in order to join the Confederate forces, an action that provoked such hostility in Meigs's Yankee blood that he apparently endeavored to obscure Rives's contribution to the project. Only recently has Rives's important role been brought to light. Meigs also requested that the bridge be officially called the Union Arch. His suggestion was adopted, but the name never gained popularity.

In June 1862 the supervision of the Washington water-supply system was transferred from the War Department to the Department of the Interior. There followed a ludicrous period of cutting and recutting of names on the tablet of the Cabin John Bridge. The name of Jefferson Davis, who as secretary of war had been responsible for the overall project when it began but who had returned to the South to become president of the Confederate States, was reputedly removed on the orders of Secretary of the Interior Caleb Smith. Rives's name may have been left off for the same reasons. In 1909 President Theodore Roosevelt ordered Davis's name restored to the bridge.

The name Union Arch was abandoned ten years after the structure was completed, and it is now officially known as the Cabin John Bridge. The longest single-span masonry arch in the world at the time of its completion, it is still the longest in the Western Hemisphere. The Ploven Arch in Europe (1904) is at present the longest stone arch in the world.

The John A. Roebling Bridge

The great suspension bridge built by John A. Roebling at Cincinnati was the first permanent bridging across the Ohio River between Ohio and Kentucky and a direct predecessor of Roebling's masterwork, the Brooklyn Bridge. Its suspension span of 1,057 feet and total length of 2,252 feet were the longest in the world at the time of its official opening in 1867. Originally called the Covington-Cincinnati Suspension Bridge, it was officially renamed in honor of its designer by the Commonwealth of Kentucky on June 27, 1983.

As early as 1810 there was talk of a bridge across the Ohio at Cincinnati. In 1839 a group of Kentuckians formed the Covington and Cincinnati Bridge Company; at their initial meeting they discussed the erection of a bridge and a railroad to the South. Nothing came of the meeting at first, however, perhaps because civic leaders listened to the protest of boatmen that a bridge would impede movement of their craft.

The 1840s were a time of intense interest in suspension bridges, with the charismatic Charles Ellet among their foremost advocates. Ellet had built the Fairmount Park Bridge, America's first major wire suspension bridge, over the Schuylkill River in 1842 and was an acknowledged leader in the field. John Roebling had less of a reputation at the time than Ellet but was also a strong advocate of suspension structures. Ellet, after his success with the Fairmount Park Bridge, had proposed that the Ohio River be spanned at either Louisville, Maysville, Marietta, or Wheeling.

Ellet's proposal activated Cincinnati-area bridge supporters. On October 21, 1845, Covington leaders passed a resolution for a bridge, and in February 1846 the Kentucky General Assembly enacted the chartering of the Covington and Cincinnati Bridge Company. Both Ellet and Roebling were consulted, but only Roebling submitted a plan. He suggested two possibilities, including a bridge with a clear span of 1,200 feet. This proposal aroused the river boatmen, who denounced the idea and brought a strong lobbying force to bear in the Ohio legislature;

John A. Roebling Bridge, Cincinnati. When opened in 1867, this suspension bridge had the world's longest main span, at 1,057 feet. Designed and built by John A. Roebling, it was the prototype for the Brooklyn Bridge, which followed seventeen years later. The bridge has remained in continuous service since its opening. (HAER Collection, Library of Congress)

but Roebling and others argued effectively against them. Still, the Ohio legislature would not grant the same type of charter that the Kentucky legislature had granted the year before. It did, however, grant one to a bridge company in Wheeling, where Ellet proposed another suspension bridge. Some Cincinnatians now approached Ellet for advice. He responded by proposing a single-span bridge, 1,400 feet long. A great stir consequently developed. At last in 1849 the Ohio legislature passed a bridge company charter that was closely patterned after the Kentucky charter. There were, however, significant amendments and stipulations attached—including a restriction against piers in the river and a minimum height of 122 feet above low water at the center of the span (this requirement was later changed to a more realistic 100 feet). A later amendment forbade alignment of the bridge with any street on the Cincinnati side, even though Covington's streets had been laid out as extensions of Cincinnati's.

In the meantime Ellet went ahead with the Wheeling bridge, eventually completing the 1,000-foot span, which made it the longest suspension bridge in the world at the time. This spurred on the Cincinnati-area bridge supporters. To test the validity of a suspension bridge and to develop further interest, they suggested that a small suspension bridge be built over Licking River near Covington. The bridge was completed in January 1854 by local engineers acting without Roebling's consultation. Two weeks after it opened, a drove of cattle caused severe oscillations to develop and the bridge collapsed into the river. Four months later, Ellet's Wheeling Bridge came down in a windstorm. These events proved devastating to the bridge company. In 1855, however, Roebling completed his famous Niagara River bridge by the falls, and it worked supremely well. Roebling's star was in the ascendant.

A contract was signed in August 1856. The bridge was to have a single span, with no central tower. Work began on the Covington side in September, and by November seven layers of heavy timber were in place. In December a cofferdam was completed. On the Cincinnati side, however, there was trouble

The primary suspension cables being put into place. (Courtesy of the Rensselaer Polytechnic Institute Archives)

because of the softer river bed. The clay bank had been washed away and replaced by dirt from cellar excavations as the town had grown. As the foundation was sunk, water from the city flowed into the excavation while Roebling struggled to pump out the cofferdam. City residents complained that their wells were running dry. Roebling finally designed his own pumps, powered by Amos Shinkle's tugboat *Champion No. 1,* and excavated the foundation to gravel. By the end of the construction season in 1858, the cable tower on the Cincinnati side was 47 feet high, the one on the Covington side 77 feet. And there they remained for five years. The bridge company simply could not raise sufficient funds to continue construction.

Curiously, the Civil War hastened the completion of the bridge. Cincinnati was threatened by General Kirby Smith's advance through Kentucky. General Lew Wallace (the author of *Ben Hur*), in charge of the defense of the area, needed a way to get soldiers across the river and had a pontoon bridge put in place. The threat passed, but it had become clear that a bridge at the site was essential. In 1863 bridge bonds could again be sold. Roebling was called to Cincinnati, and work resumed.

By the end of the summer of 1865 the two great towers were finished. The tower arches were 75 feet above the roadway. At either end of the bridge, cable anchorages were constructed, consisting of large masonry blocks (primarily of limestone). Buried within these anchorages were large cast-iron anchor plates, each weighing more than 11 tons, to which chains made of wrought-iron eyebars, forged under the supervision of Roebling, were attached. Next came the cables, made of wire manufactured by Johnson and Brothers in Manchester, England, and processed at the Roebling works in Trenton, New Jersey (the war prevented Roebling from securing American wire). The cables, 2½ inches in diameter and made of many wires, were mounted on reels and suspended on a frame. After its free end was fixed to an anchor plate, the first cable was hoisted to the top of the cable tower, laid in an iron saddle, and passed down the other side. It was then placed in a flatboat fitted with a device for paying it out. On the Covington side it was hoisted over the tower and made fast to the anchor piers. This was the guide cable, which regulated placement of following wires. The average number of wires taken across daily was 80—they were carried on little wheels that ran back and forth. In all, 5,180 one-eighth-inch wires were used, making up two major cables, each 12½ inches in diameter. The wires were not twisted but laid straight, wrapped and compressed with a strand of wire, and clamped. Each cable was given three coats of linseed oil to prevent rust. Suspenders were next attached by means of flat iron bands shrunk and fitted around the cables. Sockets and bolts hung down to receive the verticals. Wrought-iron floor beams were then affixed, and covered in turn by oak flooring.

Finally, the 76 pairs of radiating diagonal stays were installed. These stays, which many now consider structurally ambiguous, descended in straight lines from the tops of the towers to the floor. Apparently envisioned by Roebling as devices for helping stiffen the bridge against excessive vibrations, the stays became a hallmark of Roebling's designs. In addition, Roebling strung eight heavy counter, or check, stays from each tower below the roadway to the cables themselves with the seeming objective of counteracting the lifting and lateral motion that might result from wind or vibrations. Further stiffness of the bridge deck was ensured by the addition of wrought-iron Howe trusses on either side of the roadway.

The later stages of the work were supervised by John Roebling's son, Washington Roebling, who had returned from service in the Civil War. The project provided him with excellent field training for what would prove to be his greatest challenge—taking over the construction of the Brooklyn Bridge after his father's untimely death.

The completed Covington-Cincinnati bridge had a main span of 1,057 feet and two land spans of 281 feet each. The towers, of limestone and sandstone, measure 52 feet by 82 feet at their bases and are 230 feet high. The bridge opened unofficially on December 1, 1866. A huge number of people walked

A close-up of the cable. (HAER Collection, Library of Congress)

across it during the next few days. The official opening was on New Year's Day. The bridge was substantially rebuilt in 1895–1896 to accommodate increased traffic and new types of transportation: a second set of cables was added, as were new anchorages and other devices such as new steel saddles; the brick turrets were removed and stairways built to the tops of the towers. During the great flood of 1937 it was the only bridge to remain open between Steubenville, Ohio, and the confluence of the Ohio River with the Mississippi. This marvelous structure continues to provide excellent service today.

The Eads Bridge

Originally called the St. Louis Bridge, the famous Eads Bridge was later named after its designer and builder, James Buchanan Eads. Completed in 1874, the bridge represented an unprecedented use of long-span tubular steel arches and entailed the deepest submarine construction work done in the world at that time. It was also one of the first major bridges to utilize cantilever construction entirely, eliminating the need for falsework during the erection of the three great arch spans.

With the expansion of the railroad westward, the need for a bridge at St. Louis had become acute by the time Eads submitted his plan in 1867. St. Louis was already a thriving metropolis and a busy terminal for the Illinois Central Railroad. Connection with the Union Pacific was possible if the Mississippi could be spanned. There had been four proposals to connect St. Louis to the Illinois shore prior to Eads's own proposal for an arch span; two were suspension bridges, suggested by John Roebling and Charles Ellet.

In April 1866 Eads was appointed to chair a committee set up by the Chamber of Commerce to ensure that marine interests were considered in the design of a bridge over the Mississippi. The committee's report proposed that any bridge have a clear height of 50 feet or more over the main channel of the river, that it have a single clear span of 600 feet or two spans of 450 feet each, and that no suspension bridge or drawbridge of any type be allowed. The report clearly reflected Eads's own confidence that a long-span structure was indeed possible and that a rigid structure would work the best.

In the same year the St. Louis and Illinois Bridge Company was formed and secured a charter from the legislature that called for either a main span of 500 feet or two center spans of 350 feet each. Eads became interested in the idea and submitted a proposal for a bridge with a center span of 515 feet, resting on two piers, and two end spans of 497 feet each, which joined to shore abutments.

Eads had no specific training as an engineer and no formal education after the age of thirteen. He was, however, both imaginative and practical. He had made a fortune in the salvage business after inventing a diving bell and had constructed a fleet of Union gunboats for use on the Mississippi during the Civil War. His knowledge of the river was unparalleled and a major factor in his appointment as chief engineer for the project. Eads organized a team that included accomplished engineers, among them Charles Shaler Smith, Colonel Henry Flad, and Charles Pfeifer. Smith was a particularly good choice because of his training under George McLeod and Albert Fink of the Louisville and Nashville Railroad. The chancellor of Washington University, William Chauvenet, became the mathematical consultant.

Eads's design did have some precedents. The famous Coalbrookdale Bridge in England (1775–1779) had an iron arch system. Richard Delafield had constructed a modest iron span at Brownsville, Pennsylvania, between 1836 and 1838 (see Dunlap's Creek Bridge). Montgomery Meigs had used iron water mains as arch ribs in his innovative Rock Creek bridge in Washington in 1858. Tubular ribs were used in the 185-foot Chestnut Street bridge over the Schuylkill River in Philadelphia, constructed between 1861 and 1866. But the scale of Eads's proposed bridge was far beyond that of these limited experiments.

The immediate inspiration for the design may have been the triple-arch bridge completed in 1864 across the Rhine River at Koblenz. A description of the Koblenz bridge with its three arched spans made of rectangular wrought-iron lattice ribs had appeared in *Engineering,* a British publication that Eads knew well. Eads corresponded with Jacob Linville, a well-known bridge engineer, and sent him rough plans of his proposed design. In one of his letters Eads notes that the bracing is "somewhat after the plan of the bridge at Coblentz." His recently hired assistant, Charles Pfeifer, must also have been familiar with the Koblenz bridge, since he was trained in Germany and had only recently arrived in the United States. Drawings of the proposed St. Louis bridge that were put on display in July 1867 showed close similarities—including rectangular ribs—although greater spans were illustrated. By April 1868

Eads Bridge, St. Louis, completed in 1874. This celebrated bridge is named for James Buchanan Eads, its designer and builder. To found the midriver piers on solid rock, Eads used pneumatic caissons. Their sinking represented the deepest subaqueous construction work in the world at the time. Steel was used in the arch ribs—its first extensive use in a bridge. (HAER Collection, Library of Congress)

plans sent for review to Julius W. Adams, bridge engineer and vice-president of the American Society of Civil Engineers, indicated that tubular ribs were now proposed rather than rectangular ones. The tubular ribs were similar to those used by Meigs in his 1858 arch bridge in Washington, a bridge that may well have been seen by Eads during one of his frequent trips to that city.

Pfeifer worked with Colonel Flad on the detailed design of the bridge, Pfeifer doing many of the calculations involved (including determining what were then called pressure lines and determining member sizes). Apparently Pfeifer's proposed European-style cross sections for the ribs, based on rectilinear plates and angles, were overruled by Eads in favor of tubes. Eads was said to have been involved in the minutest details of the tubes and connections, but always paying attention to the effective areas for members established by Pfeifer, communicating his ideas verbally and with rough sketches.

Eads also began exploring the use of steel for the tubes. He was quite familiar with the properties of iron and steel. During the Civil War he had constructed a fleet of ironclad gunboats and river monitors for use on the Mississippi. In the process he had been in close contact with engineers in the Naval Ordnance Bureau and the Bureau of Steam Engineering, engineers who were as up-to-date on iron and steel as anyone in America. Eads developed the idea of using steel staves encircled by a steel hoop (a form that can be seen as reminiscent of the ribs of Meigs's bridge). It was not until the end of 1870 or the beginning of 1871 that the design of the 18-inch tubes had been finalized.

In the meantime Eads had to defend his proposals. Jacob Linville had come out against the bridge plans he had seen. The design had also been attacked, indirectly, in 1867 in a report issued by a group of civil engineers organized by one L. B. Boomer of Chicago—Boomer having in the previous year vied unsuccessfully with the St. Louis and Illinois Bridge Company for authority to build the bridge. This group included such well-known figures as Ellis S. Chesbrough, who had recently finished the tunnel under Lake Michigan for the Chicago waterworks (see chapter 6); Charles L. McAlpine, who had helped build a bridge over the Harlem River that involved pneumatic processes in sinking the foundations; and William Sooy Smith, builder of a bridge across the Missouri that also used pneumatic foundations. Their report challenged the concept of using an arch form for a 500-foot span and proposed alternative solutions. It noted, further, that the foundations would be safe from undercutting by scouring from the river if they were built down into the sand to 45 feet below low-water level, where they could be supported by piles. Eads was compelled to write a rebuttal, in which he expressed his particular determination to locate the bottoms of the foundations lower than 45 feet, since his own experience indicated that scouring, and hence undercutting, could occur at much greater depths. Only by resting the piers on bedrock, he maintained, could he make them safe. Eventually Eads's proposal went ahead.

Eads sank the west abutment first. Of the four foundations needed to support the triple arch, this was the shallowest—only 40 feet below mean high water. But with accumulated debris from an 1849 fire, it was no easy task. The other foundations proved even more of a challenge. The river at this crossing had a bottom profile shelving away steeply from west to east, so that on the Illinois side an abutment had to penetrate perhaps 40 to 80 feet of mud to reach secure bedrock at a total depth of well over 100 feet.

At that time the process for constructing deep underwater foundations was not well developed in the United States. The problem of increasing pressure, as well as the logistics of taking a work crew to the bottom of a river, seemed an enormous barrier. In 1869 Eads went to France to recover from an illness and to show his plans to an interested French engineering firm. Learning of his dilemma, they introduced him to the pneumatic caisson, as used at a deep-foundation bridge construction site at Vichy. The French and British had taken foundations as deep as 70 feet, solving some of the problems inherent in this type of construction. Thomas Cochrane had first demonstrated a pneumatic caisson for this purpose in 1830 in England. A cast-iron cylinder with a cutting edge was patented in England in 1843. Several other types and uses of pneumatic caissons followed. The caisson developed in Europe generally consisted of a working chamber connected to the surface by a column containing two shafts, one for men and one for materials, with an air lock at the top of each. The air lock was a small, tightly sealed room with two closely fitted doors opposite each other. The work crew entered, closed the door, opened a valve, and increased the air pressure until it equaled that of the working chamber below. Then they opened the door to the shaft and descended. Coming out, they reversed the process.

Eads adapted this method to his proposed bridge at St. Louis, even though the depths required were beyond anything ever before attempted. He had huge, rectangular, iron-sheathed boxes constructed and mounted on boat bottoms. The first such assembly was floated out to the middle of the river, where the box was freed from its false boat bottom and moored in place. Barges loaded with limestone blocks were brought beside it, and the blocks were swung atop the caisson roof to create what would become the masonry shaft of the east pier. As the weight of the stone increased, the assembly was pushed slowly to the bottom. Work crews, mostly German and Irish immigrants, then descended a spiral stairway in the middle of the stone column to dig through the mud and sand below. These workers came to be known as "sandhogs."

As sand was removed from the bottom through the use of sand pumps (designed by Eads), the men above continued to build the rising pier. The whole assembly then slowly sank. As it sank, air pressure gradually rose inside the caisson. As the pressure increased, men began complaining of stomach pains after leaving the caisson. When the shaft reached 76 feet, one man had to be hospitalized. To alleviate the problem somewhat, Eads allowed only the most physically fit to work, and then only in short shifts. Eventually, ten men died of the malady we now call the "bends," which is caused by the saturation of the tissues and blood with nitrogen. When workers left the compressed-air environment, the nitrogen formed bubbles that slowed down circulation, sometimes fatally. The workmen dealt with the bends by drinking excessively and tried to ward them off with patent medicines and amulets. More practically, they also struck for shorter hours and higher pay. It was not until after the completion of this pier that a St. Louis physician, Dr. A. Jaminet, after setting up a floating hospital and conducting numerous studies, came up with the remedy of slow decompression in a separate chamber.

The working chamber continued to descend until it reached bottom on February 28, 1870. At a depth of 93.5 feet it was the deepest submarine structural foundation ever sunk. The caisson was then filled with stone, as was the hollow core of the descent chamber. Despite ice damage and flooding, the east pier stood complete at the end of April. In similar fashion the west pier was submerged to 78 feet, and the deep east abutment to 103 feet below the river surface and 136 feet below high water. At the east abutment a newly devised decompression chamber greatly reduced the danger to workmen. Other problems slowed construction. Ice floes and a tornado destroyed the superstructure of the east abutment. Repairs were made, and eventually all four foundations were solidly planted on bedrock, ready to receive the three arch spans.

Eads's arches were composed of ribs made from sections of steel tube. No one had ever made an arch from steel tubing at this scale. Initially, plain carbon cast steel was selected for use. Eads's tough quality-control standards discouraged most contractors from offering bids for fabrication of the structure, but eventually the Keystone Bridge Company was awarded the contract for the job. Andrew Carnegie and Jacob Linville were executives in the company; Carnegie later used his influence to subcontract all bridge ironwork to his own firm, Carnegie-Kloman. The William Butcher Steel Company was to manufacture the carbon steel. Because of Eads's strict metallurgical requirements—which were perhaps beyond the capabilities, and certainly beyond the inclinations, of the fledgling steel industry—friction developed. The main braces were to be of heavy

The arches of the Eads Bridge were constructed by a cantilever method, without falsework. This was a major construction innovation for a bridge of this size. (Library of Congress)

A major problem was the joining of the cantilevered half-arches. The first closing occurred while Eads was in Europe. The engineer in charge, Colonel Flad, tried packing the ribs in ice to change their lengths and bring the projecting ends into alignment. After several nerve-racking days, the ribs were finally joined by using a special adjustable closing piece devised by Eads and left on hand for this very purpose. (*Scientific American,* November 15, 1873)

wrought-iron bars. Dozens of sample braces made by the Carnegie-Kloman Company were rejected by Eads's inspectors.

Eads turned to a new and largely untried alloy that was being produced by the Chrome Steel Company. He paid the company a royalty to allow Butcher to manufacture the steel for the bridge. Eads required that the material produced meet his mechanical tests, but he did not specify its exact chemical composition. Consequently, the composition of the material furnished and used in the bridge varied considerably. The arch ribs each consisted of four pairs of tubes, 18 inches in diameter, set side by side. The tubes were made up of 12-foot lengths of six cylindrical segments bolted together through longitudinal flanges. An extensive system of diagonal members braced the arches.

Theodore Cooper, later one of the country's leading civil engineers, was hired as inspector for the ironwork. Subsequently he took up the position of chief inspector of erection at the construction site in St. Louis.

Construction was challenging. Under normal conditions falsework would have been used to support the arches until completed, but Eads had promised to keep the Mississippi channel unobstructed. As a solution, the three arches were each cantilevered from the piers toward midspan in a counterbalanced fashion, without any centering or supporting framework. A temporary tower was erected on each pier to support cables affixed to the half-arches being cantilevered on either side. As construction progressed, half-arches from opposite piers approached each other at midspan.

Meanwhile, Eads's health had deteriorated. He was advised to take an ocean voyage. When Eads sailed for England in August 1873, Colonel Flad was left in charge of the work. The responsibility of adjusting the cantilevered half-arches and joining them at midspan—"closing" the arches—fell to him.

Although it had always been assumed that the task would be difficult, the difficulty surpassed all expectation. The stresses and, consequently, deformations in the cantilevered half-arches were not the same as those anticipated. It was also discovered that the ends of the half-arches—already deflected by

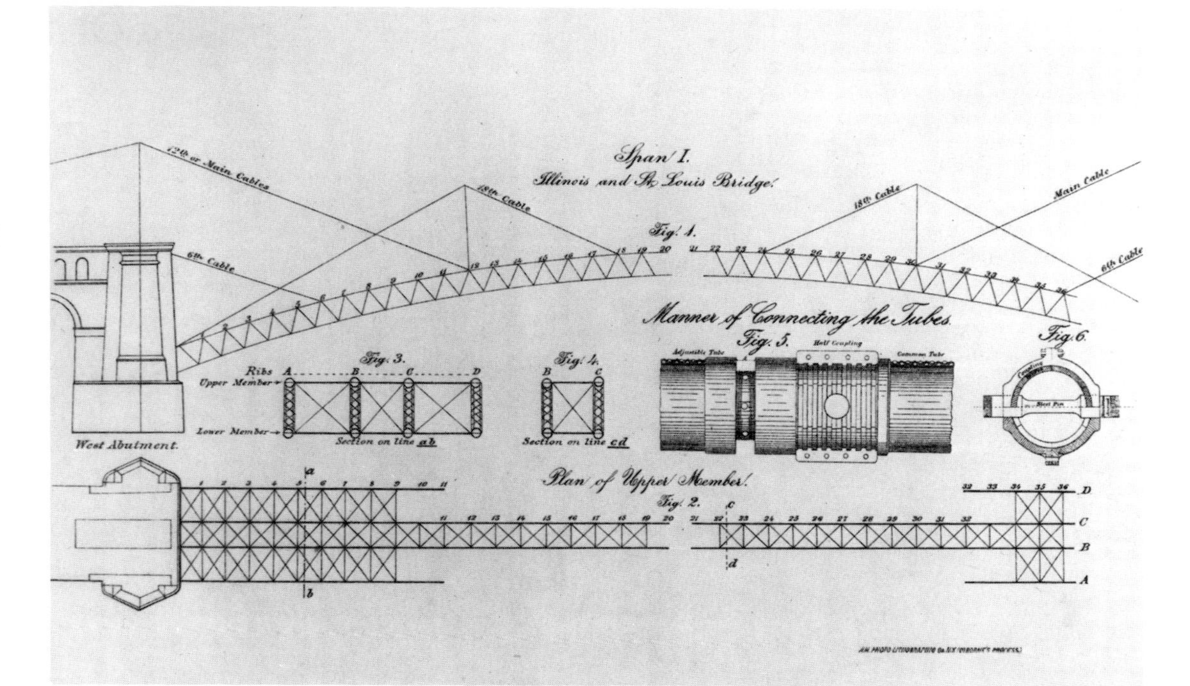

The adjustable closing piece designed by Eads to enable the cantilevered tubes to be connected is shown in this illustration from a paper published by Theodore Cooper, a well-known civil engineer who was at the time a resident engineer on the project. (*Transactions of the American Society of Civil Engineers* 3, no. 7, June 1874)

One of the great arch ribs. (HAER Collection, Library of Congress)

their own weight—rose and fell, expanded and contracted, with changes in rib length induced by daily temperature fluctuations. It was critical, however, that the arches be closed soon. A loan of half a million dollars, which had been negotiated by Eads in London, depended on the closing of the first span by September 19, 1873.

On September 14 Flad attempted to close the first two half-arches. The gap proved too narrow to accommodate the specially devised closing tubes. To change the lengths of the half-arches, and thus the critical gap spacing, Flad decided to pack the ribs in ice. On September 15 ice was placed in wooden troughs that were then fitted around the rib members and wrapped in gunny cloth. Owing to the warm weather—98 degrees in the shade at 5 p.m.—the ice had little effect, and the space between the lower tubes was still off by 2¼ inches. The night was warm, and the gap was still off by ⅝ inch at daybreak. Flad added more ice—45 tons—but at sundown the gap was still off. Again that night a warm wind blew, and the morning of September 17 showed a gap of ¾ inch. Abandoning the ice technique, Flad jacked the cantilever towers and, using special adjustable closing members devised and prepared by Eads in anticipation of just such an emergency, made the closure that night at 10 p.m. The engineers had been on constant duty for nearly sixty-five hours. The following day, September 18, a cablegram was sent to Eads in London: "Arch safely closed."

Early in 1874 a serious problem threatened the entire project. An inspector had noticed that the arch ribs had begun to fail. Two tubes in the first span had ruptured. Eads (now home from his travels) figured, correctly, that the still-attached cantilevering cables were inducing unusual stresses in the structure. The cables were released, and the arches settled firmly into place.

As if these practical problems were not enough, in the midst of construction the secretary of war, William W. Belknap, under pressure from the Keokuk Steamboat Company, demanded that a canal be built around the east abutment so that steamboats would not have to lower their smokestacks while passing under the bridge. President Grant interceded, ordering Belknap to drop the case.

On May 24, 1874, the highway deck was opened to pedestrians, and on June 3 it was opened to vehicles. A few days later, the first locomotive crossed the lower deck, and on July 2 fourteen locomotives were run back and forth across the bridge in various formations, to demonstrate its strength. At the July 4 dedication an appreciative St. Louis public placed on the bridge a medallion of Eads, with the following inscription: "The Mississippi discovered by Marquette, 1673; spanned by Captain Eads, 1874."

Eads Bridge: a typical arch abutment. (HAER Collection, Library of Congress)

A view of the interior structure of the Eads Bridge, showing the bracing system. (HAER Collection, Library of Congress)

The Kinzua Railway Viaduct

Constructed across the valley of the Kinzua Creek in McKean County, Pennsylvania, in 1882, this 302-foot-high and 2,053-foot-long railroad viaduct was among the tallest in the world at the time and remains an impressive structure today. The original structure was replaced in 1900 because of heavier train loadings; the newer viaduct is generally similar to the old one, except that it was built in steel instead of iron and was strengthened in design to carry heavier service loads.

Built to carry coal, the original viaduct was part of the Bradford Branch of the New York, Lake Erie, and Western Railroad. The railway line, pushing southward from the terminus at Bradford, moved up the steep slope of the Big Shanty in a great reverse curve and across a high plateau until the 300-foot valley of the Kinzua Creek lay in the way. The alternative of constructing four miles of tortuous grade was rejected in favor of a high viaduct. General Thomas L. Kane of the railroad joined with Anthony Bonzano of the Clarke Reeves Division of the Phoenix Bridge Company and Oliver W. Barnes, chief engineer of the railroad, in planning the great structure.

The original Kinzua Viaduct was a wrought-iron structure, 2,000 feet long and 301 feet high from the surface of the water to rail level. There were twenty towers, each made of two bents of sloped or battered columns resting on masonry piers. The towers were 99.5 feet apart, with a maximum height of 279 feet. The tallest tower was divided into ten vertical panels with horizontal struts and intersecting diagonal members. The columns were circular, of about 10.75-inch maximum outside diameter, composed of four segments of the famous Phoenix shapes. The columns had cast-iron caps and pedestals. The connecting spans were lattice girders, 6 feet deep.

Construction began on May 10, 1882. The work was done without scaffolding. A gin pole was used to erect the first tower, and a crane placed on the first tower was used to place the ironwork for the second. This procedure was repeated until all twenty towers had been raised and the connecting lattice spans placed. Iron longitudinal and transverse rods were used to support construction

Kinzua Railway Viaduct, Kinzua Bridge State Park, McKean County, Pennsylvania. Constructed in only 102 days, the original viaduct was one of the highest (302 feet) and longest (2,053 feet) in the world when it was completed in 1882. This early photograph was proudly featured in the *Album of Designs* of the Phoenix Bridge Company, published in 1885. The original structure was replaced in 1900 with a similar one capable of carrying heavier service loads.

between piers. Workers moved back and forth across the rods. A temporary track was laid in the valley floor to distribute all construction materials. Two steam hoists were used, along with some thirty miles of rope. The viaduct was completed on August 20, 1882. The work force had consisted of only a hundred men.

The structure was considered a wonder, although it swayed extensively and train speeds had to be limited to 5 miles per hour. It attracted wide interest, and excursion trains were popular. During its eighteen years of service, however, there was a dramatic increase in the weights of locomotives and rolling stock. A structure capable of carrying modern traffic was needed, and the decision was made to replace the old viaduct with a similar but strengthened design.

The reconstruction was directed by C. W. Bucholz, chief engineer for the railroad, and executed by the Elmira Bridge Company. M. R. Strong, bridge engineer, was in immediate charge. The design of the new structure was constrained by the desire to conform to the dimensions of the old viaduct. Consequently, the principal dimensions of the old and new bridges are virtually identical. The viaduct is still carried on twenty towers, each tower consisting of a pair of two-column bents. In the new design each column consists of two built-up channel members, spaced 3 feet apart and connected by a latticework of steel plates and angles. Each bent is stiffened laterally by means of deep latticed struts and massive plate-steel knee braces. The superstructure consists of two lines of plate-girder spans, spaced 9 feet apart.

Traffic over the original bridge was stopped on May 14, 1900. Two "travelers" were constructed, each consisting of a complete 180-foot-long timber truss. These travelers were run out over the old bridge from each end. Each was long enough to span three towers. While the traveler was supported by towers at each of its ends, the middle tower was dismantled and the material hoisted up to the travelers. The pieces were then carried to one end of the bridge. After the old tower had been taken away, the material for the new tower was run out on the bridge to the

The highest towers of the Kinzua Viaduct loom 302 feet above their bases. The towers of the original structure, as can be seen in this early photograph, were made of remarkably slender members. (*Album of Designs,* Phoenix Bridge Company, 1885)

Two "travelers" were used to replace the old viaduct structure with a new one in 1900. In this photograph a column section of a new tower is being lowered into place. When a new segment was completed, the traveler would be moved to an adjacent tower and the process repeated. Note that both the old and new structures are visible. (*Scientific American*, October 27, 1900)

traveler, lowered into place, and riveted up to form the new tower. Spans were treated similarly. Traffic resumed over the completed bridge on September 25, 1900—a short four months after reconstruction had begun.

At the time of its original construction, the Kinzua Viaduct ranked with the 401-foot-high Garabit viaduct in southern France and the lower 252-foot-high Verugas viaduct in Peru as one of the tallest bridges in the world. The River Viaduct in Texas now looms higher, but the power of the Kinzua Viaduct remains unsurpassed. Now safely inside the Kinzua Bridge State Park, the 1900 version is, just like its older predecessor, an excursionist's delight.

The second Kinzua Viaduct provides an early example of the use of a rigid frame structure to carry laterally acting loads. (HAER Collection, Library of Congress)

The Smithfield Street Bridge

Smithfield Street Bridge, Pittsburgh. This project represents a unique adaptation of a contemporary European engineering structure, the lenticular truss, to American needs. Completed in 1884, it is one of the oldest extant major steel trusses in the United States and was the earliest major project of the noted engineer Gustav Lindenthal. (Courtesy of the American Society of Civil Engineers)

The Smithfield Street Bridge, built over the Monongahela River in Pittsburgh, Pennsylvania, between 1882 and 1884, introduced an innovative European engineering design—the lenticular truss—to the rapidly expanding American road system. The Austrian-born engineer Gustav Lindenthal designed the trusses (often called Pauli trusses after their German inventor, Friedrich August von Pauli) using locally manufactured open-hearth steel, thus declaring his confidence in the newly developed material. Lindenthal's bridge was, and still is, a powerful and original example of structural ingenuity.

This lenticular or lens-shaped truss bridge, which connected downtown Pittsburgh with the city's south side across the Monongahela River, had been preceded by two other bridges, both designed by the leading bridge engineers of their time. The first, by Lewis Wernwag, was completed in 1818 and replaced a ferry between the two points. It was a covered wooden bridge composed of eight 188-foot truss spans and reinforced with timber arches. It was destroyed by fire in 1845 and replaced that same year by a wire suspension bridge designed by John Roebling (this was Roebling's first highway bridge). Roebling reused the original masonry piers and ingeniously hung the continuous suspension cable from links pinned to the towers. The bridge proved extremely flexible, however, so that by the 1870s the increased loads caused unacceptably large deflections and vibrations.

In 1880 Charles Davis was commissioned to replace the Roebling bridge with another suspension structure, but with larger spans. After construction on the piers had already begun, the company responsible for building, operating, and collecting tolls on the bridge was reorganized. The new managers demanded a completely different structure, one "not subject to undulations and . . . capable of enduring the continually increasing traffic without limitation of load or speed." They awarded the commission to Gustav Lindenthal, who had proposed the use of lenticular trusses.

The original portico for the Smithfield Street Bridge. (*Transactions of the American Society of Civil Engineers* 21, 1883)

Lindenthal received his formal training in Austria (where he was born in 1850). He did some bridge design work in Europe before emigrating to the United States in 1871. It was while he was working as a stone mason at the Philadelphia Centennial Exposition that his engineering talents were recognized; he was appointed a designer with the responsibility of completing two halls. He then became a designer for the Keystone Bridge Company in Pittsburgh and in 1878 was offered the position of bridge engineer for the Atlantic–Great Western Railroad in Cleveland. Three years later he returned to private practice in Pittsburgh, where he designed the Smithfield Street Bridge and the Seventh Street suspension bridges over the Allegheny. His proposal for the Sixth Street Bridge, to replace another early Roebling structure, was not executed. He later became commissioner of bridges in New York City (1902) and was designer and consultant for the largest steel arch bridge of that time, the Hell Gate Bridge over the East River. He was also involved with the Queensborough Bridge, the Sciotoville Bridge, and Hudson River and Quebec bridge proposals.

The truss form Lindenthal proposed for the project consisted of a lens-shaped structure having oppositely curved upper and lower chords separated by vertical members and crossed diagonals. There are several ways of conceptualizing the properties of this unique form. Viewed in one way, it provides the maximum structural depth at midspan, where the bending moments caused by the loading that the structure must carry are the greatest, and then tapers toward either end in a way reflecting the distribution of bending moments throughout the structure. The bridge is thus carefully and elegantly shaped to reflect the distribution of internal forces. Another way of conceptualizing the behavior of the structure under load is to envision it as a composite arch and cable structure. In this model the outward thrusts of the upper "arch" are exactly balanced by the inward pulls of the lower "cable," with the consequence that there is no net lateral force at the bridge's piers, a result long recognized as very advantageous in foundation design. Here the interstitial verticals can be seen as serving to transfer loads from the deck to the upper and lower chords, and the interstitial diagonal members as serving to stabilize the arch-cable structure under varying load conditions. The lenticular truss is, in short, primarily a truss structure, relying as it does on triangulation for stability; but it is one carefully shaped to gain the efficiencies associated with arch and cable structures, while at the same time eliminating the problematic horizontal thrusts associated with these forms.

Lindenthal emphasized other advantages of the form. Unequal temperature effects are avoided in the truss, which exposes the bottom chord to the sun as much as any other member. Another argument for it was an aesthetic one—Lindenthal thought that the sinuous curving of the bridge's chords would be an interesting addition to the cityscape.

The plans were approved, and the bridge was built in 1883 on the piers constructed for the abandoned suspension bridge project. The approach spans were deck girders and the two main channel spans were 360-foot lenticular trusses. By building on the downstream side of the piers and by designing a detachable sidewalk on the upstream side, Lindenthal provided for the bridge's possible expansion from its original 48-foot width.

Steel proved more economical than wrought iron for this project and was used in the chords, piers, posts, diagonal ties, and pins. The rest of the superstructure was made of wrought-iron with steel rivets. Lindenthal originally specified Bessemer steel but later decided on open-hearth steel because of its higher quality and uniformity.

The south abutment is essentially the same as for Roebling's wire suspension bridge. The pier, built in 1845, is concrete with a sandstone facing. The north abutment, however, was relocated 40 feet further north. Temporary anchorage for the old bridge allowed enough clearance for the new abutment to be built. The growing river traffic required that the vertical channel clearance be increased. An extended dispute between the bridge company and the boatmen was finally resolved by an increase of 20 feet to the piers and abutments.

An early view of the Smithfield Street Bridge. (*Transactions of the American Society of Civil Engineers* 21, 1883)

The bridge's cast-iron decoration is, along with the sinuous trusses, an outstanding feature. Castellated casings embellish the steel supports at either end of the bridge. By solving the visual problem of very "slender supports . . . out of proportion in comparison with the heavy piers and higher trusses," Lindenthal created a powerful composition. Although the original portals are no longer there, the bridge's entrances are still exceptionally effective.

Like his predecessor John Roebling, Lindenthal wrote many papers about his work for technical journals. In 1883 the *Transactions of the American Society of Civil Engineers* carried his paper on the Smithfield Street Bridge and introduced this important project to the world. In the paper Lindenthal described the design and construction of the bridge, with emphasis on its ability to accommodate future changes.

As planned, the bridge has accepted many modifications in its lifetime. In 1889 an additional set of trusses and a second deck were added to carry streetcars. In 1898 these trusses were shifted upstream and the deck widened to accommodate electric trolleys. In 1911 the tolls were removed by the new manager, the city of Pittsburgh.

In 1933 the bridge again attracted worldwide attention by the replacement of the steel floor system with structural aluminum beams and prefabricated aluminum decking. The resulting reduction of the dead load allowed an equivalent increase in ability to carry live load. Indeed, the bridge continues to excel in its original role—that of a forum for demonstrating innovations in steel and aluminum.

The Brooklyn Bridge

The Brooklyn Bridge, designed by John A. Roebling and built under the supervision of his son, Washington A. Roebling, is one of the world's great suspension spans. Longer than any other suspension bridge at the time of its completion in 1883, it was—and is—considered a superb feat of structural design.

The idea of a bridge spanning the East River and connecting Brooklyn with Manhattan began to be discussed early in the nineteenth century and was a popular topic of conversation by the 1840s. But the idea was repeatedly discarded as impractical. Ferry passage, though inefficient at all times and extremely difficult during harsh winters when the river was clogged with ice, remained the only way of traveling between Brooklyn and Manhattan. August Meissner, Jr., John Roebling's nephew, remarked on the difficulties of ferry travel in a letter written after the winter of 1851–1852. In 1857 John Roebling wrote to the industrialist and political leader Abram Hewitt suggesting the feasibility of a span across the East River. Hewitt was in the iron business in Trenton and knew of Roebling's already impressive achievements (see Delaware Aqueduct and John A. Roebling Bridge). He had the letter printed in the New York *Journal of Commerce.* Though widely discussed, nothing more came of it at that time.

During the Civil War years plans for a bridge were largely dormant, although the well-known engineer Julius W. Adams did develop one proposal. In 1865 Roebling presented a set of plans to a group of civic leaders in Brooklyn. William C. Kingsley, a leading contractor in Brooklyn and publisher of the influential *Brooklyn Eagle,* became intrigued with the possibilities in Roebling's suggestions and secured the interests of Alexander McCue, another Brooklyn business leader, and Senator Henry C. Murphy in a meeting held in December 1866. Senator Murphy agreed to draw up an enabling bill for the bridge. A severe winter in 1866–1867, which virtually halted ferry traffic, brought increased public support for the project, and in April 1867 Senator Murphy achieved the passage of an act by the New York state legislature incorporating the New York Bridge

Brooklyn Bridge, Manhattan to Brooklyn, New York City. Designed by John Roebling and built under the direction of his son, Washington Roebling, the Brooklyn Bridge was the longest suspension bridge in the world at the time of its completion in 1883. This photograph was taken in 1890. (Courtesy of the Rensselaer Polytechnic Institute Archives)

Company for the purpose of constructing and maintaining a bridge over the East River between New York and Brooklyn. Thirty-nine incorporators were named, and June 1, 1870, was specified as the completion date. A month later the incorporators met to organize the company. Senator Murphy was elected president. A Committee on Plans and Surveys was convened to determine who would be appointed chief engineer. After a brief consideration of Julius Adams, John Roebling's name was placed before the committee and accepted with no dissents.

Within three months of his appointment as chief engineer, Roebling had selected an appropriate site, made initial surveys and design studies, drawn up cost estimates, and prepared a report on his findings. The proposed span was an unprecedented 1,500 feet—much longer than Roebling's Cincinnati bridge—and was capable of carrying a huge load of 18,700 tons. The specifications called for four cables, composed of steel wires. This use of steel, still a new construction material, was one of the more controversial aspects of the proposal. Roebling understood the dangers involved in a suspension structure of this magnitude. To guard against dynamic loading as well as vertical and horizontal oscillations caused by violent winds, he provided stiffening trusses and cable stays (in addition to the span's own dead weight). The lines of trusses would run the entire length of the bridge.

Although the directors of the bridge company felt confident with Roebling's proposals, public skepticism still ran high. Prominent individuals, such as Horace Greeley, expressed doubt as to the feasibility of the project. Many hearings on the bridge were held, and Roebling had to defend his plans many times. Finally the Committee on Plans and Surveys approved Roebling's report and recommended that steps be taken to raise the required capital for construction. Both public and private subscriptions for stock were secured. William Kingsley played an important role here, putting up a lot of his own money as well. In late 1868 the cities of New York and Brooklyn subscribed to the effort. William ("Boss") Tweed was in control of the political ring heading New York City at the time—a group notorious for its corruption in administering construction contracts. Tweed and his colleagues became wealthy by siphoning off funds for the construction of public projects into their own pockets. Perhaps it was hopes for private enrichment that lay behind Boss Tweed's authorizing the city to subscribe to stock. Tweed fell from power, was convicted, and went to prison before construction of the bridge was very far along; but his show of support in 1868 undoubtedly caused many members of the public to view the bridge proposal with suspicion—a suspicion that persisted for years, even after Tweed was out of office. Kingsley too was ultimately accused of questionable activities. In view of both his experience as a contractor and his efforts in financing the project, the company appointed him superintendent of construction of the tower foundations. He was paid what seemed to stockholders to be an exorbitant fee, and a committee was formed to look into the issue. This affair served to increase public distrust.

During the time capital was being raised, Roebling continued to address opposition to the technical feasibility of the undertaking. In an attempt to instill confidence in the project, Roebling invited a board of prominent engineers to review his plans. The board consisted of Benjamin Latrobe, John Serrell, Horatio Allen, William McAlpine, James Kirkwood, J. Sutton Steele, and Julius Adams—all accomplished and well-known engineers. Beginning in March 1869, this group met with Roebling many times and examined his plans in great depth. They visited his prior works at Niagara Falls and Cincinnati. After two months they unanimously confirmed the feasibility of Roebling's plans. At the same time a commission of three army engineers, under the direction of the secretary of war, became involved in the project as well. Their concern, besides the feasibility of the plan, was the possible obstruction to navigation that the bridge might create. By June 1869 they had approved both the plan and its location, fixing the clearance at 135 feet. This height became standard for future spans over navigable waters. With this endorsement, a federal bill authorizing the construction was passed by Congress and signed by President Ulysses S. Grant. The bridge was now set to be built.

Roebling was not destined to see his bridge completed, however. While he was making a survey for the location of the main piers, a boat collided with the bulkhead of the Brooklyn wharf on which he was standing. Roebling's foot was crushed, and he died three weeks later, on July 22, 1869. He left a monumental task behind him; yet he had prepared the way. In August the directors appointed his son, Colonel Washington A. Roebling, to the position of chief engineer of the bridge.

The younger Roebling had attended Rensselaer Polytechnic Institute, graduating in 1857 with a degree in civil engineering. He had then fought in the Civil War, attaining the rank of colonel. Washington Roebling was thoroughly familiar with his father's methods, having worked with him on the Cincinnati bridge. In 1867 he traveled to Europe to study new methods for sinking underwater foundations utilizing compressed air. Returning only months before his father's death, Washington Roebling loyally accepted the responsibility for the bridge's completion.

Good foundations were essential to support the 271.5-foot masonry towers serving as the main piers. Pneumatic caissons were used to allow excavation under the East River. This method of founding, new to America, was being used concurrently in the construction of the Eads Bridge in St. Louis (described earlier in this chapter).

Work began on the Brooklyn pier. A timber caisson of 12-inch-by-12-inch yellow pine provided six working compartments, each approximately 50 feet square, separated by 2-foot-thick partitions. The overall dimensions of the caisson were 102 feet by 168 feet. The working chambers, 9.5 feet high, were covered by a timber roof 15 feet thick to support the load above. The timber walls of the caisson were 9 feet thick at the top and tapered to 6 inches at the bottom to form a cutting edge. This edge was fitted with a rounded iron casting. On the inside the caisson was caulked with oakum and coated with pitch to make it tight. Outside, a protective coating of

This engraving shows the caisson used in the construction of the pier foundations on the Brooklyn side prior to its being put into place. "It may be said to be a huge diving bell from which water is excluded by forcing into it air from a series of powerful air pumps worked by steam." The enormous caisson was launched on March 19, 1870. The structure was first assembled on the shore, with its long side parallel to the river on top of seven launching ways. When completed, the caisson was launched much like a ship. The launching ways were precisely curved so that the great structure would gain acceleration as it slid down to the river. (*Scientific American*, July 9, 1870)

tin and 3-inch-thick creosoted plank encased the entire caisson. The structure was built on the shore and launched much like a ship.

Before the caisson could be towed into position, the river bottom had to be dredged to a depth of 18 feet. Explosives were effectively used in this process. A timber cofferdam was then constructed to receive the caisson. The entire assembly was towed to the site and carefully centered. Three 10-ton derricks were mounted on top of the caisson, which was then loaded down with limestone blocks. Successive courses were added as the caisson gradually sank to the bottom. Similar additions were made as excavation in the working chambers continued.

As the roof of the caisson was built up, large wrought-iron shafts were installed. There were two "water shafts" to remove dredged material, and two "man shafts" or "supply shafts" for men and eventually for sending down material to fill the caisson when it reached its bearing position. Piping was also installed to supply gas and water, as well as pressurized air for pumping out sand. The "water shafts" projected below the roof of the caisson into a shaft well. The lower ends of these shafts were sealed in water, with a water column maintained above to balance the air pressure in the caisson. Into the shaft well was pumped all excavated material, which was then removed with a clamshell dredge operating through the shaft. More than 20,000 cubic yards of earth and stone were removed from the Brooklyn caisson in this manner.

Six compressors located on the bank supplied the caissons with air, which reached a maximum pressure of 23 pounds per square inch above normal atmospheric pressure. Early in the excavation work frequent blowouts occurred because of the extreme pressure imbalance. In one case the water shaft was left unsealed, and the resulting shower of mud and stones caused the caisson to settle an additional 10 inches.

The first men entered the air lock of the entrance shaft on May 10, 1870. Working in three 8-hour shifts, using only steel bars to pry material loose, the work crews made slow but steady progress. Explosives were tested and subsequently used successfully in excavation. As there was no electrical lighting at

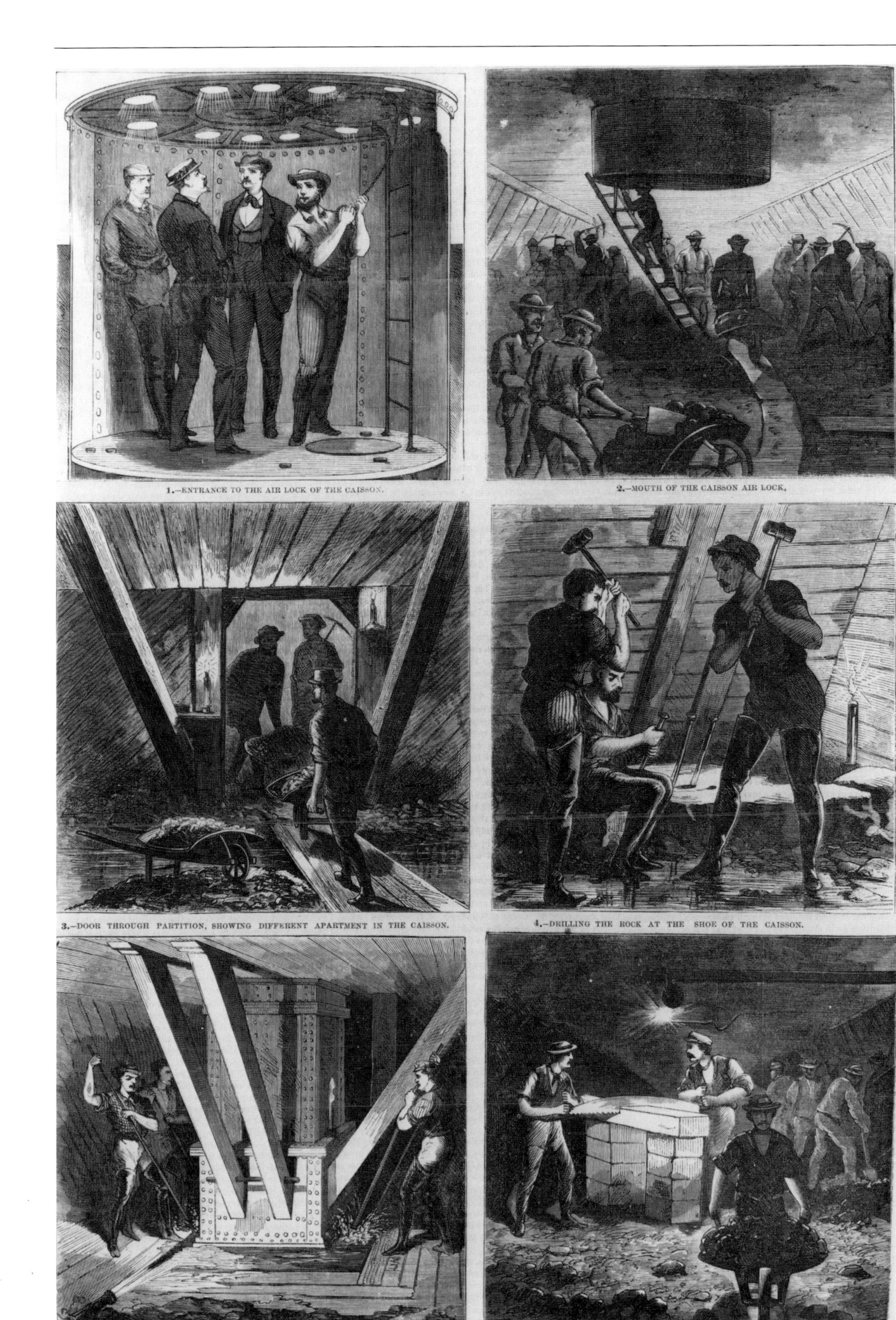

1.—ENTRANCE TO THE AIR LOCK OF THE CAISSON.

2.—MOUTH OF THE CAISSON AIR LOCK.

3.—DOOR THROUGH PARTITION, SHOWING DIFFERENT APARTMENT IN THE CAISSON.

4.—DRILLING THE ROCK AT THE SHOE OF THE CAISSON.

From the *Scientific American*, November 12, 1870: "We give herewith some engravings, showing various operations in the interior of the caisson at the Brooklyn terminus of the East River Bridge. This caisson is now only nine feet from its permanent bed, and the sinking is progressing at the rate of about one foot per week. . . . Three gangs of men—one hundred in each gang—and working eight hours each, are employed in the caisson, and the work proceeds without intermission night and day. . . . Fig. 1 represents the entrance to the caisson. It is a hollow iron shaft, having a vestibule or chamber communicating with the external air through a hatchway. . . . The lower end of the entrance shaft or 'air lock' . . . is shown in Fig. 2. The caisson in its descent requires the removal of a hard yellow clay in which are embedded large boulders which have to be broken up by blasting. Fig. 4 shows the workmen drilling one of these boulders situated under the shoe of the caisson. . . . The interior of the caisson is separated by partitions into chambers, from either of which the water may be expelled independently of the others. Fig. 3 represents the interior of one of these chambers and the door which leads from it to an adjoining chamber of similar character. Through these doors the broken stone and soil are wheeled over walk-ways to the mouth of the water shaft at the bottom, shown in Fig. 5. . . . The caisson with its load of masonry, now weight upwards of 20,000 tuns [*sic*], does not rest upon its shoe, and is only in part sustained by its floating power. The greater portion of the weight is sustained by timber frames, the uprights of which are sustained by blocks and wedges. To lower the caisson, the wedges are driven partly out, and as the impact of the enormous weight in its descent often crushes the blocks and wedges, it is necessary to supply their place by new ones. This necessitates the use of considerable timber, which is sawn by hand in the interior of the caisson, as shown in Fig. 6."

this time, various methods were tried to illuminate the work area. Candles, oil lamps, gas lights, and calcium lights were used at various times throughout the project.

The great danger was fire, and the caisson did indeed catch fire several times. The worst such incident occurred on December 2, 1870, when a candle left too close to the timber roof ignited the oakum caulking. The air pressure differential rapidly drove the fire through the caisson wall. When carbon dioxide, water, and steam had been applied without effect, Washington Roebling himself descended into the caisson and fought the flames for seven hours in the crushing pressure of the chamber. He was brought out unconscious, with the fire still consuming the timber structure. It was soon clear that the only recourse was to flood the entire caisson. This measure worked. After three days, air pumps expelled the water and repair work began. The masonry was unharmed, but it took over three months to repair the timber sections. Following this catastrophe, calcium lights and gas burners were used, and in the New York caisson interior walls were lined with boiler iron to guard against fire.

When excavation beneath the Brooklyn caisson had been completed, concrete was poured into the working chambers and laid in layers of 6 to 8 inches. The caisson was soon filled, and the setting of the pier masonry proceeded rapidly until the tower was completed. The New York pier was then begun.

Similar methods were used in constructing the New York pier, but its greater depth made the caisson for this founding 4 feet longer than the Brooklyn caisson. Its roof was 22 feet thick instead of 15, to compensate for the greater depth and weight. In this caisson the air pressure reached 34 pounds per square inch above atmospheric pressure. Special precautions were taken to avoid decompression sickness, or "caisson disease," as it was called among the workers. With a too-sudden lowering of pressure, nitrogen bubbles form in the blood and body tissues, causing pain and, in extreme cases, convulsions and collapse. It was an affliction already too well known to Captain James Eads from his work in the founding of the piers and abutments for the great Mississippi bridge at St. Louis, which had begun several months before the East River work.

A cross section of the caisson used in founding the New York pier. Workmen may be seen in the chamber at the lower left. (Library of Congress)

Pier construction. (Courtesy of the Rensselaer Polytechnic Institute Archives)

Washington Roebling worked continually in the caisson, directing the digging for all three 8-hour shifts. In the spring of 1872 he collapsed in the caisson, a victim of decompression sickness. He was left partially paralyzed and lost the use of his voice. He continued to direct the work from his home in Brooklyn Heights, where he could see the bridge. In preparing detailed drawings and written directions for the complicated proceedings, he was helped immensely by his wife, Emily Warren Roebling, who undertook a study of mathematics, strength of materials, catenary curves, cable construction, and other topics, until she was quite familiar with the demands that lay ahead. She was soon making daily inspection visits to the site. Emily Roebling became her husband's co-worker, principal assistant, and contact with others. With her help, and with the cooperation of his assistant engineers, Washington Roebling was able to maintain direction of the great project. He was often observed on the balcony of his home, examining the construction through field glasses.

When the rock ledge beneath the New York caisson had been reached and the founding prepared, concrete was poured into the caisson as before. The masonry tower was then raised to its full height, and cable construction could begin.

The manufacturing of the steel wire for the cables is illustrated in these contemporary engravings. Steel was received at the factory in South Brooklyn in the form of coiled 1/4-inch rods. Each coil was brought to a forge, where one end was heated and hammered to a point by hand. After being cleaned in vats of dilute sulfuric acid and dashed with a mixture of lime and water to arrest the action of the acid, it was ready for the drawing process. The draw plate was simply a piece of very hard steel with several carefully made conical holes of graded sizes; this was affixed to the drawing bench. The point of a coil was inserted through a hole just smaller than the rod, gripped by nippers, and attached to a great rotating cylinder. The rod was then drawn through the hole with the aid of the powerful turning cylinder. Drawing it in the same way through successively smaller holes (normally two or three times) produced wire of the right size. (*Scientific American*, March 3, 1877)

The cables used on the Brooklyn Bridge are remarkable not only for their immense size but also because they represent the first use of galvanized steel wire in cable construction. In September 1876 the New York Bridge Company adopted a resolution prohibiting bids from any company having a direct link with one of the bridge officers or engineers. Washington Roebling sold his stock in his family's wire-manufacturing plant at Trenton to allow it to bid on the project. At this time controversy existed as to whether Bessemer or the more expensive crucible steel was preferable for the purpose. Bids were received for the wire, with that by the firm of John A. Roebling's Sons being the lowest for Bessemer steel wire. J. Lloyd Haigh of Brooklyn had the next-lowest bid, but for crucible cast-steel wire. Based on a report written by Abram Hewitt, the company decided to award

After the wire was drawn, it was galvanized by being led over rollers through a bath of dilute muriatic acid heavily charged with zinc. Samples were then tested for strength. At the construction site the coils were dipped in oil, dried in air, and then dipped again and again until coated with a thick layer of hardened grease. (*Scientific American*, March 3, 1877)

Cable wire. (Courtesy of the Rensselaer Polytechnic Institute Archives)

the contract to Haigh. Hewitt, although a friend of Haigh, enjoyed a reputation as an honest and crusading liberal. Yet Washington Roebling is known to have had reservations about Hewitt. He noted that the steel to be furnished by Haigh was in fact the cheaper Bessemer steel, and that Hewitt held the mortgage on Haigh's property and had agreed not to foreclose it as long as Haigh paid him 10 percent of the payments for wire for the bridge.

The first wire cable was drawn across the river on May 29, 1877. A 4-foot-wide footbridge had been completed earlier that spring to aid in construction. A "traveler" rope was taken across the river on a scow. This rope was hoisted up to its proper position between the towers. A second traveler rope was put into position the same day. The ends of these two ropes were spliced together around driving and guiding wheels on the anchorages, thus forming a continuous line across the river. Two additional traveler ropes were then drawn across, followed by the carrier cables, three cradle cables, two footbridge ropes, one auxiliary rope, two storm cables, two handrail ropes, and four pendulum ropes. With these preliminaries complete, cable spinning could begin. From each of the continuous traveling ropes two traveling sheaves were suspended by a gooseneck pendulum bar. Every time the traveling sheave traversed the river, it carried a loop of two wires from the Brooklyn anchorage to the New York anchorage. This sheave returned empty to Brooklyn as the other sheave was traveling with two wires to New York. This process was repeated many times over.

In July 1878 Roebling discovered that despite prior testing and inspection of the wire supplied, faulty wire was nonetheless being included in the cables. The wire was normally inspected at Haigh's plant in Brooklyn and was consequently assumed to be good when it was installed. A workman noticed, however, that the pile of rejected wire never grew very large. A second man was set to watching the inspection and delivery process; he saw a wagonload of wire bearing the inspector's acceptance certificate leave the plant and go not to the bridge but to another building, where the good wire was unloaded and

The wire strands were carried between opposite shores individually by means of an engine-driven carrier wire loop. Attached "traveling wheels" paid out the wire between anchorages. To make the curvature of a newly placed wire correspond to that of the wires already present, it was passed over the great saddles, stopped, and a block and tackle attached to adjust its sag. After enough wires were in place to form a strand, a suspended "buggy" with two men traveled along the wires, binding them into bundles. (*Scientific American*, July 4, 1877)

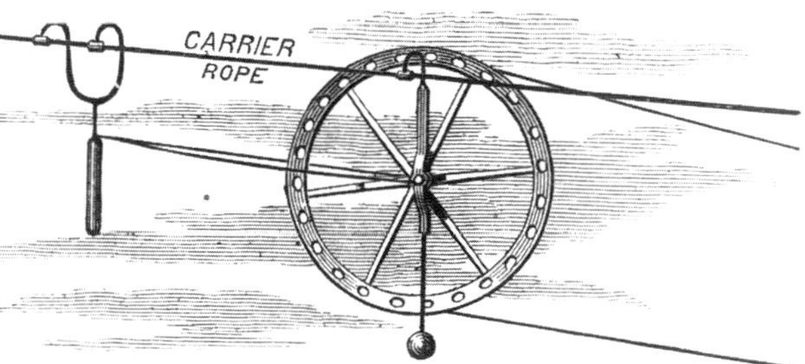

replaced with rejected wire, which was subsequently taken to the bridge and incorporated into the cables. The good wire was taken back to the plant to be recycled in front of the inspectors.

The substitution of faulty wire in the bridge was a blow to Roebling. The question then was what to do about it. Removing the wire already in place would prove time-consuming and costly. In the end Roebling decided to insist that the contractor place additional wires in the cables at his own expense to make up for the estimated number of faulty wires. This requirement meant that the cables ended up being much larger than initially envisioned. The need to have the wrapped strands form a compact circular shape dictated that either seven or nineteen strands be used. Thus the final cable was composed of nineteen strands.

Cable laying was resumed. On October 5, 1878, the last wire was run across. Eventually nineteen strands of 286 wires each (along with some additional wires needed for compactness) went into the construction of each cable. First the seven inner strands were assembled into a 9-inch cable and wrapped with wire lashings at 10-inch intervals. This facilitated final assembly and wrapping after twelve additional strands per cable were completed. Each of the four cables contains approximately 5,439 parallel, galvanized-steel, oil-coated wires, wrapped into a compact cylinder about 15.75 inches in diameter.

During this year resistance to the bridge had again mounted. Costs were running far beyond what had been authorized. A group of citizens, spurred on by the New York Council of Reform, filed a suit to halt construction in order to prevent further expenditures. Litigation followed, and in November 1878 work on the bridge stopped because of lack of funds. Roebling struggled to retain his corps of trained engineers. Six months after the shutdown some funding was again available and work resumed. But controversy persisted.

Vertical cable suspenders were attached to the great draped cables with large clamps. The lower ends of these vertical cables were then attached to the floor beams or girders that support the bridge floor. Massive anchorages for the cables had to be especially designed, as there were no natural rock formations to secure them. Constructed of limestone, the anchorages serve the additional purpose of forming part of the approaches to the bridge. Until the cables were finished, the masonry of the anchorages was left incomplete, built up as far as the last sections of the anchor chains that would be embedded in the masonry pile to secure the great cables. The approaches themselves are brick arches, faced with cut granite and resting on a limestone foundation. Like the towers, these approach ramps continue to be admired as models of elegance in design.

The floor system was completed on December 10, 1881. Composed of a single deck, it is divided to provide two elevated railroad tracks, two trolley car tracks, single-lane roadways flanking the trolley tracks, and a central walkway 15 feet wide. The entire breadth of the floor system is 85 feet. Recent renovation has expanded the roadway sections. Stiffening trusses were added to insure against the sway characteristic of the suspension structure. These trusses form a deep framework following the contour of the bridge floor as it arcs across the East River.

Owing to increased design loads, many of the floor beams and stiffening trusses had to be increased in size over the dimensions originally envisioned, which further increased costs. This led to new controversy and charges of incompetence on the part of the engineer. Newspaper articles assailed Washington Roebling and suggested that blunders had been made, or that possibly someone's pockets were being lined. The important technical magazine *Engineering News* came out with an editorial strongly defending the project—noting that fourteen years had elapsed since the original design and that the new loadings (such as thirty-ton Pullman cars) were simply unanticipated at the time. Indeed, the editorial went on, what was remarkable was that the elder Roebling had foresightedly designed a cable structure capable of meeting these new conditions. Nonetheless, criticism continued. Washington Roebling's physical impairments were used against him. At a meeting in August 1882 Mayor Seth Low of Brooklyn introduced a

Note Washington Roebling's stern admonition regarding the use of the footbridge. Several city dignitaries on an inspection tour are shown in this early, but undated, photograph. (Courtesy of the Photo Library Department, Museum of the City of New York)

The footbridge, part of the anchor bars, and a finished strand of cable ready to be lowered into position. This photograph was taken on October 25, 1878, by G. W. Pach. (Courtesy of the Photo Library Department, Museum of the City of New York)

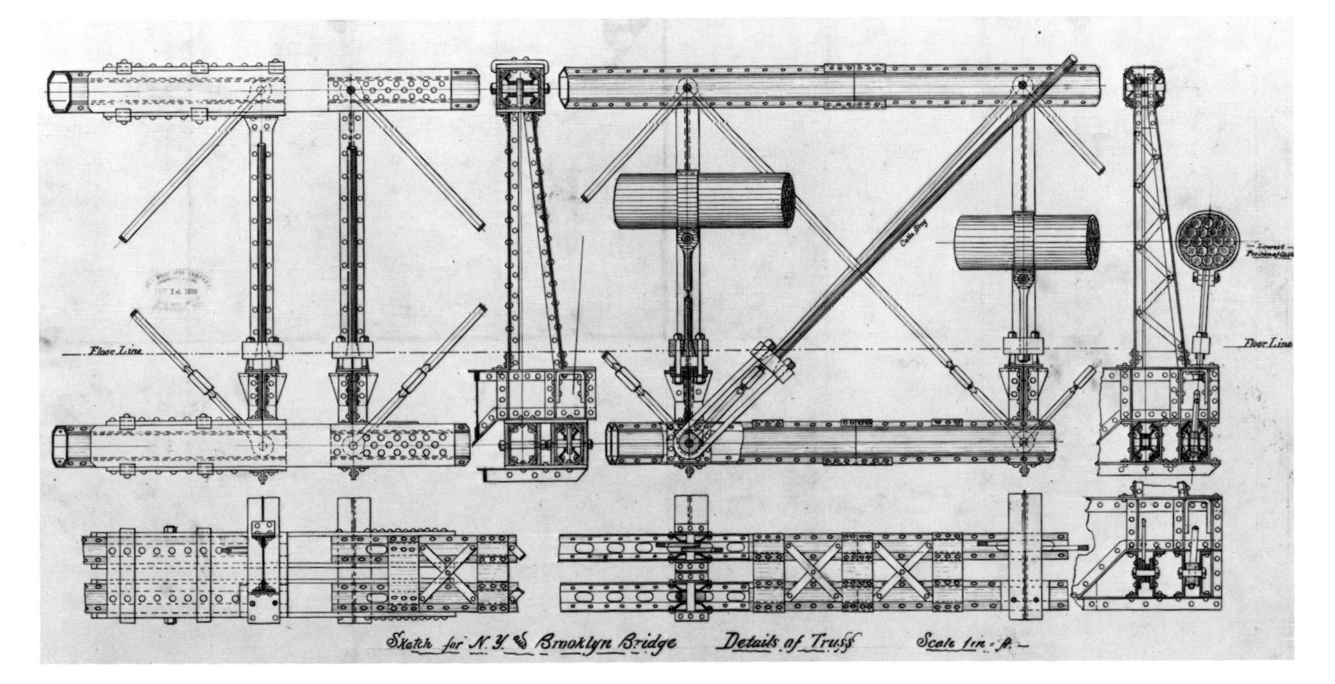

Details of the truss. (Courtesy of the Rensselaer Polytechnic Institute Archives)

resolution demanding the replacement of Roebling on the basis of his being an invalid. Though confined to his home, the engineer fought back. At a meeting of the American Society of Civil Engineers, Emily Roebling read a carefully prepared statement arguing her husband's case. (This was the first time a woman had ever addressed this body.) The presentation was highly successful and the confidence of the engineering profession gained. Public confidence was soon restored. Washington Roebling continued to direct the work until its completion.

In the spring of 1883, after fourteen years of work, the Brooklyn Bridge was completed. The central span of 1,595.5 feet, the two side spans of 930 feet, and the approaches reached a distance of 5,989 feet. The total cost of the structure was almost $9,000,000.

The bridge opened officially on May 24, 1883, amid ceremonies attended by President Chester A. Arthur and his cabinet, the governors of several surrounding states, and mayors from nearby cities. William Kingsley, now president of the bridge company, presented the bridge to the cities of New York and Brooklyn. Abram Hewitt gave an oration. The structure was applauded as a "symbol of courage, endurance, and faith." Although Roebling was unable to attend the ceremonies, later that evening a procession of citizens and officials marched to his home to honor the man who had sacrificed his health to carry his father's dream to fruition.

Today the Brooklyn Bridge is among the best-known of all civil engineering accomplishments. Indeed, the hundredth anniversary of its opening, May 24, 1983, was cause for great celebration within the city of New York. Over 750,000 people turned out for a grand parade led by the mayor, Ed Koch. Citizens donned Brooklyn Bridge hats and Brooklyn Bridge T-shirts—a species of tribute that very few civil engineering artifacts might command.

A recent view of the Brooklyn Bridge. (Library of Congress)

The Stone Arch Bridge at Minneapolis

The Burlington Northern Railway's Stone Arch Bridge at Minneapolis, Minnesota, is considered one of the finest stone viaducts in the world. It was designed and constructed in 1882–1883 under the direction of Colonel Charles C. Smith, at a time when masonry structures of this type were not common—indeed, in this particular area of the country, were virtually unknown. In carrying the railway's main transcontinental passenger line across the Mississippi, the Stone Arch Bridge played an important role in the development of the northwest portion of the United States. The bridge remains important and unique to this day.

Minneapolis in the late 1800s was a small city, but it showed promise of continued growth. A suspension bridge with a 550-foot span across the main channel of the Mississippi was built in 1855 and rebuilt in 1876 (it was eventually removed in 1889). By 1882 the need for a mainline railroad bridge had become pressing. The St. Paul, Minneapolis, and Manitoba Railroad—a predecessor of the Burlington Northern—guaranteed $3 million in bonds to furnish the city with new rail connections and to build a bridge across the Mississippi. James J. Hill, president of the company, saw tremendous potential in greatly increased east-west traffic as part of America's westward movement. The construction of the bridge was placed under the direction of Colonel Charles C. Smith, chief engineer of the Minneapolis Union Railway Company (under whose auspices the bridge was eventually built).

The site for the bridge was chosen for its proximity to Minneapolis and its centrality in relation to existing transportation routes. Just upstream from the Falls of Saint Anthony (named by Father Louis Hennepin, who discovered them in 1680), the place was then occupied by a flour mill. This particular orientation required the bridge to be built at an angle to the falls. Construction cost estimates soared as Hill decided to build it of solid stone and to incorporate a curve into the design. The unique configuration, as well as special engineering problems involving the foundations, contributed to the high costs.

Stone Arch Bridge, Minneapolis. This mainline railroad bridge over the Mississippi, constructed in 1882–1883, was a key element in the development of the northwestern part of the country, as well as of Minneapolis itself. The bridge contains 23 circular stone arch spans. (Courtesy of the American Society of Civil Engineers)

The Stone Arch Bridge under construction, 1883. (Courtesy of the Burlington Northern Railway)

The bridge is a double-track structure, 2,100 feet long, 76 feet high, 26 to 28 feet wide, and boasts 23 circular stone arches spanning lengths varying from 40 to 100 feet. The alignment of the eastern portion of the structure is tangent for 1,283 feet, crossing the line of the river current at an angle, and the remaining 817-foot western segment is constructed along a 6-degree curve. This unique and graceful combination of length and curve enables the train passenger to view the bridge while traveling on it.

The end piers are founded on a limestone ledge, the intermediate piers on solid sandstone that lies below the limestone ledge. The masonry in the foundations of the piers, up to the low-water mark, is of portland cement concrete, in which limestone obtained from a quarry along the east bank of the river was used as aggregate. Piers above the footings, extending from the low-water mark to the springing line of the arches, are of Minnesota granite. Above the springing line all of the exposed work is of magnesium limestone acquired from Minnesota, Iowa, and Wisconsin. Native limestone from the quarry at the site of the bridge was used for backing and rock filling. In all, the $690,000 bridge required 30,554 cubic yards of masonry and 18,000 cubic yards of filling for its initial construction.

The bridge was repaired and reinforced between 1907 and 1911, an operation required because of inadequate provision for drainage in the original construction. Reinforced-concrete arch rings were added to the top of the stone arch rings throughout the length of the structure, accompanied by the installation of new pipe drains. In 1925 track centers were increased 6 inches on curves, made necessary by the introduction of larger locomotives. This change also required 10 inches to be cut off from the stone copings at the top of the spandrel walls to ensure proper side clearance.

Additional repairs were made in 1965 as a result of damage caused by the record Mississippi floods of that year. On April 18, 1965, a crew taking a train across the bridge felt a jolt as a section of the bridge settled some 14 inches beneath the train. Three piers proved to be extremely weakened by the flood. Scour had washed away the footing of one pier and severely undercut another. The settlement caused the opening of many masonry joints and the falling out of a number of stones. Fortunately, critical keystones remained in place, although badly loosened. The bridge was in imminent danger of collapse. The Turzillo Contracting Company of Brecksville, Ohio, was charged with repairing the structure. Using no power tools lest the vibrations bring the bridge down, the company slowly and carefully performed the repairs. Subsurface work was carried out by divers. Holes beneath footings were filled with concrete by using a device known as a bagpipe. Deflated bags of porous canvas, custom sized for each cavity, were inserted into holes and then pumped full of grout. Excess water was squeezed through the canvas and the grout left to harden. Thirty bags were eventually used. Subsequently, each pier was ringed with drilled-in-place reinforced-concrete piers. Meanwhile cables were wrapped around loosened arch stones to keep them in place until the weakened arches could be strengthened by a 2-foot-thick underlay of reinforced concrete. This construction was performed at a cost of $1.5 million. The bridge is now in regular service.

Although it seemed extravagant at the time, James Hill's insistence on hand-laid solid stone and an artful curve, like his vision of American expansion westward, seems to have paid off. A marvel when constructed in 1882–1883, the Stone Arch Bridge with its massive masonry and graceful arches is a wonderful survivor of an ambitious era.

The Alvord Lake Bridge

The Alvord Lake Bridge in San Francisco's Golden Gate Park, built in 1889 by Ernest L. Ransome, was the first significant reinforced-concrete bridge in the United States. Although relatively small in scale, it served as a model for much larger structures. In its construction, Ransome—one of the most important American pioneers in the development of reinforced-concrete construction techniques—employed the "Ransome System," a technique involving twisted-bar reinforcement that he had invented and patented. The bridge, built for the purpose of carrying one park road over another, has an arch span of 20 feet and a width of 64 feet; the barrel of the arch is actually a flattened vault.

Ernest Ransome built many projects in the years 1880–1890. Ransome's career began with an apprenticeship in his father's ironworks in England. In 1844 his father, Frederick, patented a new type of concrete based on the heat treatment of flint stones in an alkali solution. Subsequently Frederick Ransome established the Patent Concrete Stone Company to produce the concrete. Ernest Ransome came to America in the late 1860s and helped promote his father's new product. With his knowledge of concrete, he soon became superintendent of the Pacific Stone Company in San Francisco.

In 1882 Ransome obtained a patent for a system of expansion joints to be used in plain concrete arches that spanned between iron beams (a common form of construction at the time). He used reinforcing in concrete for the first time in 1883, during the construction of a sidewalk in Stockton, California. Reinforced with smooth rods, the sidewalk soon cracked up. Believing the problem to be one of inadequate bonding between the concrete and the rods, Ransome developed a system of twisted square bars, which he patented in 1884 and used extensively throughout his later works.

In 1884 Ransome built the storage buildings of the Arctic Oil Works in San Francisco out of reinforced concrete, using his newly developed system. A flour mill for Starr and Company at Wheatport, California, followed

Alvord Lake Bridge, Golden Gate Park, San Francisco. Built in 1889 by New York's Ernest L. Ransome, this is believed to be the first major concrete arch bridge with steel reinforcing bars built in the United States. (HAER Collection, Library of Congress)

(1885). His design for the main floor of the Bourn and Wise Wine Cellar at St. Helena (1888) was extremely successful, and his reputation grew. His Academy of Sciences building in San Francisco (1889) was a major project in which he used reinforced concrete extensively. He constructed an addition to the Borax Works at Alameda (1889) that was notable because of the way he cast the slabs, beams, and joists together monolithically. This was the first example of a ribbed reinforced-concrete floor system in America—a construction approach now commonplace. By the time he built the Alvord Lake Bridge in 1889, Ransome was an accomplished designer and thoroughly familiar with reinforced-concrete construction.

Thirty years before Ransome built America's first reinforced-concrete bridge, a plain concrete arch bridge was constructed in France to carry an aqueduct of the Paris waterworks. The first plain concrete bridge in the United States was built in Brooklyn, New York, in 1871. The first European reinforced-concrete arch bridge was built in Copenhagen ten years before the Alvord Lake Bridge. As for Ransome's structure, it is concrete throughout—abutments, wing walls, and rails, as well as the arch—and its reinforcing consists of beds of longitudinal bars set near the lower and upper surfaces of the barrel and bent in approximately the same curve as the archway. There is no masonry facing on the bridge. The surface was carefully scored and hammered to imitate rough stone blocks. (In his later reinforced-concrete buildings Ransome would get away from the practice of imitating masonry and would be more honest in his treatment of the material.) A curious feature of the Alvord Lake Bridge is the artificial stalactites found hanging from the arch, perhaps to suggest a cave in the park setting. The bridge is still in use today.

After designing the Alvord Lake Bridge, Ransome continued his career with many more notable feats of reinforced-concrete construction. Two buildings for the Leland Stanford Junior University at Palo Alto were next executed in 1892—the University Museum and a women's dormitory. Later works included the factory for the Borax Company at Bayonne, New Jersey (1898); an office building for the Kelly and Jones Company in Greensburg, Pennsylvania (1904); and an office building for the Foster-Armstrong Company at East Rochester, New York (1905). While constructing these and other buildings, Ransome continued to devise new construction methods using reinforced concrete. Among his later inventions was a method of casting floor slab, window lintel, and sill pieces together, greatly enlarging window area and representing an extreme departure from earlier solid concrete slab walls. Another of Ransome's major inventions was the development of "unit" construction, whereby precast structural elements are placed in position and poured-in-place floor and ceiling slabs are integrally united with them. This evolved into a complete system of prefabricated parts to be assembled on site, called "The Ransome System of Unit Construction."

The Cortland Street Drawbridge

Completed in 1902 and still in use, the Cortland Street Drawbridge in Chicago was the first "Chicago-type" trunnion-bascule bridge constructed and opened to traffic in the United States. It became a model for this important type of urban transportation structure.

Early settlers reaching the Chicago River established ferries, which soon proved inadequate for the growing population. A crude float bridge manipulated by cables was installed in 1833; the following year a wooden drawbridge, similar to a double moat bridge, was built. The latter became a menace to navigation and was torn down in 1839. Another bridge, this time of the pontoon type that could be pulled open, was built in 1840, and soon there were several pontoon bridges in Chicago. In 1854 a pivot swing bridge with wooden trusses was constructed. An all-iron bridge followed in 1856. At the time of the great Chicago fire in 1871, there were no less than 27 movable bridges in use (the fire destroyed 8 of them). By 1890 there were 48 within the limits of the city.

The swing-type bridges in common use proved to have disadvantages—the center pier in particular was troublesome because it obstructed navigation. Attention was constantly directed toward devising new bridge types that would allow river traffic to flow unhindered but yet be easy and reliable to operate. A form of "jackknife" bridge was built in 1890 under patents issued to Captain W. Harmon. Another jackknife bridge soon followed, replacing a swing bridge at Canal Street that was only two years old, because of objections by navigation interests to the latter's center pier. Yet the Harmon bridge proved troublesome, due largely to light construction and poor execution.

In 1892 navigation interests joined forces with the Drainage Board (later the Sanitary District of Chicago) to prevent the construction of a swing bridge at South Halsted Street. After consideration of both the Harmon bridge and the Waddell lift bridge, the city decided in favor of the latter—a complex arrangement consisting of a 130-foot steel

Cortland Street Drawbridge, Chicago. Completed in 1902 and still in use, this trunnion-bascule highway bridge was the first of its kind completed in this country and became a model for this type of movable bridge. (Courtesy of the American Society of Civil Engineers)

truss span that was lifted vertically between two towers by steam-powered hoisting machinery. A Scherzer-type rolling lift bridge was built across the Chicago River at West Van Buren Street in 1894. This bridge combined the bascule—a type of moat bridge, common since medieval times, that relied on counterweights—with recent innovations in electrical and pneumatic devices. Another followed at North Halsted Street.

Thus in 1895 three new types of bridges—the Harmon, the Waddell, and the Scherzer—were in place for comparison. This was fortunate, because many of the existing bridges were in need of repair or replacement. Navigation was still hampered by the center piers of existing swing bridges. Moreover, the law creating the Sanitary District also called for a discharge capacity through the river of 10,000 cubic feet, which meant replacing many bridges in order to increase the flow (see chapter 7, Reversal of the Chicago River). In 1899 the Division of Bridges and Viaducts of the Chicago Department of Public Works organized a team of engineers to make a critical analysis of all literature on movable bridges in the United States and abroad, with the objective of identifying the type most suitable for use in Chicago. The resulting report discussed the relative advantages and disadvantages of existing types and concluded by recommending a type identified as the trunnion bascule. *Bascule* is a French word meaning "seesaw" and hence any device whose operation depends on the balance of weights at both ends of a beam supported on a fulcrum; in a bascule bridge the principle of counterbalance is of fundamental importance. The trunnion is the axle around which the bridge pivots.

Three designs were prepared by the division's staff and submitted for review to a board of consulting engineers consisting of E. L. Cooley, Ralph Modjeski, and Byron Carter. While differing in appearance and specific details, each design was based on the principle of revolving counterbalanced leaves on fixed axles, or trunnions. The board selected the design designated as number 3 with some modifications. The proposed structure was described in a 1900 report as a fixed-center, double-leaf, counterbalanced bascule bridge. Each leaf had three through trusses operated by electric motors that drove pinions geared with racks on the trusses' curved tail ends, the tail ends descending into a pocket or tail pit in the abutments when the bridge was open.

In opening, such a bridge rotates about its axle and raises its roadway to a nearly vertical position, giving a clear, open river passage. The weight of the roadway that extends over the river is generally counterbalanced by weights on the shore side of the axle. Because the river arm is longer than the shore arm, the counterbalancing weight must be fairly massive. Obviously, friction and other phenomena, including structural deformations and temperature effects, work against the achieving of perfect balance, but the counterweight principle does enable the bridge to be raised and lowered with relatively little motive power. Such a bridge is raised or lowered by the operation of a rack and pinion system. The rack is attached to the rear end of each truss. The pinions that engage the racks are operated by an electric motor and machinery controlled by an operator located in a small house near the river bank.

The proposed bridge was designed to turn through a maximum angle of 76°58′. The machinery for operating each leaf was under the approach roadway. Along the top of each abutment a shaft carried three pinions and had two sets of driving gears. Each set was driven by a 38-horsepower electric motor. On each shaft was a wheel for a hand brake that could be operated from the bridge house. There was also a worm-resistance brake at the middle of the pinion shaft. A hand-brake wheel was fastened to the shafts of the worm. The machinery was designed so that the opening of the bridge would not require more than 1 minute in calm weather or 2½ minutes in high winds. Pneumatic buffers arrested the motion of the leaf at each end. When the bridge was closed, the center of gravity of each leaf was on the river side of the pivot, but near the trunnion axis; thus it was necessary only to lock the leaf against further downward movement. This was done with bridge-closing buffers.

On August 17, 1899, the swing bridge at 95th Street over the Calumet River collapsed. This location was specified to be the site for a new bridge. About the same time a call for bids was advertised for another new bridge at Division Street over the North Branch Canal. The specifications were broadly drawn for both bridges, to allow competition among designers advocating different bridge types, but the fundamental principle of movability was there. Five designs were submitted, four of them for bridges revolving on fixed bearings. The contract was awarded to a bidder who had followed one of the city's designs. Work started in July 1900 on both of these bridges.

In the meantime the city's design number 3 was modified for a third bridge, over the North Branch Canal at Clybourn Place, later renamed Cortland Street. Contracts were awarded in the late fall. The plans reflected the modifications suggested by the consulting engineers, including the use of a full tail pit rather than separate wells for each truss. Construction proceeded at all three bridge sites, but difficulties developed at the 95th Street and Division Street sites. Consequently, the Clybourn Place, or Cortland Street, bridge was completed first and opened to traffic on May 24, 1902. The other bridges opened soon afterward.

Experience proved the Cortland Street design to be a good one. Modifications continued to be made with time, but the basic design remained largely unchanged. Since the opening of the Cortland Street bridge in 1902, the trunnion bascule has been repeatedly selected in Chicago whenever movable bridges were needed: 50 out of the 53 movable bridges since constructed have been of this type.

The Tunkhannock Viaduct

The Tunkhannock Viaduct, the largest reinforced-concrete structure of its kind in the world, was completed in 1915 at Nicholson, Pennsylvania, as part of a major readjustment in grade and alignment on the Erie-Lackawanna Railroad (later renamed the Delaware, Lackawanna, and Western). Before reconstruction the Erie-Lackawanna was a grade-ridden railway, winding along waterways through the difficult and mountainous terrain of the Appalachians between New York and Buffalo. As many as five engines were often required to haul loaded freight cars up the steep grades. The new plan straightened curves and minimized grade changes by abandoning the traditional practice of building along waterways. Instead the route was built at right angles to the natural drainage pattern of the country through massive cut-and-fill operations. By reducing the impediments of distance and grade, the railway was able to lower its operating costs.

This ambitious program was initiated by William H. Truesdale, who became president of the railroad in 1899. Truesdale put Lincoln Bush, chief engineer for the railroad, in charge of the reconstruction. Although Bush resigned from this position in 1908 to do consulting work in New York, he returned as the general contractor for the construction of the Tunkhannock Viaduct, working with Meyer Hirschthal and then with George Ray, his successor as chief engineer.

The revamping of the Erie-Lackawanna route involved two major building projects, the New Jersey cutoff and the Hallstead cutoff. Lincoln Bush planned the New Jersey cutoff, but George Ray finished the project, which was distinguished by its two viaducts at the west end. These large reinforced-concrete structures set a precedent for the Tunkhannock Viaduct, part of the Hallstead cutoff running from Clark's Summit, near Scranton, to Hallstead. The cutoff reduced the distance between these towns from 43.2 track miles to 39.6, as well as reducing curves and grades that had added to travel time; over 20 minutes was shaved from most

The Tunkhannock Viaduct, Nicholson, Pennsylvania. This structure, the largest of its kind ever built of reinforced concrete, was completed in 1915 by the Delaware, Lackawanna, and Western Railroad. Not only a great feat of construction skill, the Tunkhannock Viaduct represents a bold and successful departure from contemporary, conventional concepts of railroad location in that it carried a main line transversely to the regional drainage pattern, effectively reducing both distance and grade impediments to economic operation. (Library of Congress)

passenger schedules, and freight portage improved by over an hour. In addition, trains that would normally require five engines could make the trip between Scranton and Binghamton with only two.

The major features of the Hallstead cutoff are the two large viaducts at Tunkhannock Creek and Martins Creek; a massive cut near Clark's Summit, 2 miles long and from 20 to 60 feet deep; a fill operation near Dalton, 115 feet high; and an embankment near the south branch of Tunkhannock Creek that is 140 feet high and 2,000 feet long and contains 1 million cubic yards of fill, with a double-track tunnel 3,630 feet long.

The Hallstead cutoff involved some of the heaviest grading and concrete work ever undertaken. (To begin with, there were no government contour maps of the rugged terrain to be traversed, so the task of simply locating the new line through the hills proved tedious.) A line without grades or curves involved literally moving mountains. Crossing the valley at Tunkhannock Creek was one of the many obstacles facing the planners. The double tracks of the new line would traverse the valley 240 feet above the stream bed. The use of a concrete viaduct to span the valley was decided upon only after an extended study of the problem.

Reinforced concrete was at the time a more expensive material than steel, and hence was not often employed by railroad engineers. It was used to construct the viaducts of the Hallstead cutoff for two primary reasons: the planners believed that a concrete bridge could be more efficiently maintained, and they were also committed to building the best line possible, regardless of cost. The perceived strength and permanency of a reinforced-concrete viaduct were thus attractive. The publicity value of the undertaking was not overlooked: large advertisements appeared in the *New York Times,* characterizing the viaduct at Tunkhannock Creek as the ninth wonder of the world.

The contract for the Tunkhannock Viaduct was awarded to the firm of Flickwir and Bush in 1912. Soon a steam shovel was beginning the foundation excavation. All of the thirteen piers were carried to bedrock, some as much as 95 feet below ground level. Two of the

Construction work on the Tunkhannock Viaduct. (Courtesy of Syracuse University)

piers were especially difficult to set because of quicksand. In one case the problem was solved by dividing the last 12 feet of the excavation (to a total depth of 62 feet) into three pockets and taking each pocket down separately. For the other pier this tactic was unsuccessful, and it became necessary to use compressed air. A caisson was constructed inside the excavation sheeting, with its cutting edge 45 feet below the top of the excavation and 32 feet above the bedrock. Two lower sets of bracing timbers were built into the caisson and were carried down with it. Concrete was then placed on top of the caisson as it sank in the usual manner. The difficulties in excavating for this pier set the project back several weeks.

One of the more striking aspects of construction was the utilization of a cablecar system to carry material across the valley to the piers and archways once construction had progressed above the reach of ground-level derricks. In a few instances derricks were erected on pedestals or pile clusters about 25 feet high, but most of the high work was carried on by means of the cableway. This consisted of two lines of $2\frac{1}{4}$-inch main cable, spaced 20 feet apart; it was supported by two end towers approximately 160 feet high and an intermediate tower nearly 300 feet high. The end towers were 3,028 feet apart, the intermediate tower centrally located between them. There were four engines, which operated independently of one another. The towers were of timber, with the central tower located and designed to facilitate the arch ring and floor construction of those final centers that passed through it. Adjacent piers were carried up to the top of the floor before these centers were erected, and derricks were located on each pier. After the last steel centers were erected, all construction material was raised above them by derricks and then transferred to the cableway for transportation.

The ten 180-foot arches of the viaduct utilized five sets of three-hinged steel arch centers during construction. Each center was composed of four ribs weighing 47 tons each, spaced 3 feet 10 inches center to center, and provided support for one of the two concrete ribs composing each span. After serving their

A recent view of the viaduct. (HAER Collection, Library of Congress)

purpose under one rib of the span, the centers were moved to the next of the twin ribs. When the span was completed, the centers were rolled back under the opening between the concrete ribs and transported to the next arch position. The two 100-foot arches at the ends of the bridge used timber centers resting on a timber tower. The two arch ribs of each arch span are 5.5 feet thick at the crown and 12 feet wide. They are spaced 22 feet on center and leave a 10-foot opening between ribs. The ribs are linked together by four reinforced-concrete struts. The deep spandrel above each large arch is pierced with eleven smaller arches.

Over 162,000 cubic yards of concrete were used in the project, as well as 2,275,000 pounds of reinforcing steel. The surface of the concrete is scored to hide construction joints and to resemble stone masonry. The imitation of masonry construction included the simulation of voussoirs on the surface of the arch rings.

The bridge attracted many sightseers in the three years it was under construction. The assemblage of men and equipment was on an unprecedented scale, and the sheer size of the undertaking, which dwarfed the town at its base, was a dramatic sight. Theodore Dreiser, who visited the nearly complete viaduct in August 1915, remarked, "It is rather odd to stand in the presence of so great a thing in the making and realize that you are looking at one of the true wonders of the world." Although two men were killed in falls during construction, none are interred in the concrete piers, as popular belief held at the time. The tale, which persists in the area, seems to add a macabre allure to this already awesome structure.

The bridge opened on November 6, 1915, with a great deal of fanfare. Many dignitaries, including the governor of Pennsylvania and the mayors of New York and Buffalo, came to ride the first train across it. Still in use today, the 245-foot-high and 2,375-foot-long Tunkhannock Viaduct stands as an example of superb execution of a traditional structural design.

The Rogue River Bridge

The Rogue River Bridge in Gold Beach, Oregon, was completed in 1931 under the general direction of Conde Balcom McCullough. The seven-span prestressed-concrete structure was one of the first of its kind and was particularly innovative in its successful demonstration of the precompression technique originally developed by the famed European engineer Freyssinet.

C. B. McCullough was the bridge engineer for the Oregon State Highway Department from 1919 until his death in 1946. Educated at Iowa State College, he had been teaching civil engineering at Oregon State College (now Oregon State University) for three years when appointed to the state office. While in state service he completed hundreds of bridges, including ten major spans and many minor ones on the Roosevelt Highway alone.

McCullough was something of a pioneer in the application of emerging reinforced-concrete technologies to the design of bridges. Among his explorations were the application of Freyssinet's technique for constructing arches, the use of the Considere hinge, and the use of ferroconcrete tied arches. The Considere hinge was a short cushioning block of heavily reinforced concrete (with embedded steel hoops) that made it possible to support long-span arches on timber pilings. Its use on short columns connecting arch ribs to the deck also reduced the restraining action of the deck on the ribs. The ferroconcrete tied arch McCullough developed was particularly innovative. In this structure the deck is suspended from overhead arches and also serves as the tie for the arches, absorbing their thrusts and thus obviating the need for massive abutments. The ties consisted of steel beams encased in concrete.

The Freyssinet construction technique that McCullough employed in building the Rogue River Bridge involved constructing arches in halves, inserting jacks at the crown, and applying precompression forces. This reduced critical secondary stresses in the fixed arches that normally occurred as a consequence of rib shortening.

McCullough's bridges are as handsome as they are soundly engineered. Many exhibit the same flair for the sensitive use of reinforced concrete that characterizes the bridges of the better-known Robert Maillart. They seem to fit into the landscape better than most other bridges—as a drive down Oregon's Roosevelt Highway will surely indicate.

The Rogue River Bridge consists of seven 230-foot two-rib arch spans. There is a three-span group with anchor piers and two intermediate elastic piers, and at each end is a two-span group with an intermediate elastic pier. All piers are on wood piles driven to a penetration of about 30 feet into a sand-gravel soil. The end abutments are on rock.

Construction of the bridge by the Freyssinet method was undertaken by the U.S. Bureau of Public Roads and the Oregon State Highway Commission for the general purpose of testing the method, which held great promise for use in connection with fixed concrete arches—essentially monolithic structures with no internal hinges or pins. Until this time the fixed concrete arch had been considered by most bridge engineers as suitable only for relatively short spans, those with high rise-to-span ratios, and where relatively incompressible foundations could be found. These limitations were considered necessary because when fixed arches are constructed normally, deformations cause changes in rib length, which in turn introduce problematic bending stresses in the arch. The longer the span is, the more severe are the bending stresses, and the harder it is to design appropriate reinforcement. The deformations themselves are caused by the compressive forces induced by the bridge's dead load, volume changes in the concrete caused by temperature expansion and contraction, and movement of piers or abutments. When it was possible to use a fixed arch, however, its great rigidity was extremely attractive in comparison with other, more flexible arch forms (such as two- or three-hinged forms, which were less sensitive to the problems noted above because of the relieving actions provided by the hinges).

Various construction devices had been developed to mitigate the effects of the secondary stresses noted above and at the same

Rogue River Bridge, Gold Beach, Oregon. Completed in 1931 as an experiment in prestressed-concrete construction sponsored by the U.S. Bureau of Public Roads and the Oregon State Highway Commission, this bridge successfully demonstrated the engineering application of the precompression technique developed by the French engineer Freyssinet. Each of its seven twin-ribbed arches spans 230 feet. (Courtesy of the American Society of Civil Engineers)

time to preserve the desirable characteristics of the fixed arch, such as its rigidity under load. Among these devices was the Freyssinet method, which consisted of introducing into an arch—by means of hydraulic jacks at one or more points in the rib—deformations that would compensate for those expected to occur from elastic and plastic shortening of the rib, shrinkage of the concrete, temperature expansion and contraction, and support movements. The method was first employed in France. In 1908 a 162-foot span was constructed, and in 1920 a span of 320 feet, with a rise-to-span ratio of 1 to 8; subsequently, at Brest, a 612-foot span was attained. In all three structures there was evidence that considerable economy had been achieved. But American engineers were slow to explore this innovative technique. McCullough's Rogue River Bridge was the first in this country to adopt it.

Construction of the Rogue River span involved the use of four 275-ton jacks for each of the arch ribs in a two-span section. The jacks could be operated independently. Also, accurate Bourdon pressure gauges were used to measure applied forces. During construction, gauge pressures were constantly monitored, as were the crown openings on the extrados and intrados of each rib. Vertical movements at the crown of one rib in each span were also measured. All of these measurements were necessary to ensure that the applied forces and resultant deformations were as planned. The monitoring phase was crucial to the entire operation. After completion of the jacking operation, the jacks were removed and the openings between arch halves filled in with concrete. An essentially monolithic arch resulted.

The Rogue River Bridge is still in service today, as are a number of other remarkable McCullough spans.

The George Washington Bridge

The George Washington Bridge, which crosses the Hudson River between New York City and New Jersey, was the longest suspension span in the world on its completion in October 1931. It paved the way for further developments and longer spans in suspension bridge systems; moreover, as the first bridge to cross the Hudson at Manhattan, it became a vital link, economically and socially, between New York and New Jersey.

From the end of the Civil War until the start of construction of the final design in 1927, numerous proposals were made for a bridge over the Hudson, each representing the needs of its own time. Earlier proposals largely concerned themselves with railroad traffic, later ones accommodated the automobile. As time went on, the development of property and the construction of railroad tunnels in the downtown area pushed the bridge farther and farther upriver. At least six different bridge companies developed a total of about eighteen bridge designs. At least five locations were strongly considered. The first company, the New York and New Jersey Bridge Company, was authorized by the New Jersey legislature to construct a railroad bridge across the Hudson at 70th Street. The company proposed a cantilever structure for six railroad tracks with a central span of 2,300 feet. This design incorporated piers at a good distance out in the river. The bridge could not become a reality, however, until the state of New York passed enabling legislation similar to that of New Jersey—which it did not do until 1890 and then stipulated that the bridge could not have any piers in the river. The cantilever proposal died out.

Meanwhile, the accomplished bridge designer Gustav Lindenthal, acting under the sponsorship of the North River Bridge Company (an alliance of railroad interests), had in 1888 presented a proposal to the American Society of Civil Engineers for a 2,850-foot suspension span to be built in the vicinity of 10th Street. The bridge company later moved its location to the 23rd Street area. While Lindenthal's proposal was not well received by the engineering community, it was strongly advocated by the North River Bridge Company.

George Washington Bridge, New York City to New Jersey. This world-renowned suspension bridge, completed in 1931, was virtually double the span of its longest predecessor. Othmar H. Ammann directed the design and construction. (Courtesy of the Port Authority of New York and New Jersey)

Since the Hudson is an interstate waterway, the U.S. secretary of war appointed a board of officers from the Army Corps of Engineers to establish minimum clear spans and other specifications. The 1890 report of the board was a well-researched document. They decided that, in general, a cantilever span of 3,500 feet and without a midriver pier was twice as expensive as a shorter span of 2,000 feet with a midriver pier, but that a long-span suspension system would cost only one-third more than the 2,000-foot cantilever. And since the long span made a troublesome pier at midriver unnecessary, they proposed that a long-span suspension bridge be built. The board also proposed that the bridge accommodate increasing vehicular traffic over the years, a task to which the suspension bridge was particularly suited.

Despite this support for a suspension bridge, Lindenthal's proposal was scrapped in 1893 owing to poor financing and difficulties in gaining the cooperation of railroad companies. The New York and New Jersey Bridge Company changed its proposal from a cantilever to a suspension bridge and moved its proposed location to the vicinity of 59th Street. Some borings were made to evaluate foundation conditions, but no construction ever started.

In 1906 the state of New York appointed the New York Interstate Bridge and Tunnel Commission and the state of New Jersey appointed the New Jersey Bridge and Tunnel Commission to study the best way to cross the Hudson. After four years a recommendation was made that a bridge be located at 179th Street. This site was of interest because the river is at its narrowest here and because foundation conditions were thought to be better than elsewhere. The latter speculation was disproved by borings in 1910, and interest in the site waned.

Lindenthal redeveloped his designs and eventually submitted a proposal for a span near 59th Street capable of carrying sixteen railroad tracks and many lanes of motor vehicles; but it did not meet the waterway clearance requirements at midspan that were stipulated by the U.S. Army Corps of Engineers.

During the first decade of the twentieth century, plans were also being formulated to build a vehicular tunnel under the Hudson River in the downtown area (the Holland Tunnel eventually resulted from these efforts). These plans, along with the advent of electric trains that could be used in such a tunnel, had a tremendous impact on those involved in building a bridge anywhere along the shore of Manhattan. In 1910 two pairs of electric railroad tunnels were built at 33rd Street. The tunnels virtually stopped any plans for a railroad bridge, and with the start of World War I all investigations into Hudson River bridges came to a complete halt.

After the war a bi-state commission, the New York–New Jersey Port and Harbor Development Commission, was established. This commission recommended an agreement between the two states that was finally adopted in April 1921. The Port Authority of New York and New Jersey was born of this agreement and was charged with planning and financing bridges and tunnels. By this time vehicular traffic was increasing at an extremely fast rate, which both renewed interest in a bridge crossing and necessitated the revision of many older bridge designs to accommodate better the needs of cars and buses rather than trains. The Port Authority's decision to build the vehicular Holland Tunnel also led to renewed emphasis on an uptown site for a bridge.

In 1923 Governor Silzer of New Jersey, who had gone on record as advocating a bridge, received a proposal from the engineer Othmar Ammann. Ammann suggested a bridge at 179th Street, serving vehicular traffic only. In August 1923 Silzer and Governor Al Smith of New York joined in their support for a suspension bridge. Smith was one of the Port Authority's six commissioners at the time. The design advocated by Silzer was based on Ammann's and consisted of a 3,400-foot central-span suspension bridge with several lanes of traffic and four railroad tracks, with the required 200-foot clearance at midriver. In 1925 the states of New York and New Jersey authorized the Port Authority to construct, operate, and maintain a bridge at 179th Street. Ammann was appointed bridge engineer of the Port Authority, Leon S. Moisseiff and Allston Dana were consultants, and Cass Gilbert was the architect. Ammann was a Swiss-born engineer who had worked on a number of bridges prior to securing this commission. He had worked with a consulting engineer, John Mayer, who had made plans for a railroad bridge across the Hudson some time earlier. In 1907 he had headed the inquiry into the collapse of the huge cantilever bridge in Quebec. Later he had worked with Gustav Lindenthal on the great railroad arch over Hell Gate, and he was knowledgeable about Lindenthal's proposals for a Hudson River crossing. Interestingly, Ammann and Lindenthal seem to have remained friends after the former won the bridge job. Ammann used Lindenthal as a special consultant on the design.

Over the next several years the Port Authority held a complete investigation of every aspect of the bridge's design, construction, and long-range effects. In 1925 boring tests were made to determine the best location for the piers. It was found that the exposed rock on either shore fell off sharply toward midriver, indicating that it would not be economical to build the piers too far away from the shore line. Studies of toll revenues and construction costs were conducted in order to assure the two states of the bridge's future success. By the time construction began in 1927, the design of the George Washington Bridge had been thoroughly scrutinized, to say the least.

The bridge, as finally built, consisted of a central span of 3,500 feet and end spans of 650 and 610 feet, making a total length of 4,760 feet. The towers stood 635 feet above the water, and the deck allowed for a 213-foot clearance underneath. The four 36-inch-diameter cables each consisted of 26,474 parallel wires, totaling 105,000 miles from anchorage to anchorage. The basic design of the bridge, including its mass characteristics, gave it stability with respect to wind forces. While the anchorage on the New York side was straightforwardly constructed of monolithic concrete, that on the New Jersey side entailed blasting two sloping funnel-shaped tunnels out of the palisade. The cables were then anchored on the rock itself.

The visual aspects of the intricate exposed steelwork of the towers caused controversy at the time—and are still debated today. The architect, Cass Gilbert, had originally proposed sheathing the steel framework of the towers

Construction of the George Washington Bridge. (Courtesy of the Port Authority of New York and New Jersey)

The George Washington Bridge with construction well under way. The main suspension cables are in place and the deck is being constructed. (Courtesy of the Port Authority of New York and New Jersey)

with granite. His intent was not only visual but also to preserve the steel from corrosion and to add some stiffening to the towers. The Port Authority, however, determined that periodic repainting of the exposed towers would cost less than the proposed masonry sheathing and that the added stiffening was of marginal value. Thus, the steel was left exposed. Had this been the original intent, it is probable that a simpler tower design would have been adopted.

Ammann's design was an extremely well calculated one, not only in terms of its great load-carrying capacity but also in its response to economic, commercial, and physical needs. Financially, the bridge was a great success—the revenue from tolls quickly covered its initial cost. And as a result of the increased traffic flow, industries and housing developments surged in the areas near the bridge. The physical setting was very well planned in terms of both construction and traffic flow. Being one of the narrowest points in the river, it was a natural place to build, and since the ground on both sides is high, the need for long approaches was eliminated. With the Holland Tunnel in operation downtown, the bridge served the uptown traffic; in addition it kept the traffic between New England and points south out of the downtown area. It also created, through connections with the Washington Bridge over the Harlem River and the Triborough Bridge, easy access between New Jersey and Long Island.

Most significant is the foresight Ammann showed in designing a structure whose capacity would increase with the area's growing needs. The original six lanes were increased to eight in 1946, using the central area left open for just that purpose; and a lower level was added in 1962, creating the world's first fourteen-lane suspension bridge. The weight of this lower deck had been calculated and incorporated into Ammann's original design.

The bridge as it is used today, however, has its problems. Even fourteen lanes cannot adequately handle the incredible volume of traffic that flows through New York each day. Nonetheless, the George Washington Bridge must be highly praised for the lasting effects it has had on both the transportation system and the economic development of the region.

A recent view of the George Washington Bridge. (Courtesy of the Port Authority of New York and New Jersey)

The Pattison Tunnel

Pattison Tunnel, Cheatham County, Tennessee. The forge it provided with power is long since gone, but Harpeth River water continues to pour from this tunnel built circa 1818 by Montgomery Bell. It is the earliest known rock tunnel of significant size in the United States. (Courtesy of the American Society of Civil Engineers)

The Pattison Tunnel, also known as Montgomery Bell's Tunnel, is located on the Harpeth River in Cheatham County, Tennessee. The tunnel was constructed about 1818, a date that makes it the first significant rock tunnel known to have been executed in the United States. It was a private project undertaken by a pioneering southern entrepreneur in the iron industry, Montgomery Bell, for the purpose of harnessing the power of the Harpeth River to maintain a large iron-smelting furnace.

Because of its early date, little information about the tunnel exists. But more is known about the personality and personal history of Montgomery Bell, its instigator. Born in Pennsylvania in 1769, he was apprenticed out at an early age, first to a tanner and then to a hat maker. He migrated south in 1789. Driven by a restless nature and fascinated by the potential for establishing an iron industry in the South, he bought a furnace and succeeded in increasing production several hundred percent within a year. By the end of a decade Bell had a string of furnaces and was producing iron in large quantities.

In all of his business dealings Bell can be characterized as a man of great enterprise and great ruthlessness. In the early nineteenth century the making of iron, from the breaking down of ore containing rock by sledge hammer to the manning of the furnaces (whose temperatures sometimes ran as high as 3,500 degrees F), was a very labor-intensive industry. Slaves were Bell's source of labor, and he was known to be a brutal master. His relations with business associates and friends tended to be guided by the principle of deception for profit.

The building of the Pattison Tunnel was the most remarkable of Bell's undertakings. The Harpeth River makes several sharp turns and bends when it reaches the Highland Rim in central Tennessee, and at certain points it turns back on itself for several miles. It was at the neck of one of these bends that Bell decided to cut the tunnel, hoping to develop a head of water sufficient to power a large forge operation. At the location of the tunnel the loop of the river completes a circuit of around six miles and is separated by a limestone ridge 300 feet wide. By tunneling through the neck and short-circuiting the loop, Bell created a natural waterfall of around 18 feet, to be used to power waterwheels that would, in turn, power the hammers and bellows of the forge.

Not much is known about the actual construction of the tunnel. The equipment used was very primitive for such a project—primarily hand drills and black powder. It is known that many difficulties were encountered. Drilling through the rock found under the limestone proved an arduous task with the equipment at hand. Bell had slaves to spare, however, and this obstacle was overcome by sheer toil. Nonetheless the undertaking required great skill, planning, and fortitude on Bell's part.

With the tunnel cut, the river furnished about twice the water power it had provided before. Bell had financial problems after completing it, however, and the forge was not finished until 1832. The Pattison Forge, one of his largest, has long since gone, but the Harpeth River continues to pour from the tunnel.

The method of diverting a river to generate additional power that Bell pioneered was used in the construction of the Great Falls Dam and Powerhouse on the Collins and Caney Ford rivers east of Nashville in 1915–1916. Bell's system contributed to the growth of industry in the southern United States.

The Union Canal Tunnel

For a short time after its completion in 1827, the Union Canal, with its tunnel connecting the Schuylkill and Susquehanna rivers, was the only commercial link between eastern and western Pennsylvania. The tunnel, near Lebanon, is the oldest American transportation tunnel still in existence.

Impetus for its construction was provided as early as 1690, when William Penn suggested a canal between the two rivers. Discarded then as impractical, the idea retained its appeal. In 1762 a survey conducted by Dr. William Smith, provost of the University of Pennsylvania, and David Rittenhouse, an astronomer, laid out a route for such a canal. Serious commercial interest surfaced in 1791, when two companies were chartered by the state to carry out construction.

The first of a long series of setbacks occurred when funds were soon exhausted and attempts at financial rescue by staging lotteries were thwarted by scandal. Both companies managed to survive, limping through many years during which no construction occurred. Finally, the two merged, with renewed funding to be provided by a group of Philadelphia bankers, and work proceeded around 1815 under the guidance of Loammi Baldwin, Jr., son of the builder of the Middlesex Canal. He left the project when his insistence on dams and feeders for water supply was opposed. It was precisely the lack of such supply that later seriously impaired the canal's success. Canvass White, lately an engineer on the Erie Canal, took over the headship in 1824.

What is often called the eastern portion of the Union Canal, the 108-mile stretch along the Schuylkill River that terminates in Philadelphia, was completed in 1826. Sixty-two of these miles were canal, the remainder accomplished by slack-water pools—long stretches of the river made navigable by dams built across the flow at various points and traversed by locks, which were 80 feet long and 17 feet wide.

A unique attraction of this eastern section was the first canal tunnel ever to be built in the United States. Contracted out to a trio of brothers named Job, Samson, and Solomon Fudge, this 400-foot drive through a ridge was hailed as an engineering triumph upon its completion in 1821. Sightseers flocked to see it and ride through it. Yet it can be suspected that little beyond the desire for publicity could have forced this tunnel, because a 100-foot shift of the canal line would have avoided the ridge altogether. Moreover, the cut required was only 40 feet. But the tunnel generated intense public interest, making many potential users aware of the canal—a fact of doubtless importance to the canal's promoters. In 1853–1856 the tunnel was converted into an open cut.

The western portion of the canal, completed in 1827, connected the Schuylkill at Reading with the Susquehanna at Middletown by means of an 82-mile waterway. A truly necessary 729-foot tunnel through the watershed ridge about a mile west of Lebanon, Pennsylvania, was built on this section. The tunnel is 18 feet wide and 14 feet high. The work of Simeon Guilford, it was excavated between 1824 and 1826 through argillaceous slate.

Also built were 93 lift locks (accomplishing a 311-foot ascent and a 209-foot descent), 43 waste weirs (to carry off excess water), 49 culverts (for passage over obstacles), 135 bridges, and 14 aqueducts (chiefly for transport of supply water). The supply water was fed to most of the canal by gravity, from a pool near the summit, and to the summit level by a system of pumps driven by huge water wheels assisted by 120-horsepower steam engines.

The first traversal of the Union Canal was accomplished by the *Fair Trader,* which made the Philadelphia-to-Middletown trip in five days. The canal was soon heavily in use, but profits were eaten up by construction blunders. The channel and the locks were built too narrow for most of the vessels in use on the Schuylkill upon the canal's completion. More difficult to comprehend, in view of the two branches' being built by the same company, is the fact that the locks on the Schuylkill were built 15 feet longer and the channel twice as wide as those of the western section. (The latter channel was 24 feet wide at bottom, 36 feet wide at surface, and 4 feet deep. The locks were 75 feet long by 8.5 feet wide.)

Union Canal Tunnel, Lebanon, Pennsylvania. The oldest existing transportation tunnel in the United States, it is part of the canal that connected the Susquehanna and Schuylkill rivers, an important route to the West. It was built between 1824 and 1826 and has been restored to its original condition. (Courtesy of the American Society of Civil Engineers)

To counter this problem, the company encouraged the building of narrower vessels and was moderately successful in this effort. But tonnage inevitably suffered. And because the locks had to be used almost continuously, heavy water losses resulted, augmented by the leaky limestone soil lying beneath the canal. Such water-supply difficulties had been predicted years earlier by Baldwin. By 1830 a dam on the tributary Swatara Creek and a pumping station had been built to put more water where it was needed. Widening of the channel and locks was also begun, but it was not completed until 1856. More positively, a navigable branch line was built to reach within four miles of an important anthracite mining area, coupling the canal with a major new Pennsylvania resource.

Despite early problems, the Union Canal of the 1830s substantially contributed to the commercial development of Pennsylvania west of the Susquehanna and, by 1834, linked Philadelphia to points as far west as Pittsburgh via Pennsylvania's counterpart to the Erie Canal, the Pennsylvania Main Line Canal. The panic of 1837 and the depression of 1839 did not help matters, however, and after 1839 the heretofore-unique Union Canal had competition to face as well. That year saw the completion of the Susquehanna and Tidewater Canal, which, together with the earlier Chesapeake and Delaware Canal, provided another route between Philadelphia and the West. The Susquehanna and Tidewater employed large locks that could handle vessels of 150-ton capacity, while the Union Canal's narrow dimensions limited its use at first to boats bearing 25–28 tons and, following the widening of the canal, to loads of 75–80 tons.

Railway construction led eventually to even stiffer competition, since the relative speed of rail travel lured both passenger and freight traffic away from the canals. The early Columbia and Philadelphia Railroad also linked Philadelphia to the Susquehanna, following its completion in 1834, but hardly to the detriment of the Union Canal in those initial days of railroading when a railroad often consisted of but one track and anyone could put a car on it, hitch up a horse, and travel. Later, horses were banned, steam locomotives were improved, coal was burned instead of wood, telegraph lines were installed along the right of way, and rail operations gained tremendously in capacity and efficiency. Soon the competition from the railroad was overpowering. The Union Canal suffered severe additional competition beginning in 1857—one year after the completion of channel and lock widening—when a new railroad linked Harrisburg to Reading.

A more immediate calamity struck when the great flood of 1862 destroyed the Swatara Dam and ruined the coal-bearing branch. The Union Canal Company was soon bankrupt, and the canal was sold to the Philadelphia and Reading Railroad at a sheriff's sale. In 1885 the canal, by then of very limited usefulness, was abandoned.

The Union Canal and its tunnel, though a historic part of the canal era, never really attained full stature, because the planning for them was either shortsighted, or hampered by insufficient financing, or both. This is especially unfortunate when one considers the 729-foot Lebanon Tunnel—one of the very first, and the longest at its time in the United States—and the 45-foot dam over the Swatara, the highest at its time. All the makings of a masterpiece were there. The Lebanon Tunnel is still in existence, at any rate. In 1934 it was restored to nearly its original appearance.

The Crozet Tunnel

The 4,273-foot Crozet Tunnel, completed in 1858, represents the culmination of a tunnel-building technology that was based on manual drilling methods. At the time of its completion it was the longest railroad tunnel in the United States. Often referred to as the Blue Ridge or Rockfish Tunnel, it is located at Rockfish Gap, Virginia, near Waynesboro, and was designed and built by Claudius (Claude) Crozet, a French-born civil engineer. The Crozet Tunnel is the largest of four single-track railroad tunnels that together comprise the first penetration of the Blue Ridge, a formidable mountain barrier that had until then prevented the development of commerce between the Ohio River basin and the seaports of Virginia and other southern states.

The nearly insurmountable problems encountered in the tunnel's construction were overcome through the persistence, intelligence, and vision of its chief engineer. Born in France in 1789, Claudius Crozet was a graduate of the famed Ecole Polytechnique and subsequently served in Napoleon's army. In 1816 he and his wife emigrated to the United States, where, with the help of the marquis de Lafayette, a friend of his wife's family, Crozet obtained a professorship at the United States Military Academy at West Point. There he introduced the first American college course in descriptive geometry, writing his own text since none existed in English. His translation of Sganzin's *Elementary Course in Civil Engineering* contributed enormously to the technical literature available in America at the time (Crozet had been a student of Sganzin's at the Ecole Polytechnique). Curiously, Crozet is also credited with introducing the use of chalk and blackboards to colleges in the United States.

His successes at West Point led to Crozet's being named the country's best mathematician by a congressional committee. Yet when in 1823 the Virginia Board of Public Works offered him the post of principal engineer, Crozet eagerly accepted it. Former president Thomas Jefferson, an influential Virginian

A portal of the Crozet Tunnel, Rockfish Gap, Virginia. This 4,273-foot tunnel, built between 1850 and 1858, represents the high point of tunneling technology based on manual drilling methods. When completed, it was the longest railroad tunnel in the United States. By linking the Ohio River basin with southern ports, it facilitated the development of commerce in the South. (HAER Collection, Library of Congress)

and amateur engineer, was probably a moving force in this appointment. While in Virginia, Crozet helped form the Virginia Military Institute and gave courses there.

Virginia (which until 1863 included what is now the state of West Virginia) was primarily an agricultural state. But in addition to planters, there were businessmen, traders, and politicians, all clamoring for canals, turnpikes, and railroads to tie its vast territory together. The magnitude of the required projects made it difficult for private companies to obtain adequate resources for undertaking them. It therefore became state policy either to provide financial assistance to companies engaging in such projects (by underwriting bonds or subscribing to stock) or to build them itself. The Board of Public Works had been established in 1816 to superintend all internal improvements.

Crozet's first duties with the board included road building and map making. By 1830 he was in charge of the James River and Kanawha Canal, a project the state had inherited from an insolvent private contractor. Crozet considered the canal an inappropriate means for traversing the sharp ridges and deep valleys of the Appalachian Mountains. Instead he advocated building railroads, with tunnels through the mountains. The resulting rift between Crozet and the pro-canal legislators led to his resignation in 1831; he subsequently accepted an appointment as state engineer of Louisiana.

The canal project failed, as Crozet had predicted, and he was rehired by the board in 1837. Continued political differences, however, together with the economic depression resulting from the panic of 1837, held back rail expansion. The state legislature abolished the post of principal engineer, although Crozet continued to work as a consulting engineer to the state.

The Virginia legislature gradually came to accept the notion of railroads. Early in 1849 the position of principal engineer was reestablished, and Claudius Crozet once again filled it. At age sixty, the most spectacular part of his career still lay ahead.

In March 1849 the state of Virginia incorporated the Blue Ridge Railroad. This modest 17-mile line promised to improve the economy of the entire state, for it would extend the railroad westward by tunneling through the mountains. The new line was to originate at Mechums River, where it would connect with the Louisa Railroad, and continue to a point just west of Waynesboro. There the Louisa would begin again, extending to Covington, Virginia, where another state-chartered railroad, the Covington and Ohio, would start its route to the Ohio River. The new railway system, it was hoped, would open the region for settlement, provide transportation for agricultural and mine products to the Ohio Valley (and thence further west and south), and unify the state, which then sprawled to the Ohio River.

By the end of the summer of 1849 Crozet had completed the location surveys and specifications for the project. His plan included four tunnels and an unusually deep cut in the west slope of the mountain at Waynesboro. Three of the tunnels, the Brooksville, Greenwood, and Little Rock, would be under 900 feet in length. The westernmost Blue Ridge Tunnel was to be 4,273 feet long, by far the world's longest railroad tunnel at the time. It must be remembered that the first railroad tunnel in the country, the Allegheny Portage Tunnel, was only 900 feet long and had not been completed until 1833. Tunneling technology was little developed, and progress was slow in the days of hand drills, pickaxes, and black powder. Crozet's tunnels were all completed before the first use of the pneumatic drill in America, and dynamite had yet to be discovered.

The first step in construction was to build temporary tracks around and over the tunnel sites. Not only did this provide rail service to Staunton and Waynesboro as quickly as possible, but it also allowed construction materials to be hauled to the sites. The track over the Blue Ridge Tunnel was the most ambitious. With a ruling grade of 5.7 percent, it was at the time among the steepest mainline tracks in the world.

In February 1850 the Louisa Railroad was reincorporated as the Virginia Central Railroad. The same month, the Board of Public Works entered into a contract with Christian Detwald, John Kelly, and John Larguey for construction of the Blue Ridge Tunnel and its approaches. Laborers had been assembling in anticipation of railroad work; among them were "fardowners" from the north of Ireland, who had come to work on the Virginia Central, and Irishmen from County Cork, who had come to work on the tunnel. Their mutual resentment led to a riotous fight and burning of quarters, with the militia finally called out to quell what came to be known as the Irish Rebellion.

Labor problems were just the beginning of the troubles that plagued Crozet's undertaking. Exceedingly hard rock was encountered, causing delays; wages rose from 75 cents to a dollar per day; rock slides delayed work and endangered the workmen; building materials had to be transported from a considerable distance. All of these difficulties were described to the board by Crozet in November 1851, when he reported that his men had drilled and blasted into the two ends of the main tunnel only a total of 755 feet in the first twenty months.

The Brooksville Tunnel, although only 869 feet in length, presented its own particular difficulties. The mountain at this point was a mixture of limestone and clay, which tended to break apart upon contact with air. Until the bore could be lined with timber, tunnel crews were in constant danger of being buried alive.

In March 1854 the temporary track across the Blue Ridge reached Staunton. Cave-ins and rock slides continued to plague the operation at Brooksville. A cholera outbreak killed a number of workers and caused others to flee in fear. A general financial crisis in 1855 caused a drop in purchases of bonds, which made the board unable to meet its payments to the contractors. Fortunately, the contractors advanced their private resources, and work was able to progress. By December 1855 the headings on the east and west sides had advanced 1,718 feet and 1,809 feet, respectively.

The formidable task that Crozet had taken upon himself was brought to completion only through his characteristic blend of practicality and inventiveness. For example, Crozet had determined that the use of intermediate shafts between the portals was unfeasible. He was thus faced with the problem of boring a long tunnel with only two working faces. He had designed the tunnel with sufficient height for draft and ventilation of a locomotive's smokestack; the constant slope of the tunnel would permit drainage of water from the lower east end, while smoke would be carried out the higher west portal by convection currents. Until the two headings met, however, ventilation and drainage were serious problems.

In order to provide ventilation, particularly needed because of the noxious fumes resulting from the use of black powder explosives, Crozet devised a system of pipes and valves connected to inverted tubs. The tubs, alternately dipping in water, trapped the air and expelled it through the exhaust pipelines. Drainage, against the grade in the western heading, was effected by hand- and horse-powered pumps. Of great help was Crozet's use of the longest siphon on record, a 2,000-foot device of 3-inch-diameter cast-iron pipe, which discharged water at the rate of about 60 gallons per minute.

The two headings met on Christmas Day, 1856. So well had Crozet calculated their alignment that it deviated from true center by only half an inch. The "holing through" did not, however, signal the beginning of train service. There were delays in work due to heavy snows, and there was still a great deal of bottom to remove. Arching, trimming, and laying of the track occupied the workers for more than a year. The 150,000 bricks required for arching the tunnel (where it was not bored through solid, stable rock) had not yet been manufactured when they were needed, owing to a shortage of state funds. A total of 1,479 feet of the tunnel was eventually arched.

As for the trimming of the tunnel to the proper size and shape, a sentence from an article in the *Lexington Gazette* entitled "Brilliant Achievement at the Tunnel" indicates the continued criticism to which Claudius Crozet was subjected: "It now turns out by actual experiment that the Blue Ridge Tunnel is too small to admit the passage of a single car, much less a full train!" What the article neglects to mention is that the car in question was a clearance car, specially designed

Inside the Crozet Tunnel. (HAER Collection, Library of Congress)

by Crozet for finding narrow points in the tunnel cross section. This was the first known use of such a device. Once again Crozet had to defend himself in a report to the Board of Public Works, in January 1858. He explained the use of the clearance car in determining tunnel width. (In no place was it less than 14.5 feet; 14 feet was then considered standard for rail tunnels.) The entire tunnel was soon to be widened to a minimum of 15 feet, which proved an excellent size, serving for 86 years.

Crozet concluded his report by requesting that, if the board could dispense with his services, his connection with it be severed at year's end. The board reluctantly accepted his resignation.

On April 12, 1858, the first regular train made its six-minute passage through the Crozet (Blue Ridge) Tunnel. The tunnel had been built at a total cost of $488,000—an average of $114 per foot over its 4,273-foot length—and the average rate of progress had been 26.5 feet per month at each face. In 1867 the Virginia Central Railroad absorbed the Covington and Ohio and became the Chesapeake and Ohio Railroad. Two years later the Chesapeake and Ohio purchased the Blue Ridge Railroad from the state of Virginia for barely a third of its $1.69-million construction cost.

Little remains today of Crozet's original work. The Blue Ridge Tunnel was replaced by a new tunnel with a larger bore in 1944. The tunnel is located near the junction of U.S. 250, I-64, and the Blue Ridge Parkway. One remaining testimony to the engineering feat of Crozet is the eastern approach to the Blue Ridge Tunnel at Rockfish Gap. For almost a mile the track clings to the mountainside, which slopes steeply to the valley 500 feet below. Emerging from the western end of the tunnel, today's traveler has a breathtaking view of the valley and distant mountains, just as they appeared to the visionary Claudius Crozet.

The Hoosac Tunnel

The driving of a tunnel through the Hoosac Mountain in northwestern Massachusetts, which took place between 1851 and 1876, began with tunneling methods that had been used for centuries and ended with an almost totally mechanized process. This project became the source of modern rock-tunneling technology. Techniques that were developed during the building of the Hoosac Tunnel are still in use today.

The Hoosac Mountain was a major obstacle in any proposed northerly connection between Boston and the western markets in New York. The idea of a tunnel was first advanced in 1819 as part of a proposed Boston to Troy (or Albany) canal. In 1825 a commission was formed to study statewide transportation; it hired the well-known civil engineer Loammi Baldwin, Jr. (see chapter 13, Naval Dry Docks) to survey potential westward routes. Baldwin's report to the legislature of 1826 favored a northern route through Fitchburg and Greenfield because of better river conditions for a canal. Only the great bulk of the Hoosac stood in the way. Cost estimates for the tunnel and canal proved prohibitive, and the project died. The Western Railroad from Boston to Albany, by way of Springfield and Worcester, opened in 1841 and immediately generated demands from merchants in the north for a northerly route to New York. Alva Crocker formed the Fitchburg Railroad in 1841, and construction of the line from Boston to Fitchburg began immediately. It was opened on March 5, 1845. The Troy and Greenfield line was subsequently proposed by Crocker to complete the link to Troy. The line was chartered in 1848, despite lobbying efforts by the Western Railroad. The charter located the tunnel needed to pierce the Hoosac Mountain.

Groundbreaking occurred in January 1851, near North Adams. Chief Engineer A. F. Edwards experimented with a "full area" boring machine designed to excavate the entire face of the work in a single operation by cutting concentric grooves in the rock. The rock between the grooves was then to be blasted out. The first machine bought by the Troy and Greenfield was a "Wilson's Patented Stone-Cutting Machine" built by Munn and Company of South Boston. This machine weighed 70 to 75 tons and was designed to cut a 24-foot-diameter tunnel. After progressing some 10 to 12 feet, however, the machine broke down. Afterward, resort was made to traditional hand-drilling techniques. Progress proved very slow, and Crocker was constantly lobbying for state loans to maintain the work.

In 1856 a contract was signed with a new engineer and contractor, Herman Haupt. An accomplished engineer and author of the respected work *A General Theory of Bridge Construction,* Haupt brought additional credit to the project. He proceeded with traditional hand-drilling techniques.

An advance heading or adit was carved by pick and hammer. This shallow tunnel was then deepened into a "top heading" that had enough height to permit hammer drilling or hand jacking. One method had a two-man team in the top heading use "double-jacking" or two-handed hammers while a third man held the steel driving rod. The top-heading plan allowed the bulk of the rock to be drilled downward, which was most efficient for hand drilling. Blasting, with black powder and similar commercial variants, completed the operation. In general the heading was kept between 400 and 600 feet in advance of the blasting, so that the latter would not interfere with this work. A "bench carriage" was used to facilitate the handling of blasted rock. It was rolled back during blasts.

While the hand drilling and blasting progressed, Haupt also experimented with boring machines in the hope of speeding up progress. In August 1857 he tested a 40-ton, 650-horsepower steam-powered machine made by the Novelty Works of Philadelphia that was designed to cut an 8-foot heading. It never worked properly and was eventually retired from use. Haupt also experimented with several mechanical rock drills. One of these, the Fowle drill of 1851, consisted of a drill rod attached to, and reciprocated by, a double-acting steam piston. Developed for quarry work, it allowed drilling independent of the direction of the drill—a useful feature in mining—but the machinery was too cumbersome to find true application here. In another experiment, with the aid of his assistant Stuart

East portal of the Hoosac Tunnel, near Florida, Massachusetts. Completed in 1876, this was the longest transportation tunnel of its time in the Western Hemisphere. Many major construction innovations, such as the steam drill and the use of nitroglycerin for blasting, were first successfully used on this project. The central ventilation shaft, which provided two additional work faces, was still another remarkable engineering feature. (Courtesy of the American Society of Civil Engineers)

Gwynn, he developed a drill that could bore hard granite at a rate of ⅝ inch per minute; but the drill would not hold up in service.

By 1861 only a small portion of the total bore was completed. Hard mica-slate with quartz veining, the predominant material found, made for slow progress without successful drilling machines. On the west end a loose mica-schist, called porridge stone, was encountered, and a brick archway had to be erected to support the tunnel in this area. A shaft was sunk to grade 2,500 feet from the west portal, to a point where more solid rock was encountered. In this manner the time-consuming arch construction could be worked from two ends.

Progress was not fast enough, and there was doubt whether Haupt was equal to the task before him. The project was soon in intense financial difficulties. In addition, the tunnel was becoming a political issue. The Western Railroad management, feeling their monopoly potentially threatened, continued to lobby against the tunneling effort. Frank W. Bird, sympathetic to the tunnel's opponents, created a furor in the legislature with his attacks on the tunnel. He published anti-tunnel pamphlets such as *The Road to Ruin, or The Decline and Fall of the Hoosac Tunnel.* Haupt responded with *The Rise and Progress of the Hoosac Tunnel.* Nonetheless, financial difficulties and the failure to get necessary governmental support for loans caused Haupt to leave the job in midsummer of 1861. He subsequently became brigadier general in charge of construction and transportation for the North during the Civil War, and won commendation for achieving a high degree of efficiency among the railroads.

In 1862 the state foreclosed on the property and formed a commission to supervise the project. A prominent civil engineer, James Laurie, was called in to review the work done. He reported that completion of the tunnel was feasible and would bring economic benefit.

The commission next sent engineer Charles Storrow to Europe to observe tunnel-driving techniques there. At the Mont Cenis tunnel between France and Italy, Storrow met chief engineer Germain Sommeiller, who was

Excavation work in the tunnel: two contemporary engravings. (Library of Congress)

working with mechanical drills powered by compressed air instead of the steam employed in previous drill experiments. Problems with exhausts and transmitting steam over great distances caused steam drills to be impractical. Air-compressing machinery at the tunnel portal, with a piping system to the work face, eliminated many problems encountered earlier; but mechanical failures plagued Sommeiller's drill, and it was never extensively developed.

Storrow's report to the state commission in 1862 favored the adaptation of Sommeiller's techniques to the Hoosac project. Thomas Doane, the new chief engineer, pushed efforts further toward developing a completely dependable compressed-air drill based on Sommeiller's. In 1865 Charles Burleigh, a mechanical engineer at the Putnam Machine Works of Fitchburg, Massachusetts, invented a workable machine that fulfilled all the requirements. Designed as a separate, relatively light element attached to a movable frame or carriage during operations, it could drive a drill rod with great force and automatically feed and rotate the rod as the hole deepened.

Early Burleigh drills were tried in the tunnel as early as June 1865. By November 1866 a more reliable, easily repairable model was in operation there. Once the workers had got used to them, the machines saved half the cost of hand drilling and greatly speeded progress. Soon four to six drills were mounted on a carriage and used in the tunnel to make blastholes.

The Burleigh drill was the first practical mechanical model in the United States, and perhaps the world. Dependable, efficient, and of simple construction, it would become the prototype for all succeeding piston-type drills. The importance of this innovation to rock tunneling would be hard to overestimate. Burleigh's drill was an immediate success, and its use rapidly spread elsewhere in tunneling and mining operations.

Advances in blasting techniques were also furthered by Doane's efforts. The early methods using black powder became problematic as the tunnel advanced, because of the toxic fumes released by the blasts. By 1866 Doane had imported trinitroglycerin, or "nitro," the liquid explosive recently introduced by Alfred Nobel. Its powerful blast

The nitroglycerin plant located near the west portal. (Courtesy of the North Adams Public Library)

Trinitroglycerin was first used in tunneling at the Hoosac. This contemporary engraving shows blasting in progress. (Library of Congress)

as well as its smokeless quality seemed to solve many problems, but it was dangerous to handle as it tended to explode with or without provocation. Doane hired chemist George W. Mobray to develop techniques for the bulk manufacture of the substance and for its safe employment in tunnel work.

Mobray was able to increase stability by maintaining purity in the manufacturing process, freezing the liquid on transport to the headings, and using extreme caution in handling. At the headings the nitro was poured into cylindrical cartridges for placement in the drill holes. New fuses and an electric ignition system provided simultaneous detonation, replacing the old powder train and cord fusing of black powder. Nitroglycerin permitted fewer blasting holes over a given area of work face, and it could blow deeper holes—40 inches versus the previous 30 inches for powder. Forty percent more tunnel length was advanced per cycle of operations. Between 1868 and the completion of the tunnel, Mobray produced over a million pounds of nitroglycerin. Its successful use developed a completely new blasting practice.

With the new techniques, a bottom-heading system was adopted. Carriages mounted with up to six Burleigh drills would drive the blast holes. In three stages the full 24-foot width of the heading was blasted and cleared away.

Even with the advanced techniques now being used in the tunnel, by 1868 only a little over a third of the work was complete, and $7 million had been spent. Harassed by political difficulties and tired of apportioning piecemeal work, the state commission contracted out the remainder of the job to Messrs. F. Shanley and Company of Montreal. At a sum of $4,595,268, they were to have the tunnel completed by March 1874. With this incentive, Shanley and Company pushed the boring forward with great speed. In three 8-hour shifts, 800 to 1,000 men were employed, and some 500 were at work at any given time. Drilling stopped only for running the carriages back, blasting, and mucking or clearing the rock away.

Not only did the work progress from the east and west portals, but by 1870 a central shaft had been sunk 1,028 feet to grade,

The centering for the placement of the finished stonework on the west portal, circa 1874. (Courtesy of the North Adams Public Library)

The east portal. (Courtesy of the North Adams Public Library)

where a notch in the mountain afforded access. This shaft provided two additional work faces as well as admitting needed ventilation. Begun in December 1863, its construction had been plagued by difficulties. Not only did all materials need to be hauled out by elevators, but water collected in the bottom and had to be pumped out at a rate of 15,000 gallons per hour—and at a great expense of time and money. In 1867 a fire in the shaft killed thirteen men and halted work for more than a year. Still, Shanley and Company were able to complete the work, and in December 1872 headings from the center were joined with those from the east. A triumph of engineering skill, this shaft was found to deviate in alignment from the tunnel by only 7/16 of an inch!

On November 27, 1873, the headings were joined. In 1875, after the removal of more than a million tons of rock, the first train of cars passed through the Hoosac Mountain. Thomas Doane was at the controls. Without his energy and his technical and management skills the project would probably never have been completed. When the last details were finished in 1876, the entire project, including the 44 miles of track for the line, had cost an estimated $17 million. That enormous sum and the twenty-five years of effort had been well spent. Not only had an economical route to the West been provided, but the science of tunnel engineering had been vastly expanded. With the use of pneumatic rock drills and nitroglycerin, invention need only be followed by improvement. Experimentation and determination had produced the greatest piece of tunnel engineering in the Western Hemisphere to that time; and the Hoosac Tunnel remains a crowning achievement of American ingenuity.

The Hudson-Manhattan Railroad Tunnel

The completion of the Hudson-Manhattan Railroad Tunnel in 1908 culminated a daring attempt to develop a new method of tunnel construction based on the use of compressed air, as well as provide a long-sought link between the New York and New Jersey shores at Manhattan. In the best spirit of nineteenth-century civil engineering adventures in the United States, its driving force, DeWitt Clinton Haskin, gambled on a new technology and held to his belief that it would work. It almost did. Subsequently the tunnel was completed by William McAdoo and Charles Jacobs.

The idea of building a tunnel under the Hudson River was first considered in the 1860s. Manhattan had become a congested metropolis, and ferry service was a marginally adequate way to transport freight and passengers. A tunnel would enable people to live in New Jersey and commute into the city. The construction of a bridge was discussed but remained only a tenuous possibility. Not only was the Hudson wide and deep, its bottom was known to be deep mud in many places.

DeWitt Clinton Haskin, a successful engineer who had helped build the Union Pacific Railroad, formulated an ambitious plan for the tunnel project. Believing he could apply to tunnel building the principle of using compressed air to expel water from an underwater chamber—a principle then being applied in the construction of underwater bridge foundations, in the form of the pneumatic caisson—Haskin sought funding for the project in New York.

A British engineer, Sir Thomas Cochrane, had devised the first pneumatic caisson in 1830. James Eads, in constructing the foundations for the famous arch bridge at St. Louis in this country (see chapter 4, Eads Bridge), employed a caisson with a working chamber at its base that was kept free of water by injecting compressed air into the space. The caisson was sunk into position and constantly lowered as material was excavated and removed through locks. It was during Eads's project that decompression sickness—the "bends"—was encountered.

Hudson-Manhattan Railroad Tunnel, New York to New Jersey. Begun in 1874 and not completed until 1908, it was the first railroad tunnel under a major river in the United States and introduced the shield system of subaqueous tunneling in the United States. The installation of the iron plate linings is shown. Initially, air pressure alone was used to support the sides of the tunnel during construction. (*Scientific American,* May 8, 1880)

Still, this construction method made possible unprecedented feats in underwater building.

Haskin sought to employ a similar concept in the building of the Hudson River tunnel. The tunnel itself was to be pressurized during construction to keep the water out and to help keep the initial tunnel liners in place. The work face would be exposed so that excavation could take place. As the concept was finally implemented, a bulkhead, or thick concrete wall, was built at the mouth of the tunnel. Getting men and materials through the concrete structure was accomplished by fitting an air lock into the wall. The lock was a small, airtight steel cylinder with a door at each end, one opening into the tunnel and the other to the outside. To get to the exposed work face, men first entered the air lock, where the air pressure was gradually increased until it reached the higher pressure already created inside the tunnel; they could then open the inside door and enter the tunnel. Removal of the excavated silt was done in a unique way. The mud was first mixed with water and then fed into a 6-inch pipe. The pressure in the heading forced the liquid up through the pipe and to the surface.

In 1873 Haskin incorporated the Hudson Tunnel Railroad Company and, with the aid of financier Trevor W. Park, raised $10 million. The following year he obtained a patent for improvements in tunneling based on the use of compressed air acting directly on excavation walls. On October 29, 1874, he gained permission from the Jersey City Council to sink a shaft at the foot of 15th Street for the construction of a rail tunnel to New York. The project had hardly begun when it was halted by the first of a series of injunctions filed by the Delaware, Lackawanna, and Western Railroad Company, which operated the ferry near the route of the proposed tunnel and had much to lose from its success. Although the railway company was eventually defeated, it did manage to hold up construction of the tunnel for five years. Work resumed on September 18, 1879.

By the end of 1879 the shaft had reached its full depth of 65 feet. An air lock was installed, and work on the tunnel got under way immediately. The first tunnel plates went into place on February 9, 1880. The tunneling moved forward in steps of about 45 degrees, with the forward lining projecting over the slope. After a forwarding of the tunnel so that an additional 10 feet of tunnel shell was complete, bricklayers added 2 feet of lining to the shell. In May 1880 a news release anticipated completion: "It is expected that three years from now trains arriving in Jersey City will run directly through to New York, landing their passengers somewhere near the Metropolitan Hotel in six minutes' time." Describing the construction methods, this anouncement declared that "no expensive Brunel shields will be required." Haskin had at first considered using compressed air behind a cylindrical tunneling shield but had decided against using a shield of any kind. He believed the air pressure in the heading, together with ordinary timbering, would be sufficient to hold a shell of iron plates. These, in turn, would support the wet silt until the permanent masonry could be built.

The working crews consisted of approximately thirty-five men, three or four of whom were assigned to watch for leaks. Once a leak was suspected, a lit candle was held close to the spot. The escaping air drew the flame into the hole, locating it so that it could be plugged with sandbags.

At first the tunnel advanced at the rate of a foot per day. Soon, however, the men had grown accustomed to the job and were progressing at five times this rate. On July 21, 1880, at which time 300 feet of the tunnel had been completed, a leak was discovered. Since leaks were fairly common, this did not seem serious. As one of the men approached the leak with a sandbag, the hole suddenly widened and the compressed air began to escape with a hiss. The roofing plates, no longer supported by the air pressure, collapsed, and within moments the tunnel was half filled with silt. The foreman, Peter Woodland, directed his men into the air lock. Just as some of them had entered, a falling plate partly closed the lock door, preventing Woodland and the others from getting into the lock. Inside the lock the men panicked. The farther door could not be opened against the air pressure, which, although reduced somewhat by the leak, was still being fed by the compressors. Unselfishly, Woodland ordered the men to break the glass window in the door. This would, in effect, release all the air in the heading, causing it immediately to fill up; but it would enable the men in the lock to open the door and escape. In the end, twenty-one of twenty-eight men in the tunnel were killed, including Woodland.

Several months were spent blocking the leak and pumping out the heading. Realizing that he could not rely solely on air pressure to hold up the roof, Haskin sought a new method. His superintendent, John Anderson, suggested the use of a "pilot tunnel" extending a short distance ahead of the main structure. It was to consist of a small bore formed by a steel plate just large enough to provide working space for excavation. This proved so successful that it was patented as "Anderson's Pilot Tube." Soon progress reached 5 feet per day again. In early 1881 William Sooy Smith and Son took over the engineering responsibility.

On March 31, 1882, a blowout and flooding occurred once more. Although no lives were lost this time, the accident proved fatal to the tunnel project. Smith lost confidence in working with compressed air in sandy soil and, fearing further accidents, tried unsuccessfully to convince Haskin to use more conventional (but more expensive) methods. Smith resigned shortly afterward, and Haskin formed the Hudson Tunnel Construction Company, with Trevor Park as president and himself as manager, to complete the work. Haskin's funds were running out, however, and he found it impossible to gain further backing for his hazard-ridden project, particularly after the death of Park, who had been a major financial contributor. On November 7, 1882, all work on the tunnel ceased. Haskin did manage to fund a resumption of work in April 1887, but soon various difficulties, including flooding, brought another halt. It was probably just as well. As the tunnel extended farther under the river, the air pressure inside had to be increased. The incidence of caisson disease had risen sharply and would only increase with increasing pressure, since little was done to mitigate the problem.

Haskin's difficulty lay in supporting the tunnel walls, and the difficulty could be overcome by the use of a shield. In 1889, with

On July 21, 1880, the air pressure supporting the roof plates was reduced by a leak. The roof plates collapsed, and soon the tunnel was half filled with silt. Some men escaped into an air lock, but others were prevented by a fallen plate and were trapped in the outside heading. The men inside the air lock, however, could not get out because of residual air pressure that kept the exit door sealed. From the heading the foreman, Peter Woodland, directed the workers in the air lock to smash a glass window in the heading door in order to release the air pressure—which would cause the heading to fill up but would allow the men in the air lock to escape. Seven workers survived. Woodland and twenty others were drowned. This engraving is a contemporary conception of miners escaping into the air lock. (Library of Congress)

British capital behind the scheme, another attempt was made to complete the tunnel. The well-known British engineer Sir Benjamin Baker, who along with John Fowler had constructed the great Firth of Forth bridge, had visited the tunnel in 1888 at the instigation of English financiers and had been impressed with the project. In the new effort the British engineering firm of S. Pearson and Son introduced a tunnel shield that in its original form, developed by Peter Barlow, had been used in building the London Underground in the 1860s. Improvements in its design, by J. Greathead, had been affected by Haskin's approach. While Haskin had trouble supporting the tunnel walls, the problem encountered by Greathead was the frequent collapse of the wet-ground work face in front of his shields. Greathead followed with interest Haskin's use of compressed air alone, and it occurred to him to combine Haskin's technique with his own. He tested this configuration in 1886 in the building of a tunnel for the City of London and Southwark Subway. It became the prototype for most subsequent shields.

The next attempt at boring the Hudson River tunnel, then, involved using a Greathead shield. Cast-iron rings replaced the brick linings. Hydraulic rams forced the shield forward. Displaced silt entered through doors in the shield and was then shoveled into carts and removed. Following each forward movement of the shield, a mechanical arm set the cast-iron rings in place. There were further collapses, as well as weaknesses in the cast-iron linings and other difficulties, such as a troublesome reef of rock. Nonetheless, work progressed. Sir Benjamin Baker and John Fowler were jointly in charge of the work, and Haskin's son was the superintendent.

In the middle of 1891, however, with the north tube just 1,600 feet from completion (the tunnel consisted of two parallel tubes), a financial crisis developed in England and no more money could be raised. Work had to be shut down and the tunnel flooded. Although this second attempt to tunnel the Hudson had failed, a considerable improvement had been made in the compressed-air process, and the additional progress made on the tunnel length was very important. Progress could be seen on other fronts as well. The death toll

Construction workers who helped build the Hudson-Manhattan Tunnel. (Courtesy of the Port Authority of New York and New Jersey)

Work on the tunnel was stopped and restarted several times before the project was finally completed. Improved shields were introduced. (Courtesy of the Port Authority of New York and New Jersey) from caisson disease having become quite substantial, E. E. Moir, one of the site engineers, devised a "hospital lock" above ground. Men showing signs of the disease would enter the lock, wherein the pressure would be increased to that used in the tunnel. The patients were then decompressed at half the rate used in the regular lock.

In 1895 the English engineer Charles M. Jacobs was invited to determine the feasibility and probable cost of completing the tunnel. Jacobs concluded that there were solutions to the difficulties with the weak cast-iron rings and the reef of rock. Construction was not resumed, however, and on June 15, 1899, the entire works were sold to a group of businessmen headed by Frederick Jennings. The project was resurrected through the efforts of the Georgia-born lawyer William G. McAdoo (later to become secretary of the treasury under Woodrow Wilson). In 1901 Jacobs and McAdoo inspected the rusty and slimy shield and found it still serviceable. The following year McAdoo gained control of the project from the bondholders and incorporated the New York and Jersey Railroad Company to carry it to completion. The work was placed under the direction of Charles Jacobs. In October 1902 the first of the redesigned cast-iron rings was put in place.

As the tunnel proceeded toward New York, rock began to appear at the bottom of the heading. Blasting rock at the bottom of the tunnel would disturb the wet ground above, bringing the danger of blowout or collapse. Jacobs decided to apply intense heat to the wet silt, thus hardening it to the stiffness of dry clay; he achieved this with the use of kerosene torches. The hardening, combined with the pressure of the compressed air, held the silt in place long enough for the workers to excavate the rock.

The bulkhead at the New York end of the north tube was reached in March 1904—thirty years after the work was first begun. Construction continued on the south tube for some time longer. The first train ran through in February 1908: the building of the first tunnel under the Hudson River had finally been accomplished.

The Gunnison Tunnel

The Gunnison Tunnel, near Montrose, Colorado, was constructed between 1905 and 1909 as the main component of the Uncompahgre Valley Project. An irrigation tunnel, it was designed to divert waters from the Gunnison River through the Vernal Mesa into a 12-mile canal for the eventual irrigation of 146,000 acres of western Colorado territory.

Since the arrival of Brigham Young's Mormon pioneers in Utah in 1847, the young West had seen various attempts to tap the agricultural resources of areas rich with streams and rivers, but prevailingly arid. Sporadic local and state-financed ditchings and canals had led to modest successes and eventually, by the 1880s, to a speculative irrigation frenzy involving large acreages, stocks sold in various ventures, and, not rarely, project failures. The Desert Land Act of 1877 and the Carey Act of 1894 were the primary federal attempts to regulate a highly volatile situation, but these laws were hampered by widespread fraud and abuse; the 1893 depression did more to control such ventures. From 1888 on, the U.S. Geological Survey proceeded to survey and map much of the West, while pressure built for federal planning and implementation of large-scale water diversion efforts. With the election of Theodore Roosevelt, a firm believer in such a program, the Reclamation Act of 1902 found its prime mover and ushered in a new era in western agricultural development.

The Uncompahgre Valley Project was one of the first undertaken by the U.S. Reclamation Service as part of the Reclamation Act. The first two years following project approval were spent surveying, making topographic maps, researching, and planning water supply in the Uncompahgre Valley; analyzing tributary flow rates; determining advantageous sitings for the tunnel portals, tunnel line, and canal line; and searching for possible dam and reservoir sites. Contracts were let in October 1904, and construction began in January 1905.

Construction proceeded at four headings: the east and west portals and two headings at the base of the main shaft, sunk about one mile east of the west portal to tunnel depth at that point (262 feet). The two western tunnels were driven to meeting in July 1906, the eastern pair in July 1909; the first irrigation water was delivered through the tunnel on July 6, 1910.

Many setbacks were encountered during construction. The first occurred in May 1905, when the financially distressed Taylor-Moore Construction Company abandoned the job. The Reclamation Service took over. Three days later a cave-in killed six workers. Exonerated from blame in court on grounds of "unforeseen and unavoidable conditions," the service continued its direct supervision of the project.

The "unforeseen conditions" had been soft ground in the first few thousand feet of tunneling at the west end. Later hazards, exacerbating greatly the difficulty, danger, and expense of the Gunnison Tunnel, included pockets of combustible gas, seams of hot and cold water under intense pressure, a 2,000-foot-long fault zone, ground swelling, and flooding. Nonetheless, the overall average progress in the Gunnison Tunnel measured approximately 255 feet per heading per month. At one point an American record was achieved: 449 feet of heading advance in one month through hard granite. In 1905–1906 a three-shift gang achieved 7,500 feet in one year through shale.

Major equipment used included piston drills for granite and various augers for clay, coal, and other soft materials; and electric lighting, ventilation, and pumping machines. Power was provided by two 320-horsepower boilers, ventilation by two compressors and four cycloidal blowers feeding 16-inch riveted iron pipe. A 65-ton steam shovel was used at the west portal; 6-ton tramming locomotives, 35-cubic-foot dumping cars, and 54-cubic-foot nondumping cars (unloaded by an electric derrick) were used elsewhere.

It is very apparent that the completion of the Gunnison Tunnel held a high priority in the nascent Reclamation Service. Three private companies that had submitted proposals for completion of the project in 1905, following the failure of Taylor-Moore, were all rejected in favor of "forces working under the direct supervision of the engineers of the

Powerhouse of the river portal of the Gunnison Tunnel, Montrose, Colorado. The key to the first major transmountain irrigation system in the United States, this was the longest irrigation tunnel in America when completed in 1909. Water is diverted from the Gunnison River to the Uncompahgre Valley Project, one of the first undertaken under the Reclamation Act of 1902. (Courtesy of the American Society of Civil Engineers)

Excavation in progress for the west portal, 1905. (Courtesy of the U.S. Bureau of Reclamation)

Horse-drawn wagons hauling coal to the west portal, 1906. (Courtesy of the U.S. Bureau of Reclamation)

One of the electric locomotives used during construction, 1907. (Courtesy of the U.S. Bureau of Reclamation)

Reclamation Service." The South Canal, the other major component of the Uncompahgre Valley Project, was far less complex and was let to bid. It was satisfactorily completed in 1908.

Civil projects on the scale of the Gunnison Tunnel entail such dangers that loss of human life during their prosecution often seems inevitable. Yet only six lives were reported lost. This of course compares favorably with the cost at the Hoosac Tunnel—seventy-five men.

The completed tunnel runs 30,583 feet, roughly east to west. It is 10 feet wide, 10 feet deep at the sides, and 12.5 feet deep at the vertical center of section. The tunnel is concrete lined for about two thirds of its duration. At the eastern portal the bottom of intake is placed 7 feet below the low-water line of the Gunnison River, which at this point is passing through a steeply walled, 2,000-foot-deep canyon, with many surging rapids. This placement, while necessitating cableway construction apparatus for material transport to the east portal, was adopted because it greatly shortened the required tunnel length from that in previous plans. The west portal emerges in the Uncompahgre River valley, connecting to the 12-mile South Canal, which performs the remainder of the transfer at an estimated flow rate of 1,300 cubic feet per second.

The Holland Tunnel

When completed in 1927, the Holland Tunnel, with its unprecedented length of over 8,500 feet, was a bold step forward in vehicular tunnel construction. Its 29-foot-diameter twin tubes had been shield driven through extremely difficult river bottom conditions; its unique ventilating system for drawing off exhaust-laden air was a major innovation. Linking Manhattan and Jersey City, it was originally called the Hudson River Vehicular Tunnel but was later renamed the Holland Tunnel as a memorial to its builder, Clifford M. Holland.

The tunnel was initiated by two state-appointed bodies, the New Jersey Interstate Bridge and Tunnel Commission and the New York Bridge and Tunnel Commission, who began their deliberations in 1906 and were considering a vehicular tunnel by 1913. The original idea of building a bridge was replaced after investigations indicated that its cost would be prohibitive. The all-weather nature of a tunnel was attractive, in addition to its lower cost. Authority was granted in 1919 by the two states for the commissions to proceed with the construction of a vehicular tunnel between a point in the vicinity of Canal Street on the island of Manhattan and a point in Jersey City.

The able Clifford Holland, a young tunnel engineer for New York's Public Service Commission who was in charge of the construction of all subway tunnels under the East River, was chosen by the commissions to design and build the tunnel. Chief Engineer Holland took office in July 1919 and gathered a staff consisting largely of individuals who had worked on the East River subway tunnels. A number of traffic-engineering studies determined the dimensions and the best location for the tunnel. The planned separation of entrance and exit plazas was unique at the time. Eventually the planners settled on two tubes, each of which would accommodate two traffic lanes in a single direction. Each tube would have a diameter of 29.5 feet, the largest in the United States at the time. The north tube would be 8,558 feet long, portal to portal, and the south tube 8,371 feet.

Holland Tunnel, Manhattan to Jersey City. This twin-tube vehicular tunnel beneath the Hudson River, with its unprecedented length of 8,500 feet, was a major step forward in underwater crossings when completed in 1927. The 29-foot-diameter tubes were shield driven through extremely difficult river bottom conditions that were overcome by the ingenuity and determination of engineers Clifford M. Holland, Milton H. Freeman, and Ole Singstad. (Courtesy of the Port Authority of New York and New Jersey)

Holland faced many design and construction difficulties with a subaqueous tunnel of this length. A principal design problem was the ventilation system. With an expected constant stream of motor vehicles passing through, the tunnel would soon become lethal if some way of removing the fumes were not devised. Holland studied this problem very carefully, beginning by attempting to ascertain what quantities of exhaust were to be dealt with, what were the components of the exhaust, and what would be their effects on tunnel occupants. He initiated a series of studies conducted by the U.S. Bureau of Mines, Yale University, and the University of Illinois. The goal was to provide a tunnel as safe as any roadway.

This phase of the project included a series of tests on motor vehicles in closed chambers as well as on the road. Exhaust fumes were measured and analyzed. Volunteers were exposed to fumes to determine their effects on people. The tests attracted much attention among engineers, such research having never been the basis of a tunnel design before. It was looked upon as of the greatest importance in determining the future development of vehicular tunnels.

The final design consisted of four vent buildings with a total of 84 fans (42 blowers and 42 exhaust fans). Fresh air was to be drawn in and blown by the blower fans into a duct running beneath the roadway. The air would enter the tunnel proper through a series of slots spaced 10 to 15 feet apart, located above the curb on each side. The steady stream of fresh air emanating from these louvers would mix with the exhaust fumes from tunnel traffic, and the mixture would be drawn up through grills in the tunnel ceiling by powerful exhaust fans in the ventilation buildings, where it would then be discharged into the open air. A complete air change could be accomplished in 90 minutes.

As for the method of constructing the tunnel itself, Holland considered trench, caisson, and shield techniques. The need to reduce interference with river traffic, as well as the silty composition of the river bottom, influenced his choice of the shield method rather than an alternative such as dredging a

At the heading of the tunnel, December 13, 1923. (Courtesy of the Port Authority of New York and New Jersey)

trench, floating tube segments over the trench, and sinking them in place (a process that would necessarily obstruct river traffic). By now the shield-driving technique was an accepted method of tunnel construction and known to be particularly suited to the subsurface conditions encountered in this case. Originated by Isambard Brunel for the Thames River Tunnel at London in 1825, the shield in various forms had been subsequently used in America several times, including in Alfred Beach's pneumatic subway in New York and in the Hudson-Manhattan Railroad Tunnel. The shield used in Holland's time was a steel cylinder whose forward end acted as a cutting edge and whose rear end overlapped the tunnel lining of inserted cast-iron rings. Inside the shield, hydraulic jacks pushed against the tunnel lining to drive the shield forward. As the shield thrust forward, the encountered material was either pushed aside or admitted into the shield through special openings and then removed. In subaqueous tunnels, compressed air prevented the entry of water into the shield.

Construction began on October 12, 1920. The first shield was built on the New York side, and tunneling was begun on October 26, 1922. Work progressed from both shores. Each shield was some 30 feet in diameter and a little over 16 feet long. The upper half had a 2.5-foot projecting hood. The shields were divided into 13 compartments. The thirty 10-inch jacks used in each shield had a total thrust available to drive the shield forward of about 6,000 tons. The cast-iron rings that lined the tunnel consisted of 2.5-foot-wide, 6-foot-long segments, bolted together. These were put in place by a hydraulic erector. A special grout lining was introduced under high pressure into the void created by the difference between the diameter of the shield and that of the rings. Hemp rope and lead wire were used to make junctures watertight.

The work was difficult. Holland was ever-present. Sadly, he did not live to see the completion of the great tunnel. He died on October 19, 1924, at the age of forty-one, and was succeeded by his assistant, Milton H. Freeman. Less than two months later, on December 7, 1924, the south tube headings were joined. Freeman himself died on March 24, 1925. Ole Singstad, Holland's engineer of design, succeeded Freeman and carried the work to its conclusion. On November 13, 1927, the tunnel was opened to traffic. In 1928, the first full year of operation, it handled 8,744,600 vehicles and became an indispensable part of the transportation network in the area.

The Moffat Tunnel

The Moffat Tunnel, a 6.1-mile rail passage under the Rocky Mountains 60 miles west of Denver, was the first tunnel to cross the Continental Divide and the longest railroad tunnel in the Western Hemisphere when completed in 1927. It eliminated 23 miles of twisting, winding track that cross the divide on the surface. The tunnel made possible the creation of a competitive and efficient transcontinental rail line through Denver, which previously had not been adequately served with rail transport. It opened up western Colorado to markets in the Midwest and generally aided in the development of the region. In the construction of the Moffat Tunnel a number of new tunnel-building techniques were demonstrated; as a by-product of one of them, the rapidly growing city of Denver benefited from a major new source of water.

In the 1860s it was commonly expected that the Union Pacific Railroad would pass through Denver, the largest city in the area. Surveys by Grenville Dodge of the Union Pacific had identified mountain passes in Wyoming and Colorado suitable for the expanding railroad. Near Denver only the Rollins Pass seemed a possible route over the Continental Divide, but it was still felt to present an uncompromising barrier, and the main line of the Union Pacific eventually passed north of Denver—to the chagrin of many Denver citizens and merchants. Agitation for a line through Denver persisted, however, and surveys over Rollins Pass were made by several railroad companies in the late 1860s. The Chicago, Burlington, and Quincy Railroad acquired part of the grade in 1884–1885 and ran additional surveys over the pass, as well as proposing two tunnels. Under the leadership of David H. Moffat, a rail line was eventually built over Rollins Pass in 1904. The line, which came to be known as the Moffat Road, was the only one going west from Denver at the time of construction. As of 1923, when tunnel construction began, the line extended about a third of the way to Salt Lake City (about 130 miles). Denver itself was served by spurs of the Denver–Rio Grande and Union Pacific railroads, whose main lines were both about 150 miles away.

Moffat Tunnel, 60 miles west of Denver, Colorado. Opened in 1928, this was the longest railroad tunnel in the Western Hemisphere at the time. It demonstrated new tunnel construction techniques and the innovative concept of using the pioneer bore as a permanent aqueduct. (Courtesy of the American Society of Civil Engineers)

The Moffat Road crossed the Continental Divide at Rollins Pass, at an altitude of 11,600 feet. At the engine-changing station at the pass, snowsheds were maintained to enable trains to operate during the winter months, when enormous drifts formed there. In spite of this measure the road was blocked up by snow 30 to 60 days each winter. It was estimated that 41 percent of the cost of operating the line went into fighting snow blockades, mostly around Rollins Pass.

Another major operational problem was the steepness of the grades in the section of the line around the pass. Engine changes had to be made three times: before the divide, at the pass, and on the other side. The grade between the changing stations and the pass was 4 percent, which of course added considerably to the effective burden of a train. Four or five specially designed heavy engines were needed to haul each train across the divide. The 90-mile trip from Tabernash (the western changing station) to Denver often took 14 to 16 hours.

When finished, the Moffat Tunnel eliminated all the 4-percent grades and reduced the maximum grade on the road to 2 percent, saving 2,400 feet in elevation. Instead of the four or five heavy engines, a single normal engine could haul a typical train from Tabernash to Denver in half the time previously required. The eastern portal of the tunnel is located at an elevation of 9,198 feet, the western one at 9,085 feet, both far enough below the timber line to avoid large snowdrifts on the tracks. The tunnel was constructed with a slight peak or high point in its center in order to facilitate drainage. On the eastern side of this peak the grade is 0.3 percent, and on the western side it is 0.9 percent.

Boring proceeded from both ends simultaneously—both ends of the main tunnel heading and of the so-called water tunnel, which served during construction as a conduit for crew, power lines, compressed-air feeders, water supply and drainage, and ventilating systems. The water tunnel was dug 75 feet away from the main tunnel and at a grade 7.5 feet higher than the floor of the main tunnel. The two-tunnel system was new in the United States, although it had been used in Europe and Canada. It enabled only actual excavation and construction to occur in the main tunnel, while the other was used as a service tunnel. The two-tunnel setup also permitted an extremely efficient use of manpower. The drilling crew and the mucking crew simply alternated, and one never had to sit idle while the other was working. Machinery was adapted to this system by being made mobile. Crews changed positions by moving through the nearest crosscut connecting the two tunnels; these were placed every 1,500 feet or so. Another innovation was the use of electric muck-moving machines, which saved considerable time over hand loading.

The general drilling procedure consisted of driving the railroad tunnel's center heading (8 by 8 feet) each way from the crosscuts from the water tunnel (8 by 9 feet) and later enlarging around the center heading in the main tunnel to a full 16 by 24 feet, standard American Railway Engineering Association dimensions. At the west portal the procedure differed somewhat. Because of soft ground the top heading was excavated first, and once that was supported, the bench was dug out to the full enlargement dimension. The heading was drilled with 26 holes, 9 feet deep; this took about 2.5 hours. Enlargement from the 8-by-8-foot heading to the 16-by-24-foot standard was done, in the hard rock, by the ring-shooting method: rings of 27 holes each were drilled 4 feet apart in the direction normal to the center line of the tunnel. The holes were filled with explosives and the ring segments blasted away.

Of the various difficulties encountered during the boring, one of the more serious was water leakage into the tunnel. This was worst at the east portal, where on one occasion 1,800 gallons per minute flowed in from a seam leading from Crater Lake, 1,300 feet above. On another occasion a seam 1,800 feet under Ranch Creek caused a complete flooding of the tunnel. Both leaks materially hampered progress.

Another particularly difficult problem was the result of the different geological conditions found in the east and west portals. Contrary to the predictions of the geological surveys, at the west portal large seams of a soft, water-laden gneiss schist were found, instead of the solid granite that was found in the eastern bore. When the talc-like schist was exposed to air, and especially in the presence of underground water, it had a tendency to collapse into the bore. Geologists had predicted that there would be less than 2,000 feet of questionable rock that might need support. In the end, 2.5 miles of such seams had to be supported, adding considerably to costs and delays. Bench excavations were especially difficult through the soft ground. By the time the bench crew started work, the roof had been exposed long enough to become treacherous, because of cracks and pressure buildup. Cave-ins occurred, one of them killing six workmen.

A special piece of machinery, the Lewis Cantilever Girder, was invented by the project's chief engineer, George Lewis, to enable construction to proceed through the soft seams. This device supported the roof timbers and wall plates (as a cantilever extending back from the heading) while the bench was excavated and the posts set. It was designed to be slid from one position to the next. In the soft rock the timber support systems had to be adapted and some steel-reinforced concrete used where the timber failed altogether. In all, 916 linear feet of steel lining was buried in concrete and a total of 12,000 yards of concrete lining was used, including that around the wooden sets.

The water tunnel, once it was no longer needed as a service tunnel for construction, became an aqueduct, transporting water from the Colorado River basin on the western slope to the dry eastern slope and to Denver. It was lined with concrete in order to prevent loss of water through seams and seepage into the main tunnel.

The Moffat Tunnel project was perhaps unique in its time in the amount of attention given to safety and comfort considerations for the workers. Both camps, east and west, were equipped with decent lodgings, schools, hospitals, and recreation facilities. The men were instructed to report sick at the slightest sign of a cold, and consequently the incidence of pneumonia was greatly cut down. Hot meals and an unlimited supply of coffee were provided to the crew in the tunnel. All in all, both efficiency and worker morale were very high. Progress was rapid, and casualties were kept to a minimum. On the human level the project was a definite success.

When the final holing-through blast, touched off by President Coolidge on February 10, 1927, broke down the last granite barrier between the east and west headings, it was said that the Moffat Tunnel was both the highest and the lowest tunnel in United States history: the highest because it was more than 9,200 feet above sea level, and the lowest because the bore was 2,800 feet below the crest of James Peak on the Continental Divide. Twenty-three miles of treacherous mountain tracks had been bypassed. But the tunnel was without value as long as the rail line did not continue to Salt Lake City, and for some years financial problems prevented the extension of the road. Even a relatively inexpensive connection to the Denver–Rio Grande line at Dotsero proved to be out of range. Other problems that continued to haunt the tunnel project included the financing of the increased costs due to the unexpected rock conditions. It was not until 1932 that things began to fall into place. In that year the Denver–Rio Grande Railroad obtained control of the Moffat Road. The earlier plan to connect the two via a cutoff west of the tunnel at Dotsero was finally realized in June 1934, putting Denver and the Moffat Tunnel on a major transcontinental line.

The Detroit-Windsor Tunnel

This subaqueous highway tunnel, built between 1928 and 1930, is a remarkable engineering feat. Its construction involved three distinct tunneling techniques: cut and cover, a shield with compressed air, and a trench and sunken tubes. Of particular note, its 32-foot-diameter main channel section required 65 miles of arc welding, the first major use of arc welding in tunneling history. The helical ramps at the Detroit approach were another innovation.

The Detroit-Windsor Tunnel was the third great subaqueous tunnel built for vehicular traffic in America, preceded by the Holland Tunnel in New York and the Posey Tunnel at Oakland. It is larger in diameter than either of these, but not as long as the Holland Tunnel. The Detroit-Windsor Tunnel extends from a terminal in the business district of Detroit under the Detroit River to a terminal in the center of the Windsor, Ontario, business district.

As early as 1871, groups in Detroit were arguing the merits of a bridge versus a tunnel across the river. Responding mainly to shipping interests and their fear of river impediments, attention focused on a tunnel. In 1871 ground was broken for a railroad tunnel intended to have a 15-foot bore surrounded by masonry. Work proceeded slowly, until at 135 feet sulphurous gas was encountered. Workmen became sick and refused to enter the tunnel. The project was abandoned. A second railroad tunnel was started in 1879 to connect Grosse Ile with the Canadian shore. The workmen soon encountered limestone formations, which caused excavation costs to soar. This project too was abandoned.

The completion of the Grand Trunk Railway Tunnel at Port Huron, Michigan, in 1890 caused another flurry of activity, as Detroit businessmen feared a loss of trade to Port Huron. Although nothing immediately came of it, the eventual result was the Michigan Central Railway Tunnel at Detroit, begun in 1906 and completed four years later.

Detroit and Windsor finally had a tunnel. But a new form of transportation, the motor

Detroit-Windsor Tunnel, connecting Detroit, Michigan, and Windsor, Ontario, constructed 1928–1930. This subaqueous single-tube highway tunnel was an exceptional engineering achievement, using three distinct tunneling techniques: cut and cover for the land sections, compressed-air shield for the channel approaches, and trench and sunken tube under the main channel. The 32-foot-diameter main channel section involved 65 miles of arc welding, the first major use of arc welding in tunneling history. The helical ramps at the Detroit approach were another engineering innovation. One of the tubes to be submerged is shown. (Courtesy of Parsons Brinckerhoff Quade and Douglas, Inc.)

vehicle, soon became well established and required its own river crossing. A vehicular tunnel was proposed in 1919 but failed to materialize because of the Canadian government's stated inability to offer financial aid for the project. One tunnel proponent, however, would not quit so easily. Fred W. Martin, a Windsor Salvation Army captain, visited the engineering firm of Parsons, Klapp, Brinckerhoff, and Douglass in New York and suggested that a Detroit-Windsor tunnel would be not only a feasible but also a profitable undertaking. A group of Detroit bankers agreed to back the project with some help from Chicago and New York bankers and with the Parsons firm's guarantee of the costs. Thus it was that the Detroit-Windsor Tunnel became one of the few tunnel projects at the time to be privately financed, built, and operated.

Construction began in the summer of 1928, on both sides of the river. Completion of the tunnel was accomplished by three separate tunneling methods. The work was divided into five sections. The two land approaches were built by the cut-and-cover method, which simply involved cutting a trench and erecting in it a standard box-type structure with steel vents and concrete jack arches, which were subsequently covered over. The shield method was used for the two sections extending from the open-cut operations to the harbor line. Muckers dug ahead of the shield, slicing away the earth with power-operated knives and passing it back through the shield. After a few feet had been excavated, the shield was forced ahead by thirty hydraulic jacks to occupy the space. After each of these great shoves, a hydraulically operated erector arm picked up segments of the lining and swung them into place. When a "ring" was completed, the muckers again advanced.

The river section was constructed by the trench-and-tube method. This involved digging a great ditch across the river and then sinking nine huge steel tubes into it. The tubes were built in Ojibway, Ontario. The shapes were welded watertight and then launched like ships, to be towed by large tugboats to the crossing. Interior and exterior concrete was poured. Then, almost submerged, each section was swung into its proper position over the trench, fastened to a barge by steel cables, and lowered into place by applying weights to each end. Alignment of the tubes was accomplished through the use of tall masts attached to each tube as it was sunk. Divers joined the tubes together, and each joint was sealed with a collar of tremie cement. The trench was then back-filled.

When contact was made with the shield-driven sections, the alignment was only one inch off. The sections were attached, and watertight bulkheads removed. A concrete roadway, ventilation towers, and other ancillary facilities were constructed. The tunnel opened on November 1, 1930.

Chapter Six

Water Supply and Control

Introduction

In 1754 a community water supply system was built by Hans Christopher Christiansen in Bethlehem, Pennsylvania. A plentiful supply was provided by a hydraulic pump forcing spring water through wooden pipes into a wooden reservoir, from which water was distributed. Although there had been earlier experiments with water distribution from central sources to nearby locations, the Bethlehem system represented one of the first successful trials of a new technology on a large scale. The new technology was that of water impoundment and distribution by means of piping systems, reservoirs, mechanical pumping equipment, and, later, tunnels and aqueducts. The emerging nation required more than the provision of water for consumption by communities, however: the need for hydropower, flood control, and irrigation also demanded the creation of systems of water supply and control.

Natural springs initially provided water for many of the nation's settlements. Private wells, such as those sunk in Boston by 1640 with the permission of the selectmen, were also common early sources. These wells often proved inadequate or became fouled, however, and other means were sought to supply water. The Massachusetts general court incorporated a company in 1652 to provide water for Boston. A 12-foot-square reservoir was built, which was linked to nearby springs and wells by gravity feed through a system of connected bored logs. This supply, used for both drinking and fire fighting, never proved completely successful. By 1817 some 40 miles of wooden water mains served about 800 Boston families, but reliance continued to be placed on several thousand private wells. It was not until the building of the Cochituate Aqueduct, completed in 1848, that the city made a major step in solving its water problems.

Other cities also moved to deal with their water problems. Schaefferstown, Pennsylvania, constructed a gravity-feed system in 1732. Salem, North Carolina, built a citywide distribution system using bored wooden logs in 1775. Cincinnati was the first of the cities in the West to develop a citywide system, with construction beginning in 1819.

Far more ambitious, however, was Philadelphia's famous waterworks project, prompted by a series of disastrous yellow fever epidemics in the 1790s. Under the able direction of Benjamin Latrobe, a system unique in America—particularly in its use of steam-driven pumping engines—was built and put into operation in 1801. The replacement of the wooden main system with a cast-iron system in 1818 also marked the first extensive use of this material for pipes in the United States. The system was a marvel of its day and a prototype for other urban systems.

New York was also struggling with its water supply system at this time. Again relying initially on wells, the city eventually constructed a covered wooden reservoir and limited distribution system in 1774. It was soon abandoned, and reliance was once again put on wells. But this solution was known to have shortcomings. For example, the cholera epidemic of 1832 was attributed to polluted wells, and not enough water was on hand to contain the disastrous fire of 1835. The city's problems were finally solved through the construction of the renowned Croton Aqueduct. Built largely under the direction of John Jervis, the aqueduct was opened in 1842 and was immediately successful. The original Croton Aqueduct was later expanded as the city grew, and new aqueducts added to tap other watersheds, but the first one served the city well in its burgeoning years.

Chicago relied at first on Lake Michigan for its water needs. With population growth and increasing shoreline pollution, however, the city soon felt the need for a more sophisticated water system. Shortly after the severe typhoid epidemic of 1848 Chicago granted a charter for a municipally owned waterworks, but by the early 1860s shoreline pollution had seriously contaminated the intake areas for even the new system. To respond to this problem, a unique 2-mile underground tunnel was built under the direction of Ellis Chesbrough out into Lake Michigan to a timber intake crib in deep, clean waters.

The complex water supply systems created for Philadelphia, Chicago, and New York were among the wonders of their day. Not only did the new systems prove a great convenience to the populations of these rapidly growing cities, but they dramatically reduced deaths and sickness due to diseases caused by contaminated water supplies. The ability of cities to deal with another constant threat—fire—was also greatly increased.

The knowledge of techniques necessary for building these systems depended upon self-teaching and remote precedents. Certainly European prototypes, such as the great London waterworks, were important; but as with so many other European innovations, these were not necessarily well known to engineers in the United States, particularly in the eighteenth and early nineteenth centuries. The rapidly developing industrial technology of the nineteenth century made many of these systems possible, as evidenced by the introduction of steam-powered pumps that could raise well water to the height required for sufficient gravity flow, and by the use of cast iron for the making and reinforcing of pipe networks. Although systems of this type were built to serve existing city fabrics, they also became highly influential in shaping the course of further urbanization. Indeed, without them the city as we know it today would be unthinkable.

These great urban systems were oriented toward solving problems of quantity and access. The quality of the water was also highly valued, and felt to be of importance in controlling the spread of disease, but initially improvements in water quality were achieved for the most part by tapping distant watersheds or removed water bodies. With time, however, the need for purification techniques that would treat the water before it even reached distribution networks became evident. The growing body of knowledge resulting from the work of such scientists as Snow, Pasteur, and Koch in the period from the 1850s to the 1870s brought a better understanding of the role of water in the transmission of germs. This in turn led to improvements in filtering and treatment techniques for water. Methods such as slow sand filtration—employed in Richmond in 1832—had been explored earlier, but it was not until the end of the nineteenth century that more sophisticated techniques based on scientific principles were established. Work by James Kirkwood and by the Lawrence Experiment Station in Massachusetts was critically important in these developments. Water processing thus became yet another component of the comprehensive water supply systems used in American cities.

In the field of water supply and control for irrigation, most significant developments occurred in the western part of the country. The irrigation ditches built on the west bank of the Rio Grande in 1598 are among the country's first constructions to serve irrigation needs, as are the acequias of San Antonio, which date from 1718 and were built to serve the fields and missions of the original Spanish settlement. The mission at San Diego built an irrigation system, and around 1776 irrigation systems were developed near the present site of Santa Cruz. On a larger scale, the Mormon pioneers in the valley of the Great Salt Lake placed lands under irrigation as early as 1847. Small diversion structures and short canals were built. As important, however, to the development of later water practices in the west were the Mormon practice of attaching water rights to the land and the creation of companies to build irrigation works. The California gold rush of 1849 also led to increased irrigation needs, and by 1857 important irrigation developments were occurring in southern California.

Most of the irrigation ventures at the time were privately sponsored. These endeavors were not always successful. Conflicts over control and governance of water resources also developed among private interests and local and state governments. It was soon argued by many that the issue of water rights and the development of water resources would benefit from federal involvement. Such involvement began in the late nineteenth century with experiments such as that at Embudo, New Mexico, in which members of a team sponsored by the U.S. Geological Survey developed stream-gauging techniques that added significantly to the existing scientific knowledge about water flow. The project was led by John Powell, a naturalist and proponent of federally assisted irrigation programs for developing the West.

Major federal intervention came about through the 1902 Reclamation Act, which provided for federal sponsorship and construction of water resource projects. A series of multipurpose projects that included meeting irrigation needs were developed, among them the construction of many notable dams for water impoundment, such as the Roosevelt Dam in the Salt River Project and the Elephant Butte Dam. Subsequent reclamation acts increased the number of irrigation projects in the West.

Flood control was the chief objective in the development of many water control systems. On the Mississippi River, for example, which has a long history of major floods, a levee over a mile long and 3 to 4 feet high was built in 1727 on the site of the present New Orleans under the direction of the French engineer Blond de la Tour. By 1803 levees of one type or another extended about 100 miles upriver from New Orleans. The state of Mississippi passed legislation authorizing the construction of levees in 1819. Nonetheless, devastating floods continued to occur periodically. The Swampland Acts passed by the federal government after the floods of 1849 and 1850 were significant in setting the basis for later flood control projects. The Mississippi River Commission was created by Congress in 1879 and, although primarily responsible for navigation, proposed the building of levees. In 1882 the Rivers and Harbors Act placed primary construction responsibility in the hands of the Army Corps of Engineers, but the Mississippi River Commission still undertook the building of some well-designed levees. The commission's mandate was extended by the first Flood Control Act in 1917. The 1925 Rivers and Harbors Act provided for waterway planning. Disastrous floods, however, continued to occur. One of the worst, in 1927, precipitated a comprehensive federal flood control plan for the Mississippi. A series of engineering efforts involving levees, revetments, floodways, channel stabilization, cutoffs, reforestation, and other techniques were brought to bear—approaches still used today.

Other river basins were also the subjects of extensive flood control programs, notably the Missouri River basin, where the measures included the building of many dams, reservoirs, and other facilities. The Ohio and Tennessee rivers were the targets of major programs as well. One of the most interesting projects to be carried out by a local district was that in the Miami River basin in southwestern Ohio, involving the construction of a variety of dams and other facilities.

Many of the federal flood control programs in the West were undertaken as a consequence of the influential 1902 Reclamation Act. Other acts of Congress that involved the U.S. government in flood control include the 1920 Federal Water Power Act, creating the Federal Power Commission; the 1928 Boulder Canyon Act; the Tennessee Valley Authority Act of 1933; and the Flood Control Act of 1936, which established flood control as an appropriate federal activity and set in motion a large number of activities by the Corps of Engineers. Other agencies, such as the Soil Conservation Service, have also been extensively involved in projects related to flood control.

The great systems of water supply and control that serve municipal water supply, power, irrigation, flood mitigation, drainage, industrial, and other needs are truly significant engineering contributions to the social and economic fabric of our country.

The Acequias of San Antonio

The acequias, or irrigation canals, located in the present city of San Antonio, Texas, are one of the earliest recorded water supply and irrigation systems in North America. Sections can still be seen of the first acequia, dating from 1718, which was built to supply water to the mission of San Antonio de Valero, now known as the Alamo. The later Espada Acequia, with its dam and aqueduct, is still operational, as are parts of the San Juan Acequia. Traces of the San Jose and Upper Labor acequias can still be found.

These humble waterways formed an essential part of daily life in the old Spanish settlement. Their course greatly influenced the physical pattern of the community's development into a major city. It was only toward the end of the nineteenth century, with the advent of a modern water supply system, that the importance of the acequias diminished. Moreover, they typify the early Spanish irrigation systems that were the prototypes for the extensive systems that evolved in the West in the midnineteenth century. The introduction of these huge irrigation works was a major factor in the development of many parts of the West.

Few considerations were more important in determining the location of Spanish settlements in the Southwest than the availability of water. Estimates of the costs of building irrigation canals and dams on potential sites were carefully considered by provincial authorities. The decision to locate a settlement in the area now encompassed by the city of San Antonio was significantly influenced by the presence there of several gushing springs.

In 1716 a party of Spanish soldiers under Martin de Alarcon, governor of Coahuila and Texas, met with the party of the missionary Fray Antonio de San Juan Buenaventre Olivares in the gently sloping valley to establish a presidio and mission. Father Olivares chose a spot on the west bank of the San Antonio River for his mission, the San Antonio de Valero. About a year later he moved the mission to the east bank, where a new structure, now known as the Alamo, was built. Digging began in 1718 on a long irrigation canal from

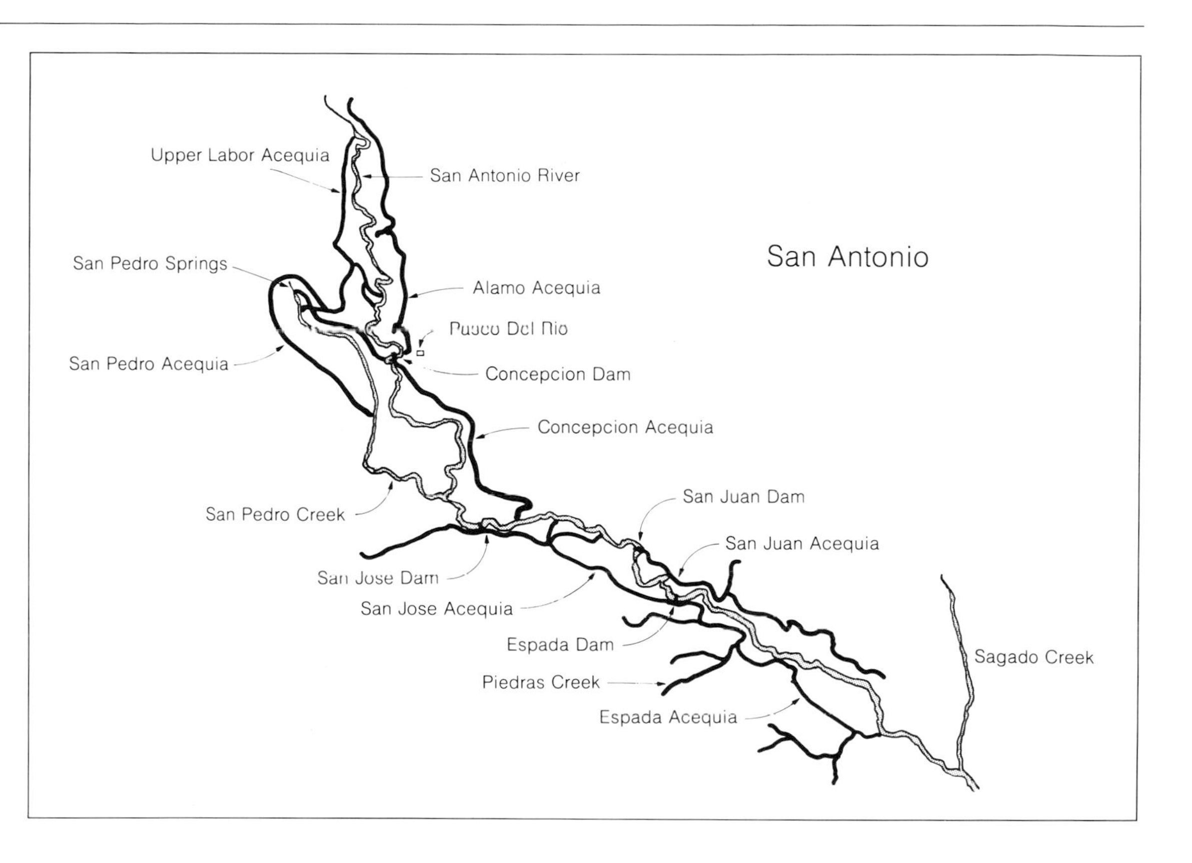

Acequias of San Antonio, Texas. Built in the eighteenth century to serve the Spanish missions clustered along the San Antonio River, these canals represent one of the earliest recorded water supply and irrigation systems in the country. Construction of the first acequia began in 1718. (Courtesy of the American Society of Civil Engineers)

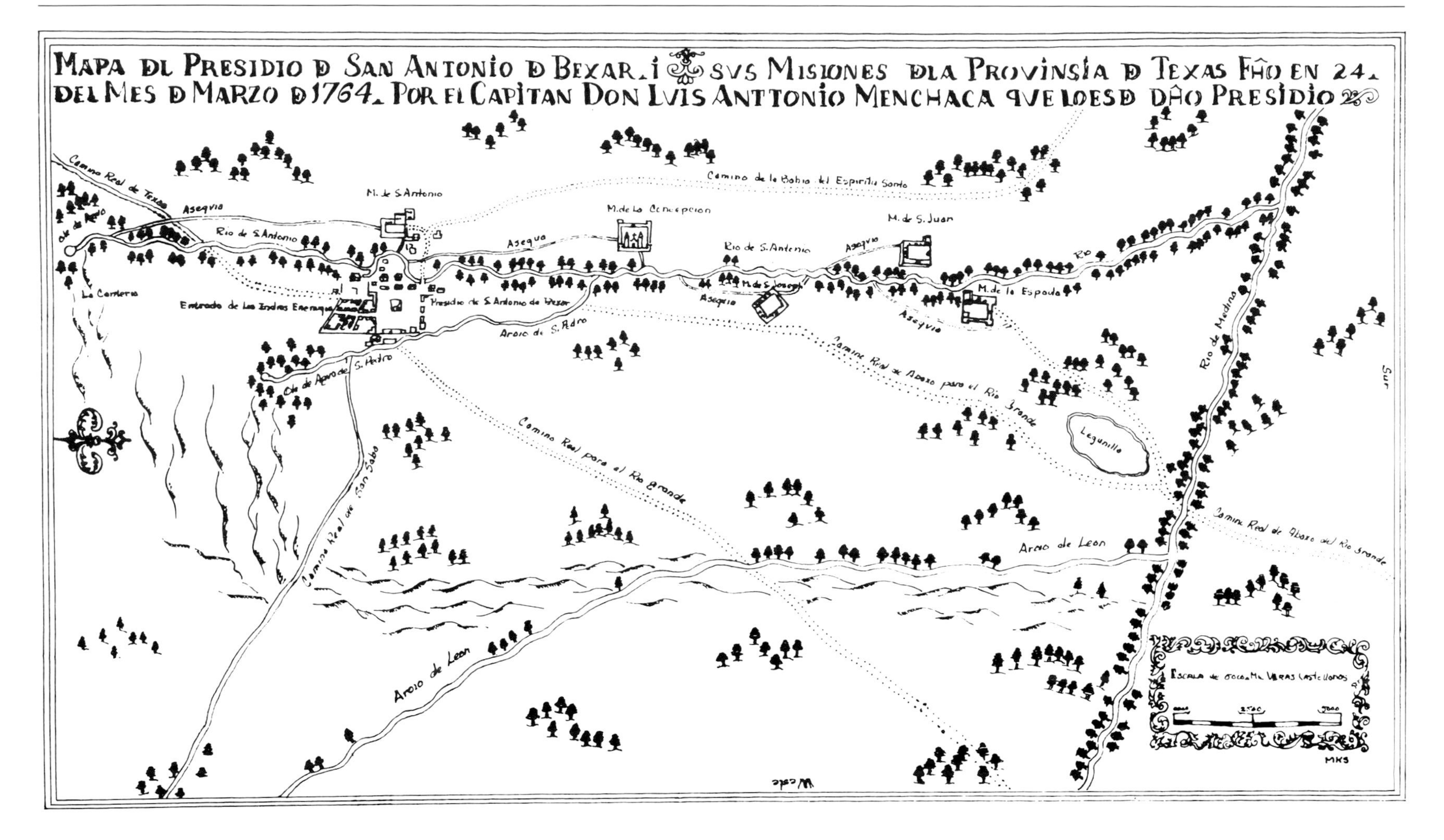

This 1764 map shows San Antonio's missions and several of the acequias. (Courtesy of Brown University)

The Espada Acequia, one of the original eight canals. (Courtesy of the American Society of Civil Engineers)

the headwaters of the San Antonio River to the fields adjacent to the mission. By 1727 the mission enjoyed a working irrigation system known as the Alamo Madre Acequia. The system continued to be extended until 1744. Over 600 acres of the mission's lands were fully watered.

The Alamo Madre Acequia was the first of eight major acequias to be constructed for the new colony. The locations of several of these are indicated on a map drawn in 1764 by Captain Luis Menchaca of the Presidio de San Antonio de Bexar. A canal serving the San Juan mission, started in 1731, watered about 500 acres for over two centuries. A portion of the San Juan Acequia has been renovated and is operational today. The acequia serving the mission of San Francisco de la Espada, also started in 1731, still flows vigorously. This canal drew its waters from behind the Espada Dam on the San Antonio River. The Piedras Creek was a barrier to the 5-foot-wide canal, but the creek was vaulted by an arched masonry aqueduct that still carries the waters of the acequia.

The Concepcion, or Pajalache, Acequia, which served the Concepcion mission and other parts of the colony, was actually finished before the Alamo Madre Acequia, even though it was begun later. Some accounts of the period suggest that it was finished about 1729 or 1730, although it should be noted that the Concepcion mission was not established on the San Antonio River until 1731. The San Jose Acequia served one of the most beautiful of all the Spanish missions in Texas, the mission of San Jose. It provided water for herds of domestic animals and for a grist mill, as well as irrigating 600 acres of land. The San Jose Acequia was completed in 1730. Other canals include the Upper Labor Acequia, known to be under construction in 1777, and the San Pedro Acequia, which was watering fields by 1739.

The last major acequia to be built was known as the Alazan Ditch. (During the United States period in Texas history, the canals were typically referred to as ditches rather than acequias.) Now virtually obliterated, it was opened on June 9, 1875, but not completed until late 1876. Also constructed in the same period was the Valley Ditch, actually an extension of the Alamo Madre Acequia.

The identity of the engineers who built the acequias is not known, although most were probably Spanish army officers. What construction guides they used or what technical knowledge they possessed is also largely unknown. There were major European prototypes for canal construction, however, such as the Canal du Midi, built by Pierre-Paul Riquet in France near the Spanish border in the 1680s, and numerous smaller-scale irrigation works. It is likely that at least some of the methods employed in constructing one or another of these were known to the Spanish engineers. The design principles involved, however, were probably largely empirical and not theoretically based. Belidor's classic *Architecture Hydraulique* was not published until 1740, and Chezy's important equation for the velocity of flow in a channel was not enunciated until 1775.

One section of the Alamo Madre Acequia was excavated in 1966 and examined to determine the nature of its construction. The walls were found to be lined with soft, quarried limestone blocks that varied in depth from 10 to 14 inches and in length from 11 to 14 inches. Originally, five courses of stone were used, set with a dry, sandy mortar. The height of the walls was 5.2 feet, and the width of the canal 6.3 feet. This same segment has since been completely restored and bears a historic marker placed by the state of Texas.

The Bethlehem Waterworks

Regular operation of the first pumped waterworks in North America began in Bethlehem, Pennsylvania, in 1755. The system's water-powered machinery pumped water to every house in the town.

Gravity-feed water systems had already been tried in colonial America by the time the Bethlehem system was developed. The Massachusetts general court incorporated a company that in 1652 constructed what was known as "the circuit"—a small reservoir linked to nearby springs and wells by gravity feed through a system of connected bored logs. The supply, used for both drinking and fire fighting, proved inadequate and fell into disuse. Another gravity system was constructed in Schaefferstown, Pennsylvania, in 1732.

But it was the Moravian community of Bethlehem that provided the country's prototype for a sophisticated and practical water supply system. In 1754 Hans Christopher Christiansen, a millwright from Denmark newly arrived in Bethlehem, began to build the waterworks for the small community. His project turned out to be much more successful than any predecessor. The townspeople had long obtained their water from a well near the town gate at the bottom of the hill on which Bethlehem sat. Delivery kept one man busy with buckets and carts from morning till night. Water power was already in use, running an oil and bark mill nearby, and it seemed feasible to adapt this mill's wheel to the task of furnishing water to the town. Christiansen decided to undertake the project.

The two most important considerations were the material for the pipes and the construction of the pump. Fortunately, one of the city's pioneer founders, John Boehner, who was then serving as a missionary in the West Indies, was visiting Bethlehem. An ingenious and practical man, he had some familiarity with pumping mechanisms and was able to construct a model of a pump and its connections. Christiansen then began to build a full-scale machine. In anticipation of the next phase of the project, hemlock logs were floated down the Lehigh River by raft from Gradenhuetten in March 1754, to be made

Pump house for the Bethlehem waterworks, Bethlehem, Pennsylvania. Built in 1754–1755, this was the first significant pumping system to provide drinking and washing water in North America. A wooden waterwheel, driven by the flow of Monocacy Creek, drove wooden pumps that lifted the water through wooden pipes to the top of a hill, where the water was then distributed through pipes by gravity. The existing building dates from 1761; it was preceded by a frame structure built in 1754. (Courtesy of the American Society of Civil Engineers)

into water pipes, and a pump house was built near the oil mill where the water was to be supplied for Christiansen's first experiments. On the evening of June 21, 1754, the mechanism was demonstrated. To everyone's astonishment and delight, Christiansen's pump forced the water to the top of the hill. Christiansen proceeded to perfect the machinery, and when it was ready the pipes were laid and a shingle-roofed wooden water tower, crowned with a weathervane, was constructed in the middle of the town. Small distribution tanks were constructed there also. On May 27, 1755, water was forced into the 70-foot tower, and on June 27 it flowed through the pipes for the first time. The source of supply was a spring from magnesian limestone near the banks of Monocacy (then Menagassi) Creek. From there the water was conducted underground for 320 feet to a cistern, where it was raised by a 5-inch lignum vitae pump via the bored hemlock logs.

Bethlehem, like all the other communities that used wooden pipes, had trouble with leakage. Christiansen tried to remedy the problem by substituting 1.25-inch pipes fabricated from sheet lead and soldered along the edges. The pipes were bedded in a cement of pitch and brick dust and laid in a gutter. Apparently this innovation was unsuccessful.

In 1761 Christiansen built a larger works, using an 18-inch undershot wheel, 2 feet clear in the buckets, which drove three single-acting iron force pumps of 4-inch bore with an 18-inch stroke. The pumps were made by the community blacksmith, Brother Blum. Christiansen was paid 30 shillings "for ye water running." The main was made of gumwood and the distribution pipes were of pine. The latter were replaced in 1769 and a lead main was installed in 1786. Lead distribution pipes were also added in some places in that year, and cisterns were constructed around the town. In 1791 the last of the pine pipes were replaced.

An English traveler named Isaac Weld wrote of Bethlehem in 1796: "Every house in the town is supplied with an abundance of excellent water from a spring, which is forced through pipes by means of a hydraulic machine worked by water, which is situated on the banks of the creek. Some of the houses are supplied with water in every room. The machine is very simple, and could easily raise the water, if necessary, several hundred feet." The Moravians apparently developed similar systems when they moved elsewhere, for a visitor to the Moravian community of Salem, North Carolina, in 1786 commented with like admiration that all houses were supplied with running water that was brought to town by a conduit 1.5 miles long.

Bethlehem's wooden tank stood in the square until September 1, 1802, when it was replaced by a 15-foot octagonal stone tower constructed on Market Street. This new structure distributed the water from an elevation of 112 feet above the spring. Bethlehem was undergoing a building campaign at the time, and in an effort to ensure the construction of a new church, it had been suggested that the water tower be incorporated into the belfry of the church. Apparently this idea was too revolutionary for the citizens, who eventually raised the money to build a church separately.

The next innovation was the introduction in 1813 of iron pipes, joined with leather fittings held secure by iron clamps. In 1817 the increased demand for water was met by building a reservoir, 17 feet by 10 feet by 7 feet deep. A second reservoir was added in 1832. In the same year the triple pumps, which had been used for 70 years, were replaced by a double-acting pump of greater capacity made by Rush and Muhlenberg of Philadelphia. It was still in use in 1890. In 1845 the Bethlehem Water Company took over the works from the water committee, which had controlled them from the start. In 1868 steam power replaced the water-power machinery, and in 1871 the town purchased the entire system.

The early works were a source of widespread admiration and were undoubtedly known to Benjamin Latrobe when he designed the Philadelphia system. Those who submitted proposals for the New York City waterworks may also have been familiar with the Bethlehem works. A register of the names of visitors to the Bethlehem works includes Governor Penn and Lord Charles Montague (April 1768), Baron von Reefsdorf (1774), John Adams (1777), and General George Washington and adjutants (1782).

The Philadelphia Municipal Water Supply System

The Fairmount Park waterworks graces the eastern bank of the Schuylkill River in Philadelphia. Its Greek revival buildings form part of the first truly large-scale municipal water supply system in the country. William Penn's green and rocky "Faire-Mount" rises dramatically behind the cluster of buildings. The hill, now the location of the Philadelphia Museum of Art, was the site of the original reservoir. The Fairmount works were used from 1812 to 1909, when a filtration system was added, the forebay filled in, and much of the waterworks machinery discarded.

By the 1790s Philadelphia's water sources —wells, cisterns, and springs—were inadequate. Philadelphia was the largest city in America; 28,522 persons lived in the city proper, and those on the outskirts brought the total number to 42,520. The purity of the Philadelphia water was a matter of civic pride as early as the 1750s, when the city began buying private wells. By 1771 the city owned 120 of the 498 wells in the city. By the 1790s water from many of these sources (particularly shallow wells) had developed an offensive taste and smell. In addition, the amount of water available from pumps and wells was inadequate to fight the fires that easily developed in such a large and densely packed city.

Another very serious problem was that of disease. A three-month yellow fever epidemic in 1793 resulted in 4,000 deaths—10 percent of the population. Yellow fever hit again in 1794, in 1797, in 1798 (that year leaving 3,800 dead and forcing three quarters of the population to the countryside), and for a fifth time in 1799. The causes of the disease were a source of extended public debate. One account attributed it to filth from the streets and marshes, another to infection from foreign ships. There was general agreement, however, that cleaning the city would help combat the occurrence and spread of disease. Fire companies were instructed to flood the streets daily with clean water. Whether or not clean water was available became the subject of another controversy. Many believed that all the water was contaminated because Philadelphia wells were placed so near privies and graveyards.

During the 1790s the Delaware and Schuylkill Canal Company proposed to bring fresh water to Philadelphia from Norristown, 16 miles away. The company failed and the plan was never accomplished. The attempt did, however, bring to the forefront the question of whether or not a municipal water supply should be owned by private corporations or by the public. By the end of the eighteenth century the need for fresh water was acute, and Philadelphia, unlike most other American cities, had committed itself to public control of its water systems. Several plans for a new supply system were debated. Completing the aborted canal from Norristown was considered, but the two years it would take to finish was deemed too long. Although no extensive municipal water supply system existed in the United States when Philadelphia decided to initiate one, adequate technology for building such a system had been available in this country for several years. Bethlehem, Pennsylvania, had completed a system in 1755 consisting of water-powered pumps and gravity-feed distribution pipes, and several other cities had experimented with various types of distribution systems.

Benjamin Henry Latrobe, whose plan was finally used by Philadelphia, moved there from Virginia in 1799 to begin designing the Bank of Pennsylvania. Latrobe, born in England in 1764, was gaining a reputation as an architect and engineer. He had built several grand homes in England and had made a study and design for the Chelmsford Canal in 1793–1794 before emigrating to the United States. While in England he had trained as an architect under Samuel Pepys Cockerell. It was from Cockerell, and less directly from other English architects including the Adam brothers, that Latrobe had adopted the Greek revival style. He had trained as an engineer with John Smeaton. In America he had been commissioned to survey the possibilities for navigating the Appomattox River, had worked for the Dismal Swamp Company in Virginia, and had designed a penitentiary in Richmond. The latter, long a Richmond landmark, embodied Thomas Jefferson's notions of a progressive and humane prison

The Fairmount waterworks on the Schuylkill River, part of Philadelphia's historic municipal water supply system, constructed between 1799 and 1822. This innovative system was a prototype for water supply systems in other major metropolitan areas. (HAER Collection, Library of Congress)

system. Latrobe had also planned improvements for the Norfolk fortifications and designed several elegant houses elsewhere in Virginia.

Latrobe had to be an active campaigner for his plan in Philadelphia. A strong political opposition to a municipal system was fired by the Delaware and Schuylkill Canal Company and its supporters, who were unwilling to lose the chance to complete their project. In the end Latrobe's arguments, often advanced through pamphlets, won out. His plan had been worked out in only a few weeks, barely two months after his arrival in Philadelphia and at a time when he was busy with the design of the Bank of Philadelphia. Latrobe's plan, efficient and revolutionary for the time, was to be in operation by July 31, 1799. It called for the drawing of water from the Schuylkill rather than the Delaware River. The latter, Latrobe said, was contaminated by wharves, ships, marshes, and sewers, whereas the Schuylkill water was "of uncommon purity," flowing on a rock bed from limestone sources. A drawback of his plan in the public eye was the size of the machinery to be used in the works. The employment of steam power to raise water from the river, an unheard-of idea, engendered tremendous controversy.

Attacks on the Latrobe plan were vicious and often directed at Latrobe personally. In the era of isolationism that followed the revolution, distrust of foreigners ran high. Regionalism, with roots in the colonial period, persisted as well. Latrobe's British-Virginian background was highly suspect in Quaker Philadelphia. The attacks continued throughout the construction of the works. Latrobe's overly optimistic estimates of cost and completion time caused some to doubt his honesty, and this helped to prolong the controversy.

Philadelphians were especially skeptical about the use of steam engines. Latrobe, although familiar with the waterwheel pumps used in London, was convinced that steam engines, which had been used successfully for other purposes in this country and in Europe, would better serve this purpose as well. Where his knowledge of steam engineering originated is something of a mystery. Later in his life he related that his father had been a friend of Matthew Boulton (1729–1809), the English steam engine manufacturer; yet he himself seems not to have worked with steam engines before the Philadelphia project.

Latrobe's plan provided for one steam engine to be located on the Schuylkill at Chestnut Street, about a mile below the surviving Fairmount works. Water would be pumped from the river to a settling basin at its edge and from there through a tunnel to Centre Square, the present site of Philadelphia City Hall, on high ground and precisely in the center of William Penn's original plan for the city. There the water would be pumped to a height sufficient to allow it to flow by gravity to all the settled districts of the city, which at that time lay largely east of Centre Square. In addition, a masonry culvert was to bring drinking water from Spring Mill, 12 miles away. Most of the culvert was to be sunk below the frost line; it would cross valleys on segmentally arched viaducts. Latrobe's plans went beyond a system for the mere supply of water. He foresaw Philadelphia as a city of public fountains and bath houses; the former to cool the air, the latter to ensure the health of the populace and provide the city with income.

The Spring Mill proposal was too costly for the public to accept, but the rest of the plan—though still inadequately funded—was adopted. Latrobe was to be paid $6,350 plus expenses. The drawings were done quickly and well, by Latrobe and his chief draftsman, Frederick Graff (1774–1847). Now owned by the Historic Society of Pennsylvania and the Franklin Institute, they are exquisite examples of engineering drafting.

The steam engines were to be made by Nicholas Roosevelt in his Soho works outside Newark, New Jersey. Roosevelt was an acknowledged master engine builder, one of the few in the country. He was possibly the only person at the time who could handle the contract for the machines. A New Yorker, Roosevelt had formerly been in partnership with John Smallman, an Englishman who had once been a foreman in the Boulton and Watt engine works in England. Latrobe chose to use the conservative Boulton and Watt engine design, with a newly designed and improved valve system. (Later Latrobe and Roosevelt increased the efficiency of the Boulton and Watt design by doubling the air pump.) The engines were designed to raise 3 million gallons a distance of 50 feet per day. Roosevelt agreed to keep the machines in repair for five years. Because the city thought the capacity of the engines was well beyond its present needs, it leased the surplus power to Roosevelt. As it turned out, demand soon met the supply.

Separate contracts went out to four builders for the approach to the canal and basin, for the lower tunnel, for the vertical well at the lower pump house, and for the upper aqueduct and a three-arched portion to carry water across a gully in Chestnut Street, respectively. On March 12, 1799, work began on the Chestnut Street tunnel between the two engine houses. Progress on the waterworks was not smooth. The excavation, through solid rock, took longer than expected. In addition, logs large enough to be made into pipes were scarce. Some of the pipes rotted before the project was completed. When water was flowing in them, they lasted; problems arose whenever they were allowed to sit empty and dry out. Terra cotta or iron pipes could have been used, but these were economically unfeasible.

Financing conditions were unfavorable, and although a tax was finally levied, few funds were collected because of a new epidemic of yellow fever and the resulting dispersal of the population. After 22 months' work the project had exceeded its $127,000 budget by more than $220,000. But on January 27, 1801, the Centre Square works opened. On the night of the 26th, Latrobe, workmen, and friends had lit the fires under the boilers. The hydrants in the streets had been left open, and by morning water was flowing through the gutters. Latrobe was proclaimed a genius.

An open basin, 84 feet wide and 200 feet long, ran east along Chestnut Street from the river. Tide-lock gates enabled the system to shut out brackish high-tide water and admit only the fresh ebb-tide water. From there the water flowed through a 300-foot oval tunnel laid in rock to a well, 10 feet in diameter and 39 feet deep. Above the well was the lower engine house, which took the water by means of a 6-foot-diameter, 3,144-foot brick tunnel down Chestnut Street to Broad Street and the Centre Square, or upper, engine house. The Centre Square engine pumped the water up 50 feet into two wooden tanks, which held more than 20,000 gallons. From there water flowed down into an iron chest outside the engine house and thence into wooden mains for distribution to the rest of the city: two 6-inch mains for Market Street, one 4.5-inch main for Chestnut Street, and another for Arch Street. From these the water branched off into 3-inch and 4-inch logs for distribution through the cross streets.

Latrobe did not differentiate the purely functional from the aesthetic in his design for the Centre Square pumping station. In spite of the suspicion that he was an unrealistic visionary, he was permitted to design a building of true architectural distinction. In both appearance and mechanics, it was a masterpiece. The structure was classical in style and symmetrically composed. Like other Latrobe buildings, it was characterized by simple, plain, geometric forms. Its square, white marble base, 60 feet by 60 feet and 25 feet high, was decorated with entablature, pilasters, and arched windows set into arched and recessed wall panels. The portico entrance was supported by two large Doric columns. Atop the base was a 40-foot-diameter drum capped by a flat dome with an oculus/chimney. The planted grounds, soon decorated with William Rush's *Water Nymph and Bittern* fountain—the first decorative fountain in America built with public funds—became a popular spot for strollers and out-of-town visitors. Inside the pumping station were the engines, sunk halfway into the basement and connected to the water mains. The first floor housed the city water office and the engine keeper's apartments. Near the latter was the access to a coal cellar below. The wooden reservoirs fit neatly within the circular drum, and smoke from the engines escaped through the center of the dome.

The popularity of the waterworks' delightful setting did not help to bring it immediate financial success. At first there were almost no customers. Water was still available at public pumps and from private wells and cisterns, the use of which was considered unsafe by public officials but not prohibited. At the end of a year's operation there were only 154 subscribers, and about half of those received their water free because they had purchased bonds for the waterworks' contruction. Total revenues amounted to only $537. Plans to add 3,000 feet to the distribution system were defeated, even though that expansion would have increased the number of subscribers. Pipes leaked, the operation ran large deficits, and the city was again hit by a yellow fever epidemic. Because the habit of filling buckets at public pumps was strong, there was little incentive to pay $5 per year for a private, household pump. In addition, Philadelphia adopted a policy of providing free water for the poor, which virtually made free water available to all. On top of all these drawbacks, the tepid Schuylkill water was considered unpalatable by many who were used to chilled water. Gradually business grew, however. Fireplugs appeared in the streets in 1803, providing another demand on the system.

As the population grew—which it did dramatically over the next decade—the administrative problems associated with the unprecedented activity of running a large municipal water system began to surface. The flat-rate subscription system then in use was questioned, and the city tried to determine a procedure for dealing with the "willful waste" of water. What had proved in every way satisfactory was the ownership of the system by the city rather than by private companies.

By 1811 it was clear that the system could not supply sufficient water to the city. The search for a solution began. An overhaul of the current system, it was estimated, would cost $7,333 plus $18,000 per year in operating expenses. An alternative plan was to move both engines to the Schuylkill while continuing to use the Centre Square reservoir. A third plan involved drawing water from the Wissahickon, a smaller stream near the Schuylkill.

The plan finally adopted was the work of John Davis and Frederick Graff. Fittingly, Latrobe was the background progenitor of the new system. Both Davis (who later was responsible for the Baltimore water system and other canal systems) and Graff (who had already worked on several independent commissions and would later become the best-known waterworks authority in the country) had been trained by Latrobe. In 1805 Graff had become superintendent of the Centre Square operation; he remained in charge of the Philadelphia waterworks until his death in 1847.

Davis recommended building a reservoir on what was then called Morris Hill. Previously known as Faire-Mount and located about one mile above the lower engine house, it rose directly above the river to an elevation 96 feet above the city's datum line. The fine view afforded by Faire-Mount had made it a popular spot ever since William Penn's time, and Penn had hoped to build his house there. Davis and Graff agreed that a new and larger reservoir was needed and that Morris Hill was the ideal site. Their plan recommended moving the entire waterworks operation to the Schuylkill—both the pumps and the reservoir itself—and abandoning the Centre Square station entirely. One new pump engine was to be installed; the second was to be moved from Centre Square.

The estimated yearly operating cost was only $8,360, a bargain compared with the other options. Since the high operating cost of the current system was one of its chief drawbacks, the Davis and Graff plan was understandably attractive. Although it was not accepted without public controversy, the commotion was considerably smaller than that which had nearly destroyed Latrobe. The names of neither Latrobe nor Roosevelt (who had manufactured Latrobe's steam engines) were popular in Philadelphia following the financial entanglements and disputes that had grown out of the first waterworks construction. It was fortunate that Latrobe had found new practitioners in Davis and Graff, and it was also fortunate that there was now a rival for Roosevelt's position of premier steam engine manufacturer. Oliver Evans of Philadelphia had built a high-pressure engine and was rapidly supplanting Roosevelt.

The new pumping station was under construction in 1812. The engine house was completed by 1813 and the reservoir was in operation in 1815. The works buildings, like Latrobe's Centre Square engine house, were not merely utilitarian structures. They were built in the Greek revival style. The central engine house was a three-story symmetrical structure with elegant architectural detailing, including windows set into recessed arched panels, string courses, a cornice, and a colonnaded porch. Inside, the rooms were large and handsome. The Great Room, where refreshments were served to visitors to the project, was an enormous space with a vaulted ceiling, arched wall niches, and decorative wall and window moldings. The wings of the building were apparently used as living quarters for the two engine keepers and their families.

The system was viewed in its day as an engineering marvel, and little heed was paid to its architectural merits. Graff was responsible for the engineering and is generally credited with the design of the buildings as well. The engine building and the subsequent temple-pavilion additions to the complex are distinguished architectural achievements; and since there is no evidence that Graff had any architectural training—and no evidence besides this complex that he had any architectural ability—it has been suggested (though not substantiated) that either Latrobe or his pupil Robert Mills or both had some hand in the design of the buildings. Like the Centre Square station, the new waterworks was decorated with the work of Philadelphia sculptor William Rush, and over the years new Rush pieces were added along the buildings.

When the reservoir was completed in 1815, the city installed a cast-iron Boulton and Watt–type engine, built in Philadelphia. The machine soon developed leaks in both the boiler and the pump. The planners then made arrangements to purchase, at a price of over $30,000, a modified Boulton and Watt machine developed by Oliver Evans. The high-pressure Evans machine, ready in December 1817, operated under more than 200 pounds per square inch of pressure and pumped 3,072,656 gallons every 24 hours. This engine later exploded and killed three workmen.

By 1817 Philadelphia was nationally and internationally acclaimed for its water supply system, which had a pumping capacity of over 3 million gallons per day and a reservoir

This engraving was made by W. H. Bartlett around 1835. The view is from the northwest and shows the reservoir. Note the famous "Bridge of Sighs." (HAER Collection, Library of Congress)

that could hold over 3 million gallons. That part of the system, efficient and adequate, served to point out the inadequacies in the other parts, however. The mains, even though they now numbered six, could carry only a million gallons per day. Theoretically, their capacity was much greater; but, Graff conjectured, the friction of the wooden pipes and the sharp angles of the side-street distribution system lessened the volume capacity. In 1818 it was decided to replace the system with iron pipe. Iron had not been used before because of its scarcity and its troublesome leaking joints. By 1822 two miles of pipe had been laid with satisfactory results. Philadelphia continued to replace the wooden pipes at the rate of 10,000 feet each year for the next 10 years, which made the city an enormous buyer of iron. As late as 1832 wood pipe was still being laid, and it was not until 1849 that a complete conversion was made.

The conversion to iron pipe was not the last problem Philadelphia faced. It became evident, about the same time iron pipe was first being installed, that steam power was still a far too expensive way to pump water. The decision was made to start to convert the steam system to water power, at an initial cost of $346,000 but with a reduction in yearly operating costs from over $30,000 to $23,000. The conversion also promised a significant increase in water supply. The new plan was engineered by Graff. Housing for eight waterwheels and pumps was to be added. Three machines would satisfy the current demand, with five more to be added in future years. The Schuylkill River would be dammed with earth, logs, and stone diagonally along its width (a total of 1600 feet), backing up the river for six miles. A forebay was to be blasted out of the rocky base of Faire-Mount behind the engine house—an enormous task. From the forebay the water would flow through the millrace onto wheels about 15 feet in diameter and 15 feet wide, and then through stone arches back into the river. Graff himself designed the "double forcing pumps" that supplied the reservoir.

Everything was in place and working by July 1, 1822. The dam held—to the surprise of many—and the Philadelphia system now had margin to expand. The waterwheels proved efficient. Each of the three wheels could pump 1.25 million gallons every 24 hours. Moreover, Philadelphia had a hedge against future costs. Increasing the pumping capacity to 10 million gallons per day, it was estimated, would increase the operating cost of the waterwheel system by only $10 per day. A comparable expansion under steam power would have increased the daily operating cost by $550.

In 1830 the waterworks made a profit for the first time. Philadelphia had a system that, with few changes, served it well through the rest of the century. The quality of the water remained high for a long time. Although a filtration system had been discussed as early as 1812, none was installed until 1909, long after many other municipal systems had begun filtering their water. In 1827 Latrobe's Centre Square building was pulled down, and in 1850 the waterwheels were replaced by more efficient water turbines. Additional reservoirs were built and the distribution system was expanded.

The new waterworks, like the old, served as the social promenade for the city. Charles Dickens found the site in 1842 to be "no less ornamental than useful," and Philadelphians agreed. The 253-foot by 26-foot terrace was often filled with pedestrians strolling along the river. Each change in machinery brought alterations to the buildings, but throughout the century they retained their unified and distinctively classical appearance. By the latter part of the century Philadelphia recognized the need to preserve the purity of the Schuylkill River water and did so by establishing Fairmount Park in the west end of the city. The park, which now extends over 4,000 acres, is a visible legacy of the Fairmount waterworks to the present-day city of Philadelphia. Less evident are the growth potential and high standard of living this system afforded the city. The early example of Philadelphia's municipal system presented the rest of the country with technical, political, and administrative expertise to be used in the development of later systems all over the United States.

The breast wheel is shown in this view of the interior of the wheel house made in 1853. (HAER Collection, Library of Congress)

This view of the Fairmount waterworks from the west was executed by T. W. Bovell in 1871. (HAER Collection, Library of Congress)

The Croton Aqueduct

The New York City water supply system, completed in 1842, was in its time the model for large-scale municipal water systems throughout the United States. Its chief feature was a 40-mile aqueduct running from a dam near the mouth of the Croton River to receiving and distribution reservoirs on Manhattan Island. The history of the system's construction reflects not only an unprecedented engineering achievement but also half a century of political wrangling between private investment companies and municipal government, both of which sought to control the project. The system that resulted stood as an international example of engineering skill in an era of rapid growth and formed a cornerstone in the creation of the modern city.

In the closing years of the eighteenth century New York was experiencing the same water problems that Philadelphia faced. There was neither an adequate supply of good water for drinking nor a sufficient quantity for fighting fires. The tragic results resembled those in other rapidly expanding cities. In 1776 New York lost 493 houses to fire; in 1795, 1798, and 1799 epidemics attributed by many at the time to unclean water took as many as 2,000 lives each summer.

By 1696 sixteen wells had been dug with public and private funds. But in the 1720s the city stopped building wells and allowed residents to build their own wells in public streets. Water was normally drawn from these wells with buckets or balance poles, but in 1741 the city required that pumps be installed on all wells at the expense of the users. By the 1750s most well water was foul and bad tasting. Before the revolution a thriving private enterprise grew up around the Tea Water Pump, so named because it sold water good enough for making tea. The Tea Water Pump gardens became a resort and social center. By the 1780s, however, even this water was far from pure. The well was located near the Collect, or Fresh Pond, then a "sink and common sewer." By that time also, the cistern water used for washing was dirty and in short supply. Unfortunately, the Hudson and East rivers were tidal, and their salinity ruled out the option of drawing water from them.

Croton Aqueduct, New York. Completed in 1842, the New York City water supply system was the most outstanding municipal system in the United States and was the prototype for many large-scale projects elsewhere. John B. Jervis was the chief engineer for the project. The aqueduct at Sing Sing Kill is shown. (HAER Collection, Library of Congress)

The city, concerned about the diminishing supply of good water, took action in 1774 and commissioned Christopher Colles, an Irish engineer, to build a waterworks. The system was to be located within the city at the Collect, which was to be cleaned and kept free of future pollution. A Newcomen or atmospheric steam engine (a type long used to pump water from mines in England) would force the water through a wooden tube to a quarter-acre reservoir on top of a hill. The project was well under way by the time of the Revolutionary War, but the British occupation of New York put a stop to construction. At the end of the war little remained of the works. Proposals for new waterworks began to appear in the 1780s, but no work commenced for many years. Plans came from, among others, Samuel Ogden (1785), Zebiner Curtis (1794), Amos Porter (1795), Benjamin Taylor (1795), Joseph Browne (1796), Christopher Colles (1797), and Nicholas Roosevelt (1797). The delay in new construction was due in large part not to a lack of technology, enthusiasm, or finances but to extended debate over the public or private ownership of the works.

A yellow fever epidemic in 1798 brought public dissatisfaction over the lack of a system to a head. Philadelphia, Boston, and Baltimore were all either planning or building their own systems, and New Yorkers believed they should follow suit. As a result, the New York Common Council decided on December 24, 1798, to accept the plan of Joseph Browne, a physician, scientist, and engineer. He proposed that water be diverted from northern streams outside the city by means of a dam across the Bronx River, then flow through a canal to the Harlem River, from which a waterwheel and pumps would raise it to a high reservoir. The well-known English engineer William Weston, then working on canals in Pennsylvania and New York, was hired to make the survey. The waterworks, it was decided, would be publicly owned, and a tax was levied to finance it. The public supported the plan to build a system, but just what kind of system should be built was still debated. Critics of the Browne proposal charged that the water supply would be insufficient and tepid, and further argued that a cheaper means was available—the Collect could be cleaned and used to supply New York.

Weston's report of February 1799 advised against using the Collect. He estimated that the source springs, although not yet contaminated, soon would be. The Bronx River, Weston argued, would supply more than enough water. Moreover, Weston redesigned Browne's system to eliminate the machinery. He proposed running a canal from the Bronx River to a reservoir on the shore of the North River. From there the water would pass through iron cylinders to a distributing reservoir north of the city. Weston also recommended that an advanced filtration system be added to the works.

The need for a steady supply of good water was by this time urgent. It was obvious to some that much money was to be made from the system. Desire for personal gain kept the controversy raging. Who should supply the city and by what means? The solution was the result of a series of complicated political maneuvers, perhaps masterminded by Aaron Burr. Burr's real concern in the matter was banking, not water supply, but he saw the waterworks as an opportunity to establish a bank controlled by his political party. He established the legal groundwork for his scheme in the New York state legislature and organized the Manhattan Company. The company was given broad powers: to use the land, to dam and divert water with almost unlimited control, to lay pipes and dig canals. Further, though required to provide free fire-fighting water, the company was not obliged to repair streets it destroyed in laying pipes, nor was it limited in its power to fix rates. Most important of all for Burr and his colleagues, it was allowed to invest surplus capital, and thus the door was opened for it to become a great investment business (it eventually evolved into the Chase Manhattan Bank), unchecked by the restrictions placed on regularly chartered banks.

Although many citizens objected, they were stuck with this company, which advertised yet again for a waterworks plan. Nicholas Roosevelt, William Weston, and Christopher Colles all resubmitted their proposals. The company decided to use a Roosevelt steam engine at the original site near the Collect to pump water from that source into a reservoir, from which it would flow into long pipes for distribution through the city.

In 1800 the reservoir at Chambers Street was in place—a 132,600-gallon stone, sand, clay, and tar structure fronted with four Doric columns and a reclining figure of Oceanus. The distribution system followed the geography of the island down Broadway and laterally from there through the streets to the Hudson and East rivers. The pump was originally operated by horses and in 1803 converted to steam power.

Public support, good at first, did not last. The wooden mains broke frequently, and there were no public hydrants. Revenues fell short, precluding repairs. In addition, streets where pipes had been laid went unrepaired. Since there were no hydrants, water for fires was obtained by drilling holes in the pipes, which, although later plugged, continued to leak (the term "fireplug" originated from this practice). Relations steadily deteriorated between the city and the Manhattan Company, whose primary concern proved to be investments. Each year brought a new round of bitter complaints. The city was forced to build 40 public cisterns between 1817 and 1829 in order to maintain an adequate supply of water for fire fighting. The population growth—from 64,489 in 1800 to 202,589 in 1830—placed tremendous additional demands on the system.

In 1821 the city undertook a thorough exploration of the problem. Canvass White, an engineer who had worked on the Erie Canal, was hired to survey sources and distribution possibilities. Once again the proposal was to go northward out of the city for a water supply, this time to the Bronx River and the two Rye Ponds. At this point New York planners also consulted Frederick Graff, who had helped design and now supervised the Philadelphia water supply system. That system was envied by all those urban centers that were feeling the effects of urban growth.

White filed his report in 1824. He proposed linking the Saw Mill River with the Bronx River in order to supply 9,600,000 gallons per day. He suggested four different plans for conveying the water, recommending one that, because it began high enough, would enable the water to be carried across the Harlem River on a bridge without the use of pumps. This proposal initiated a new battle among the Manhattan Company and other groups for the right to build and control Manhattan's water supply system. The victor was a new group, the New York Waterworks Company, which in 1825 received a charter to build a system. Canvass White and another Erie Canal engineer, Benjamin Wright, headed the company. Nothing constructive came of it, however, and the company was forced by political pressures to liquidate in 1827. Meanwhile, new drilling technology gave New Yorkers hope that water could be obtained on the island itself. They were mistaken, however, in their belief that Manhattan Island held limitless supplies of underground water. Although one well was driven for years, no water came from it.

In 1828 the Fire Department, in desperate need of water, persuaded the city to build a wooden reservoir, to be filled by horse-driven pumps and connected to iron mains attached to hydrants. The Fire Department argued that the iron mains could later be taken over by the waterworks company for its distribution system.

It was not until the 1830s that the problems became so acute as to force the city to take the steps that led to the creation of the Croton Aqueduct. By that time it must have seemed to New Yorkers that the very basis of their society was in jeopardy. People so commonly spiked water with spirits to make it palatable that drunkenness had become a widespread problem. In addition, the New York beer business was losing out to Philadelphia's, because the latter had better-tasting water for the beer's manufacture. Philadelphia's system permitted private bathtubs all over the city, while New Yorkers were still having their drinking water delivered to them in buckets.

Dr. P. S. Townsend of the Lyceum of Natural History of the City of New York dashed the hopes of the well-digging faction when he published a report indicating that there were no significant underground water stores on Manhattan since the island was composed of solid gneiss. Perhaps the final blow was the 1833 Asiatic cholera epidemic, which struck North America and Europe. The cause of cholera was unknown at the time, but since foul air and uncleanliness in general were thought to aggravate it, a well-washed city was considered a preventive necessity. New York spent $110,000 in an attempt to clean the city—which was so dirty and smelly that travelers had long since been avoiding it. In spite of this massive effort, 3,500 people died and 100,000 fled to the country. Proof of the importance of clean water was presented to all by Philadelphia, which suffered only 900 deaths. After studying the Philadelphia system once again, New York at last committed itself to spending whatever was necessary to solve its problem once and for all.

Benjamin Wright, now the city's street commissioner, was ordered to do a new survey of the Rye Ponds and the Bronx River. He put Timothy Dewey and William Serrell in charge of the field work. Colonel De Witt Clinton, a civil engineer with the federal government, was appointed to do a special investigation into possible supply sources. Clinton based his recommendation on the large collection of reports and proposals the city had accumulated over the last several decades. The Croton River in Westchester County, he said, was the only possible source of water for New York. Although the cost of bringing that water to the city would be high, Clinton's recommendation was persuasive. The next year, 1834, a former army engineer, Major David B. Douglass, was sent on a survey of Westchester County. He supported Clinton's choice of the Croton, concluding that the Bronx watershed, while of excellent quality, was insufficient to supply the rapidly expanding city. The Croton promised both good water and an adequate supply in all seasons. In addition, its elevation was high enough to permit the water to reach the city by gravity alone: the steam machinery that Philadelphians had found so expensive could be eliminated. Douglass's survey led to a plan to dam the river at the Muscott rapids, 11 miles north of its mouth, and to run the water from there through Westchester County, across the Harlem River, and into the city via a closed masonry aqueduct. He offered two alternative routes at prices that made the 1824 estimate of $2,000,000 for the Saw Mill River plan look desirable—$4,550,237 and $4,718,197.

The shockingly high cost of the system provided an occasion once again for political opponents of the plan to begin lobbying campaigns. In retrospect it is hard to imagine how New Yorkers could have maintained the hope that they could solve their problem in a piecemeal way, yet the abundance of alternative plans, all considered seriously, testifies that this was so. Faced with widespread opposition to Douglass's proposal, the Water Commission asked him to review his cost estimate. Douglass managed to make a few cuts by relocating parts of the system, but the price reduction was insignificant. In the meantime yet another survey was done, this on an independent venture by John Martineau, who lowered the cost to $3,800,000 by moving the dam to the mouth of the Croton and carrying the water across the Harlem River not on an expensive aqueduct but by means of an inverted iron-pipe siphon, supported on a massive embankment. An arch through the embankment would permit navigation and tidal movement.

The Water Commission consulted Frederick Graff of Philadelphia, deliberated, and made its recommendation, accepting aspects of both plans. It proposed "that a dam of sufficient elevation be erected near the mouth of the Croton River, and from thence the water to be conducted in a closed stone aqueduct to the Harlem River. The river to be crossed by inverted syphons of wrought-iron pipes of 8 feet diameter, formed in the manner that steam boilers are. From the south side of the river, a line of stone aqueduct will again commence and proceed across Manhattan valley to the distributing reservoir at Murray's Hill [site of the New York Public Library]." The cost for the supply system was estimated at $4,150,000, and the distribution pipes were estimated at $1,262,000, for a staggering total of $5,412,000. Revenues, based on 30,000 subscribers, were put at over $310,000 annually. The system, the commission decided, should be municipally owned. The voters were well supplied with information on the commission's proposal: the report was published and distributed, and summaries appeared in the papers. The election of April 16, 1835, decided the issue by an unexpectedly large majority: 17,330 in support of the new system, 5,963 against it.

The project was beset with problems almost immediately. Major Douglass was appointed chief engineer and was instructed to make the first surveys. Painstaking care was needed to fix the exact location of each piece of land that was to be flooded or built upon, so that negotiations for purchase could begin. Douglass was given only sixteen assistants, although he insisted he needed sixty or seventy and that more than the allotted time was required for accurate surveys. The Water Commission refused his requests. Relations between Douglass and his employers continued to deteriorate until he was fired on October 11, 1836.

John Jervis was chosen to take Douglass's place. Whereas Douglass was professionally trained—a Yale graduate and a former professor at West Point—Jervis was a self-taught surveyor who had received on-the-job training. Benjamin Wright had been his supervisor and mentor on the Erie Canal, and by 1819 Jervis had been promoted to the position of resident engineer on the canal. Next he had served as principal assistant to Wright on the Delaware and Hudson Canal and then as chief engineer. Jervis had afterward turned his interest to railroads, taking charge of the Mohawk and Hudson and then the Schenectady and Saratoga. By 1836 he was again on the Erie Canal, at work on enlargements.

Jervis is credited with intelligence, honesty, and a good head for business. Despite public debate over the firing of Douglass, Jervis soon began to make steady progress on the water system. One of the reasons for his success was his selection of excellent assistants—Horatio Allen, another self-taught engineer, who had worked with Jervis on the Delaware and Hudson Canal, and Edward H. Tracy, an Erie Canal veteran. Jervis realized that the reliability of the system depended on the quality of its design and construction. He was also economically minded. In principle he preferred embankments to bridges and siphons to either of these alternatives, provided they would yield the appropriate effects. He was well aware of the problem of water spillage and freezing in masonry structures.

Jervis began work in 1837. He organized the Croton project into four divisions, with 96 subdivisions called sections. For each section Jervis prepared minute specifications, and for each division the contract was awarded to the lowest responsible bidder. Jervis established strict standards and inspected the project closely. He even developed a standardized system for culverts. Jervis's plan for the aqueduct was published in the January 1838 Water Commissioners' report:

The foundation of the aqueduct is stone, upon which is laid a bed of concrete, composed of broken granite and hydraulic cement; the side walls are of hammered stone, laid up with cement; the floor is composed of an inverted arch of hard brick, eight inches thick; the lining of the side walls and the upper roof arch are of the same thickness and materials, all laid with hydraulic lime mortar. No common mortar is permitted in the whole structure. The culverts and bridges are of dressed stone, of great strength and suitable dimensions; all laid with hydraulic cement, which undergoes the usual tests before it is passed by the engineer.

The construction proceeded steadily through Westchester County, but not without difficulty. Residents of the county did not like the aqueduct's passing through their neighborhood, fearing loss in the value of the land and limitations on its future use. They also resented the work crews who camped several thousand strong on their farms, often only a few yards from the farm house. The laborers formed an army of 5,000, and although large-scale damage to property on their part is doubtful, certainly minor infringements occurred. The farmers helped to delay the project by sabotaging the surveyors' efforts. Surveying stakes planted one day disappeared overnight, and nothing could be done except to resurvey and replace them.

Work was nearly complete on the dam, a 90-foot granite structure with large earthen embankments, when disaster struck. On January 8, 1841, a heavy rain began to fall on frozen ground, already covered with 18 inches of snow. The 400 acres of water behind the dam rose at the rate of 14 inches an hour. The waste gate was too small for the flooding waters, which rushed over the masonry overfall. The structure held fast, but when the water rose over the embankment, the earth gave way between the structure and the north bank of the river. The ensuing flood wiped out four bridges, three mills, and six houses downstream. Three people died.

Jervis decided to rebuild the dam on a radically different design. The earthen embankment was discarded and replaced by a stone barrier. The overfall was expanded threefold to 260 feet. The first step in construction was to clear the bed of mud and then to construct a concrete armature for the structure. The concrete was layered with 1-foot hemlock piers on either side, and that in turn was wrapped with alternating layers of white oak and stone. The entire structure was then covered with cut granite laid with hydraulic mortar, the joints kept to a maximum of 3/16 inch. Downstream a stilling basin was formed by a secondary dam. The basin would break the body of water coming from the lip of the dam and would also help protect the wood from rot by keeping the timbers wet. The primary dam was 50 feet high and held an estimated 500 million gallons, discharging 35 million gallons daily.

The water followed a 180-foot tunnel through the rock bed beneath the reservoir to the gatehouse, where it passed through a screen frame of oak boards with 1-inch interstices covered by a fine brass screen that excluded all but very small fish. (There was early suspicion about the water, "all full of tadpoles and animalculae," so that one was "in dreadful apprehension of breeding bull frogs inwardly.") Two sets of cast-iron control gates manipulated by iron rods guarded the entrance to the aqueduct, where the water began its 41-mile journey south, falling an average of 15 inches each mile. Because of the elevation of the site, no steam machinery would be needed for distribution in Manhattan. When on level ground, the aqueduct was laid on a foundation of 3-inch to 12-inch concrete, depending on soil conditions. Brick was laid on top of this in the form of an inverted arch, and the side walls were carried up to the springing point. This U-shape was then roofed with another brick arch. The tunnel interior measured about 7.5 feet by 8.5 feet. Hydraulic mortar was thickly applied to each brick, and the brick was then squeezed into place so that excess mortar oozed out, forming a tight connection. The workmanship was careful and the structure, where it remains today, is still sound. A wide variety of other structures were also built. They included 16 tunnels between 160 feet and 1,263 feet long, 114 culverts, 33 ventilating shafts, 6 waste weirs, a major bridge at Ossining, an inverted siphon at Manhattan Valley, a bridge across the Harlem River, and several other smaller valley crossings.

The first major crossing coming south from the dam is the Sing Sing Kill Bridge at present-day Ossining. The valley was 536 feet wide and 70 feet deep, with only a small stream of water below. Jervis's design was a simple arch with an 88-foot span, 20 feet wide and 80 feet above the stream. In order to protect the masonry from frost, Jervis took pains to make the bridge watertight. In the spirit of work done by the noted English engineer Thomas Telford, he lined the aqueduct conduit with bolted cast-iron plates. He also provided for copper drains within the structure to carry out any water spills. Jervis believed that the bridge would have a longer life if he reduced the loads; but since the water conduit needed a heavy earthen cover to protect it from frost, the load on the bridge was abnormally high. Therefore, he reduced the dead load by omitting the rubble fill normally placed between the outer walls. Instead, he tied the walls together with interior spandrels, which also supported the deck. The air space provided additional insulation and good drainage. This soundly engineered and painstakingly constructed bridge, still in existence, has settled only 3.5 inches since it was built.

The Mill River crossing was the next important structure, located 12.75 miles from the dam. Jervis planned to build a double culvert there, because it served his purpose more economically than did a bridge. A single culvert was eventually built, however, 87 feet high, 25 feet wide, and 172 feet long. The culvert arch was covered with a dry wall, which was capped with concrete for the conduit

The Croton Aqueduct at Yonkers. (HAER Collection, Library of Congress)

bed, and each side was banked with earth. The conduit was itself buried in dirt, and the embankments were buttressed.

The aqueduct proceeded from there to the ravine crossing at Irvington-on-Hudson, where Jervis designed a masonry arch 62 feet high, 14 feet wide, and 148 feet long. It, like its four sister ravine crossings, is in good repair today. In Yonkers the aqueduct crossed the Saw Mill River via a 648-foot tunnel cut through earth and rocks, and then over a stone arched bridge of two 25-foot spans. Tibbett's Brook Valley necessitated another tunnel through solid rock, this one 810 feet long. The water then ran 2 miles to the eastern bank of the Harlem River.

The High Bridge, designed by John Jervis for the crossing of the Harlem River, was one of the greatest engineering challenges in the building of the Croton system. This view from the northeast shows the bridge during construction of the large main. (HAER Collection, Library of Congress)

The crossing of the Harlem River was one of the two great engineering challenges of the system. It proved to be a political challenge as well. Bitter debate arose over whether the crossing should be made by a high or a low bridge. At the crossing point the river was 620 feet wide. This, together with the depth of the sloped banks, totaled 1,420 feet. Jervis advocated a low bridge carrying an inverted siphon on top, a plan that was much less expensive than a high bridge.

An inverted siphon is not a true siphon, but its shape is close enough to suggest that name. In fact, it is no more than a pipe that runs down one side of a valley, across the bottom, and up the other side. The water is moved on the uphill slope by the pressure head created by the height of the water on the opposite side. The great advantage of the siphon was its low initial cost; its great drawback was loss of elevation, since it was not fully efficient in raising water. Yet every foot of elevation was precious, for the higher the water when it reached the distribution reservoir, the farther it could be distributed. Jervis was forced to plan for siphons sparingly in spite of their economic attractions.

For this crossing Jervis proposed the low bridge as a responsible and thrifty solution; Frederick Graff, who was consulted in the design, concurred. Yet residents of Westchester attacked Jervis for trying to cut off navigation north of Manhattan Island. Even when it was pointed out that the Harlem River was not

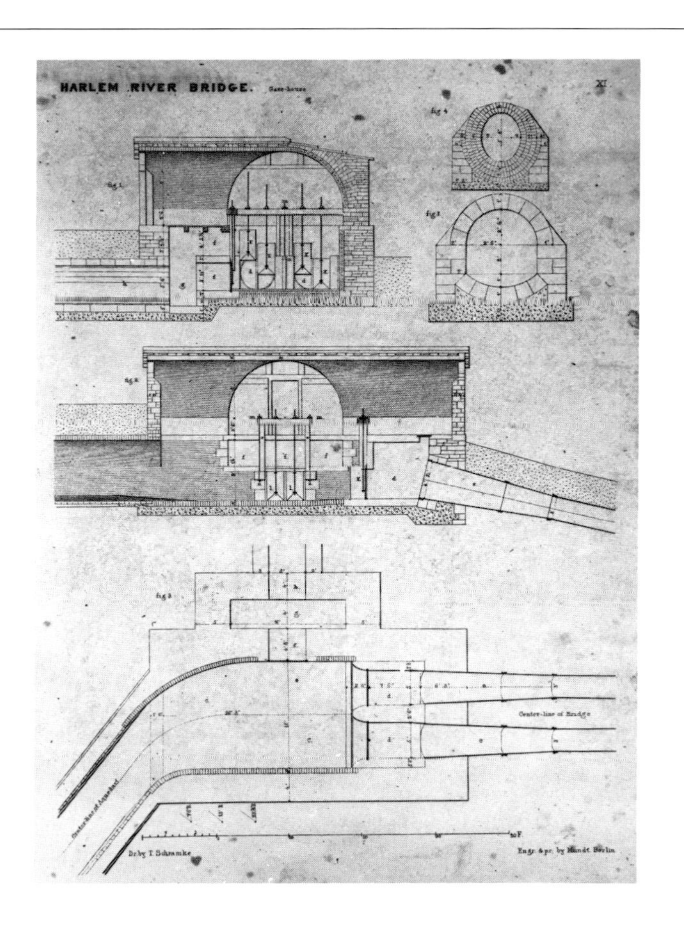

Construction details of the gatehouse of the High Bridge. (HAER Collection, Library of Congress)

In 1937 several of the original arches of the High Bridge were removed and replaced with a single steel arch. (HAER Collection, Library of Congress)

currently navigable, land speculators who hoped that it might eventually be made so refused to give up the battle. Others who argued that only a high bridge would fit the city's image joined the fray. Eventually Jervis was ordered by the state legislature to prepare two alternative solutions, one for a high bridge that would not interfere with shipping and one for a tunnel under the river. Jervis concluded that a bridge meeting the legislature's specifications, with a height of 100 feet or more above high water, would cost $838,613. A tunnel would run $636,738. Nevertheless, he recommended the bridge. He regarded the tunnel as a "very uncertain venture" and cited the problems with Brunel's Thames tunnel. He also feared that salt water would damage the pipes. Moreover, construction of the tunnel would take four years or more, and New Yorkers were impatient for Croton water. Jervis's recommendation was accepted. He wrote a friend, "I cannot say by any means that I regret this. As you know, engineers are prone to gratify a taste for the execution of prominent works."

In designing the Harlem River High Bridge, Jervis's goal was to combine "stability, permanence, symmetry, and economy." Although his preference was for small arches and closely spaced piers, the expense for each pier was very great. To complicate things further, the condition of the riverbed was largely unknown. Keeping in mind the state dictate of an 80-foot pier spacing, Jervis compromised, spacing the piers over the water at 80 feet and those over land at 50 feet, the two connected by a transition zone of one 70-foot and one 60-foot arch. Six of the sixteen piers rested on the riverbed, the remainder on land. Each was 20 feet by 40 feet at the base; their height ranged from 1 foot to 84 feet at the springing point of the arch. Originally Jervis had planned an even higher bridge, but he reduced that by 12 feet, saving $100,000. The height reduction necessitated a shallow inverted siphon on top of the bridge, since the bridge abutted Manhattan Island below the grade line.

The foundations were difficult. Because Jervis believed that the aesthetics of a bridge located in a populated area were important, he used local stone and a system of parapets and other features to decorate the structure. Like the Sing Sing Kill Bridge, the High Bridge was well protected from water damage. The conduit was lined with cast-iron plates, and insulating spaces were provided in the parapet walls. Interior spandrel walls were again used instead of rubble fill, and the piers were left hollow so that the differential settling of the dressed stone and the rubble fill inside would not force the face to bulge out. This precaution allowed an earlier and more uniform hardening of the concrete.

The water was carried by two pipes in order to avoid a discontinuation in service if a breakdown occurred. The two could carry 60 million gallons of water a day, and though planned to be 40 inches in diameter were reduced to 36 inches as an economy. Five feet of earth protected the pipes from frost. For six years, until the bridge was completed in 1848, the pipes were carried on a temporary crossing to Manhattan. This arrangement worked so well that the city once considered abandoning its plan for the more permanent structure.

Once on the Manhattan side of the river, the water entered the Jumel Tunnel, cut through solid granite. At 150th Street and 10th Avenue a tunnel was bored 1,215 feet through rock. The passage down 10th Avenue was by an inverted siphon of 36-inch pipes across a 4,180-foot depression having a maximum depth of 105 feet. This part of the route, together with the 1,400-foot masonry embankment crossing at Clendinning Valley, was the most difficult engineering problem on the island side of the system.

The water flowed into the Yorkville Receiving Reservoir, between 79th and 86th streets and 6th and 7th avenues. (This reservoir was later incorporated by Olmsted into Central Park.) The reservoir, which soon became a fashionable gathering place, held 150 million gallons in a 37.5-acre lake. It was divided into two sections by an earthen embankment, permitting the draining and repair or inspection of either side. Two 36-inch cast-iron mains connected the Yorkville Reservoir with the Murray Hill Distribution Reservoir, located between 40th and 42nd streets and 5th and 6th avenues. The Murray Hill structure was designed by Jervis in the Egyptian style, then popular. It measured 386 feet by 420 feet and was 49 feet high. It held 20 million gallons in its two divisions. Its double wall system was composed of outer walls 4 feet thick, inner walls 6 feet thick, and an interior vault bricked and supported with spandrels.

The population of New York had easier access to this part of the Croton system than to any other, and though it was not the most impressive of all the structures, it elicited a great public response and helped symbolize the entire project. A stone staircase led to a promenade along the top, which afforded a view from 119 feet above sea level over all of Manhattan. Many claimed that this was one of the finest views in all the world.

Jervis completed the entire system at a cost of $8,575,000 plus $1,800,000 for the distribution pipes, a figure within 5 percent of his estimate. The first water reached New York in June 1842, with the *Croton Maid,* a small boat carrying officials, riding its crest. A great water celebration occurred on October 14. One hundred cannons sounded, church bells rang, and fountains burst forth all over the city. New York had, at last, an adequate supply of water and a system that was praised worldwide for its engineering accomplishments. A citizen wrote:

> Nothing is talked about or thought about in New York but Croton water; fountains, aqueducts, hydrants, and hose attract our attention and impede our progress through the streets. Political spouting has given place to water spouts. . . . All parties hail the advent of pure and wholesome water after its journey on the earth, and under the earth, and across the watercourses of miles, as a proud event for our city, and one which enables the Knickerbockers to hold their heads high among the nations of the earth.

With the rapid growth of the city in the latter part of the nineteenth century, the water supply system had to be expanded. This resulted in the construction of the New Croton Aqueduct, built between 1885 and 1893. This aqueduct included a 30-mile water tunnel, the longest and largest in the world then and for a long time afterward. Another dam was also built. The New Croton Dam was constructed

The New Croton Dam, about 1910. (Courtesy of the American Society of Civil Engineers)

in two phases between 1893 and 1906. Its impressive height of 291 feet above its rock foundation, which is 124 feet below the bed of the river, made it among the highest of all cut-stone masonry dams. With its curved, stepped spillway, spanned by an arch bridge and situated at the edge of a park, it has become a favored visiting spot in the metropolitan area. The Old Croton Dam is still intact, although submerged except in times of severe drought.

Perhaps the boldest of all the engineering feats in the expansion of the system was the construction of a 7.25-mile inverted siphon under the Harlem River, the first of such a size. It recalls the original intention of Jervis and others to use an inverted siphon to carry water over the Harlem River.

The Chicago Water Supply System

The water supply system designed for Chicago by Ellis S. Chesbrough and completed in 1869 served as a model for other cities bordering large bodies of fresh water and set a precedent for future expansion of the Chicago system. The Chicago Water Tower and the Chicago Avenue Pumping Station stand today as symbols of a daring achievement that involved building a two-mile tunnel under Lake Michigan. Heralded as the eighth wonder of the world in its time, this tunnel, with its remarkable intake structure (the Two-Mile Crib), brought Chesbrough international fame.

Lake Michigan provided the early settlers of Chicago with one of the largest natural freshwater reservoirs in the world. Since those settlers were few until the 1830s, little thought was given to providing a public water supply. Water was drawn from wells, the lake, or the river; where necessary, vendors carried it by bucket to outlying districts. This system could not last for long, however. With rapid population growth came contamination of these sources.

In 1836 the private Chicago Hydraulic Company was chartered by the state legislature to supply the new sections of the burgeoning city. The resulting system, completed in 1842, began with an iron intake pipe extending from a crib 150 feet out in Lake Michigan to a 15-foot-deep pump well on shore. A 25-horsepower steam pump forced water from there to a reservoir at Lake Street and Michigan Avenue, whence it flowed through about 2 miles of 10-foot cedar logs, fitted with special wooden joints, to the users. Pressure from the reservoir was adequate to force water to the second stories of buildings, but it reached only a small portion of the southern and western sections of the city. The rest of the population, which by then had reached 4,500, relied on wells, the river, or water carts.

The population was growing at an alarming rate and by 1851 had reached 35,000. The private company could no longer supply the needs of Chicagoans, so the legislature incorporated the Chicago City Hydraulic Company, a public system run by three elected commissioners, which began by purchasing the rights of the private company. William J. McAlpine was named chief engineer. McAlpine designed Chicago's new waterworks, which were completed in 1854. The city's population was by then 70,000, and the works were designed to supply the needs of a city of 162,000, the estimated 1875 population. (Chicago, in fact, reached that size in 1865.) The new system had a 3-foot by 4-foot timber intake pipe laid in a trench in the bottom of the lake. The intake terminated in a brick suction well (located 600 feet out into the lake), whose mouth was protected by a semicircular timber-and-stone breakwater. The breakwater, however, provided a shelter for more than the intake mouth. In cold weather millions of small fish sought the enclosure near the shore, from which the water was pumped. In spite of every effort, these fish would enter the reservoirs of the city, and come out in scores from every hydrant, alive and swimming. A contemporary observer noted that "every drop of water drank [*sic*] in the city was highly flavored with fish, and one was obliged to look twice in his goblet to see that he did not swallow one alive."

The pumping station was located at Pine Street (now Michigan Avenue) and Chicago Avenue and was built of brick in the Italianate style. It housed an 8-million-gallon-per-day, vertical-beam condensing steam engine, which became known as Old Sally. The water pumped from the lake into a well equipped with a high standpipe. The smokestack and standpipe were enclosed in the station's tower. The distribution system consisted of about 9 miles of iron pipe and an iron reservoir located 80 feet above lake level. Old Sally operated for nine hours each day except Sunday, when no water was pumped unless it was needed for fire fighting. The engine was used by the Chicago system for 50 years. Between 1854 and 1856 three more iron storage tanks were added. Within this time the new supply had become insufficient, and a second pumping station with a 12-million-gallon-per-day capacity was added in 1857.

By the late 1850s, though, supply was not the only problem. The quality of the water was becoming a more immediate concern. The population had reached 150,000, but low per-capita consumption helped to keep the system adequate for several years. Mean-

The 1869 Chicago Water Tower, a familiar landmark and symbol of Ellis Chesbrough's ambitious and successful undertaking to supply safe water to the citizens of Chicago. The system included a two-mile tunnel under Lake Michigan, leading to an intake crib. (Courtesy of the American Society of Civil Engineers)

while, industry and commerce were increasing. Miles of sewers were constructed along the Chicago River, and these, along with the distilleries and slaughterhouses, dumped their wastes into the river, which gave off an almost unbearable odor and made Chicago the butt of jokes all over the country. The river, which more resembled a currentless bayou, during most of the year did not spread its pollution into Lake Michigan; but the spring floods carried the river water out into the lake, contaminating the water supply. Typhoid and dysentery were common; it became clear that the intake must be moved out to purer waters. The Civil War drew the talents of engineers to other tasks, however, and the problem continued to worsen.

Several solutions were proposed during these years, but none was considered satisfactory. They included a series of intercepting sewers for purification of the water, as in London; the diverting of the waters of the Calumet and Des Plaines rivers into the Chicago River by means of a feeder and pumps; a ship canal with a reversal of the flow of the river (this was later carried out: see chapter 7); and a covered aqueduct, 10 feet in diameter, to extend from the lake to the river to the South Side. The idea of constructing a tunnel under the lake bed was also proposed but initially defeated by the many engineers who declared it impractical. Nonetheless, Ellis Sylvester Chesbrough, the city engineer who was brought in to investigate the problem, recommended just that.

Chesbrough was born in Baltimore in 1813. Because his family had suffered financial reverses, he received little formal education. At thirteen he was employed as a chainman with an engineering party in the preliminary survey of the Baltimore and Ohio Railroad. He was later hired by the Allegheny and Portage Railroad and in 1831 became associated with William McNeill on the construction of the Paterson and Hudson. He was appointed senior assistant for the building of the Louisville, Cincinnati, and Charleston Railroad in 1837 and worked from 1844 to 1846 on the construction of the Boston and Providence. His first job with water supply systems was in Boston in 1846, when he was appointed chief engineer of the West Division of the Boston Water Works, planning structures such as the Brookline reservoir. He was appointed sole commissioner in 1850. Chesbrough left Boston for Chicago in 1855 to take the post of chief engineer for the Chicago Board of Sewerage Commissioners. No city in the country at that time had what could be called a sewage system, and Chesbrough was sent to Europe to study systems there. Upon his return he devised a system for Chicago; he was later a consultant on sewage systems in New Haven, Indianapolis, Milwaukee, Memphis, and other cities. An authority on waterworks as well, he was consulted by Cambridge, Toronto, Detroit, Memphis, Milwaukee, Pittsburgh, and Jacksonville, to name a few.

In 1863 Chesbrough made his recommendations for the Chicago water supply system. He proposed driving a 5-foot-wide brick-lined tunnel through the clay bed of Lake Michigan to a distance of 10,567 feet. Although this drastic proposal had many critics, experiments were begun to test its feasibility. In June borings were taken in the lake bed, beginning 20 feet from shore. These showed blue clay under sand in varying proportions about 200 feet from the shore, and, at 2.25 miles from shore, sand-covered, soft, marshy clay, which grew firmer the deeper it went. The water at 2.25 miles, about 30 feet deep, was clear enough to reveal small objects at 18 feet; its temperature was 60 degrees at the surface and 51.5 degrees at the bottom. These findings convinced leading engineers that the plan was sound. Chesbrough set to work supervising the drawings and specifications, and bids were received and opened on September 9, 1863. The range in the bids—from $239,548 to $1,056,000—was due to differing opinions about the soil conditions. Several of the firms expected to find sand, gravel pockets, and boulders, which would make tunnel construction difficult. Dull and Gowan, who were willing to take all financial risks of the potentially dangerous work on themselves, were awarded the job. Their cost estimate of approximately $315,000 turned out to be far too low.

Work on the tunnel began with the construction of the shore shaft, on the site of the old pumping works at the east end of Chicago Avenue. Ground was broken on March 17, 1864. Cast-iron cylinders, manufactured in Pittsburgh, were sunk 26 feet into the ground, with a lower brick shaft continuing to a depth of 69 feet. A temporary building was erected over the shaft to house the steam engine, the contractor's office, and bricks and tools for the workmen. An elevator was constructed within the shaft to bring up excavated clay and to send down workers and brick. When water began to ooze in at the bottom of the shaft, a pump was added. When the shaft was completed, a drift was made 50 feet west of the bottom chamber, in which to mount a Pike transit, which would be used to give the line of the tunnel.

Construction of the tunnel proper began on May 26, 1864. Its bell-shaped mouth of 6 feet tapered off to the 5-foot standard for the remaining distance. The tunnel was formed from two semicircular top and bottom arches. The nearly circular form was adopted to ease the problem of striking the centers, and it proved effective enough to be used in all the succeeding tunnels. The lower arch was built to templates customary in sewer work and was kept about 6 feet in advance of the upper arch, which was built on a ribbed center of boiler iron. The inside dimensions allowed just enough room for two men to work side by side, with miners in the front followed only 10 or 20 feet behind by masons, who laid the lining brick. Both teams were relieved regularly so that the work could continue without stop 24 hours a day, 6 days a week.

After the tunnel had progressed 750 feet, the first of several side chambers was built, allowing for the construction of a turntable for the small railroad cars that were used to transport clay out and building materials in. These cars were propelled by workmen at first, but as the distance grew, the use of animals was considered. After some difficulty, two mules were found who were small enough to go side by side down the shaft and into the tunnel. These mules were "tractable animals," according to contemporary reports, who "after a little experience and training, learned their work, and performed it well." They could pull several cars at a time, and wore small lamps on their collars to announce their presence. A timetable was prepared to avoid tunnel collisions. Side chambers identical to the first one were incorporated into the tunnel every 1,000 feet. Each was lined with brick, carefully groined to the tunnel lining. On the tunnel's completion these chambers were bricked up. The increasing length of the tunnel caused the air to grow impure, but this was solved by installing in the shaft a steam bellows attached to a perforated pipe. The bellows drew off the foul air, which in turn drew fresh air down the shaft and into the tunnel.

Work began on the lake shaft with the construction of the crib, a 40.5-foot-high iron and timber pentagonal structure that could be inscribed within a 98.5-foot-diameter circle. Its three separate walls were bolted together and braced, each caulked and tarred like a ship. The whole was constructed of timbers, bolted with square iron rods. The bottom was formed of three 12-inch layers of timbers, connected by 36-inch bolts. The crib contained 15 separate watertight compartments surrounding an open well in the center for the tunnel shaft. Each exterior angle of the structure was protected from ice and ships by 2.5-inch-thick iron armor that extended 2 feet on each side of the edge and 12 feet down from the top. The mammoth structure was nearly as large as the Chicago Court House. It was launched on July 24, 1865, and while thousands watched, seven tugs pulled it into place and cut it loose. The gates were opened, and as the water entered the 15 compartments, the crib settled evenly to the bottom. The compartments were then filled with stone and the entire structure fastened to the bottom of the lake with marine mooring screws of a type that had been used before on the Thames tunnel. Before the crib could be fastened, however, a violent storm blew it slightly off mark, so the line of the tunnel had to be altered in order to meet the new position. Later the stones were removed and the compartments filled with concrete so that even if the timber rotted away, the structure would remain in position.

Several feet of the crib rose above the water, and a house with a cupola for a light and fog bell was mounted on this. Winter was coming, and with it ice, so a large quantity of brick and mortar was stockpiled there, along with provisions for workers. Seven huge cast-iron cylinders, 9 feet in diameter, were then

The original Two-Mile Crib. This great pentagonal box was built on shore, floated two miles into place, and sunk into about 40 feet of water. Stone masonry was built inside to sink it even deeper into the muddy lake bottom. An iron intake tube was fitted into the masonry and connected to the brick tunnel under the lake bed leading to the pumping station. (*Frank Leslie's Illustrated Newspaper*, November 3, 1866)

taken out to the crib and bolted together along their heavy flanges. Each weighed 11 tons. When the whole shaft was together, it was placed by means of a block and tackle and sunk 27 feet into the soft lake bed. At the base of the intake shaft a second 50-foot drift was constructed, this time eastward, for alignment of the work and the handling of cars at the foot of the shaft.

The tunnel had progressed 4,815 feet from the shore shaft when tunneling began from the lake shaft on January 1, 1866. Work proceeded in the same manner as on the land shaft side, except that the clay from the excavation was dumped into the lake. As both tunnels progressed, workmen began striking pockets of gas, which caused delays in the operations. After some experience with these pockets, miners learned to tell if they were about to hit one by the sound the pick made as it bit into the clay. Work was stopped if a pocket was found, a hole was drilled into it with an auger, the gas was lit, and it was allowed to burn itself out. Only when gas was encountered in large quantities did workmen leave the tunnel. Safety lanterns were employed in order to avert explosions. As for the large stones anticipated by some of the bidding contractors, few were encountered.

One day in September 1865, miners struck a crevice in the clay that brought a flood of water into the tunnel. The workmen fled but were soon able to return to repair the leak. This was the only major incident during the three years of tunneling. Conditions were safe enough for a number of dignitaries, including General Ulysses S. Grant, to visit the tunnel while construction continued.

Only a thin wall of clay stood between the two tunnels on November 30, 1866. The distances had been carefully marked every 5 feet on the walls of the finished tunnel, and they read 8,275 feet in the shore tunnel and 2,290 feet in the lake tunnel, with 2 feet remaining to complete the total length of 10,567 feet. Dull, Gowan, and Chesbrough, along with a few others, descended into the tunnel, some on each side, to remove the last of the clay separating the two. At 3:40 p.m. the final connection was made. The alignment of the two tunnels was off by 7 inches. The final masonry stone, a marble slab filled with American coins, was placed by the mayor on December 6, 1866, where the two halves met.

In early March lake water was allowed to enter the tunnel for the first time and flow toward the shore. On March 8 the tunnel was partially pumped out to enable the engineer to inspect it by boat. Since everything was in order, the dedication ceremonies were scheduled for March 25, 1867. With great fanfare pure water flowed into the city through the structure that was named the eighth wonder of the world in the popular press. Chesbrough's controversial project had brought him international fame.

The new pumping station and water tower were built on the site of the old at North Michigan and Chicago avenues. Completed in 1869, the structures were designed by the architect William W. Boyington in a Gothic style. The lower wall sections of the water tower are capped with cut stone battlements, and a cupola tops the tower, which was constructed in five sections and ornamented with battlemented turrets and parapets. Narrow slit windows are cut into the shaft. The whole is built of Lemont limestone, then common but now rare. Its foundation consists of 168 wood piles capped with 12-inch oak timbers; concrete fills the interstices. Massive stones bring the foundation up to 6 feet below grade. The pumping station housed a new engine, designed by De Witt C. Crieger, the engineer at the old waterworks, who was later elected mayor. The tower housed the vertical cast-iron standpipe, 138 feet tall and 3 feet in diameter, which relieved the excess pressure in the distribution lines caused by the pulsating action of the early pump. Oscar Wilde, who visited the city in 1882, praised the machinery as "simple, grand, and natural" but described the water tower as "a castellated monstrosity with pepper boxes stuck all over it." The tower and pumping station did prove to be fire resistant—they were among the few buildings to survive the great fire of 1871, even though the complex stood at the heart of the blaze. Both served as landmarks after the ashes cooled, guiding Chicagoans who searched for the remains of their homes. Today these period pieces stand as a memorial to one of the nineteenth century's more remarkable and successful engineering projects.

The tunnel bringing clean water from Lake Michigan to the pumping station in Chicago lay about 60 feet below the lake surface. It was made of two courses of brick. A second tunnel was installed in 1880. (*Frank Leslie's Illustrated Newspaper*, November 3, 1866)

The Embudo Stream-Gauging Station

The stream-gauging station established at Embudo, New Mexico, in 1888 for measuring the flow of the Rio Grande was the first of its kind in this country; the training camp associated with it furnished the newly established (1879) U.S. Geological Survey with a team of hydrographers who made some of the earliest hydrological studies in the western United States. What was started at Embudo led to the gauging of many of the nation's streams and a more accurate evaluation of its surface water resources. John Wesley Powell, the head of the project, was convinced that it was necessary to know the differential flow of seasonal water in western streams and rivers if hydraulic structures were to be built to regulate them. In an area such as the southwestern United States, agricultural production depended absolutely on access to limited water resources. Not only did his team of hydrographers develop the engineering equipment and skills to measure stream flow, but Powell also endeavored to set the groundwork for a land-use policy that would make the best use of limited water resources in the West. Although never implemented as he envisioned, his recommendations were thoughtful and provide an alternative view of how public lands in the West could have been developed.

The Spaniards who first settled the southwestern United States, having come by way of Mexico, understood how to adapt the harsh, water-starved area to farming. They divided the small areas suitable for agriculture in such a way that each plot contained some desirable land and no owner could control the entire water supply. Thus, farms radiated out perpendicularly in narrow strips from a river's edge, with the land decreasing in fertility from the water outward to the mountains. Vegetables, which required the best land and most water, were planted in the area closest to the river, grain was planted beyond that, and the rocky land that extended up into the mountains or out into the desert was set aside as cattle pasture. New Mexico in particular still shows evidence of this method of land management. The Spanish were aware

The first group of student hydrographers at the stream-gauging station at Embudo, New Mexico. Established in 1888 for measuring the flow of the Rio Grande, this station was the first of its kind in the United States and served as a training ground for hydrographers. (Library of Congress)

that the possibility of community growth in the region depended upon irrigation. The irrigation laws that persist in some areas of the Southwest today were originally introduced by the Spanish, who had themselves inherited them from the Arabs. Because systems of irrigation ditches required a communal effort, their introduction led to the formation of ditch companies, which amounted to a type of political unit. The master of the ditch held a position equivalent to that of mayor. Exact methods of surveying were not well understood, and therefore land came to be reckoned not by an arbitrary physical measurement but by its productivity—that is, the amount necessary to support one cow or the amount that would produce two bushels of wheat.

The Anglos who came to settle did not possess the same understanding of the limitations of the desert terrain, nor were their customs of land development communal. The U.S. Land Ordinance of 1785, enacted to facilitate the measurement and settlement of the vast areas in the western territories, set the general rule for the surveying of all future western acquisitions. Land, no matter what its topographical or geological characteristics, was typically marked off by a national grid, six miles square. A few people argued when the Land Ordinance was adopted that such a uniform grid, unresponsive to the particularities of local terrain, was not the most suitable tool for settling arid portions of the West. In areas where water was scarce, they pointed out, one section might control the only water source for miles, while the neighboring sections would be left with no access to water at all. This piecemeal attitude toward the land would render much of it worthless for supporting life. Almost a hundred years after the 1785 Land Ordinance John Wesley Powell came to the same conclusion himself.

Powell spent much of his youth between 1867 and 1878 traveling through the arid western states, studying the American Indians and their languages and at times making expeditions by boat down the Colorado River. He also served as Director of the Geographical and Geological Survey of the Rocky Mountain Region. His observations led to the writing of one of the most important books ever produced on the West, *Report on the Lands of the Arid Region of the United States,* in 1878. The report was highly controversial. In effect, it recommended an entirely new method of surveying and settling all the land west of the 100th meridian—approximately 1,300,000 square miles of territory running from the middle of the Dakotas. The methods Powell proposed were much more appropriate to the productive use of the western territories than the grid method was, and were aimed at reclaiming large areas of the land by irrigation. The report was well detailed, with specific and innovative recommendations. Powell's proposals included the following: an inventory of the flow of all streams in order to evaluate their potential for irrigation; a determination of the average amount of water required to irrigate an acre of land; the preparation of topographical maps outlining stream drainage areas and the division of those areas into individual irrigation districts of politically independent entities; the determination of the cost of impounding dams and canals to be constructed and owned by those districts (all settlers, before receiving government lands in any of the districts, would have to agree to participate in the cooperative arrangement); and the abandonment of the current method of laying out land subdivisions.

The main thrust of these recommendations was to incorporate into every land grant the right to a fair share of the water in the area, with the farmer automatically belonging to the political unit that could settle any dispute that might arise concerning his water rights. Powell also proposed that Congress consolidate all geologic and geographic expeditions into one bureau. Of all his suggestions Congress acted on the latter only, forming the Geological Survey in 1879. Clarence King was appointed director, but when he resigned in 1881 Powell succeeded him. Powell's other suggestions were ignored, and surveying continued to be based on the grid system. Settlers moved in, accepting the popular and unfounded theory that "rain follows the plow"—a notion Powell's opponents worked hard to encourage. A record cold winter hit in 1886, however, and was followed by a drought lasting nearly a decade. As bankrupt farmers abandoned their land, Congress came to reconsider Powell's suggestions. By now, however, many smaller streams were already privately owned. Only unclaimed areas were available for Powell's proposal, and they had to be surveyed before settlers reached them if the government was to avoid the legal complications of squatters' rights and vested interests. In March 1888 a resolution to implement Powell's plan was passed, and in October the project was funded. The Irrigation Survey was turned over to Powell, who set up mapping, engineering, and hydrometric units and established a camp at Embudo, New Mexico, to train engineering-school graduates in stream gauging. Hydrographers were rare and desperately needed.

Embudo, a small and isolated village in northern New Mexico, was selected by Powell for his training camp for several reasons. It was west of the 100th meridian, which was the cut-off line for project funds. The area was served by the Denver and Rio Grande Western Railroad. The Rio Grande at Embudo would not freeze over during the winter; moreover, it was ideal for gauging at that point, and Powell thought the results of the work might be immediately useful for a large irrigation dam planned downstream near Sante Fe. Finally, Mexican farmers were becoming concerned about the amount of water being diverted from the Rio Grande before it was available for their agricultural purposes.

Conditions were far from ideal at the camp. Tents were provided for the men's housing, but the temperatures at that high altitude were colder than had been expected. Some men dug trenches in which to sleep, while others excavated a cave in the side of a hill. The twenty-one permanent members included, in addition to the fourteen students and instructors, a cook, two laborers, and a packer. Salaries ranged from $50 to $100 per month.

The instruments were primitive at first. No current meters were available, so float measurements were used. Later, levels were run alongside the river to ascertain its slope and used in computing river discharges. Daily readings of the barometric pressure, the temperature of the water, and evaporation were also taken—the latter from the cook's bread pan. When more sophisticated equipment arrived in January 1889, the measuring section was moved from its first location, about a mile from the camp, to a point downstream across from the railroad station. Here a steel cable was stretched across the river, with a separate tag line above it for locating the measuring points. Data were gathered from a boat that was held in place by ropes from the cable. The recording gauge, placed 75 feet upstream from the cable, was a horizontal-cylinder type, similar to the tide gauges used by the Coast and Geodetic Survey. It probably resembled the type designed by Joseph Saxton in 1845, when he was in charge of the Office of Weights and Measures. A slope gauge was constructed directly over the intake pipe leading from the river to the recorder's still well. The cook's bread pan was replaced with a floating evaporating pan that had a precision scale for measuring the loss of water. Studies were made to determine the amount of sediment carried by the river, and bedload movement was also recorded. The researchers hoped to determine whether there was a connection between the size of the pebbles or sand at the river bottom and the velocity of the current that deposited them.

A current meter was eventually obtained from the U.S. Navy. It was a cable-suspended Haskell Direction-Indicating Meter. When it proved too large to measure the very shallow flow of water at Embudo during the winter, engineers at the camp developed a smaller version to suit their needs. About the same time a rod-suspended meter became available. It, too, was modified by the Embudo engineers for their specific purposes.

After the engineers had become familiar with the Embudo equipment, they were sent to neighboring streams to select sites for gauging instruments and to take measurements. By April 1889 the training period was over, and ten students received the title of hydrographer or assistant hydrographer (four others had dropped out of the program). They were then sent out to other areas in the Southwest to carry on the work of the Irrigation Survey. The Embudo station was left in the hands of the railroad station agent, who was taught to read the gauges. Readings have been taken continuously ever since (except for a break between 1904 and 1912), at various times by railroad personnel, the state of New Mexico, and the Geological Survey. Improvements have been made over the years to ensure the permanence of the gauges. The station can be seen by travelers on Highway 64 between Espanola and Taos, New Mexico.

Even though the development of techniques for stream gauging and the training of hydrographers were successfully under way in 1889, political realities were growing that would soon put a stop to Powell's plan for a government policy that would make efficient and fair use of the limited supply of water. The "rain follows the plow" creed continued to lure people, and despite the unpopularity it caused him, Powell continued to insist that only 1 to 3 percent of the arid land available could be profitably farmed. Even then, he argued, irrigation would have to be provided. The remaining area, he maintained, should be used for forests and grazing. President Cleveland and his attorney general supported Powell's plan to prevent homesteaders from settling on any of the remaining lands of the entire public domain in arid regions until the topographic maps were completed, the hydrographic studies made, the plans for dams and irrigation canals formulated, and only as many farms laid out as the volume of water available could support. The plan, which would delay western settlement, of course drew numerous opponents. A move was made to repeal the entire 1888 act, and a congressional committee was told, "Every representative of the arid region would prefer there would be no appropriation [to the Irrigation Survey] to having it continued under Major Powell."

Powell's great power as head of the Irrigation Survey had long been controversial. The proposal of new legislation that would have given Powell even more power enabled his opponents to obtain approval for a Senate investigation into his activities. The hearings were held from January 17 to March 28, 1890, and attention focused on funds used for the mapping program. Powell had always insisted that without good topographical maps the Irrigation Survey could not properly carry out its work. One of his administrative assistants testified that an engineering and hydrographic survey could be satisfactorily completed without a topographic survey. When it was learned that $60,000 of the original $100,000 allotment had been used for mapping operations alone, some congressmen felt that Powell's actions constituted a misuse of public funds. Powell's supporters were reduced in number sufficiently to permit the western congressmen to control the future of the Irrigation Survey. In 1890 funds for the survey's work were reduced from $720,000 to $162,500, and this was available for use only west of the 101st meridian. The controversy led to Powell's resignation as director of the Geological Survey in May 1894.

Although Powell's sound proposals for the settlement of the West were never to materialize, his pioneering work in the field of stream gauging was carried on by his former engineer-in-charge at Embudo, Frederick H. Newell. Newell was transferred from the Irrigation Survey to the Geological Survey, and when the Geological Survey was empowered by Congress to continue work in the field of stream gauging, Newell was put in charge of these operations. His work in the Geological Survey earned him the title "father of systematic stream gauging" in this country.

Powell is now commonly acknowledged to have been correct in his conviction that an irrigation system cannot be planned without a knowledge of how much water will be available from each channel during periods of low, normal, and flood flows, and that such knowledge depends upon accurate stream-gauging measurements. The work of the Embudo station, which is still in operation, led to a broad evaluation of natural surface-water resources and to the establishment of criteria for the design of the immense water systems needed to support large and concentrated populations.

The Marlette Lake Water System

The system that conveyed water from Marlette Lake to Virginia City, Nevada, was the first one in America developed to sustain the high pressures required by mountain topography. The first of its three pipelines, constructed in 1873, included an inverted siphon that sustained a 1,887-foot pressure head, twice that of the next-highest-pressure line in the world at the time.

Virginia City, the most famous of the Nevada mining towns and the site of the Comstock Lode, was the greatest producer of silver and gold high-grade ore in history. The first serious prospecting in the region was under way in 1850. By 1857 water was brought up to the Gold Hill prospecting area via a crude wooden flume. The territory was rough and wild, just as legend has it. John Jessup was one of many who staked claims in the Gold Hill area. He was killed in a fight over a card game, and his claim was jumped by the man who was to discover the ore-rich lode in 1859. It was in the same year that Virginia City's first substantial dwelling was built. Until then the population of between 200 and 300 slept outside or, if lucky, in tents. In winter many inhabitants dug holes in the mountainside in which to sleep. Water problems were serious from the very beginning, for in addition to the domestic needs of the population, water was required in great quantities for the operation of the mines.

The earliest settlers made use of natural springs to supply their wants, but these proved adequate for only a short time after the discovery of valuable ores. Wells were dug in an attempt to ease the situation. Eventually a water company was formed, which took its supply from the water discovered west of Virginia City when prospecting tunnels were dug into Mt. Davidson in the Gold Hill area. When the tunnels were pushed farther into the mountain, a new stratum of water-bearing rock was struck, temporarily easing the problem. At last the whole top of the mountain was riddled with tunnels, and although much money was spent on efforts to tap new sources of water, the mountain was running dry.

Marlette Lake Dam, constructed to provide one of the two water sources for Virginia City, Nevada. Completed in 1875, this water system was the first in America developed to cross rugged mountain terrain. The 1,700-foot pressure head sustained by its inverted siphon made it the highest-pressure pipeline in the world at the time. (Courtesy of the American Society of Civil Engineers)

By 1862 there were two suppliers, which in May of that year consolidated their operations, and their names, to form the Virginia and Gold Hill Water Company. The company bought or leased water from seven tunnels, bringing it via wooden flumes and ditches into cisterns from which it was distributed to Virginia City and nearby Gold Hill. The distribution mains in Virginia City began as roughly made wooden boxes that ran along the surface of the ground, with lead tubing used for the subsidiary branches; in August 1863 the water company began laying iron pipes. Although a satisfactory distribution system was now in place, the amount of water available was scanty. Each spring the mains ran full with the water from melting snow, and each fall the town scraped by with half-empty pipelines. As the population grew, the problem became more severe.

The water company's primary supply was the Santa Rita tunnel until the Cole Silver Mining Company's prospecting tunnel, sunk in 1867, hit a new quartz seam and an abundant water supply. Apparently the latter was fed by the same springs that supplied the Santa Rita tunnel, for the Santa Rita immediately dried up. The water company leased the new source from the mining company, and because this was still insufficient, supplemented it with waste water from the Virginia City mine shafts, a practice that brought complaints from the inhabitants of both Gold Hill and Virginia City. The water company managed to keep complaints at a minimum by convincing each town that it was the one receiving pure water.

In 1870 the Cole Mining Company refused to renew the water company's lease and instead laid a pipeline of its own. In retaliation the Virginia and Gold Hill Company dug an adit from another tunnel, intercepting and drawing off all the water before it reached the Cole tunnel. The mining company took the matter to court, where the water company denied that its intention was to obtain water, insisting rather that it was mining for ore. The court decided in favor of the mining company and ordered the water company's tunnel blocked so that water would flow back into the Cole tunnel. Cole was therefore able to continue furnishing the area with domestic water. The Virginia and Gold Hill Company was not ruined by the decision, however, for it continued to furnish waste water from the mines for use in the Gold Canyon mills; its revenues, in fact, exceeded those of the Cole Company.

This legal skirmish did nothing to ease the water crisis. In 1871, however, the Virginia and Gold Hill Company, recently reorganized by new owners, decided to go ahead with a daring plan that had been put forward and rejected years ago. The idea was to conduct water from the Sierra Creek across the Washoe Valley. An engineer named S. M. Buck had been consulted and in June 1864 had submitted a discouraging report. According to Buck, "To bring water across Washoe Valley at a sufficient height to make it available to supply Virginia City would, to say the least, be one of the most arduous undertakings of engineering and mechanical skill in modern or ancient times." He added in a later letter, "It is an undertaking in which no prudent capitalist would ever invest his money; and I hardly need observe that without capital, and that in great abundance, this undertaking could never be accomplished." Buck's pessimism is understandable when one considers that no pipeline had ever been laid that had to contain a pressure head of over 1,000 feet, which was what would be required by this plan.

The water company's new directors included a group of major developers and financiers of the early San Francisco building boom. It was their backing that turned the Sierra Creek project into a reality. They consulted Herman Schussler, chief engineer of the Spring Valley Water Works in San Francisco, a graduate of the Prussian military academy at Oldenburg who had attended civil engineering schools in Zurich and Karlsruhe. He had emigrated to San Francisco in 1864 and was responsible for a number of waterworks on the Pacific Coast. In his report of October 1871 Schussler reviewed the proposed pressure pipeline route from Sierra Creek—known at that time as Dall Creek and later as Hobart Creek—and made some minor adjustments. Further, he reported that he had inspected Marlette Lake and noted it to be a natural reservoir. He recommended that a dam be constructed there to impound water, adding that water could be brought around from Marlette Lake either to the north or to the south by flumes, but that he preferred the northern route to Dall (Hobart) Creek—this would include a tunnel—and then across the Washoe Valley.

Schussler's plan involved the construction of a diversion dam on Hobart Creek, high on the eastern slope of the Sierra Nevada, from which a wooden flume 18 inches deep and 20 inches wide would convey water 4.6 miles to the inlet of an 11.5-inch-diameter pipeline. The pipeline—a 7-mile inverted siphon at whose lowest point, the Lakeview Saddle, the pipe would be 1,997 feet lower than at the inlet—would bring the water to another wooden flume, 471 feet lower than the inlet. This flume would carry the water 4 miles to the Five Mile Reservoir, from which a third wooden flume, 5 miles long, would convey it to tanks above Virginia City. The system was designed to carry 2 million gallons daily.

The iron used for the pipeline was shipped from Scotland in plates 3 feet by 10 feet. At the Risdon Iron and Locomotive works in San Francisco these were cut and rolled with a cylinder into 11.5-inch-diameter, 36-inch-long pipe sections; the edges were lapped an amount sufficient to permit two lines of rivets; and the sections were joined to form lengths of pipe a little over 26 feet long. The ironworks was furnished with a diagram of the elevations and the course on which the pipe was to be laid. Each 26-foot section was accordingly made to fit a certain spot. When the route lay around a point of rocks or other obstruction, the pipe was made to the required curve. Other curved sections were required when the line crossed deep and narrow ravines. Ten different gauges of metal were used in the 7 miles of pipe, depending on the pressure that had to be contained at each location along the route. At the Lakeview Saddle the static head (the vertical distance between the lower pipe and the hydraulic gradient of the pipeline) was 1,887 feet, or 819 pounds per square inch, almost double the pressure withstood by the next-highest-pressure pipeline in the world at the time.

The pipe was shipped to Reno on the Central Pacific Railroad and thence to Lakeview by the newly constructed Virginia and Truckee Railroad. The sections were joined in the field by means of wrought-iron rings, sealed with lead. The pipe was laid in shallow trenches. A blow-off valve was placed at each depression along the pipeline for the removal of sediment, and on top of each ridge was a valve for blowing off the air when water was let into the pipe. The air valves were also designed to let air in, should the pipe break below them, and thereby to prevent the formation of a vacuum and collapse of the pipe. In all, there were 14 air valves and 15 blow-off valves.

The 7 miles of pipeline, over very rough terrain, were laid in just six weeks—a remarkable feat. The Sierra Nevada water first reached Virginia City on August 2, 1873, only five months after the start of fabrication of the pipeline. It was soon apparent, however, that the Hobart Creek water supply was not sufficient for the needs of Virginia City. Therefore, as had been suggested by Herman Schussler, a small dam was constructed at the Marlette swamp, on the western slope of the Sierra Nevada, to create Marlette Lake. Work was started on May 1, 1875. A 4.3-mile wooden flume conveyed the water from the lake along the western side of the range to a saddle, where a 3,994-foot tunnel was constructed to the Hobart Creek drainage area. A wooden flume, 4.7 miles long, then conveyed the water to Hobart Creek. A second pipeline, 10 inches in diameter, closely followed the alignment of the 1873 pipeline.

A third pipeline was constructed in 1887. An additional supply of water was obtained by constructing an 8.25-mile wooden flume northward from the west portal of the 1875 tunnel, along the mountains rimming the east shore of Lake Tahoe. The new 12-inch pipeline was laid alongside the other two.

The second pipeline, laid in 1875, still carries water to Virginia City. The pipe from the first and third lines was dug up and used to replace the flumes. The tunnel caved in many years ago, and the Marlette Lake supply ceased. On some recent occasions when the Hobart Creek water was low, however, pumps were installed and water pumped over the divide.

The water system supported an oasis of life and industry in the Virginia City area. A local newspaper, the *Territorial Enterprise,* extolled this engineering marvel in its June 17, 1875, issue: "One of the boasts of Virginia City, Gold Hill, and Silver City is, that in this land of barrenness—of shifting sands and burning alkali, they have the purest and best mountain water and plenty of it. Nor is the boast lightly made. There is no place in the world where so many natural difficulties have been overcome and so many triumphs achieved as in bringing the pure, fresh, and soft water of the Sierras across Washoe Valley and into the places above mentioned."

The First Owens River–Los Angeles Aqueduct

The 240-mile aqueduct constructed between 1907 and 1913 to carry water from the Owens River valley to the metropolitan Los Angeles area made it possible for that city, located in semiarid terrain, to continue the spectacular growth that had begun in the 1870s with the arrival of the railroad.

Southern California was not inviting to early colonizers. With its scarce water and extreme temperatures, it remained largely unpopulated long after other areas of California had been settled. Founded by the Spanish in 1781, El Pueblo de Nuestra Señora de los Angeles de Porciuncula drew its water from the Los Angeles River, which carried a small amount of surface water but a considerable underground flow. Archeological evidence shows that the ancient Pueblo Indians drew from it to meet their limited needs. As the area was settled by the Spanish from the south and the Anglos from the east, water shortages developed.

The public supply of water was begun in Los Angeles in the early 1860s, when the city gave the privately owned Los Angeles City Water Company a franchise to provide water to its inhabitants. After the franchise expired, control of company property fell to the city, and in February 1902 municipally owned works were initiated. During that time the population of the city had increased dramatically. In 1860 there were 3,700 inhabitants; in 1880, 11,000; in 1900 they numbered 102,000; in 1905, 200,000. By 1913, when water from the Owens River finally reached Los Angeles, over half a million people anticipated its arrival.

By the turn of the century the demands of the rapidly growing population had severely lowered the water table in the area. In one place, a municipal study commission discovered, the water table had dropped a startling 29 feet. Investigations by the U.S. Geological Survey concluded that further drawing of the underground water was hazardous. Furthermore, farmers in the areas surrounding the city were forced to drill ever deeper wells for irrigation. The search began for a new supply

Dedication day at the Los Angeles Cascades, part of the first Owens River–Los Angeles Aqueduct, Kern and Los Angeles counties, California. Built between 1907 and 1913, this aqueduct system was unprecedented in size and scope. It provided Los Angeles with a large flow of water that stimulated its growth. (Courtesy of the American Society of Civil Engineers)

of water both to accommodate the city's expansion and to preserve the agricultural productivity of the area.

William Mulholland had immigrated from Ireland at the age of twenty and had first worked in the Los Angeles area digging artesian wells. After six months he had gone to work for the Los Angeles City Water Company as a ditch cleaner. He had stayed with the company, advancing to the position of superintendent and chief engineer in 1882. When the city took over the operation of the waterworks, Mulholland retained his status as superintendent. One of his principal tasks was to find new water sources for the expanding city.

Knowing Los Angeles needed water badly, Frederick Eaton, retired superintendent of the Los Angeles City Water Company, was developing a proposal to bring the water of the Owens River to Los Angeles. This was a distance of 240 miles, much of it over the Mojave Desert. Although the plan was unprecedented in its size, Eaton, who had become a rancher in the Owens Valley, was familiar enough with the topography to feel that his idea merited serious exploration. In 1904 Eaton studied the route an aqueduct might take and began to acquire land and water rights along it. He presented his plan to the city. The aqueduct would carry a specified amount of water for domestic use by Los Angeles, with the rest going to Eaton and his associates for use outside the city. Los Angeles would build and finance the aqueduct. In return Eaton would provide land and water rights. Mulholland was by now faced with a desperate need for water for the city. He soon set out on a forty-day survey of the Owens Valley and an aqueduct's possible route.

While Eaton was beginning to develop his aqueduct plans, the U.S. Reclamation Service was investigating the possibilities for putting 100,000 additional acres of the Owens Valley into cultivation. An engineer, J. C. Clausen, from the Reclamation Service came into the valley to investigate the feasibility of the project. He was acting under the instructions of J. P. Lippincott, supervising engineer for the Reclamation Service in California. According to the plan developed, the land would be sold to local farmers at the government cost of

$1.25 per acre, plus the cost of bringing water to it—estimated at about $23 per acre. The Reclamation Service, in turn, would require that the local farmers relinquish all power and reservoir rights. Eight reservoir sites had been previously located by the inhabitants, and these were released to the Reclamation Service. A reservoir with a 140-foot-high dam was to be constructed at the head of the Owens River, and canals would draw water from the Sierra and the White Mountains on either side of the valley. Most of the local citizens welcomed the plan as potentially benefiting the valley. As a step in securing the area for the project, the Reclamation Service withdrew from sale all public lands in the area and filed on the water rights.

In 1904, while Clausen was developing his plans, Lippincott wrote to the chief engineer of the Reclamation Service that there was a possibility of not constructing the project, "but of our stepping aside in favor of the city of Los Angeles." During this time, evidence suggests, Lippincott not only was holding his post in the Reclamation Service but was employed by the city of Los Angeles as well (he officially resigned from the Reclamation Service in March 1906 and went to work for Los Angeles as an assistant to Mulholland). It was at this time that Eaton was starting to purchase land in the Owens Valley. In making some of these purchases, he has been accused of misrepresenting himself as an agent of the Reclamation Service (with the power of eminent domain).

A board of engineers met in San Francisco in July 1905 to review the plan finally developed by Clausen, who strongly favored the project in his report. At the meeting Lippincott agreed to the feasibility of the plan but urged that the city of Los Angeles have access to the water for domestic supply. The board followed his advice and did not approve the project.

In the meantime, Mulholland completed his survey of the proposed aqueduct route and recommended it to the Board of Water Commissioners, who in turn ordered a preliminary cost estimate on which to base a bond issue. The board, the mayor, the city attorney, and Mulholland made another trip to inspect the route. Eaton by this time had obtained several of the water rights and options along the route, which he was willing to sell to the city. The water board approved the plan and agreed to buy Eaton's property, which totaled 22,670 acres plus rights to land having 16 miles of frontage on the Owens River. In addition, easements were procured for perpetual storage at the Long Reservoir site. All of these operations were kept from the press in order to prevent land speculation in the valley and the subsequent rise in cost to the city.

The proposed aqueduct at last became public knowledge in July 1905, when the *Los Angeles Times* made the official announcement. The city was wildly enthusiastic. The Chamber of Commerce sent a committee to investigate the quality of the water, and when a favorable report was received, urged voter approval of a $1,500,000 bond issue to pay for preliminary work—surveys, land purchases, water rights. In September 1905 the bond issue passed by a fourteen-to-one majority. The Reclamation Service (with its own project killed by the engineering review board) joined forces with the city to get President Theodore Roosevelt's approval of a federal bill confirming Los Angeles's right to public land. Roosevelt concluded that the use of the land to supply Los Angeles with water was acceptable after receiving assurances that the water would not be used for private irrigation purposes for private gain. Nevertheless, 100,800 acres of desert in the San Fernando Valley had just been annexed to the city, and Mulholland owned land there. This area was to serve as the terminus for the proposed aqueduct. In point of fact, aqueduct water was used for years to irrigate land just north of the city, often at the expense of the city itself. Land rose in value from $20 per acre to $200 per acre in anticipation of the $1,000 to $1,500 per acre it would command if it could be used for citrus production.

The water board chose three of the foremost civil engineers in the country to design the project: John R. Freeman, Frederick P. Stearns, and James D. Schuyler. Freeman was a past president of the American Society of Mechanical Engineers, a consulting engineer for the water supply system of the city of New York, and a consulting engineer on the Panama Canal. Stearns was president of the American Society of Civil Engineers and, like Freeman, a consulting engineer for both the Panama Canal and New York City's water supply system. Schuyler was familiar with southern California, having served as assistant state engineer from 1878 to 1882. The designer of the Sweetwater and Hemet dams and a consulting engineer for domestic irrigation works in Hawaii and arid areas of America, he was also a past vice-president of the American Society of Civil Engineers and a member of the Institute of Civil Engineers of London.

Schuyler made two preliminary trips over the route of the aqueduct before the others joined him for an eight-day tour to prepare plans and estimates. Their final estimate was $24,485,600. The quality of the water was found to be good; the flow would be minimally affected by drought; the completion time was estimated at five years; and no insurmountable engineering difficulties were anticipated. Further, it would be possible to supply the city with hydroelectric power from the project—a most attractive prospect, since industry in the area was growing as rapidly as the population.

Even though the project promised to pay for itself eventually, money for construction was needed at once. While bond issues were common practice for municipal projects, the amount to be borrowed here—$25 million—was not in the least common. This amounted to $88 per capita, and no city of comparable size had ever attempted to obtain such a loan. Later, it was expected, the city would vote on an additional bond of $6.5 million for the development of hydroelectric power. The people of Los Angeles approved the bond issue by a majority of ten to one, and New York banks were successfully solicited to underwrite the loan. Two factors may have been decisive in persuading the latter to take on such a risk: first, the assurance from nationally prominent engineers that the project was technically feasible as well as practical, and second, Los Angeles's immaculate reputation in matters of municipal waterworks financing. The city, after purchasing the waterworks in 1902 from the private company, had rebuilt the entire system to supply the expanding city without increased taxation; all costs had been met by income from water sales alone. This excellent record of financial management convinced eastern bankers that a loan to the city would be a worthwhile investment. With its financing guaranteed, work on the mammoth aqueduct project was ready to begin. The system, which would be a prototype for many of today's, was described in the popular press as "the greatest of all municipal projects."

Construction work on the aqueduct. (Courtesy of the American Society of Civil Engineers)

The proposed aqueduct was impressive not so much for its size or technical innovation as for the character of the land it would traverse—the waterless and mountainous Mojave Desert. Working conditions would be difficult. Lack of water, lack of power, high temperatures in the desert, and low temperatures in the mountains were formidable practical considerations. The plan was for water to be diverted from the Owens River 35 miles north of Owens Lake in order to avoid contamination from alkaline deposits. It would be carried by open canal 60 miles to a large reservoir at Haiwee and from there, through 128 miles of tunnel, conduit, and siphon, would travel to the Fairmont Reservoir. The water would drop rapidly over the next 15 miles, providing power to generate electricity. After collection into the Dry Canyon Reservoir and then the San Fernando and Franklin reservoirs at the upper end of the San Fernando Valley, the water would be distributed to the city. Along the 235-mile length of the aqueduct there were to be 24 miles of unlined canal, 53 miles of tunnels, 37 miles of lined canal, 98 miles of covered conduit, 12 miles of inverted siphons, and 10 miles of power waterways. The water would flow by gravity alone over the entire length of the aqueduct. The project was expected to take 5,000 men five years to complete.

Work began with a sectioning of the project. The line of the aqueduct was divided into 11 parts of 6 to 23 miles each, depending on the type of construction necessary. An assistant engineer was put in charge of each section and given large latitude to solve problems and make decisions within that section. Bids were collected from private contractors for work on the line, but when they came in 50 to 100 percent over the estimate of the city engineers, water board superintendent Mulholland persuaded the city to do its own work.

In October 1907, a year before preliminary work started anywhere else in the system, the Elizabeth Tunnel was begun. This was the longest tunnel—26,780 feet through solid granite underneath the San Fernando Mountains—and one of the most difficult construction tasks. It was undertaken first because it was thought to be the only part of the project that might hold back the five-year completion timetable.

Water, transportation, and power were the three problems. Not only did the workers need water to drink, but great quantities of it were needed for the concrete construction itself. Since no water was available in most of the desert and since such large quantities could not be hauled in by the barrel, the city laid a 268-mile pipeline, connected with large storage tanks, along the aqueduct route. This auxiliary pipeline ran from the intake to the San Fernando Valley.

Transportation of the enormous loads of construction materials and equipment was initially intended to be carried out by mule teams, but desert conditions made this impracticable. There was little water available for the teams and no natural food supply. Bids for freight teams were astonishingly high: 25¢ to 28¢ per ton per mile. Further calculations revealed that the city could build and operate its own railroad for 10¢ a ton-mile and could buy its own mules for a mere 12¢ to 15¢ a ton-mile. The city hoped that it could persuade a private firm to build the track connecting the Owens Valley agricultural and mining region with Los Angeles. This would guarantee a permanent rail connection instead of a temporary, city-operated one. The freight estimates were large enough to attract one private company: the Southern Pacific built the line and carried construction materials for the city at a special rate of 4.5¢ per ton-mile. The track was completed in 1910.

Railroad construction solved only part of the transportation problem. In addition, 505 miles of roads were necessary. It was soon decided that mule teams were too expensive to use for this short-distance hauling, so the planners invested in a recent development, the caterpillar tractor. After one of these machines had successfully completed a trial period of a few months, the city bought 28 more. Unfortunately they proved to be cost effective only when new. Soon they began to break down, and maintenance was expensive. The city tried making replacement parts in its own shops from patterns purchased from the manufacturer, but the repair bills remained excessive, and animals were brought in to do the work. Some machines were sold to local farmers, and others were disassembled for their steel, which was used in the formwork for casting the concrete. This was certainly one of the last times that animal labor won out over the machine in a large construction project.

Power for construction was generated locally by means of two hydroelectric plants built especially for this project. The Cottonwood and Division Creek plants supplied power over a 218-mile, 30,000-volt transmission line. Transformer banks had to be installed at each supply branch of the line. The power supplied the construction machinery, cement mill, machine shops, and lighting for the camps. In addition to the power line, 240 miles of telephone and telegraph line, strung on poles, was installed. Once these lifelines had been established, construction became possible. Because the distances were too great to permit daily commutes, workers had to reside in temporary towns. The buildings provided were transportable, easily knocked down and reassembled elsewhere. In all, 2,291 buildings in approximately 57 camps were built.

The supply of cement for concrete construction was another hurdle. No city had ever gone into the cement manufacturing business simply to supply a public works project, but the Los Angeles engineers decided that municipal ownership in this area, too, would provide the most economical results. Therefore, Los Angeles created the town of Monolith, whose population numbered 250 laborers and whose only reason for existence was the production of 1,000 barrels of cement every 24 hours. The cement mill was supplied with limestone and clay from deposits in thousands of acres of land the city had purchased in the Tehachapi Mountains for that purpose. Nearby were also discovered deposits of volcanic ash, sometimes called tufa or pussuolana. This material was ground and mixed with equal amounts of portland cement, producing a substance named Tufa cement. Its biggest advantage was that it cost almost 30 percent less than portland cement. Eventually this new product was used with excellent results for all concrete work throughout the project. The demand for cement was too great for the city's mill, however, even though it was producing round the clock. An additional 200,000 barrels were obtained from private manufacturers.

As the work proceeded, the city continued to purchase land for reservoir sites and to protect its water rights. Much of the land along the aqueduct's path was eventually set apart as national forest to protect the water quality. In all, Los Angeles purchased 135,000 acres of land, twice what was then the area of the city.

The intake for the aqueduct was constructed on the Owens River 15 miles north of Independence, the county seat of Inyo County. The intake is a diagonal weir and sluiceway, 325 feet long, with four radial control gates of concrete construction. Between these gates is a 240-foot concrete overflow weir. A steel sheet was driven down about 20 feet on each side of the intake and extends 75 feet into each side of the aqueduct bed to prevent underflow. Heavy rubble pitching protects the two sides of the weir from undercutting by the backlash of water.

The next section of the aqueduct consisted of an unlined canal, 38 feet wide on the bottom, 64 feet wide at the water line. The following 35 miles consisted of open, lined canal (lined because the ground was porous in the area), following the ancient beach line of the once-larger Owens Lake. The side slopes were made relatively flat so that the concrete could be placed without formwork. Power shovels did the excavation. After this channel came the Haiwee Storage Reservoir, made by a fill earth dam at the lower end and a levee at the upper. Below the reservoir the channel was covered; it remains so for the rest of its journey south, through a combination of concrete-lined conduits, concrete-lined tunnels, and riveted-steel inverted siphons, 135.5 miles to the Fairmont Reservoir.

In addition to the 70 miles of cut-and-cover work through the Mojave where the concrete covered conduit was used, part of this 135-mile passage ran along the eastern Sierra face. This stretch was of such topographic severity that many tunnels had to be constructed—164 in all. As there was considerable concern about completing the difficult Elizabeth Tunnel on time, a bonus was awarded to each team of men for every foot of tunneling by which they exceeded the average footage established for that particular area. The men gladly accepted the challenge and worked so quickly that their daily wage increased by 30 percent. The dramatic jump in wages precipitated an official investigation into the matter to determine whether the system was fair to the city. The results showed that in spite of the increased wages the cost per foot of the tunnel had fallen by 10 to 15 percent. The city continued the bonus system throughout the project.

Two work crews were used in the tunnel construction, alternating between shoveling and light work during the entire shift. Trains with electric locomotives were used in the tunnel to take out muck and bring in lining materials. Both techniques enabled the crews to keep up a terrific pace on the Elizabeth Tunnel. Because the north end was driven through decomposed granite, frequent timber or steel shoring was put in place. The south end of the tunnel was driven through solid granite and required no timbering. On February 28, 1911, after driving 13,370 feet from the north and 13,500 feet from the south, the two bores met at the center. The Elizabeth Tunnel had been completed after only 40 months of work instead of the predicted 60, and at a cost only two-thirds the original estimate. The crews had broken the American record for hard-rock tunnel driving with a figure of 604 feet in one month.

In addition to the tunnels and covered conduits, the water was carried by a series of 23 inverted siphons. Most were constructed of riveted steel, manufactured in Pennsylvania. Because of the extreme isolation of the site and the difficulty of adjusting defective parts, all steel pieces were test-assembled by the manufacturer, then disassembled before being shipped west. The largest of all the sections was 36 feet 10 inches long and weighed 52,000 pounds. It was hauled into place by 52 mules.

Just how to use the siphons was a matter of debate and experiment at first. Early attempts to bury them were failures: when leaks occurred, they reacted with the mineral deposits in the soil and rusted the pipes. The next siphons constructed were carried on piers. The piers were built under empty siphon sections, but the large pipes altered their shape when filled with water. Often a carefully constructed pier had no pipe to support after the pressure of the water in the pipe caused it to rise several inches. Another difficulty was caused by the longitudinal expansion and contraction along the pipe before it was filled. After water was running inside, changes in length were not a problem, because the temperature was held more or less uniform by the bulk of water. Pipes were eventually laid in trenches about one-third their circumference in depth. The city, having profited from the bonus plan for tunnel diggers, used the same incentive system for its riveting crews.

Some siphons were as long as 15,597 feet. The Jawbone Siphon, while not the longest (only 7,096 feet), carried water under the strongest pressure, about 368 pounds per square inch or 850 head-feet. The Jawbone was economically designed to taper in diameter and plate thickness to reflect the distribution of pressures in the pipe. Not all of the siphons were steel, however. Eleven were constructed of concrete, which was estimated to be cheaper and better for heads under 75 feet. For heads higher than this, water begins to seep through the concrete. The pipes were constructed over a collapsible form that could be dismounted and loaded into a car, which was pushed along inside the pipe.

After this chain of tunnels, siphons, and conduits the water was to enter the Fairmont Reservoir, formed by a concrete core wall and earth fill dam, 115 feet high and 1,516 feet long. The reservoir was not intended primarily for reserve storage purposes, although it could serve that function as well, but instead was built to act as a forebay, regulating the flow of water to the envisioned power plants.

An additional 11 miles of conduits and tunnels were constructed, leading to the wheels of two large hydroelectric plants. From these a powerway to a third plant was built. The Dry Canyon Reservoir was then put in place, created by another fill dam. This was intended to act as an equalizing reservoir, compensating for the variations in draft caused by power plant and city use of water. Beyond the San Francisquito Canyon and in the area that is now the Santa Barbara National Forest, the aqueduct came to an end in the San Fernando and Franklin reservoirs, 1,465 feet above Los Angeles. That elevation afforded perfect distribution to all parts of the city.

Water from the Owens Valley first reached Los Angeles on November 5, 1913. The aqueduct was a model of efficiency and economy, and for eight years operated with great success. From the beginning, however, inhabitants of the valley were anxious about the project and future developments. Despite repeated requests for information on the policies the city was to pursue, little was forthcoming. Finally in April 1913 a meeting with Mulholland was held in Bishop. Agreement was reached that the city would admit rights to existing ditches, not interfere with underground water, and withdraw opposition to the reopening of public lands for settlement. A period of relative calm followed until 1921–1922, when a severe drought again initiated tensions between Owens Valley farmers and the city. Both the farmers and Los Angeles were in desperate need of water. The city was forced to choose between building flood storage dams and buying up additional rights to surface water. When it decided to pursue the latter course, the death knell sounded for agriculture in the valley.

Los Angeles already owned significant land and water rights in the area, which had originally been acquired by Eaton. The city renewed its acquisition activities. All of the agricultural irrigation ditches had long been controlled by community ditch companies, cooperative organizations that allowed each farmer to have a say in the distribution of water. Several influential local citizens began taking options on the important McNally ditch. They soon gained control of the ditch and turned it over to Los Angeles for a handsome profit. The city ignored the hapless minority owners of the ditch, who could not obtain loans for their drought-stricken farms and were forced to sell to the city. Meanwhile, the farmers formed an irrigation district in 1922, in the belief that a single organized body would have greater strength in dealing with the city. Rights for individual ditches were transferred to the district.

Los Angeles continued its campaign of acquisitions, using a "checkerboard" system: every other ranch was bought and allowed to dry up. This depreciated the value of the remaining ranches. Hard-pressed ranchers still suffering from the drought could not obtain loans—not even from friendly organizations, because of the uncertain future of the valley and the unannounced intentions of the city. The farmers, not giving up, claimed that the water from the ditches did not belong to the city and insisted that if the city did not irrigate its property the water should go to neighboring irrigation ditches. Los Angeles responded by attempting to build bypass channels.

On May 21, 1924, a small charge of dynamite exploded against the aqueduct wall near Lone Pine. Newspapers reported the incident, and a Chamber of Commerce committee from Los Angeles came to investigate. For many inhabitants of Los Angeles this was the first sign that anything was at all wrong in the Owens Valley. In an attempt to call statewide attention to their plight, the farmers then engaged in relatively peaceful demonstrations and kept the gates of the aqueduct open at a point near Lone Pine. The city responded by filing an injunction against the demonstrators. Newspapers throughout the country followed the story and most strongly condemned the actions of the city in the valley. A state engineer sent by Governor Richardson also condemned the city's actions.

Despite some efforts at arbitration, conditions continued to deteriorate and more ranchers were forced to sell out. The city began digging wells on their property, ostensibly for irrigation purposes. On April 3, 1927, a well near Bishop was dynamited. Another was blown up the next day. On May 13 the aqueduct was again dynamited. Some valley people believed the dynamiting was done by city workers. During the late spring and summer of 1927, six additional dynamite blasts occurred against the aqueduct wall. The city hired armed men and detectives.

The last of the dynamiting was in July 1927, for financial disaster struck many farmers in the next month. Several banks in the valley were forced to close. The banks were owned by Mark and Wilfred Watterson, influential businessmen and property owners who had for years provided leadership and financial support to the embattled farmers. Their own businesses were now in financial trouble. The Wattersons tried to get outside help for their businesses and banks but were not successful. Clerks in the office of the State Superintendent of Banks noticed some irregularities at the Inyo Bank, and investigations followed. The Wattersons were charged with embezzlement and later found guilty. Many farmers lost their own deposits in the bank failures. With this calamity, active resistance in the valley came to an end.

On March 12, 1928, the San Francisquito Dam collapsed. Five hundred lives and millions of dollars' worth of property were destroyed. Mulholland came under a great deal of criticism. In 1929 a change in policy occurred on the part of the city of Los Angeles with the election of a new mayor, John Porter, who began to ease relations with the remaining valley inhabitants. By now many of the latter believed that it might be best if the city simply bought out the entire valley. Committees were formed and methods for determining reasonable prices devised. The plan was set in motion. A 1929 court case established the precedent that the city must buy all ranches in the valley whose owners desired to sell, not just those with critical water rights. Eventually the buyout plan was largely accomplished. Soon the city was looking to engage in a similar program northward in the Mono Basin. The Owens Valley had become virtually barren—a sad epilogue to the history of a remarkable construction feat.

The Miami Conservancy District

The Miami Conservancy District in southwestern Ohio was the first major regionally coordinated flood control system in the United States. Dams were combined with levees and channel improvements in this large project. Engineers trained in the Miami River valley eventually went all over the world to undertake flood control projects utilizing techniques initiated here. Since completion of the system in 1922, cities in the region have not been damaged by flooding.

The Miami Valley drains 4,000 square miles of gently rolling territory. The three main branches of the river system—the Stillwater, Miami, and Mad rivers—join one another at Dayton to form the Great Miami River. Also at Dayton, Wolf Creek empties its waters into the Miami, and between there and Hamilton the waters of Twin and Four-Mile creeks are added. Not far south of Hamilton, and just above its mouth on the Ohio River, the Great Miami is joined by the Whitewater River from Indiana. The soil of the region is largely clay, which sheds water quickly.

As early as 1805 serious floods were recorded. Nevertheless, towns grew up along the unpredictable streams because they provided transportation, water supply, and a source of power. Thus the cities in the area, some of which were highly industrialized by 1913, were located largely in the flood plains of the rivers. Levee building to control flooding began in the Dayton area after the 1805 flood. Particularly destructive floods were recorded again in 1814, 1828, 1832, 1847, 1866, 1883, 1886, 1897, and 1898. Each brought renewed control efforts, which were usually wiped out by the next flood.

A flood in the spring of 1913 proved more damaging and greater in scope than any of its predecessors. A tremendous storm inundated the entire Ohio River valley, and the Miami Valley area received the worst of it. On March 25, streams in all of the Miami Valley cities overflowed their levees. Most small residential buildings were flooded to their eaves, and Dayton's Main Street was under ten feet of water. Ruptured gas lines started fires, which could not be fought. Thousands of

The Miami Conservancy District, Dayton, Ohio. This was the first regionally coordinated flood control system in the United States. The project consisted of the construction of five dams, levee and channel improvements in nine towns, the relocation of four railroad lines, the removal of one village, the lowering of water and gas mains, and many other works. Since its completion in 1922 the system has protected the Miami Valley from flooding. An aerial view of Huffman Dam, part of the system, is shown. (Courtesy of the American Society of Civil Engineers)

people were stranded without food on the top stories of their houses for days. The current was too strong for rowboats or rafts to reach them, and lives were lost in rescue attempts. A spring snowstorm and cold weather then beset towns that were without heat or adequate shelter. The railways were out of service, so relief supplies could not be brought in. Martial law was declared and an emergency government set up to carry out relief work.

Estimates of the death toll were high during the flood, but when the waters subsided it was found that many presumed lost had survived. Property loss, however, was enormous. The prospect of another such disaster in the future drove the citizens of Dayton to immediate action. Within sixty days they had raised a fund of $2,136,000 to initiate a solution to the flood problem. They hired the Morgan Engineering Company of Memphis, headed by Arthur E. Morgan, to develop prevention plans. It soon became apparent to the engineers that a piecemeal attempt at protection would not succeed as well as a comprehensive, regional project.

The legal and political apparatus for such a project did not exist, so while the engineers began their work, law drafters developed the appropriate mechanisms. On February 17, 1914, the Conservancy Act of Ohio, which would permit the formation of the Miami Conservancy District, was executed by the governor. A battle over the constitutionality of the act ensued, postponing the establishment of the Conservancy District until June 28, 1915. The conflict pitted farmers in the areas where flooding had not occurred against the populations of the towns and cities lining the waterways, who suffered most of the flood damage. The farmers could see little reason why they should pay higher taxes to finance the safety of the townspeople; further, they feared the loss of their land. Those south of Dayton were mostly in favor of the act; those north were opposed.

A campaign of shows and talks with question-and-answer sessions conducted by the directors and engineers of the Morgan Company, who had by then developed preliminary plans, was begun in order to persuade the general public of the need for regional flood control. Although the campaign was eventually successful, the directors and engineers had to spend a great deal of energy fending off counterschemes developed by those opposed to a regional solution. It was suggested, for example, that a huge concrete conduit be built under the river bed to carry the flow for 125 miles and also to provide electric power. One of the more outlandish schemes submitted for consideration was the drilling of holes down to underground caverns—which were hoped to exist—so that flood water might be received and stored therein. A third scheme advocated the enclosure of the entire river in a concrete flume. None of these plans was feasible, and their proponents provided no technical information or cost estimates. Yet these bizarre suggestions had their effect on the public, and organized efforts to repeal the Conservancy Act continued well into 1915.

The work of the engineers had actually begun right after the flood; field surveys were initiated on May 29, 1913. There were no U.S. Geological Survey topographic maps for most of the Miami Valley, and most of the knowledge of the region came from railroad and highway profiles. Therefore, surveyors were sent from house to house over the entire valley to record flood level marks and to estimate the volume of flow and the rate of travel of the flood crest. From that information a topographic survey was made. All records of rainfall and of previous floods were studied; 1913 flood discharges were calculated; the frequency of floods was determined; the capacity of the present channels was computed; foreign international flood prevention plans were reviewed; and rainfall and flood records from other areas were compared with those from the Miami drainage basin. The engineers concluded that the Miami River was well suited to control by dams. Extremely high and low flows were found, the great variation being due to the flash floods typical of the region. Retarding basins were considered ideal for the control of flash floods; therefore the engineers proposed the construction of giant dry dams.

The idea of dams without water behind them, massive constructions that would stand unused during most of the year, appeared shockingly wasteful to a part of the populace. There was no precedent for such a system, although the engineers did manage to find a few smaller dry dams that, while not intended for flood control, did serve that purpose. Two such dams were 200-year-old structures in the Loire Valley of France. One of them, in Pinay, was still in the middle of a farming area. The flood plain here showed a 30- to 50-percent greater crop yield than in the unflooded areas. Moreover, an average of only 11 percent of the crops had been lost to flooding in the history of the area. The engineers proposed that the same practice of planting in the retard basins be practiced in Ohio, a particularly convenient plan since the greatest danger from floods existed from January through March, when no crops would be affected. The accumulation of silt from flooding would improve the productivity of the soil and counterbalance any possible drawbacks due to flooding. This plan was officially adopted on November 24, 1916.

The Miami Conservancy District governing board appointed a committee of hydraulic engineers and flood control and dam experts to oversee the engineering portion of the planning. The first decision the committee faced was to what extent the valley should be protected and at what expense. Priority was given to safeguarding human life, since the area to be protected was heavily populated. After considerable study, the committee decided to provide protection against a flood up to 20 percent greater than that experienced in 1913, by means of a combination of retard basins and channel improvements. The amount of control provided by each was determined by economic considerations—that is, by the cost of constructing a dam in a given location versus that of channel improvements.

Five earthen dams were to be placed across the river valleys. Concrete outlets, founded on rock and passing through the base of each dam, would allow the normal river flow to pass through. Outlets would be sized so that only the amount of flood water that could be handled safely by the river channels below the dams would be permitted to pass through. Any amount beyond this would collect in the valleys behind the dams and be let off gradually after the flow decreased. Each dam was to have a concrete spillway constructed slightly below its top level, so that water would never reach the top and overflow. Eventually it was calculated that a flood 40 percent greater than that of 1913 would bring the waters up only to the level of the spillways and not to the top of the dam. Since a flood of that magnitude was considered virtually impossible, it was debated whether the additional expense of providing for spillways was reasonable. The loss of life that might result from failure of a dam decided the matter. The spillways were kept in the plans, even though no one expected they would ever be used.

Channel improvements were planned for the cities only. Therefore, the land outside the cities would be only partially protected. The improvements, to take place in nine cities, included widening and deepening the channels, raising levees, and removing trees, bridge piers, and other manmade objects that provided resistance to the river flow. In most of the small towns the improvements were to be relatively simple and inexpensive. But in Dayton and Hamilton, the two most industrialized areas, urban growth had encroached upon the river to such an extent that changes in the channel were expected to be difficult and costly. A number of bridges would have to be removed or rebuilt, and gas, water, and sewer lines would require alteration. Four railway lines, with about 50 miles of track, would need relocation, along with several highways and power lines. All of these were provided new rights of way on higher ground. The complete removal of the small village of Osborne from the flood area had to be undertaken.

Legal measures were taken to provide for the project's financing. The Conservancy Act granted the Miami Conservancy District the power to levy benefit assessments, borrow money, condemn land, and perform the other actions necessary to ensure the project's success. In August 1915 the District set about determining the value of benefits to be gained from the project. Half of the benefits were assessed to the cities and counties, the other half to the individual properties subjected to flooding. The properties were assessed according to their initial value and benefit, the degree of protection needed and provided, and the depth of the 1913 flooding on that land. Forty thousand owners controlled the 60,000 pieces of property that were appraised for benefits. The construction was to be financed by the proceeds from the sale of bonds, which were secured by the appraised benefits. The bonds were to be retired by 1949, a period of 34 years. The money to take up the bonds, pay the interest on them, and maintain the system would come from an assessment placed on the benefited property. Approximately 30,000 acres of land were purchased outright for the basins.

The call for construction bids went out on November 15, 1917. The Morgan Company, dissatisfied with the usual quality of specification writing, resolved to provide a model for future engineering projects. The specifications that they distributed contained an extraordinary amount of advance information. Definite times and dates were cited, and the quality of the work required was stated plainly. The entire document was organized as a self-indexing system tied and referenced to the unit price items of the bid. All of these steps were innovative.

In spite of the care the engineers had taken in their preparation of specifications, the bids that came in were filled with qualifications and protection clauses for the contractors. The project was large, and not one of the contractors had undertaken anything of its size. Other factors that confused the bids were the three- to five-year construction period and the inflationary conditions brought about by World War I. Both of these made accurate prediction of future costs difficult. After analyzing the bids, the Conservancy District ruled that all but one were either too high or too irregular to be accepted. On December 3, 1917, the District decided that the only solution was to form its own construction company, hire its own workers, and purchase its own equipment. Charles H. Locker was hired to head construction. Locker had been involved with several large projects in the past, including the Chicago Drainage Canal, the Shoshone Dam, the Wachusetts Dam, the Weston Aqueduct, the Western Maryland Railroad, the Catskill Aqueduct, the West Neebish Ship Canal, and the New York Barge Canal.

Meanwhile, the engineers were refining their plans and organizing their own administrative system to control the design of this vast project. At the head of the organization was Arthur E. Morgan. In later years he served as chairman of the Tennessee Valley Authority, president of Antioch College, and vice-president of the American Society of Civil Engineers. Sherman M. Woodward did most of the experimental and theoretical work on the hydraulic-jump energy-dissipating device. He was also responsible for overall planning and balancing of the system. Morgan placed engineers joining the program into a general pool. Each was expected to understand the whole project and its rationale as well as the particular area to which he was eventually assigned. The project was divided into twelve areas or divisions, with one supervising engineer for each. The assignments for division engineers were made as soon as the general design for each area was finished but before the detailed specifications had been worked out. The division engineer at that point became the design engineer as well.

This procedure developed out of Morgan's concept of dynamic design. In this model each engineer is familiar with the overall project and with the detail of a particular part of it. The design can therefore remain fluid, modified if necessary at any point by a designer who is intimately familiar with the problems of his part of the project. Morgan believed this led to an efficient method of construction that would produce a more fitting structure than if all design decisions were made in advance by an engineer who was not working in the field. Another notion of Morgan's that was used on the project was what he termed "conclusive engineering analysis." Each possible approach to a problem was assigned to a different engineer or group of engineers, who analyzed it and made recommendations; final decisions were made only after weighing all the alternatives. Morgan credited this procedure with many innovative developments.

The construction team was gathered in 1917–1918. In order to start construction during the war, the District obtained "essential industry" status. The workers, most of them locals, varied in number from 750 to 2,000. Railroads were already in place where they were needed, but new highways had to be leveled and paved. A central warehouse and repair shop were constructed, with subdivisions at the various work sites. Workers were housed in model towns; the houses, after the project was completed, were to be sold off by the Conservancy District. The five towns comprised 230 major buildings and 200 sheds, and had running water, baths, five mess halls, and five stores. These towns were self-governing, and they developed into true residential communities, with public schools (used for adult aducation as well as for teaching children), community groups, and family gardens.

Much of the excavation work was in rock or below water level. A thorough study was made of the potential use of steam shovels, draglines, and other excavation equipment. Dragline equipment was practically unknown at the time, but it had the advantages of long reach, the ability to handle heavy loads, and the capacity to excavate under water while mounted on a barge. It was therefore selected over the more common steam shovel and, despite the doubts of many, proved itself remarkably well suited to this task. Electricity was to be the primary source of power on the project. This, too, was an innovation. Because machinery was difficult to obtain during the war, and because the electric dragline was a new type, the District was forced to track down the only ten draglines in existence, requisitioning them from all over the country. Eight of them were steam draglines and had to be converted to electricity. The purchase of the draglines was the largest of all the project's equipment purchases.

The District set up a daily flood-warning and river-stage-forecasting service for the safety and convenience of the construction crews. From the measurement of rainfall, saturation of the ground, and temperature were developed river-stage predictions that were accurate within a few inches. When necessary, machines and workers were removed from the flood area until the danger was past. In addition to these accomplishments, the District established its own labor code, which enabled it to work with one rather than many labor organizations and their regulations.

Five dams were constructed: the Germantown, the Englewood, the Huffman, the Taylorsville, and the Lockington. All have variable-slope sides. At each change of face is a 10-foot-wide berm, either equipped with gutters or paved with rock. The berms, intended to break the force of rain water running down the face, also act as inspection roadways. The cross sections of the dams are heavy, "to insure safety beyond any doubt" to the densely populated area. Each outlet channel expands into a stilling basin in order to reduce erosion of the river channel below due to the water's increased velocity. The basin is designed to create a hydraulic jump to further reduce water velocity.

Construction began on January 27, 1918, at the site of the Huffman Dam. All five dams were of earth construction, with concrete conduits permitting normal river flow through the dams. All were identically constructed by means of the hydraulic fill method, with only minor variations. There were two types of outlets. Those in the Germantown and Englewood dams had separate conduits and spillways. The other three had the conduit and spillway in the same structure. Concrete plants were set up at each dam site, and gravel was mostly obtained from the excavation for the outlets. Collapsible sectional forms were used in tunnels, and movable panels elsewhere. Where possible, equipment and forms were standardized.

The hydraulic fill method of construction and the hydraulic jump are two of the important features of the dams. The latter was a device used to reduce the velocity of water as it left the conduit pipes. The jump is created by a concrete structure designed to cause the water to form itself into a standing wave. This uses some of the energy the water carries; the rest is dissipated when the water swirls around the stilling basin.

The hydraulic fill method originated in "placer mining" in the western United States and was used there long before it spread to other parts of the country. The process produces a solid dam with an impervious clay and silt core, unbroken from top to bottom, which is stabilized by banks of sand, gravel, and boulders. The process begins with the use of a cannonlike hose that emits a stream of water under pressure. This hose furnishes material for the dam from a borrow pit, either by tearing down hillsides or by breaking up excavated rock with the force of its blast. The material is then carried down the hillside by the stream of water in open sluices or, if the borrow pit is below the level of the dam, lifted by sump pumps and transported under pressure in pipes. In either case it arrives at the dam site traveling at a high velocity. Two parallel pipes are stretched across the dam site, on the outside edges of what will become the embankment. The water and its contents are dumped out of the pipes along the entire length of the dam, and the contents of the water are distributed as the water flows from each side of the dam inward to what will become the core. Thus the largest rocks drop off as soon as the water leaves the pipe, the gravel is carried farther in toward the core area, the sand is dropped off next, and finally the water flows into the pool that occupies the core area. By now the water has slowed down considerably and carries only small clay and silt particles. As more and more water is added, the clay and silt settle to the bottom. The gradually increasing weight of the accumulated mass compacts the lower levels, squeezing the water to the top. From here the water, having deposited all its foreign matter, is drained off. The dam is gradually built up, carrying the pool with it on top of the water-impervious core. The sand, gravel, and rocks form stabilizing shoulders that decrease in density as they extend outward. The result is watertight, free from settlement, and simply produced by readily available materials.

Flood control during the construction of the dams was critical. Since a dam could not be constructed in a single year, there was a genuine possibility that a flood could occur during construction. Its path must not take it over the partially finished dam, wiping out the work to date. The outlets, therefore, were constructed first. Because the dam would not be able to resist any back-up water while under construction, the outlets had to be oversized at first, to handle all the flood water that might need to pass through. Only after the dam was completed were the outlets filled in with concrete to their precisely designed size. The work on the dams was staggered so that not more than one dam would be at a critical stage during flood time.

Germantown Dam is located on Twin Creek, two miles northwest of Germantown. The valley here is narrow with steep sides, providing a good site for a dam, with relatively little concrete yardage required. The outlet conduits were positioned at one side of the valley floor at the foot of a side hill, some distance from the river channel, to which they were connected by new channels. They were therefore built with no interference at all to the river channel. As with the rest of the dams, a concrete girder bridge originally spanned the spillway but was replaced with a new steel and reinforced-concrete bridge in 1983. Lockington Dam is located on Loramie Creek, two miles north of its junction with the Miami River. Its outlet and spillway, combined in one structure a little east of the river channel, consist of two retaining walls 27 feet apart at the bottom and flaring outward toward the top. Over the gap, 16 feet below the crest of the dam, is an ogee overflow weir, 62 feet high, which forms the spillway. Through the base of the weir are two outlet conduits, 9 feet by 9 feet, with arched roofs. Englewood Dam is on the Stillwater River, ten miles above Dayton. It is the highest of the dams and, like the Germantown Dam, has a separate outlet and spillway, the latter an ogee weir 100 feet long. The Taylorsville Dam, ten miles above Dayton on the Great Miami River, required the largest concrete structure of all. Its retaining walls are 90 feet high. The spillway is 132 feet long, and the outlet has four conduits, 19 feet high and 15 feet wide; all are in one piece. The Huffman Dam, on the Mad River eight miles east of

Dayton, uses a combined outlet and spillway. The spillway crest is 100 feet long, and the outlet contains three 16.5-foot-high and 15-foot-wide conduits.

While work was in progress on the dams, channel improvements began in the towns. What improvements were made differed according to the needs of the area and the cost involved. Hamilton and Dayton, because of their riverside industrialization, proved the most difficult. In both cases channel excavation was the major work, in combination with minor amounts of levee raising. This was also true in Troy and Piqua. Cut-off channels were constructed in Troy and Middletown. In West Carrollton, Miamisburg, and Franklin construction was restricted to levee building.

Since improvements caused increased water velocities, sod or concrete revetment was used to protect critical points in the river bank where the flow was in the direction of the bank or where erosion might cause damage to valuable property. The revetment attached to concrete walls consisted of a mat of concrete blocks held together with galvanized iron wire cable. The edge of the mat was held in place by large, cast-in-place concrete blocks.

The standard channel design was one with a depressed bottom, one-quarter the width of the entire channel, sunk 8 feet below the rest of the channel bed. Where the river was straight, the depression was located at the center of the river; where curved, at the outside bend of the curve. In Dayton the channel depression swung from side to side as the curve of the river changed. A 30-foot-wide berm often stretched from the channel depression top to the base of the levee. Where this was not possible, the base of the depression continued up with the slope of the levee. The standard levee was designed to be 8 feet wide on top. Many of the levees were faced with 6 inches of concrete with wire mesh reinforcing formed in 8-foot by 12-foot slabs with expansion joints. Lamp black was added to darken the concrete and reduce glare. In many places sewers had to be equipped with floodgates that would remain open under normal circumstances but would shut in case of flood so that the sewers would not back up. Numerous gas and water lines were also adjusted to the new river channel dimensions, and occasionally a bridge was replaced by one whose piers did not impinge on the river bed. The construction of the retarding basins, however, obviated the wholesale bridge reconstruction that would have been required were channel improvements to have handled all the effects of flooding.

The first test of the completed system occurred in April 1922, when a severe storm brought four days of heavy rain. Under previous conditions the amount of runoff would have caused the river in Dayton to rise at least 18 feet above normal, well above flood level. In fact, the river peaked at 9.6 feet above normal. Half of the reduction was attributed to the channel improvements and half to the dams. The rivers were not turbulent in the cities, and the hydraulic jumps functioned as they were designed to do. Almost no damage occurred anywhere in the district.

In addition to protection from floods, there were several byproducts of the completed project. One was the conservancy farms, which were leased to local farmers. A second was the development of recreational areas, dotted with man-made lakes formed from the borrow pits. Dumping in the streams was now prohibited, so their quality improved markedly. And finally, since the continual dredging of the rivers resulted in a source of gravel, the District was able to make money by leasing the dredging rights to commercial businesses.

The influence of the Conservancy District eventually spread well beyond the Miami Valley. The Ohio Conservancy Act was a pioneer undertaking that inspired the adoption of similar laws in many other states, enabling communities elsewhere to complete similar projects without the support of state or federal funds. The project introduced the dry dam, a new method of flood control, and in the process helped perfect the earth-core, hydraulic-fill dam. Three of the technical reports the District published, including *The Hydraulics of the Miami Flood Control Project,* were historically significant contributions to their respective fields. The dragline came into use for the first time as a major construction tool on a large-scale project. In short, the Conservancy District project was the first well-planned project of its kind. Not only did the ideas it developed concerning labor relations, legal organization, and engineering management serve as models for the later Tennessee Valley Authority, the TVA project itself was largely carried out by men trained on the Miami Conservancy District construction operation.

Chapter Seven

Environmental Engineering

Introduction

The building of America was a social and technical accomplishment that has many clear symbols in the form of bridges, tunnels, buildings, and other physical artifacts. The rise of modern environmental engineering, with its impacts on the health of America's citizens, was an essential element of this accomplishment, but its specific symbols are less visible and its results now frequently taken for granted by the public. Yet if one were to picture any of our cities without the benefit of the many contributions of the field of environmental engineering, the vision would be frightening.

The late nineteenth and early twentieth centuries were the nascent period in the development of a branch of engineering that applies scientific knowledge to the problems of water quality and waste disposal. The field has more distant antecedents, of course. The earliest civilizations explored techniques of water supply. Ancient peoples, however, primarily viewed the water supply problem in terms of quantity and accessibility. Attention to the quality of water came much later, and even more recent has been a true understanding of the relation between water quality and the public health. That such a relation does exist was appreciated in a vague way by the creators of some of the great nineteenth-century water supply systems of Europe and America—notably in major urban areas such as Philadelphia and New York, which in their early years experienced great epidemics that were thought attributable at least in part to "the water." Responses initially took the form of avoiding the use of fouled well water or contaminated springs by bringing in water from outlying districts. The 1840s saw the rise of a movement, led largely by the English, toward "cleanliness" as a way of controlling diseases such as typhoid and cholera. Specific techniques for improving the quality of the water supplied to a community were also explored. The slow sand filter was first used just prior to this period by the Chelsea Water Works Company in London in 1827; it was introduced in this country in 1832, in Richmond, Virginia. The Richmond experiment, by Albert Stein, designer of the city's waterworks, was not fully successful, and it was not until James Kirkwood's design of a slow sand filter system for Poughkeepsie, New York, in 1872, that a truly effective installation was made. Another system followed at Hudson, New York, two years later. Kirkwood's innovations had been generated by a review of European systems sponsored by the city of St. Louis in 1865.

Before 1850 only a handful of communities in America had sewerage facilities, with drainage systems being only marginally more common. Boston, Philadelphia, New York, and other cities shaped newly paved roads with crowns and gutters alongside or down the center for drainage. Some of the wooden drains built by urban dwellers to carry off water in cellars were being connected to sewers beneath the streets by the early 1700s. In 1704 Francis Thrasher built a sewer in Boston that became so useful that others soon were constructed. Boston became one of the best-drained cities in the world. Other cities were less progressive. Philadelphia had some underground sewers and culverts by the 1750s. New York began constructing trunk sewerage lines in 1769. Such developments were not widespread, nor were they necessarily carefully thought out. In 1857 Julius Adams prepared plans for the sewerage system of Brooklyn, New York, based largely on European models and employing engineering principles. Ellis Chesbrough proposed a sewerage system for Chicago in 1858, after a study of European systems; and Herbert Shedd prepared plans for a system in Providence, Rhode Island, in 1874.

A historic moment for environmental engineering came with the famous Broad Street Pump case in London in 1854, when John Snow demonstrated that cholera was transmitted by water polluted with excrement from those already infected with the disease. Snow's work, the development of the germ theory of disease by Pasteur (announced in 1857), and subsequent work by Koch and others finally established a scientific explanation for the transmission of disease. The new understanding did not immediately affect American practices, which did, however, continue to stress the vague notion of cleanliness as a way of controlling disease. Eventually the study of germs led to a series of investigations, largely in England, that spawned modern methods of filtration of both drinking water and waste water, the use of chlorine as a disinfectant, and bacteriological techniques for measuring water quality. American engineers of the time borrowed extensively from these investigations.

The year 1887 saw a turning point in the development of American waste-water treatment practices with the establishment of the Lawrence Experiment Station by the state of Massachusetts for the study of water quality. Massachusetts had been a leader in the evolution of sanitary engineering since the report of Lemuel Shattuck in 1850, which had led to the creation of the Massachusetts Board of Health. This was the first state agency to seriously investigate water pollution and treatment methods. The agency's research was conducted at the Lawrence Experiment Station, where a remarkable multidisciplinary staff, under the direction of Hiram Mills, helped to lay the basis for the field of environmental engineering as we now know it. By the turn of the century a number of formative approaches to the water treatment problem had been developed. Slow sand filtering had been improved on the basis of scientific principles elucidated at the Lawrence Experiment Station. Between 1895 and 1897 George Fuller's experiments at Louisville, Kentucky, established the principles of mechanical filtering systems.

Before 1900 few sewage treatment plants existed in the United States; most sewage went directly into streams, rivers, lakes, or oceans. Intermittent sand filters were used in some locations, contact beds in others. Worcester, Massachusetts, and Providence, Rhode Island, boasted chemical precipitation plants. The trickling filter, first developed in England, was introduced in Reading, Pennsylvania, in 1908. The Imhoff tank, which combined clarification and sludge digestion, became widely used after 1910. Disposal of sludge was usually done through drying beds or sludge lagoons. The activated-sludge process, developed largely in England as a consequence of work done at the Lawrence Station, soon came into use. San Marcos, Texas, received the first installation in 1916, but the system at Milwaukee, constructed between 1919 and 1929, was more innovative. Milwaukee also developed a process for transforming the sludge into a salable product.

The collection of garbage had been more or less systematically addressed by urban areas for some time because of existing beliefs about links to disease. The reduction of garbage became relatively widespread in the first part of the twentieth century as well. Incinerators, used in some urban areas prior to 1900, were common thereafter. The turn of the century was also a time of emerging interest in the reduction of stream pollution, although little was done to rectify such problems before World War I.

The period after the First World War saw continued improvements and refinements in water and waste treatment. New methods of coagulation were developed and new mechanical techniques explored. Later decades saw renewed attention to the reduction of pollution of streams, rivers, and lakes. The years immediately before and after the turn of the century, however, continue to stand out as the period of greatest development in the field of environmental engineering.

The Louisville Waterworks

The rapid growth of American cities in the nineteenth century necessitated finding not only greater water supplies but also new methods of purifying water for domestic and industrial use. The larger cities turned to the rivers and lakes on which they were built to supply their residents with water. But in many cases this water was filled with germs and organic matter known to cause typhoid, cholera, and other diseases, and was not usable. In the late 1800s the city of Louisville, Kentucky, conducted experiments in filtration that were to have tremendous significance for the purification of municipal supplies across the United States.

In the late 1850s the Louisville Water Company, under the direction of Theodore Scowden, chief engineer, began the construction of its waterworks along the Ohio River. The first pumping station was built between 1857 and 1860, and in the latter year the works went into operation with a 10-million-gallon reservoir. Charles Hermany, who had supervised the construction of the pumping station, was appointed chief engineer of the works in 1861. At first the particles present in the river water did not pose a major problem, but after a few years it became evident that some form of filtration system would be required. Turbidity from runoff, rather than contamination from human wastes and other city-generated pollutants, remained the primary concern until the late 1800s, when the construction of a sewer system brought the direct discharge of wastes into the Ohio.

In 1876 the water company recommended experiments in filtration of the turbid Ohio River to remove sediment and organic impurities. Meanwhile, the new 100-million-gallon Crescent Hill Reservoir was being readied for operation 2.5 miles away from a new standpipe with a capacity of 18,330 gallons. This supply system went into operation in 1877. It was not until 1884, however, that Hermany began his tests of slow sand filtration.

Hermany's filtration tests took eight months, using two 12-foot-diameter filter tanks similar to those recommended by James P. Kirkwood in his reports on filtration of the muddy Mississippi River water at St. Louis. The experimental system did indeed

The Number One Pumping Station, Louisville waterworks, Louisville, Kentucky. Here in the 1870s Charles Hermany first demonstrated the practicality of rapid sand filtration of water on a municipal scale. This was a milestone in American environmental engineering. (Courtesy of the American Society of Civil Engineers)

remove impurities, but because of the high turbidity of the water it was very inefficient. The filters required cleaning far too often, and the output of clean water was only 1.5 million gallons per acre daily, much too low for the growing city of Louisville. It was evident that a rapid sand filtration process would increase the capacity of the system, but the kind of filter it would require did not yet exist. In the midst of these tests, progress on the waterworks was continuing. In 1888 Hermany constructed a new pumping station to replace the old Cornish beam engines and boilers, which had become inadequate for the increased capacity.

In 1893 the famed Lawrence Experiment Station in Massachusetts, a leader in the development of environmental engineering techniques, succeeded in devising a slow sand filter that was capable of removing typhoid fever germs present in the water supply of the city of Lawrence. This was a significant development in water purification techniques, and the first time one of the major disease-causing germs was eliminated from a public water supply. But because it was a slow filtration process, it was impractical for use in a large metropolitan area such as Louisville. During the same period St. Louis was also a leader in confronting problems associated with drawing water from rivers.

In 1895 the city of Louisville came up with a plan to employ rival filter-manufacturing companies to compete in experimental tests of their products. Chemist George Warren Fuller, Hermany's associate, was put in charge of the experiments, which involved many types of rapid sand filters, the Harris magnetoelectric filter, an electrolytic apparatus, the MacDougal polarite filter system (which used a settling tank and a clay extractor), and several of the water company's own devices. Two of these competitors had devised very novel methods of water filtration: a sand filter designed by William Jewell used chlorine gas to assist in the reduction of harmful bacteria, and the Harris magnetoelectric filter employed magnets with a high-voltage current to achieve similar results.

Unfortunately, none of these filtration systems proved suitable. The electric systems and the polarite system were deemed inapplicable, and the sand filters proved inadequate, because before successful filtration could occur the water had to be free of suspended particles and thoroughly coagulated as well. George Fuller set about devising a system that incorporated both precoagulation and presedimentation. He conducted experiments with several types of coagulants and concluded that alum should be used in the system at Louisville. In 1897, after the ability of the new apparatus to remove from the Ohio River water all four types of impurities—turbidity, color, organic matter, and bacteria—had been demonstrated, plans were made to put it into operation.

The new system consisted of several elements rarely used before the tests at Louisville. First, the filter tanks were much larger than any employed before—they covered a rectangular area of 0.1 acre and had an output capacity of 37.5 million gallons daily. Second, there was a new sand support and strainer complex, with layer upon layer of wire netting. Third, a sand agitator was used to stir and move the sand beds to permit easier maintenance. Finally, the coagulation apparatus consisted of new, more reliable elements, including three concentric tanks for preparing the coagulant and a feed pump that was more precise than those previously used.

In 1906 the waterworks passed into the hands of the Board of Works (trustees of the Louisville Water Company), and two years later Charles Hermany died, leaving the unfinished filter plant to be completed under the direction of Theodore Liesen, the new chief engineer. After problems developed in the filter cleaning system, George Fuller was given the authority to propose changes to keep the sand layers from becoming clogged by mud accumulation. He decided that the filter units should be divided into two, the sand agitator eliminated, the strainer equipment remodeled after Cincinnati's to give a 20- to 24-inch-per-minute rise of wash water, and a separate wash water tank provided, instead of taking wash water from the distribution supply. With these changes in effect, the system went into operation on July 13, 1909; but it was not until 1912, after further changes in the wash water waste system and the coagulant supply, that the full procedure of presedimentation and precoagulation was begun.

George Fuller went on to Cincinnati, where he developed a rapid filtration plant following the direction provided by Charles Hermany. Everything that was learned at Louisville about the sand filtration of turbid inland waters was eventually practiced in many other American cities and is now standard procedure in municipal water purification.

The Chain of Rocks

Chain of Rocks, St. Louis, Missouri. The settling basins of this water treatment plant are shown, circa 1920. Here in 1904 a civil engineer and a chemist cooperated to develop a chemical coagulation process for purifying the turbid waters of the Mississippi River. (Courtesy of the American Society of Civil Engineers)

The problem of creating clean water supplies for use in American cities occupied many of the great engineering minds of the mid–nineteenth century. The challenge was formidable in cities that relied on the muddy or befouled waters of rivers. The city of St. Louis, on the Mississippi, was the site of many significant developments in the field of water purification. Its Chain of Rocks plant was the first significant full-treatment plant in the United States.

St. Louis had used Mississippi River water for drinking and residential purposes ever since it was founded in 1764. The first waterworks was established in 1829, at which point the problem of sediment in the water was officially recognized. From 1830 until the mid-1860s the operation of the works was impeded by excess sedimentary deposits in the system. At first, river water was pumped to a brick and stone reservoir 60 feet above the river, from which it was distributed to the city. Within a few years deposits of silt and mud had risen to such a depth that a wooden tank was built on top of the original reservoir to accommodate the overflow of water supply. In 1849 a new reservoir was built of masonry 120 feet above the river, but four years later it had accumulated about 4 feet of sediment. This time the waterworks removed the sediment and continued using the reservoir. In 1855 another masonry reservoir was built adjoining the second one, but the sediment problem was so great here that in 1860 the outlet pipe was completely blocked by mud. A 17-foot vertical addition had to be built to allow for extra water capacity. The sediment collected at the rate of 2.5 feet per year.

At the end of the Civil War James P. Kirkwood, chief engineer of the works and an eminent civil engineer, recommended that a system of filtration be employed. He was sent to Europe to study filtration systems in use in major cities there, only to find that the nature of their rivers made their filtering plants inappropriate prototypes for St. Louis. Meanwhile, the city of St. Louis had decided not to adopt filtration anyway, and when Kirkwood returned he recommended that a system of sedimentation be developed, to allow the impurities to settle out of the water before it was stored in the reservoir. He proposed that a series of settling tanks be used on a rotating basis, that these tanks be designed to allow for frequent cleaning, and that a well for measuring output capacity be included in the system. Having completed his report for St. Louis, Kirkwood went on to work in upstate New York, where he developed a sand bed filtration system for Poughkeepsie in 1872.

The water plant, the Bissell's Point pumping station, that was built on Kirkwood's recommendations included only the settling tanks and pumped water directly from the river—at a depth of 5 feet below the surface, where it was fairly free of floating particles as well as bottom sediment—to the four 100-million-gallon tanks. While one tank was being filled, two were settling (or one was settling and another being cleaned) and the fourth was drawn from to supply the standpipes and 60-million-gallon reservoir. The tanks were cleaned once every four months, when about 16 inches of sediment had accumulated.

Later in the century many cities and their engineers began experimenting with methods for further purifying turbid river waters such as that of the Mississippi. In 1884 L. H. Gardner, then superintendent of the New Orleans waterworks, conducted experiments using chemicals as water-cleansing agents. He proposed the use of perchloride of iron, following the theory that additives such as ferric chloride produce flocculent precipitates, or large aggregations of the suspended particles, which settle out much more rapidly than the original materials. Later experiments at Rutgers University convinced civil engineers that precoagulation and presedimentation were both necessary for successful filtration. In 1888 James A. Seddons, while testing designs for new settling basins at St. Louis, developed several principles for sedimentation that governed the design of water systems for years to come. Among these principles were the following: the basins should not be covered; the size and shape of the basins has little to do with controlling internal water movement during filling; because of this movement the time needed for filling cannot be included in the time allowed for sedimentation to occur. Then in the early 1890s the Anderson process of precoagulation with tiny particles of metallic iron was tested at St. Louis with a slow sand filtration system. The process worked well, but unfortunately a slow system was inappropriate for St. Louis.

During this period of experimentation the Chain of Rocks water treatment plant was being constructed eleven miles north of St. Louis. The plant, the largest of its kind at the time, was completed in 1894 and consisted of a new intake, a new primary pumping station, larger settling tanks, and a new seven-mile conduit. The system pumped river water at the Chain of Rocks to the new 132-million-gallon settling tanks. Then the water was gravity fed seven miles south to the Bissell's Point pumping station and from there was repumped to the distribution system. On May 1, 1960, the Bissell's Point station was taken out of service and a new distributive pump was put into service at the Chain of Rocks plant.

Experiments at the Chain of Rocks led to many new elements of water treatment technology that are still in use today. First, the European idea of sand and gravel filter beds, as outlined by Kirkwood, was eventually adopted at St. Louis. Then, at the turn of the century, the upcoming World's Fair led the city to adopt a system of coagulation to aid both sedimentation and filtration in purifying the water. St. Louis became the first large city to adopt ferrous sulfate and lime—tested at Quincy, Illinois—as coagulants, paving the way for Cincinnati, New Orleans, and other cities along the Mississippi to follow. This new coagulation system was influential in controlling the typhoid and cholera epidemics of 1903. A few years later a system of double coagulation, with milk of lime and sulfate of iron, plus sulfate of alumina, was used to provide floc for more rapid sedimentation, as demonstrated in the 1880s by Gardner. Finally, in 1915 a chlorine disinfection chemical filtration system with a capacity of 120 million gallons a day was added in order to kill additional bacteria after the filtration process.

These later improvements—sand filtration, flocculation, and chemical disinfection—in addition to the earlier system of sedimentation, represent all the major elements of water purification in use today. The Chain of Rocks was also the first plant to combine sedimentation with the advanced European filtration processes. For these reasons the system in operation in 1915 at St. Louis must be considered the first modern full-treatment facility in the United States.

The Lawrence Experiment Station

The Lawrence Experiment Station, established in 1887, was a pioneer research facility dedicated to the study of problems associated with water supply and sewage and waste disposal. The station was born out of landmark legislation enacted in 1886 by the state of Massachusetts, entrusting the supervision of the water supply and the treatment of sewage to the state board of health. The Lawrence facility was to conduct experiments for the board of health, chiefly on methods of purifying sewage and disposing of industrial pollutants. Over the past century it has served as a model and leader for research facilities of its type, not only in the United States but throughout the world.

The station was constructed on the banks of the Merrimack River in Lawrence, Massachusetts, and was first placed under the leadership of Hiram Mills, a civil engineer. The farsighted Mills initiated a series of studies in sanitary engineering and treatment methods that were all pathbreaking in their time. His major contribution to the station was his realization that because the science of sewage treatment was in its infancy, both engineers and scientists from disciplines such as bacteriology, chemistry, and biology would have to work together as a team to advance the field. Mills assembled a talented staff at Lawrence, and these scientists and engineers began, in effect, to invent modern sewage and waste disposal techniques. In taking this initial attitude Mills established a tradition of strong research work that helped the station solve new sewage and environmental problems as they were brought about by new industrial methods, new chemicals in refuse, and new home appliances such as garbage disposers.

After Mills's tenure a series of distinguished scientists and engineers headed the facility. The scope of the research was expanded to include exploring ways to recover substances from wastes and assessing the impact of pollution, as well as its treatment, on the air, water, and land environments. In later years a laboratory was established for research into the hydraulics of plumbing. The station became a teaching center, disseminating the knowledge its investigators had attained to visiting scientists and engineers.

Among the Lawrence Experiment Station's more important accomplishments are an 1887 study recommending a system of sewage disposal for the Mystic and Charles river valleys in Massachusetts; the design of a filtration plant for the city of Lawrence in 1893; the establishment of a rainfall observation network; the completion of several metropolitan sewage system networks in Massachusetts; the construction of a shellfish purification plant in 1928, benefiting the New England fishing industry; studies on fluoridation; numerous pollution control studies; and the first studies to detect radioactive materials in the water supply. The station was the first organization to address industrial waste disposal as a problem distinct from the disposal of other types of sewage. The efforts of the station were also directly responsible for the reduction in the incidence of waterborne diseases such as typhoid fever.

The Reversal of the Chicago River

The reversal of the Chicago River, carried out in the last decade of the nineteenth century, represents an innovative engineering solution to problems of water supply and pollution in an era of rapid American urban development. The direction of flow of the water that carried Chicago's wastes was changed so that instead of entering Lake Michigan it flowed into the Des Plaines River and thence into the Illinois, the Mississippi, and finally the Gulf of Mexico, thus preserving Lake Michigan as a source of clean water for the city of Chicago. The project, which involved the construction of a 28-mile channel through a glacial moraine and bedrock ridge, was unprecedented in scale and introduced earth-moving techniques later employed elsewhere. More earth was moved in this undertaking than in the digging of the Panama Canal.

The rapid expansion of Chicago as a gateway to the West in the nineteenth century resulted in a growing pollution problem. From the time of its incorporation in 1837, Chicago relied on Lake Michigan as its sole source of fresh water. A supply system of intake cribs constructed one to two miles from the shoreline in the lake and connected to pumping stations through underground tunnels furnished the city's burgeoning population with water (see chapter 6, Chicago Water Supply System). From 1856 waste disposal was accomplished through a sewer system that fed into the Chicago River—which, though virtually stagnant most of the year, carried untreated sewage into Lake Michigan and the intake cribs during the spring floods. This unfortunate interface between the two systems resulted in periodic outbursts of typhoid fever, amoebic dysentery, and cholera.

In an attempt to relieve the increasing domestic sewage and industrial waste problem, work was done in 1871 to deepen the existing Illinois and Michigan Canal and create a gravity flow away from the lake toward the Des Plaines River, thus carrying away from the lake a portion of the untreated sewage. The canal had been constructed much earlier in the century in response to demand for a southern outlet to the Great Lakes—a possibility because of the unique geology of the area. At the time of the formation of the Great Plains, a vast waterway connected Lake Michigan with the Gulf of Mexico through the Chicago, Des Plaines, Illinois, and Mississippi rivers. A slight tilting of the lake plateau, however, occurring approximately 10,000 years ago, virtually closed this southern outlet, leaving a shallow water trail and large swamp in its place. The explorers Joliet and Marquette, first passing through the region in 1673, noted the importance of a southern waterway to the Gulf of Mexico in their reports. After American independence the northern outlet from the Great Lakes through the St. Lawrence River was controlled by the British, and demand for a southern outlet grew. Construction of the Illinois and Michigan Canal began in 1836 and, although its planners took advantage of the residual pathway that already existed, proceeded with great difficulty, owing to underlying bedrock formations and a slumping national economy. Work was finally completed on a modified "shallow-cut" design in 1848, thus realizing the first man-made southern connection to the Gulf of Mexico.

The deepening of this canal in 1871 provided only partial relief of the waste problem, as changing water levels rendered the gravity flow unreliable. Moreover, the growth of towns along the Des Plaines River produced further problems as sewage from Chicago polluted their drinking supply. In 1876 a pumping station and conduit from Lake Michigan were built to flush fresh water down the canal, but this proved of little help; additional pumping installations in 1881 brought only a marginal improvement. Outbreaks of water-borne disease continued to afflict the metropolis.

The crisis escalated in August 1885, when 6.19 inches of rain fell in a two-day period in and around Chicago. Flood waters overwhelmed the pumping stations and sewer pipes, sending the scourings of catch basins and inlets into the river and lake. An immense mass of sewage and bacteria spread across the lower portion of Lake Michigan, fouling the intake crib for the city's water supply system. Within months almost 12 percent of Chicago's 250,000 inhabitants died of cholera and other diseases. A drainage and water supply commission was established in

Completed in 1900, the construction of the Sanitary and Ship Canal, between Lake Michigan at Chicago and the Des Plaines River at Lockport, resulted in the reversal of the Chicago River, so that water from Lake Michigan drained into the Gulf of Mexico. This multipurpose project, involving water supply, pollution control, transportation, and power, required more earth moving than did the Panama Canal. (Courtesy of the Chicago Historical Society)

January 1886 to develop a comprehensive, long-term solution to Chicago's water problems. The group's report, submitted the following year, recommended that a sanitary and ship canal be constructed connecting Lake Michigan at Chicago with the Des Plaines River at Lockport; that this canal be of sufficient size to permit a flow of 24,000 cubic feet of water per minute for every 100,000 inhabitants of Chicago, based on an expected future population of 2.5 million; and that the canal be connected to the Chicago River and deep enough to permanently reverse the river's flow. These recommendations were formally adopted in November 1889 by the Illinois legislature, and the Sanitary District of Chicago was created. This autonomous agency, known today as the Metropolitan Sanitary District of Greater Chicago, embraced an area of 185 square miles and had its own elected officials and independent taxing authority.

The 28-mile channel between Lake Michigan and Lockport that engineers accordingly devised for the Sanitary District brought about the second reversal of the Chicago River in 10,000 years. The design relied on dilution and natural biological processes to render the sewage harmless to downstream communities. (Treatment plants had to be added in later years.) In addition to diverting contaminants from Lake Michigan, the Sanitary and Ship Canal provided drainage margins to allow for storm runoff under flood conditions, thus preventing a recurrence of the 1885 disaster. Navigational facilities for barges were also provided, so that this waterway could take the place of the obsolete Illinois and Michigan Canal. Finally, land acquisition along the channel's route, authorized under the Sanitary District's charter, created sites for future industrial development.

Ingenious construction techniques, particularly with regard to earth moving, were employed during the eight-year course of the project, which involved canal building, river dredging, bridge and road building, and other activities. More cubic yardage of rock and

Excavated material was run up the embankments for removal. (Courtesy of the Chicago Historical Society)

The Needle Dam, shown under construction, was opened on January 2, 1900. (Courtesy of the Chicago Historical Society)

earth was displaced than in any other single earth-moving project to date. Fifteen miles of the route were dug through solid rock; almost 8 miles through earth, mostly clay; and an additional 5 miles through combined earth and rock. At the project's completion in January 1900 a force of 8,500 men had been working without halt for eight years, blasting and removing 29,559,000 cubic yards of earth and 12,261,000 cubic yards of rock. The Lockport control structure contained sluice gates and the Bear Trap Dam, by which the amount of water flowing through the channel was regulated. Horse-drawn graders removed and leveled the first 13 miles of earth excavation, beginning at the Chicago end. This completed section measured 160 feet wide at the canal's bottom and 225 feet at the top. The final 15-mile section to Lockport, cut through rock, was accomplished with dynamite and steam-operated cranes and steam shovels mounted on rails in the trench. These digging techniques, particularly the manner in which steam shovels were semiautomatically operated, were later adopted in the Panama Canal dig. The channel section in the rock cut had a profile of uniform 160-foot width at top and bottom. The depth throughout was 24 feet.

Extensive dredging of the Chicago River was required in order to bring about its reversal. In addition, a 13-mile section of the Des Plaines River was rerouted, the former riverbed being employed as a portion of the new channel. Finally, new bridges were constructed to maintain existing roadways across the new canal.

On January 17, 1900, the gates at Lockport were opened for the first time. A great crowd gathered along the river's edge in Chicago, full of doubters as to the predicted reversal. But the Chicago River did indeed change direction, and sewage stopped flowing into Lake Michigan. In the following decades Daniel Burnham's Chicago Plan of 1909, as constructed, created one of the world's most beautiful and inviting urban shorelines along this formerly polluted lake's edge. By 1922 the death rate from typhoid fever was down to 1 per 100,000 of the city's inhabitants.

Minor modifications were incorporated over the years. In 1910 the 8-mile North Shore

Chapter Eight

Dams

Introduction

Small dams were commonplace in early America. Built to impound water to feed grist and manufacturing mills, they could be found almost anywhere fast-flowing streams or rivers promised a source of power. Many small dams were also built in conjunction with the country's canal networks. In 1834, however, a more ambitious structure, one of the first high masonry arch dams to be built in North America, was completed in Vermont at Ascutney. The Ascutney Mill Dam represented a project conceived by entrepreneurs for the generation of power for private enterprise. Other large dams were soon built to supply water power in many places—there were notable examples at Paterson, New Jersey, and Lowell, Massachusetts—reflecting the industrialization of the nation.

Dams were built not only to supply hydropower but also to provide adequate water in areas of rapid urbanization. The famous Croton Aqueduct project, for example, completed in 1842 to supply water to New York City, involved the construction of the Croton Dam, a major masonry structure that formed a large reservoir on the Croton River (see chapter 6). During the 1880s increasing water demand in New York brought about the construction of the New Croton Aqueduct and the New Croton Dam, built three miles below the older one. In 1871 the Druid Lake Dam, the first major earth fill dam in America, was completed in Baltimore; though built without sophisticated knowledge of soil mechanics, it served as a model for the future construction of earth dams. Two notable western structures built to meet urban water needs are the Marlette Lake Dam, constructed in the 1870s as part of the system that supplied water to Virginia City, Nevada (see chapter 6), and the Cheesman Dam, built in 1905 for the burgeoning city of Denver.

The settling of the western states during the nineteenth century brought a demand for water sources for irrigation of the land. The many important western dams constructed to meet this need were often designed to serve the purposes of flood control and hydroelectric power supply as well. The harnessing of water resources in the West gave rise to questions of water rights and thus to federal government involvement. The 1902 Reclamation Act reflected the federal government's commitment to the irrigation of the West. The act provided for federally led planning, construction, and development of irrigation works as part of a major public works program, to be financed by the sale of public lands and to be carried out by a new agency, the Reclamation Service (known as the Bureau of Reclamation from 1907). During the reclamation period, between 1902 and the First World War, no less than seventeen earth, two masonry, and seven concrete dams were built as the major elements of regional projects. The Minidoka Dam on the Snake River in Idaho, 86 feet high with a crest length of 4,475 feet, was an important early earth dam. The Buffalo Bill Dam in Wyoming, completed in 1910, was the first major concrete gravity-arch dam built in America. In 1911 the Theodore Roosevelt Dam was completed as part of the Salt River project in Globe, Arizona; 280 feet high, it is the highest rock masonry dam in the world and was the first of its kind to yield electric power on a commercial scale. The Elephant Butte Dam, an important concrete gravity-arch structure, was completed in 1916 as part of the Rio Grande project in Truth or Consequences, New Mexico.

In 1928 the construction of the famous Hoover Dam was authorized as part of the Boulder Canyon project. Built during the years of the Great Depression, this structure, with its unprecedented height of 726 feet, soon came to have a symbolic value as a great public works project that helped improve the economy of the West. Like most,

Boulder Canyon was a multipurpose project. Not only was the Hoover Dam built to impound water for irrigation, to mitigate flooding on the Colorado River, and to provide hydroelectric power; the overall project also included a dam and power plant in nearby Black Canyon and the All American Canal leading into Imperial Valley in California.

The Grand Coulee Dam was another major project undertaken in the period between the wars. Funding was authorized in 1933 after a series of reports indicated the potential for water power, irrigation, navigational improvements, and flood control in the Columbia River basin. The Grand Coulee Dam was one of ten recommended and the uppermost in the chain; the Bonneville Dam was the major downstream structure. A 530-foot-high, 4,173-foot-long gravity dam, the Grand Coulee remains the world's largest concrete structure. In addition to the long-term benefits, its construction provided a tremendous source of jobs in the region during the Depression. Other dams were subsequently built in the Columbia River basin.

Whether built for water power, irrigation, flood control, or the generation of electricity, whether financed by private entrepreneurs or by the government, such projects could only have been realized through a rapidly growing country's ardent desire for equally rapid development. They attest not only the national will for progress but also the foresight and inventiveness of their engineers.

The Ascutney Mill Dam

Erected in the town of Windsor, Vermont, in 1834 and still standing today, this was one of the first high masonry arch dams to be constructed in North America. The 120-foot-long granite structure, with a crest height of 42 feet above tailwater, anticipated the twentieth-century concrete gravity-arch dam. Despite their lack of scientific knowledge and of contact with the relatively advanced technologies that marked nineteenth-century European dam design, the builders of the Ascutney Mill Dam achieved a structure that was eminently suited to the tasks for which it was intended and that demonstrated the practicality and safety of such permanent high masonry dams for water impoundment and power generation.

The importance of navigable inland waterways in postcolonial America is evidenced by the hundreds of small towns that grew up along New England rivers, of which Windsor, on the banks of the Connecticut in southeastern Vermont, was typical. Running north-south, the Connecticut River linked trading and shipping towns such as Windsor with major cities such as Hartford, Connecticut. The 1820s, however, saw the completion of both the Erie and the Champlain canals, connecting the Hudson River with Lakes Erie and Champlain. The result was a swift decline in Windsor's importance as a port of exchange between water and overland shipping; business traffic that had formerly passed through Windsor on the way to the Boston markets was now diverted to the western side of the state and the New York and Great Lakes markets. The economic pinch was felt in Boston as well: as early as 1825, plans were drafted for a canal across the length of Massachusetts, linking Boston with the Erie Canal and the Hudson River at the state's western boundary. The scheme remained unrealized, however, and towns such as Windsor were left to seek alternative industries upon which to base their economies.

Within a few years of its incorporation in 1764, Windsor, like other colonial settlements, had its primitive, water-powered saw and grist mills. By 1795 several grist and clothing mills and a small wooden mill dam

Ascutney Mill Dam, Windsor, Vermont. Built in 1834 and still in service, this granite structure is America's earliest masonry dam of significant size. It was a structural precursor of twentieth-century concrete gravity dams. (Courtesy of the American Society of Civil Engineers)

were well established. The primitive technologies then employed, however, hampered the exploitation of this resource. Occasional droughts, as in the summer of 1825, left Windsor's mills idle for periods of up to three months. Nonetheless, water power was felt to hold the key to Windsor's future as inland water traffic shifted westward. By 1830 the town's depressed economy clearly demonstrated the need for local and private initiative to sponsor a renewed industrial base.

In 1833 a group of Windsor businessmen formed the Ascutney Mill Dam Company. Shares at $50 each were available for public purchase, although previous experience with water-powered industry in Windsor rendered their purchase a risky investment. Nonetheless, the company's strategy—based as it was upon successful examples in Paterson, New Jersey, in Lowell, Massachusetts, and in Manchester, New Hampshire—seemed sound. The company was founded upon the idea of a "great dam" with sufficient storage capacity to produce a predictable and reliable source of power year round. By attaching clearly defined water rights, measured in cubic feet per second, to salable properties, the company's founders hoped to attract large-scale industrial enterprise to Windsor and thus to reverse the community's declining fortunes. If a few local businessmen and adventurous stockholders realized their own fortunes, so much the better. From the outset, then, the Ascutney Mill Dam, the creation of private initiative, was conceived to pay its own way and eventually return a handsome profit to its investors.

The company's directors hired Ithamar A. Beard, an engineer of some prominence in New England, to survey the mill brook and indicate the best siting for a storage dam. As soon as Beard had recommended a site, defined the requisite storage capacities, and plotted future mill sites and available water rights for the downstream properties, the company advertised in Windsor's *Vermont Republican and Journal* of March 1834 for proposals to build a simple buttressed-wall dam. Five proposals were received, and in early April the contract for construction was let to Simeon Cobb of nearby Westmoreland, New Hampshire.

Cobb's influence on the dam scheme effected major changes in the structure's planned configuration. The son of a millwright and contractor, Cobb had worked on railroad projects in Boston and New Jersey and on the grading of Boston's Bunker Hill. He brought to the Ascutney Mill Dam project a knowledge of contemporary civil engineering practices in America. In addition, he was probably aware of Zachariah Allen's recently published *Science of Mechanics,* an important early source for engineering practices and procedures in nineteenth-century New England. Allen's 1829 treatise contained the following recommendation concerning mill dams: "The form of the dam, if practicable, should be made somewhat semicircular with the concave facing down the stream. In this case an arch is formed, and the pressure of the water cannot carry away the timber work of the frame unless the abutments fail." Allen went on to describe the procedures and construction methods required for the realization of a mill dam remarkably similar to that, albeit in stone, constructed at Windsor. Whether or not he had read Allen, in any event a marked revision in the form of the Ascutney Mill Dam was effected after Cobb's arrival on the scene.

By April 19, 1834, the old wooden dam on the mill brook was being removed, and on June 4 the first stones of the new dam's foundation were laid. The structure was built from west to east, thereby allowing the water to be diverted to the stream bed's eastern edge and necessitating the building of only a small, temporary cofferdam behind the dam's eastern end. Power culverts to supply regulated water pressures to downstream sites were installed once this small cofferdam was in place; upon their completion the stream's waters were diverted through the culverts as construction proceeded.

Work on the dam structure went quickly. By the end of July 1834 the diverting culverts were completed, and wooden planking was being placed on the dam's downstream face, indicating that most of the masonry construction was complete. Material shortages were nonexistent. Granite was simply cut from surrounding outcrops, trimmed to fit, and laid in hydraulic cement. The wooden planking, made in nearby sawmills, was of native hemlock from the adjacent woodlands. Octagonal pins of local ash wood were fitted between the stones, either to serve as reinforcement against tension stresses induced by high water pressures or to secure the wooden planking against the dam's stone face. By the end of November construction was essentially complete and water impoundment begun. The dam stretched 120 feet in a gentle arch between abutting stone outcrops; the 36-foot-thick base tapered to 12 feet at the peak over a height of 42 feet. This base width–to–height ratio, just under 0.9, approximates a standard established in much more modern gravity-arch dam design. A 100-foot spillway across the dam's top permitted safe water overflow, particularly during the spring high-water season. Impoundment proceeded slowly through the winter of 1834–1835; water to power several existing downstream mills had to be maintained through the power culverts during this period. By the following spring, however, a reservoir covering approximately 100 acres, seven-eighths of a mile long, had been formed, as Ithamar Beard had specified.

The dam's experimental design and unprecedented scope worried the citizens of Windsor—particularly those living and working below the dam and its growing reservoir. Sheets of ice and water cascading down the 40-foot-high spillway during the first winter sent vibrations and a constant rumbling noise through the town. The worry turned to panic, and the dam company's directors, who had no engineering expertise, themselves began to fear the collapse of the structure. Thus in the spring of 1835 the newly formed reservoir was hastily drained and a visual inspection of the dam made. Despite the absence of any evidence of distress in the structure, a decision was made to buttress its downstream face. A 150-foot-long wall, 20 feet thick at the base and tapered to a 3.5-foot-thick coping at the top, was added in April 1835. The wall was built with a battered profile, thus providing a series of steps over which the water spilled, damping both vibration and noise. A concrete lip was added along the top edge much later. The dam exists today in this modified configuration.

This photograph shows the concrete lip later added along the dam's top edge.

Confidence in the structure and in its potential for water power, however, were not restored with the 1835 revisions. Mill sites, with water privileges as defined by Ithamar Beard, were offered for sale in the spring of 1836 without result. Suits by lending banks were instituted against the Ascutney Mill Dam Company; its stock was subsequently seized and sold at public auction. By 1843 the shares originally priced at $50 sold for 5.5¢. The entrepreneurial dreams of the Mill Dam Company's directors were to go sadly unrealized.

The dam's hydropower potential did eventually benefit the town of Windsor. The Robbins and Lawrence Armory, manufacturer of rifles for the United States Army, established a major plant there in 1845. This factory's pioneering development of interchangeable component manufacture revolutionized the armaments industry worldwide. The post–Civil War period brought machine tooling and cotton milling as well to a now thriving metropolis. With the advent of hydroelectric power production, control of the Ascutney Mill Dam was taken over by a public corporation for this purpose in 1898 and remained so until 1952. Today the dam is owned by the city of Windsor, its hydropower potential once again offering the promise of a competitive and readily available energy source.

The Druid Lake Dam

Baltimore's Druid Lake Dam, which together with its storage reservoir was constructed over a seven-year period between 1864 and 1871, was the first major earth fill dam to be successfully completed in the United States. Designed by Robert K. Martin, a civil engineer for the Baltimore Water Department, the structure demonstrated the viability of earth fill design and construction for large-scale water impoundment projects in this country. Martin's achievement is magnified by the ingenious methods applied in the dam's erection. He authored a meticulous construction sequence, utilizing locally available material, to meet the dam's storage requirements and to realize a project of both strict economy and civic embellishment. Many of Martin's construction methods and design principles, with modifications to account for modern earth-moving machinery, are still in use today at earth fill dam sites. In addition, the Druid Lake Dam and storage reservoir provided a vital link in Baltimore's water supply network, enabling the city, in a period of rapid urbanization, to meet the requirements of its burgeoning population. The Druid Lake Dam has served Baltimore continuously since its completion well over a century ago.

The supply of drinking water was characteristically sporadic and uneven in early-nineteenth-century Baltimore. From the beginning the city had employed both private and public utilities to supply fresh water. The principal private contractor, the Baltimore Water Company, supplied water from nearby lakes and marshes, using a feeder network of iron pipes imported from Philadelphia and England. The city supplemented this supply base with its own network of wells and public pumps, dug and maintained by its own forces. A system of natural springs, distributed throughout the city's geography by virtue of the marshland base upon which Baltimore is founded, was also purchased by the city. Thus the early environment of Baltimore—in character with that of early Boston, Philadelphia, or New York—featured hand-operated water pumps and springs, around which structures of classical architecture were erected.

An early view of the reservoir created by the Druid Lake Dam, Baltimore. Built between 1864 and 1871, this was the first major earth fill dam to be constructed in the United States. It had a number of features unique for its time and was an important forerunner of other earth fill dams that have been built across the United States. The reservoir became a popular recreational area. (Library of Congress)

The rapid growth of the city in the early nineteenth century rendered the public-private water enterprise obsolete. The privately operated Baltimore Water Company was underfinanced and thus unable to extend piping networks to newly settled areas. Insufficient resources produced low water pressures in the existing pipes, boding disaster in the event of a major fire. The publicly operated network of wells and springs evidenced its own inadequacies. Without a centralized sewage disposal system the city was forced to permit privies on developed sites. This haphazard sewage dispersal fouled the wells and springs with untreated wastes. Periodic cholera and dysentery outbreaks menaced the population. At last in 1854, after almost twenty years of debate, the city took over the Baltimore Water Company and embarked upon a regional, publicly controlled water supply policy.

The newly established Water Department of the City of Baltimore took several major steps during the 1850s to introduce fresh supplies of drinking water. To the city's northwest, Jones Falls was dammed and preparations to create reservoirs at Lake Roland, Hampden, and Mount Royal were begun. By 1862 the storage reservoirs, with a total capacity of 480 million gallons, were completed and in service, but not without problems. Heavy spring and fall rains turned the reservoir waters, which drained agricultural areas, muddy and occasionally manure infested. Baltimore's daily consumption taxed the supply network, particularly in the summer months, causing occasional influxes of the fouled water into the city's supply lines. With the science of water purification and filtration still in its infancy, an alternative solution to the problem of fouled water was forwarded—the creation of a holding reservoir within the city itself, storing a sufficient quantity to permit natural biological and settlement processes to clear the detained water supplies of the outlying reservoirs. Moreover, the holding reservoir would provide an additional source during periods of drought and would serve to maintain sufficient water pressure for fire-fighting equipment using the recently installed hydrant network.

The water board selected Druid Park, an undeveloped ravine site within the city's western boundary, for the reservoir and appointed one of its engineers, Robert Martin, to design and supervise construction of the reservoir dam. From the outset Martin approached the project with two principles firmly in mind: economy—the water board's resources being already overtaxed—and the creation of a major civic adornment, serving as the focus of a large, open public space, to be preserved for recreation amid Baltimore's haphazard expansion. Toward these ends, Martin opted for an earth fill dam structure, utilizing material excavated from the reservoir bed, a solution without American precedent. The gentle slope of the dam's dry face, together with alterations to the ravine's natural slopes, would provide large areas of attractive parkland.

Construction of the dam began on March 7, 1864. All trees and vegetation within the ravine area to be covered by water were removed; and the topsoil, stripped to the depth at which impervious strata were reached, was stockpiled for future use on the dam face. At the dam site all subsoil was removed to the level of underlying solid rock; a 5-foot-wide groove was then cut into the rock for the length of the dam. In this groove, and extending 5 feet above the rock, was laid a stone wall, 4 feet wide at the base and tapering to a 2-foot width at its peak. Composed of stone excavated from the reservoir bed and laid with hydraulic cement, the wall had battered sides for greater strength and greater contact with the eventual overlying saturated-clay puddle core. Martin envisioned the wall as the structural "tooth" of the dam, anchoring it against lateral pressure from the impounded water, as well as providing an impervious barrier against water seepage through the dam's base. The wall extended through gravel ridges to solid rock at either end, securing the dam structurally into the ravine slopes.

On and around the masonry wall, Martin next constructed the puddle core, an impervious, full-height, vertical barrier that would later be surrounded by the earth-fill structure. Beginning with a 36-foot-wide base, centered upon the masonry wall, the core was to be built up from saturated clay, again excavated from the reservoir site. Earth banks, laid in 3-inch lifts of excavated material, each rolled and compacted with horse-drawn rollers, were erected on either side of the space the core would occupy, to serve as scaffolding for the puddling operation. Three successive earth embankments were built up, one upon the other, to reach the height intended for the puddle core. The trench thus created between the embankments was subsequently flooded, and into it clay was deposited by hand shovel. The clay would initially compact under its own waterborne settlement, preventing the formation of undesirable air pockets within the puddle core. At the completion of each 2-foot lift the water was drained from the trench and compaction equipment was introduced onto the clay, squeezing water from and thus compacting the puddle core.

At the same time, embankment operations at the earth bank's outer edges proceeded to align the earth fill dam's walls to the prescribed slopes. In this way the puddle core, eventually reaching a height of nearly 114 feet and tapered to a 17-foot width at the top, proceeded to rise uninterrupted, except in adverse weather, throughout the erection sequence. Only locally available material, excavated from the ravine's slopes and bed, was used during the dam's construction.

Martin's self-prescribed economies extended to the realm of material handling as well. Steam-powered excavating shovels removed needed fill from the ravine's slopes and bed. The material was subsequently hauled to the earth bank in horse-drawn carts on a temporary railroad track built for this purpose, and dumped in place. In this manner each portion of the dam's material was handled only once. Similarly, handling of the clay for the puddle core was limited to its final placement in a saturated state. Temporary rail track was pulled up and relocated as the earth bank rose, bringing earth fill directly to where it was needed. These methods for both material handling and construction sequencing, pioneered by Martin, served as guidelines for the building of earth fill dams across the United States well into the twentieth century. Adapted for modern machinery and earth-moving equipment, they continue to serve the construction industry today.

By January 1871 the Druid Lake Dam was substantially complete, and water impoundment for the 429-million-gallon reservoir was begun. When finished, the dam stood 119 feet high at its peak, tapering from a base width of 640 feet to 60 feet at its top. The only substantive difficulty to arise during the construction concerned the original 30-inch-diameter influent, effluent, and drain pipe installations. Laid in the inner embankment in the spring of 1865, the pipes were supported through the puddle core on a series of stone piers spaced on 6-foot centers. Stone collars rigidly attached the pipes to the piers, to prevent water percolation along the pipe lengths and thus through the dam structure itself. Differential settlement between the earth embankment and the stone piers, however, caused the pipes to crack almost immediately. Upon discovery of this defect in 1866 the pipes were removed, the puddle core sealed, and new pipes were laid through the southern ravine slope, without incident.

Druid Lake Dam exists today substantially unchanged from Robert Martin's pioneering design. A drive along its outer face was added after the original construction was complete, linking Druid Park's two entrances and thus forming a continuous promenade along the reservoir's perimeter. Druid Park was subsequently developed according to Martin's vision of a vast, open space within the dense aggregation of Baltimore's development. Modern highways, however, have diminished its recreational value. The dam, monumental in dimension for its time, continues to serve in its original role in Baltimore's water supply network.

The Cheesman Dam

When completed in 1905, the Cheesman Dam, on the South Platte River in Colorado, was the world's highest gravity-arch stone dam. It provided the first substantial and continuous on-stream storage of water for municipal use in the entire Rocky Mountain area. Located 48 miles southwest of Denver, the dam was a key element in the water supply system of the new city—a system whose creation was one of the most important corporate acts in determining Denver's development into a major urban area. Depending largely on mass for its strength, the dam was also arched in a 400-foot radius to provide additional safety, thus becoming the first major dam in the United States to incorporate the innovative gravity-arch concept in its design.

In the second half of the nineteenth century Denver was rapidly earning its early nickname as "the Queen City of the Plains." It was expanding at an enormous rate. It soon became apparent that direct stream flow and wells could not be depended upon to provide adequate water. As was common at the time, private companies were created to supply the city. Among these fiercely competing enterprises was the Denver City Water Company, founded by Walter Scott Cheesman and other civic leaders in 1870.

Cheesman was born in 1838 at Hempstead Harbor, Long Island, and attended New York public schools. He came to Denver in 1861 to manage a drugstore owned by his two elder brothers and devoted the next forty-six years to real estate development and other projects that contributed to the growth of the city. Primary among his many activities was the development of a water supply system. In 1894 the Denver Union Water Company, a successor to the Denver City Water Company, was incorporated, with Cheesman elected president. The new company immediately proceeded with plans, developed by Cheesman ten years previously, to bring water to the city from a mountainous area 48 miles away. Increasing the storage capacity of the Marston Reservoir to 8 million gallons and installing filter beds at the mouth of the South Platte Canyon were the basic elements of the initial project. The scope of Denver's water crisis, however, soon led Cheesman to consider a much grander project. The construction of a dam in the same area would create an immense reservoir and ensure an adequate and continuous supply of water.

The first attempt at construction of a dam was under way in 1897. By April 1900 the dam, a rock-filled structure with a sheathing of ⅜-inch steel plate facing upstream, had reached a height of 50 feet. On May 3 it was destroyed. Torrential rains and a massive runoff had caused the failure of the older Lake George Dam, located upstream. The released waters in turn swept away the partially completed structure below.

Cheesman, undaunted by this setback, intensified his efforts to complete the project. Charles L. Harrison, an eminent engineer who had worked on the Panama Canal survey, was engaged to design a new dam. Harrison drew up plans for a structure "so reinforced and anchored that no flood could disturb it." Common practice in dam design of the day was to rely upon large masses of material for strength and resistance against overturning effects caused by the impounded waters. In a significant innovation in American design practice, Harrison also gave the dam an arch shape, which imparted a large extra measure of strength and safety. A new construction contract was let on August 29, 1900—an amazingly short time after the first failure. The Geddis and Seerie Stone Company of Denver became the contractor.

The site of the dam presented many obstacles to construction. Added to the roughness of the terrain was the fact that the nearest railroad was 23 miles away. Work progressed steadily, however, through to completion on January 1, 1905. The construction time of less than five years is still considered an engineering marvel. The dam rises 221 feet above the stream bed and is 1,100 feet long, including the spillway. Its thickness ranges from 176 feet at the base to 18 feet at the top. A roadway, 14 feet wide, crosses the top of the dam and is flanked by 6-foot parapet walls. The structure is made of solid granite ashlar blocks, which were quarried on the site and set in cement mortar. Approximately 42 million pounds of cement was used in the building of the dam and structures at the

Cheesman Dam, Colorado. The world's highest gravity-arch masonry dam at the time of its completion in 1905, this was also the first major dam in the United States to incorporate the gravity-arch concept in its design. The Cheesman Dam provided the first continuous on-stream storage of water for municipal use in the West. It remains a key structure in the Denver water supply system. (Courtesy of the American Society of Civil Engineers)

reservoir. All material except the native granite had to be transported from the railroad over rugged mountain terrain. The granite blocks averaged 4 to 6 tons each, the largest weighing approximately 11 tons. Many blocks were floated to the dam site on rafts.

Interestingly, environmental aspects were not ignored. Native stones were used to blend with the surrounding escarpments. The dam was designed so that no outlets penetrated its face; rather, these were located in natural granite, some distance away, and were connected by tunnels more than 500 feet long to the dam. Water cascaded over the natural granite cliffs and boulders and plunged into the stream bed below.

With the completion of the Cheesman Dam a new era in the development of Denver began. The dam is still an essential factor in Denver's water supply system. When the reservoir is full, the dam retains 79,100 acre-feet. The surface area of the lake is 875 acres, its shoreline 18 miles in circumference. Surrounded by federal forests, with potential sources of pollution remote, the water in the lake remains as uncontaminated and crystal clear as it was in 1905.

The Buffalo Bill Dam

The Buffalo Bill Dam, named the Shoshone Dam until 1946, is located on the Shoshone River about six miles downstream from Cody, Wyoming, in the northwest corner of the state. The highest concrete arch dam in the world when completed in 1910, it was one of the first major structures to be built by the United States Bureau of Reclamation and the first major dam to be designed and built using the trial-load analysis technique. This method, a forerunner of today's computer-assisted dam design, marked the beginning of the exact science of large, concrete arch dam design.

The impetus behind the project came from its present namesake, Colonel William "Buffalo Bill" Cody. Wyoming's Bighorn Basin, lying downstream from the project's site, was settled by Cody and his followers in the last decade of the nineteenth century. Perceiving the potential of the basin's sagebrush flats for large-scale agriculture and grazing through irrigation, Cody formed and outfitted a party—which included Wyoming State Engineer Elwood Mead, a future director of the Reclamation Bureau—and led an expedition up the Shoshone River to publicize that potential. In 1899 Cody acquired from the state the right to appropriate water from the Shoshone for irrigation of approximately 60,000 acres of land in the Bighorn Basin. A small canal was constructed on the southern side of the Shoshone to divert the needed water. Four years later the Wyoming State Board of Land Commissioners, with Cody's approval, adopted a resolution urging the Reclamation Bureau to complete Cody's proposed irrigation development. In 1904 Cody transferred the water rights for the irrigation of 60,000 acres to the Department of the Interior, of which the Reclamation Bureau was a branch, and thus began the official involvement of the federal government.

Cody's proposal for the Bighorn Basin was limited to a system of canals by means of which the Shoshone's water could be used to make arable land. Reclamation Bureau engineers, however, conceived a large water storage reservoir and dam as a better means of utilizing the river's flow. Such a dam could provide for both irrigation and hydroelectric power generation, the feasibility of which was being demonstrated at the bureau's Roosevelt Dam and Salt River project near Phoenix, Arizona, under construction at the same time.

The engineers chose a site at the upstream end of a walled canyon on the Shoshone River. The selected area ruled out an earth fill dam—a common type at the time—and instead a thin concrete arch dam was planned (this choice, in turn, was a factor in the exact site location). The canyon's walls were to be of critical importance as abutments to which the dam's arch would transfer the superimposed forces. The thinness of the proposed structure was remarkable by contemporary standards. Construction started in 1905. Water impoundment began in 1910, a few months before the dam was finished. Modifications were undertaken in 1915 and 1922–1923; and between 1945 and 1948 the Heart Mountain power plant and the Shoshone Canyon conduit, utilizing the Buffalo Bill Dam's capacity for generating hydroelectric power, were constructed.

The dam has a structural height of 325 feet and a hydraulic height of 233 feet. It is 10 feet wide at the top and a maximum of 108 feet wide at the base, resting upon a granite foundation. The crest length is 200 feet. A total of 82,900 cubic yards of concrete was employed in the dam's construction. The spillway was of a concrete side-channel weir design, originating upstream from the left abutment and passing through the abutment in the form of an unlined tunnel. The operating capacity is 25,000 cubic feet per second. The reservoir created by the dam has a capacity of 439,800 acre-feet and a surface area of 6,710 acres. Approximately 93,000 acres of farmland are presently irrigated from this source, and the Heart Mountain plant operates at a capacity of 5,000 kilowatts.

The Buffalo Bill Dam stood in 1910 as the world's first dam with a height-width ratio greater than one. This feat was attributable to its arch design, made possible by the technique of trial-load analysis—first developed by Reclamation engineers for this site.

Buffalo Bill Dam, Cody, Wyoming. This was the first major dam to be designed and built using the trial-load analysis technique. At the time of its completion in 1910 it was the highest dam in the world and the only dam with a height-width ratio greater than one. (Courtesy of the American Society of Civil Engineers)

The thin arch design had proved feasible in America only twenty years earlier, with the Bear Valley Dam on the Santa Ana River in California. That structure's engineer, F. E. Brown, had adopted the form in response to the dam's remote site: the high cost of transporting materials and equipment necessitated the strictest economy. Brown seems to have arrived at the arch design without knowledge of precise structural techniques and without specific knowledge of earlier arch dams constructed in Spain, France, and Italy. His breakthrough at Bear Valley came about through his rejection of the conventional gravity dam concept, in which overturning effects from the reserved water's thrust were counteracted by mass alone—an impractical approach, Brown realized, for the canyons of America's western rivers. Brown proposed a conventional vertical arch simply turned on its side, employing the river canyon's vertical walls as abutments to which the water's thrust could be transferred. As a result, the dam was not required to resist the water's translational and overturning effects with its own mass; rather, it became a load transfer device. It could therefore be lighter and of substantially cheaper construction. Despite a dangerously inadequate spillway, the Bear Valley Dam performed flawlessly for twenty years. A second arch dam built by Brown in 1895, also in California, firmly established the soundness of the concept in the American West. Other arch dams began to appear in California and Australia in the 1890s, and concrete became increasingly popular as the material best suited to them.

Brown's Bear Valley Dam had a structural height of but 65 feet; Reclamation engineers were proposing a 325-foot-high dam at the Shoshone site, at that time the highest dam in the world. The Shoshone project, because of its immense scale, presented an additional challenge to its designers: How could they determine the forces present in a complex structure of this type and thus design to assure its stability? An inexact understanding of these forces might lead to disaster. Their response was to develop a new model of the dam's action—the trial-load analysis technique. This method provided a means for determining the distribution of loads, both horizontally and vertically, in arch dams for the first time. As a result of the modeling procedure, the engineers introduced the concept of coexisting systems of horizontal arches and vertical cantilevers as a means of introducing vertical stability to the horizontal arch. The water load on the dam was conceived of as divided between the two systems of structural elements so that both underwent identical, but opposite, rotational and translational movements at the points of intersection.

At the Buffalo Bill Dam a simple analytical model, employing several horizontal arches and one vertical cantilever at the dam's crown, was employed. Its success led to increasingly sophisticated techniques that firmly established trial-load analysis as an efficient method for arch dam design. Later Bureau of Reclamation projects designed on the trial-load analysis principle include the Glen Canyon Dam, the Hungry Horse Dam, and the 700-foot-high Hoover Dam. Recent computerized stress analysis systems represent the evolutionary refinement of this modeling technique, first formulated at the Shoshone project.

The Theodore Roosevelt Dam and Salt River Project

The Salt River project, incorporating the Theodore Roosevelt Dam, was the first major project undertaken by the United States Reclamation Service, under whose auspices the western landscape was greatly altered. The Roosevelt Dam, located 30 miles northwest of Globe, Arizona, on the Salt River, was Reclamation's first venture into large-scale dam construction. On its completion in 1911 the 280-foot-high dam was one of the largest in the world. Further, the Salt River project was one of the first multipurpose federal undertakings, involving farmland irrigation, flood control, recreation, and power generation. The electric power supplied to Phoenix in 1911 for its new streetcar system was the first demonstration of hydroelectric power generation on a large scale.

A large part of the area presently irrigated by the project is occupied by metropolitan Phoenix. The Salt River valley was irrigated and cultivated in historic times, but when pioneers arrived in the mid–nineteenth century it was essentially uninhabited. The settlers began irrigating the valley about 1867, constructing small diversion dams out of brush and rock. These structures, however, were no match for the river's erratic flow, ranging from a trickling stream in late summer to floodwaters in the spring. Compounding the difficulty, the washed-out structures could not be rebuilt until water levels in the river were once again at a minimum, resulting in inadequate irrigation supplies for land in cultivation. Frequent years of drought, following periods of steady water flow, caused the loss of vineyards and orchards, which require a consistent water supply over a number of successive years in order to thrive. Without adequate damming and storage capacity the new farmers could receive little benefit from mountain runoff and spring rains.

These difficulties prompted the formation in 1895 of the Farmers' Protective Association, an organization whose purpose was to put forth a united voice to both territorial and federal governments, urging legislation securing the water rights of those farmers inhabiting the Salt River valley. The association was enlarged and renamed the Salt River Valley Water Users' Association in 1903. Its first goal was to gain the help of the United States Reclamation Service in irrigating the valley; its second goal—and one of equal importance to the Salt River project's eventual realization—was the establishment of a corporate body, representative of a larger constituency, that would provide for direct contractual interaction with the newly established Reclamation Service. As a direct result of its ability to take action in unity, the association became the beneficiary of the Reclamation Service's first major undertaking.

Investigations by Reclamation engineers as to the feasibility of establishing water storage facilities on the Salt River got under way shortly after such actions were requested by the Water Users' Association. The construction of a large storage dam and reservoir, under contract with the association, was proposed by the Reclamation Service and authorized by the secretary of the interior in March 1903. The Theodore Roosevelt Dam, named after the incumbent president, was subsequently designed and approved, and construction proceeded in August of the same year. The dam, of cyclopean masonry in a thick arch design, was located 76 miles northeast of Phoenix, Arizona. At the same time a diversion structure, the Granite Reef Diversion Dam, was also started as the cornerstone of a system of irrigation canals and laterals from which the valley's farmland would be supplied. It was situated 30 miles downstream from the Roosevelt Dam, at the confluence of the Salt and Verde rivers, 22 miles east of Phoenix, and it utilized the Verde River's water for irrigation as well. The Granite Reef Dam was designed as a concrete ogee weir with two principal feeder canals, one on either side.

The Roosevelt Dam construction, headed by Reclamation engineer Louis C. Hill, took place over an eight-year period. The first stone was placed in 1906, and water impoundment began in 1910. When completed in 1911, the dam stood 280 feet high, with a hydraulic height of 234 feet. Its thick arch design resulted in a top width of 16 feet, expanding to a maximum base width of 184 feet along a crest of 723 feet. The total volume of rubble masonry included in its mass was 355,800 cubic yards. The dam's original spillway consisted of an unlined open channel

Theodore Roosevelt Dam and Salt River project, near Phoenix, Arizona. This was the Bureau of Reclamation's first project and one of the first multipurpose projects (irrigation, flood control, power generation, and recreation) undertaken by the federal government. When completed in 1911, the dam, at 280 feet, was one of the highest in the world. (Courtesy of the American Society of Civil Engineers)

through each side abutment. The side-channel spillways were later controlled by nineteen 20-by-15.9-foot radial gates. The large reservoir created by the dam, capable of storing sufficient water supplies for year-round irrigation, covers a surface area of 17,300 acres, with a conservation storage capacity of 1,382,000 acre-feet. The average annual inflow into the Roosevelt reservoir from the Salt River is 750,000 acre-feet (1913–1983 mean). The average annual diversion at Granite Reef, which includes Verde River inflow, is 992,000 acre-feet (1974–1983).

The Granite Reef Diversion Dam, completed in 1908, stands 29 feet high, with a hydraulic height of 18 feet. With a crest length of 1,128 feet, the structure contains 35,000 cubic yards of concrete. The principal diversion facilities consist of the Arizona (north side) Canal, operating at a capacity of 2,000 cubic feet per second, and the South Canal, with an operating capacity of 1,700 cubic feet per second. Each canal head works is controlled by a system of 15-by-9-foot slide gates. Some of the canals now fed by the system date back to the Hohokams (prior to A.D. 1400). The modern canals date from the 1880s, having been dug by private companies under contract with the valley farmers; but these were subsequently improved and incorporated into the Reclamation scheme upon construction of the Roosevelt and Granite Reef dams.

After several years of operation, in November 1917, the Salt River project was transferred to the Salt River Valley Water Users' Association to operate and maintain. The association in turn constructed three concrete thin-arch storage and power dams—the Horse Mesa, Mormon Flat, and Stewart Mountain dams—downstream from the Roosevelt Dam on the Salt River during the period 1923–1930. This substantially supplemented the project's hydroelectric capacity. Between 1936 and 1939 the Bartlett Dam, a concrete multiple-arch storage structure on the Verde River, was erected, and the four existing dams on the Salt River were repaired, under contractual agreement between the association and the Bureau of Reclamation (whose name had been changed from the Reclamation Service in 1907). A sixth storage structure, the Horseshoe Dam, of earth fill and rock fill design, was built on the Verde

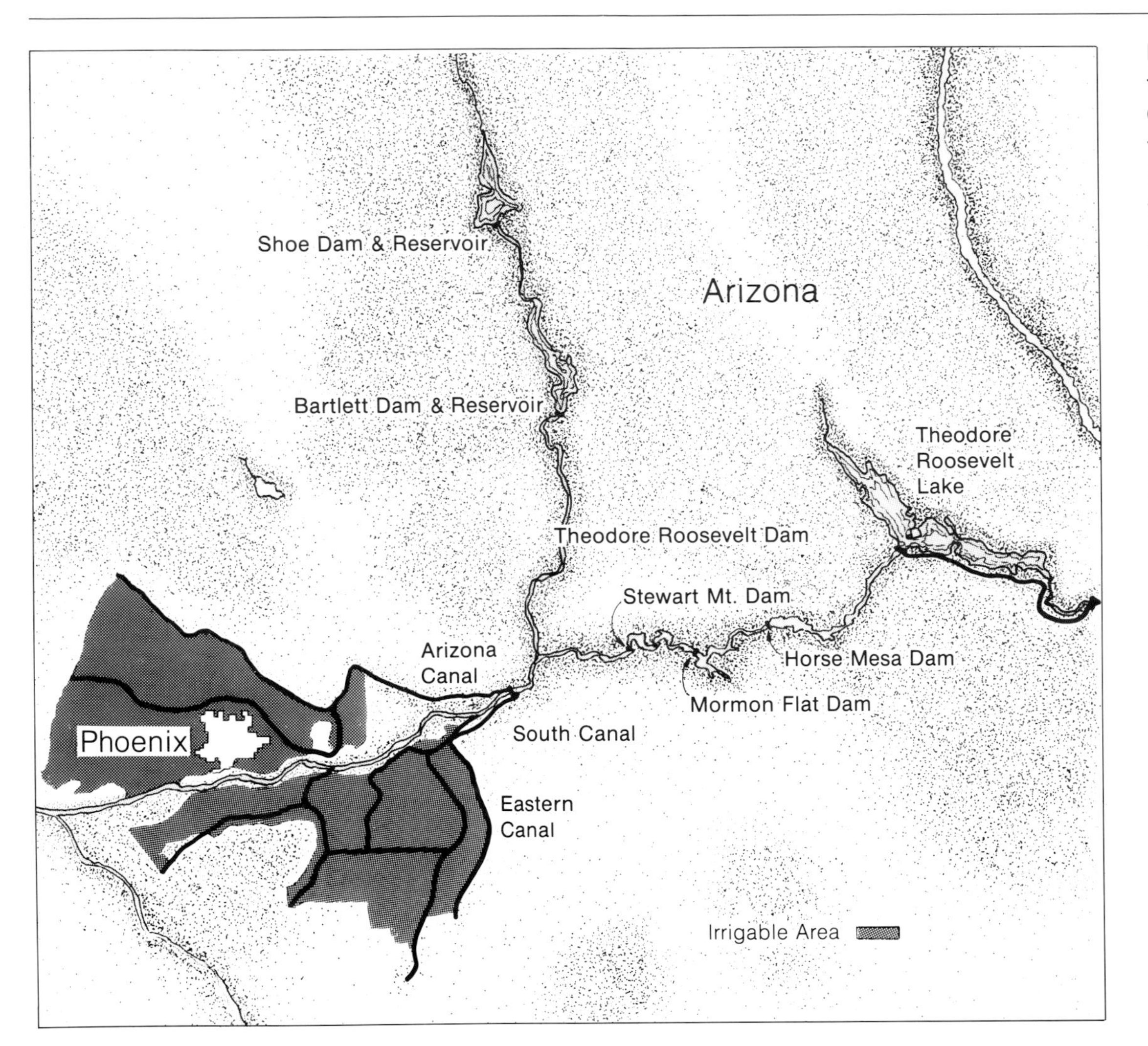

The Salt River complex.

River upstream from the Bartlett Dam during the years 1944–1946 by the Phelps-Dodge Corporation under contract with the association. At present the Salt River Valley Water Users' Association operates and maintains the network of irrigation and drainage canals below the Granite Reef Diversion Dam. The existing system of power plants—located at each of the four storage dams on the Salt River, with two small additional plants on the outskirts of Phoenix—is operated by the Salt River Project Agricultural Improvement and Power District, a governing body incorporated in 1937 and encompassing the same lands as the Water Users' Association.

The electric power produced at the Theodore Roosevelt Dam in 1911 was the first demonstration of practical electric power production on a commercially usable scale in the United States. It marked the Bureau of Reclamation's first venture into hydroelectric power as well. The Reclamation Act of 1902 had made no mention of power, and no bureau policy with regard to power production existed at the time of the Roosevelt Dam's construction. Rather, it was due to the efforts of the dam's supervising engineer, Louis Hill, that hydroelectric power became an element of a Reclamation project.

Engineer Hill, in charge of construction on the Salt River project, from the outset foresaw the advantages of locally available power for the dam's construction. Before dam building commenced, Hill planned and dug a 20-mile diversion channel, terminating at a power plant just below the dam site. This produced electricity for construction machinery. The channel, today named the Power Canal, was used at low-water periods to divert water to the Roosevelt Dam power plant until it was taken out of service in the early 1950s. At the time of the Roosevelt Dam's construction, Reclamation Bureau projects were carried out wholly or in part by government manpower, a practice continued until 1924. Thus, the economic advantages of using machinery to perform many of the heavier dam-building tasks, despite the wilderness site, were not lost on Hill. As a result of this planning, by 1906 locally produced power at the Roosevelt Dam site operated a cement mill and powered hoists and tramways for transporting giant stone blocks from the quarry to their placement in the rubble masonry dam. The advantages of hydroelectric power were immediately obvious, and the bureau subsequently authorized construction of a permanent power plant at the Roosevelt Dam. As a further step, in 1906 Congress authorized that revenues obtained from the sale of hydroelectric energy be maintained within the project's allocations, thus aiding those farmers whose lands were to be irrigated by project water by reducing their payments for such services. As a result, the Reclamation Bureau was able to partially fund its own projects with hydroelectric power revenues. This policy, further defined at the Rio Grande project in 1911, became the basis for all future Bureau of Reclamation multipurpose projects.

At present the Salt River project's watershed area includes some 13,000 square miles, supplying year-round irrigation to approximately 250,000 acres with water from the Salt and Verde rivers. Four storage dams and one diversion dam form an almost continuous 60-mile chain of lakes on the Salt River. There are two other storage dams on the Verde River, whose confluence is above the diversion dam on the Salt River. Irrigation is provided through approximately 1,300 miles of canals, laterals, and ditches. Recreational uses at the reservoirs include year-round fishing, boating, swimming, and waterfowl hunting. Hydroelectric power generation is accomplished by six power plants on the Salt River. There are two low-head hydro units installed in the canal system. The project also operates five steam plants.

The benefits southern Arizona has derived from the Salt River project are widespread. Hydroelectric power is furnished to the project's three surrounding counties, including metropolitan Phoenix, and to many expanded irrigation districts dependent on pumped well water for irrigation. Desert has been transformed into fertile farmland, most of which produces two crops a year. Seasonal flooding of the Salt and Verde rivers has been controlled. Finally, the system of reservoirs and lakes produced by the project's complex of dams and canals has spawned diverse recreational and tourist industries throughout the region.

The Elephant Butte Dam

This concrete gravity-arch dam, located at Truth or Consequences, New Mexico, is the chief element of the Rio Grande project, one of the first undertaken by the United States Bureau of Reclamation after its formation in 1902. Three of the bureau's most noted engineers, Arthur P. Davis, Louis C. Hill, and E. H. Baldwin, worked on the project as design engineers. (Both Davis and Hill were later to head the Reclamation Bureau.) The project is aimed at controlling the Rio Grande's seasonal floodwaters and providing a reliable source of irrigation water for farmland in New Mexico, Texas, and the Republic of Mexico. The completion of the Elephant Butte Dam in 1916 created one of the world's largest reservoirs, and its linked irrigation systems were the first to allocate water across international boundaries. In addition to the Elephant Butte Dam, the Rio Grande project includes the Caballo storage dam, four diversion dams, 141 miles of canals, 462 miles of laterals, 457 miles of drains, a hydroelectric power plant, 490 miles of transmission lines, and eleven substations.

The mild climate, rich soil, and readily available irrigation water of the Rio Grande valley have long attracted human habitation. Spanish explorers arriving in the first part of the sixteenth century reported that the Pueblo Indians were irrigating croplands there; the methods the Spanish described are still used by Pueblo descendants in the twentieth century. Settlers who came to the valley between 1840 and 1850 constructed simple diversion dams and canals at strategic points along the river. These structures, however, unable to resist the ravages of seasonal flooding on the river, had to be rebuilt annually.

By the 1890s extensive settlement and irrigation development in southern Colorado and central New Mexico were absorbing the normal summer flow of the Rio Grande. The result was longer and more frequent periods of dry riverbed at the southern reaches around El Paso, Texas. Several local, small-scale dam and diversion projects were proposed by private companies, but conflicting interests prevented their realization. Meanwhile, the Republic of Mexico sued the United States government for loss of water from the Rio Grande.

Under the authorization provided by the Reclamation Act of 1902, federal intervention in the Rio Grande water allocation issue began officially in 1903. Feasibility reports filed in 1904 recommended creating a large storage reservoir north of El Paso by building a dam across the Rio Grande. In the following year the Secretary of the Interior authorized construction. A 1906 treaty allocating 60,000 acre-feet of water annually to Mexico removed the final diplomatic obstacle to the project's realization. In 1907 the State Department funded $1 million for construction of the Elephant Butte Dam, and the Rio Grande project became the first engineering endeavor concerned with the international allocation of water.

The first structures to be completed in the system were the Leasburg Diversion Dam (a concrete ogee weir 20 feet high) and the 6-mile Leasburg Canal. These were finished in 1908, providing the first permanent diversion facilities. A site was selected for the Elephant Butte Dam, 125 miles north of El Paso; construction was scheduled to begin in 1908, but difficulties in acquiring all the needed reservoir land delayed it for four years. Building finally began in 1912 and was completed in 1916, with water storage under way by January 1915.

The Elephant Butte Dam (originally named the Engle Dam) was designed as a concrete gravity structure, the simplest and most economical means of creating the large storage reservoir required. It was immense for the time, rising 301 feet, with a hydraulic height of 193 feet, and the dam slopes to a maximum base width of 228 feet. A total volume of 629,500 cubic yards of unreinforced concrete was employed in erecting the dam.

The structure is situated upon hard, sound, fissured sandstone in irregular beds, containing pockets and interbedded strata of friable shale. Several small natural springs are interspersed within the foundation area, into which cement grout was injected during construction. In addition, in an attempt to prevent seepage into the fissured sandstone, a cement grout curtain was installed beneath

Elephant Butte Dam, on the Rio Grande at Truth or Consequences, New Mexico. Completed in 1916, the dam was part of the first civil engineering project in North America concerned with the international allocation of water. (Courtesy of the American Society of Civil Engineers)

an excavated upstream cutoff trench. Despite this measure a critical seepage condition beneath the dam, abetted by a rising groundwater table, occurred in 1916 and required the installation of a drainage system. Seepage has been a continual problem at the dam site, and improvements in the drainage system have been made from time to time since the dam was completed.

Mass concrete for the dam's construction was composed primarily of locally acquired crushed rock and rock screenings, blended with sand for aggregate. Differential movements due to temperature effects during the concrete setting process were controlled by pouring large blocks in an alternating, checkerboard arrangement. In addition, the faces of abutting pour blocks were coated with heavy oil to prevent mass concrete bonding and subsequent expansion and contraction cracking. Massive rocks, weighing up to 8 tons apiece, were also deposited in the green concrete upon placement. Finally, transverse contraction joints were spaced at 80- to 100-foot intervals in the dam's lower 200 feet, a spacing of 35 to 57 feet being employed in the top 100 feet.

The reservoir created by the Elephant Butte Dam was one of the world's largest at the time. Covering a surface area of 36,600 acres, the reservoir's capacity is 2,206,800 acre-feet of water. Its size was designed to allow for sediment accumulation, due to the large volume of silt carried in the Rio Grande's seasonal floodwaters. The Reclamation Bureau built an additional storage facility, the Caballo Dam, 25 miles downstream from Elephant Butte in 1938, to compensate for water capacity lost to silt accumulation in the Elephant Butte reservoir since 1916. At the same time an electric power plant was constructed at the Elephant Butte Dam, providing hydroelectric power in the vicinity for the first time. Today, water held at the Elephant Butte reservoir is used for winter power generation, then held at the Caballo reservoir for irrigation during the summer. Water is provided to Mexico under the terms of the 1906 treaty by means of the American Diversion Dam and Canal system, located 2 miles northwest of El Paso.

The Elephant Butte Dam stands as a monument to a farsighted water conservation project that combined the solutions to both local and transnational problems of water allocation. Through planning on a scale far larger than had been attempted previously in the region, the Bureau of Reclamation served the interests of both sides of the Rio Grande.

Early stages of construction of the Elephant Butte Dam. (Courtesy of the Bureau of Reclamation)

Construction proceeding on the Elephant Butte Dam. (Courtesy of the Bureau of Reclamation)

Blasting at the dam site. (Courtesy of the Bureau of Reclamation)

Note the temporary concrete batching plant in this view. (Courtesy of the Bureau of Reclamation)

Later stages of construction, Elephant Butte Dam. (Courtesy of the Bureau of Reclamation)

Chapter Nine

Buildings

Introduction

The emergence of the engineer as a professional with an independent status is one of the more significant of the interrelated themes running through the history of American building. Buildings in America, and the process of creating them, might be described with respect to many different kinds of characteristics—whether the technologies involved were preindustrial or postindustrial, whether the building types and forms were indigenous or derivative, whether the designers were craft-based or professionally trained, and so on. Essential to any such description, however, is the acknowledgment of a close connection between the character of America's buildings and builders and the character of her social and economic development.

The initial building forms of colonial America were largely craft-based and derivative of European practices. Although reference to European precedents (Spanish, English, French) continued throughout the period, most early American settlers learned mainly from each other and modified the derived forms so as to make them more suitable to local climatic conditions and resources. Some of their construction techniques evolved still further to establish a unique American building vocabulary in some regions of the country. Significantly, early settlers were normally confined to the use of locally available materials workable by hand and hand-tool techniques—although water-powered sawmills were introduced early in the seventeenth century, enabling the settlers to produce squared framing timbers. While timber construction was most common in New England, building types were quite different in the colonies farther south. In Maryland, Virginia, and Florida suitable clays and bricks were used to create more highly detailed and elaborate structures for the more affluent and aristocratic population, although timber found widespread use in these regions as well.

In all of these cases the carpenter-builder followed the pragmatic traditions of the colonial craftsman. As the level of building became more sophisticated with the growth of the colonies, however, a transition occurred. Slowly the notion of the carpenter-builder became elevated to the more specific notion of the builder as architect. As early as 1724 a body entitled the Carpenters' Company was established in Philadelphia. In this city, colonial building craft had reached a protoprofessional stage. Master carpenters who made up the membership were acting as both contractors and what we now know as architects. In 1734 the Carpenters' Company established a professional library, which became renowned for its assemblage of English treatises on the building arts. The systematic study and practical application of mathematics, geometry, surveying, and drafting eventually transformed the members of the Carpenters' Company into highly trained builders.

Abroad in western Europe the professions of architecture and engineering were also in transition. The rapid evolution of the sciences, and in particular the theory of mechanics, in the seventeenth and eighteenth centuries led to a successful challenge of the design practices of the time. French institutions such as the Académie Royale d'Architecture helped disseminate new design approaches. The establishment of the Ecole Polytechnique and the Ecole des Beaux Arts in France in the latter part of the seventeenth century marked a turning point in the development of the engineering and architectural professions. The Polytechnique assumed the role of technical teaching and the Beaux Arts that of visual concerns. Many traveled Americans of the eighteenth and early nineteenth centuries were strongly influenced by the Beaux Arts approach, which included a strong classical tradition in masonry construction. Immediately following the Revolutionary War architecture became established as a "gentleman's profession" in the young republic.

In the postrevolutionary era the classicism so prevalent in Europe found an accepting home in America as well. To a remarkable extent the guiding force in the architecture of the new republic was Thomas Jefferson. His classical designs for the University of Virginia and his own home, Monticello, directly influenced designers of the time; and through his various official positions he exerted great indirect influence as well. A number of talented architects were soon practicing in the young country—Mills, Strickland, Magnin, Latrobe, and others. These individuals were not dilettantes but technically trained professionals. Jefferson's sponsorship of Benjamin Latrobe, for example, was of far-reaching importance. Latrobe, trained as an engineer in a German university, found no difficulty in adapting classical forms to the buildings of his famed and influential Philadelphia waterworks. Latrobe went on to numerous other important building projects and became well known as an "architect and engineer."

The 1820s saw the ascendancy of the Greek revival style, particularly in the cities of the northeastern and mid-Atlantic states. In frontier areas, however, where the first wave of building was associated with traders and trappers, the building forms were indigenous. Then, as merchants and farmers began appearing on the scene, a new wave of building followed, with its more refined Greek revival forms. The products of the crude technology of the first wave often coexisted with the classical forms of the second. The South was slow to embrace the Greek revival. Cities like Charleston clung quite happily to their prerevolutionary styles well into the nineteenth century. Common craft traditions in building prevailed during this era, although handbooks, architectural guides, and carpenters' guides made architectural styles and building principles accessible to an ever widening audience. The popular works of Asher Benjamin are examples of this genre.

The shifting styles of the time were not particularly influenced by practitioners of the civil engineering profession. Preoccupied as they were with canals, roads, and bridges, civil engineers had little involvement with architecture as it was then defined. Existing craft-based technologies and rule-of-thumb construction methods were adequate for the buildings of the time. Industrialization, however, soon demanded new building forms—factories and mills—and also brought the need for new ways both of thinking about buildings and of producing them. The cultural changes brought about by industrialization also affected the architecture of the day. Many of the foremost thinkers were confident that an improved world could be built on science and technology, while others foresaw only disaster. A new eclecticism in building design began to emerge, with attacks on the remnants of the classical tradition coming from the Romantics and others.

The impact of the technical developments associated with industrialization in America would be hard to overestimate. Water supply systems, improved sanitation, better heating and lighting methods, all significantly influenced the nature of the built environment. New methods of building were also beginning to appear. Specific events—such as the development of the machine-made nail and of steam-powered sawmills, which led to the creation of the revolutionary balloon-framed house—had a profound impact on building practices. The evolution of cast-iron technology for making buildings perhaps best exemplifies the developments of the day. While this industrially based construction approach was at first mainly used to replicate building elements in the classical mode—as in the case of the cast-iron building elements made by James Bogardus—a whole new building language rapidly evolved that was a consequence of its own technological origins. The introduction of railroads and other artifacts of the industrial age led to innovations not only in bridge design but in building structures as well, owing to the need for long-span train sheds and other new types of structures. The eventual development of the skyscraper in Chicago and New York in the late nineteenth century, for example, was made possible by advances in the technology of producing building elements that were initiated during this era, as well as by theoretical developments in the field of structural engineering that occurred largely outside the context of "architecture" or even simply "building construction."

The period 1840–1860 saw the beginning of an unfortunate schism between the architectural and engineering professions that persists to this day. The burgeoning building demands of society, particularly the demand for industrial buildings, could no longer be filled by the master builders and architects of the day, and more and more buildings came to be designed and built by engineers as the nineteenth century wore on. Under the influence of critics such as Ruskin, the direction assumed by architecture—as a product of the professional architects of the day—began to have little to do with the attitudes of the efficiency-minded engineers who were engaged in their own building programs.

The apparent gap between technology and architectural ideology and form continued to deepen into the 1880s. In earlier times the prevalent aesthetic and stylistic standards had not been in serious conflict with the technology of building. This was no longer true. There was no meeting of the minds.

Further developments in technology in the more mature phase of the Victorian era, however, began to release building design from its age-old limitations of span and height. Worldwide attention was drawn to innovative structures other than buildings—the Eiffel Tower and Roebling's Brooklyn Bridge, for example—whose construction had important implications for building practices in general. The latter part of the nineteenth century also saw the development of that most indigenous of all American architectural forms, the skyscraper, and with it the need for a cooperative effort between architects and engineers.

A technological event of the late nineteenth century that was to have a profound impact on the design of buildings was the introduction of reinforced concrete as a building medium. Unreinforced concrete had long been used in construction in America, but reinforced concrete was largely a European invention. Significant early reinforced-concrete buildings in America include the Ward House in New York and Cincinnati's Ingalls Building. The twentieth century brought the development of precast and prestressed reinforced concrete, which also dramatically affected the form of buildings constructed by both architects and engineers.

By the turn of the century it was commonplace for architects and engineers to be jointly involved in the design of major buildings, yet their respective professional roles were quite rigidly segregated. The emergence of modernism, however, promised a convergence: it seemed to bring with it the attention of the architectural profession to the values already present in the engineering professions—efficiency and the like. Some buildings began to look like diagrams found in civil engineering textbooks. But a closer look at modernism in architecture surely reveals this interpretation to be highly suspect at best, and the apparent nearing of attitudes between the two disciplines to be illusory.

The increasing emphasis on scientific methods in the engineering schools of the 1920s further entrenched the separation of the roles of architects and engineers. Rapid improvements in structural analysis techniques occurred during this period as the academic branch of the profession laid more and more stress on them. With these improvements came a confidence in the ability of engineers to undertake difficult and complex projects, as is evidenced by the great multistory buildings constructed in the 1920s and 1930s.

The engineering of buildings has by now become a highly efficient process. The recent advent of computer-assisted structural analysis has enabled engineers to analyze efficiently and quickly more complex structures than ever before. Certain design activities have been greatly improved and expedited. Innumerable examples of their uses already exist, yet the potential impact of computer techniques on building design is relatively unexplored.

A consequence of the trend that developed early in this century for engineers to become preoccupied with analytical methods is that architects once again began assuming primary or even exclusive control over the forms of buildings, with engineers more and more frequently limiting their role to analysis and such restricted design issues as member sizing. In this same context, an attitude of extreme permissiveness on the part of many engineers in terms of "providing what the architect wants" also began to emerge earlier in this century, perhaps because it was increasingly true that inefficient, even ridiculous structures could be analyzed and subsequently engineered to be safe and serviceable.

A central issue becomes apparent here. On the one hand, improvements in analytical methods stemming from attention to the scientific basis of the profession have enormously enhanced the ability of engineers to address large and complex building undertakings. On the other hand, the preoccupation with analytical methods has led many engineers away from assuming an aggressive role in the shaping of building structures. Obviously, exceptions on the individual level may be found, as well as exceptions in connection with certain building types—for instance, high-rise construction. But the overall trend is there. The example of Benjamin Latrobe, "architect and engineer," is rarely followed today. While this model is surely not the only one capable of producing buildings that can be admired by both the architectural and the engineering professions—and perhaps not even the best model in today's society—the current role distinctions implicit between the professions and so aggressively maintained by them might well have puzzled Latrobe and the other great builders of the nineteenth century.

The Castillo de San Marcos

This seventeenth-century fort in St. Augustine, Florida, serves as a tangible reminder of America's Spanish past. Erected by Spanish military engineers from Havana, Cuba, the Castillo provides a rare embodiment of medieval European design principles on the North American continent. It has played a vital role in the development of Florida through successive administration by the Spanish, British, and United States governments. First completed in 1695 (a series of renovations was carried out between 1738 and 1756), the fort is the oldest extant major engineered structure in the United States.

The town of St. Augustine has traditionally occupied a pivotal position on the strategically important Florida peninsula. Its location was determined in the sixteenth century by the route of Spanish treasure ships making their way back to Europe. Loaded with gold and silver from Mexican, Chilean, and Peruvian mines, the galleons passed through the Bahama Channel, between Florida's southern tip and the heavily guarded outpost of Havana, Cuba, and were carried northward up Florida's eastern coast by the Gulf Stream. Here the treasure-laden ships, at the mercy of seasonal hurricanes, were also vulnerable to increasing forces of French and English pirates. By the mid-1500s, moreover, the Spanish empire was in decline, financially overburdened, its armada destroyed by the English naval hero Sir Francis Drake. French Huguenots, in open disregard for Spain's territorial claim to the Florida peninsula, established a military settlement, Fort Caroline, at the outlet of the St. Johns River on the northeast coast. Incensed by the French trespass and the potential threat to his treasure route, King Philip II of Spain in 1565 dispatched Captain General Pedro Menendez de Aviles to Florida to destroy Fort Caroline. Menendez set up a military base thirty miles to the south that was to become St. Augustine. Fort Caroline was duly destroyed in the following year, after a French force seeking to destroy Menendez's own outpost was eliminated. (A storm had shattered the French fleet, and the attackers were forced to surrender to Menendez, whose men then slaughtered some six

Castillo de San Marcos, St. Augustine, Florida. Completed in 1695, the fort was built by Spanish engineers to protect their country's interests in Florida. It is the oldest extant major engineered structure in the United States. (Courtesy of the American Society of Civil Engineers)

hundred of the Frenchmen one by one.) Menendez's outpost remained and became the oldest permanent European settlement on the mainland.

St. Augustine's early history was marked by hunger, poverty, and periodic military conflict. Menendez had erected a wooden fort on arriving in 1565. The semitropical climate, however, together with war and hurricanes, wreaked havoc on wooden fortifications. No less than nine such structures were constructed and abandoned in the hundred years preceding erection of the Castillo de San Marcos. Sir Francis Drake plundered Spanish settlements in the Caribbean during the 1580s, stopping off on his journey homeward to do the same to the poorly defended St. Augustine in 1586.

English designs upon the North American continent increasingly jeopardized the Spanish treasure route and tiny St. Augustine through the seventeenth century. Captain John Davis, an English buccaneer, sacked the town again in 1668. Two years later the English established a military outpost at Charleston, in modern South Carolina, only 200 miles, or two days' sail, to St. Augustine's north. Her control of Florida now threatened from both land and sea, Spain responded by planning a permanent fortification at St. Augustine, from which defensive military operations could be mounted, and to which both land and sea forces could retreat.

The project got under way with the arrival of a new governor for Florida, Don Manuel de Condoya, in 1671. Most of the funding for the fort was grudgingly supplied by the viceroy of New Spain in Mexico. The cornerstone was laid toward the end of 1672. Labor, food, and money shortages slowed the erection process over the next 23 years, delaying the citadel's completion until 1695.

The completion of the Castillo and supporting smaller forts—such as Fort Manazas, which guards Manazas Pass, a back route to St. Augustine—effected a marked improvement in Spanish control of Florida and its coastal waters. Sited to guard both sea and land approaches to St. Augustine, the Castillo served the Spanish effectively for 91 years, providing a base for operations against pirate forces and English troops and offering refuge to the townspeople. Although laid siege to several times, it was never captured.

The Castillo's invincibility did not forestall Spain's imperial decline, however. Spanish Florida was ceded to Britain in 1763, the result of an unsuccessful Franco-Spanish military campaign against the English. The citadel was used as a base of operations and military prison by the British during the American Revolution. Britain's tenure of the Castillo was short-lived: Spain resumed control in 1783, in exchange for Britain's annexation of Gibraltar. In 1821 the Castillo came into American possession, together with all of Spain's Florida territory. Renamed Fort Marion—after a Revolutionary War hero who, ironically, came from Charleston—the citadel was adapted for use as a military prison and later as an operations base during the Spanish-American War. Its military role ceased in 1900, and the fort was dedicated as a national monument in 1924. Legislation enacted by Congress in 1942 restored its original name.

The Castillo's quadrangular, four-bastioned plan derived from sixteenth-century Italian-Spanish principles of fortification design, adapted to the peninsular site by Ignacio Daza, a Spanish military engineer from Havana, Cuba. The triangular bastions located at each of the Castillo's four corners represent an evolution from the rounded towers of the medieval castle: the sharply angled walls of the bastion were less susceptible to damage from cannon balls than were the walls of the rounded tower. The general form of the fort was common in European fortifications of the time.

The Castillo's 12-foot-thick walls, battered back to 9 feet at their 25-foot-high peak, were constructed of coquina, a native rock unique to Florida's coast. Formed by natural force from millions of tiny shells cemented together under great pressure with a natural mortar of oyster shell lime, this material was quarried on Anastasia Island, across the Matanzas Bay from St. Augustine. The stone is relatively soft and thus was easily cut in the quarries by Indian stonecutters, although it hardens under exposure to air. The character of the stone rendered it highly effective against cannon fire, as blows against its surface were absorbed without splintering the fort's walls. Stonemasons, most of them convicts and captured enemies, built the Castillo's walls using a quick-setting lime made by heating oyster shells to a white-hot temperature in two kilns on the fort's site. Quarried stone and rubble to fill the fort's thick walls were ferried across the bay by Indian laborers on rafts and canoes the Indians had built for this purpose. Within the citadel's walls, structures to house troops, munitions, and food supplies were erected in timber, with foot-square timbers to support the roof, upon which guns were mounted against attack. Wells were dug in the courtyard to provide fresh water for the sheltered townspeople and soldiers during times of siege.

Despite financial shortages throughout the Castillo's difficult construction, ornamental and aesthetic considerations were not ignored. The fort's outer walls were originally covered with a white, waterproofing plaster of lime, reflecting the sun's rays. Finely detailed cornices and pilasters were added, to cast shadow lines and to visually divide and cap these expansive surfaces. Sentry towers, erected upon the corner bastions, were plastered with a mixture of lime and clay, creating a strong, dark-red tone, in sharp contrast to the brilliant white of the fort's perimeter walls. Even the iron gates were elaborately detailed in wrought iron, imported from Spain.

During the first period of construction, from 1672 through 1695, masonry work within the Castillo's walls had been avoided, in large part because of a shortage of skilled labor. Coastal Florida's humidity quickly rotted the wooden terreplein, or gun deck, rendering the Castillo's weapons useless by 1738. At this time another Spanish engineer, Pedro Ruiz de Olano, began renovation work on the fort's interior. Twenty-four massive masonry arches were erected around the interior courtyard in place of the wooden chambers. The arched spaces created beneath the rehabilitated terreplein were designed to withstand the impact of bombs lobbed into the

The Castillo was located to protect the seaward entrance to St. Augustine. (Another fort, still standing, at Manazas Pass protected the town from attack via the Manazas River.) (Library of Congress)

Castillo. Along the fort's eastern wall, against which attack from the sea was directed, ceilings above the masonry arches were 4 feet thick; the remaining three sides had 2.5-foot-thick ceilings. A stone ramp, replacing the original wooden stair, was built between the courtyard and the terreplein, providing for rapid movement of both guns and troops during battle. Finally, the moat surrounding the Castillo's perimeter was extended to 40 feet in width, with earthwork defenses beyond. Today the Castillo, now maintained by the National Park Service, looks essentially as it did in 1756, at the completion of Pedro Ruiz's renovations; it retains the essential character of its sixteenth-century design.

were timber beams used. Bogardus designed the building so that it could be disassembled and reerected elsewhere (this remarkable building was disassembled in 1859).

Bogardus continued to expand his business and remained innovative. He helped introduce the I-beam to the United States, possibly learning of it through the works of Fairbairn and Hodgkinson. Among his better-known works was the second printing plant of Harper and Brothers (1854), which used cast-iron floor beams with wrought-iron ties and 7-inch wrought-iron rail beams that helped transmit floor loads to girders. His McCullough Shot and Lead Company tower (1855) was an exploration of skeletal construction and foreshadowed later developments in multistory construction in the nineteenth century.

Daniel Badger was another innovator and enthusiastic advocate of iron. He began his career in Boston, where he constructed a store building on Washington Street with iron columns and lintels on the first story. Badger soon moved to New York, where he established the highly successful Architectural Iron Works at Duane and Thompson streets. Before long he was producing iron members for buildings across the East (including the Watervliet Arsenal storehouse) and abroad as well. He executed the building, designed by John Gaynor for E. V. Haughwout and Company (1857), in which Elisha Graves Otis installed the first passenger elevator with an automatic brake. Many medium-height commercial buildings followed, typically having long rows of columns and spandrel beams on their facades. Badger also constructed a grain elevator for the United States Warehousing Company in Brooklyn (about 1861) that was an all-iron frame in the spirit of Bogardus's shot tower.

By the time of completion of the famous Wanamaker Department Store (1859–1868)—perhaps its finest example—the era of the iron-front building was drawing to a close. New building forms were emerging for which the bolted iron-front technology was less suitable. Certainly the eventual development of the skyscraper in New York and Chicago necessitated new forms of braced and riveted construction—although these very buildings were presaged by cast-iron works such as Bogardus's shot tower.

The Watervliet Arsenal storehouse stands at the southeast corner of the arsenal, just east of the former bed of the Erie Canal. An excellent example of the products of the great iron revolution, it is detailed in Renaissance revival style, proportioned for stone but prefabricated almost entirely in cast and wrought iron. The rectangular building has a 100-foot transverse dimension, divided into two 25-foot side aisles and a 50-foot center aisle. The longitudinal dimension of 196 feet is divided into sixteen bays by two rows of double ("Siamese") columns. There are fourteen 12-foot interior bays and two 14-foot exterior bays. The side aisles each contain a gallery floor. Cast-iron stairways in each corner lead to the gallery level.

The foundation consists of cut limestone over random rubble footings on the perimeter, while interior columns have ashlar bases dressed to the sill. The sixteen longitudinal bays are defined by transverse cast- and wrought-iron Fink trusses over the center aisle (spanning 50 feet) and modified Fink trusses and composite beams over the side aisles. Both trusses are 8 feet deep. Center and side trusses share the same collinear top chord. The truss members themselves are plain wrought iron except for cast-iron cruciform compression struts. Turnbuckles are used on the 1-inch-diameter lower chord. All connections are bolted. The supporting columns are 28.5 feet high and taper from 10 inches to 6 inches in diameter. The duplex columns are integrally webbed. Gallery joists are wood. Composite gallery beams are principally cast iron but use a lower wrought-iron rod.

The walls are cast-iron panels connected by flathead, countersunk machine screws through flanged and lipped surfaces. The paired cast-iron pilasters support one end of the roof trusses and stiffen the wall. The corner pilasters are built-up box columns. The non-loadbearing fenestrated panels between pilasters are generally $5/16$ inch thick. The walls on the end and side elevations are topped by stiffening members. The end-wall gables are sheathed with corrugated iron, framed with different structural sections above the top plates of the end walls. These walls have been further stiffened recently by the addition of welded frames.

The gabled end elevations are divided into eight bays containing cast-iron personnel double doors in the end bays and, originally, rolling iron vehicular doors in the interior bays. Only two of the latter remain, and they are inoperative. The easterly rolling doors have been replaced by double, wooden, half-glazed doors. There are circular windows on the gable ends as well. The openings on the side elevations are either glazed or covered with fixed iron plates. Three combination ventilation monitors and skylights are located at approximately quarter points on the roof ridge. Artificial lighting, plumbing, and heating, for which there was no original provision, were installed at a later date.

Truss members are of wrought iron, except for the cruciform compression elements, which are made of cast iron. (HAER Collection, Library of Congress)

Two rows of double ("Siamese") columns run the length of the building. (HAER Collection, Library of Congress)

The William Ward House

The William Ward House, built between 1873 and 1876 in Port Chester, New York, was the first in North America to be constructed entirely of reinforced concrete. It synthesized two decades of efforts by engineers in France, England, and the United States to develop a fireproof material for building construction that was adaptable to a wide range of architectural purposes. Its owner and builder, William E. Ward, planned the house as a showpiece for the manifold possibilities inherent in the new material, choosing as the vehicle for this demonstration a remarkable and unlikely architectural type—the "picturesque" suburban villa, popularized by A. J. Downing and the Hudson River School at midcentury.

The Ward House brought to fruition simultaneous but largely independent developments occurring in reinforced-concrete technology in both Europe and the United States since the middle of the nineteenth century. The Industrial Revolution in Europe had introduced cast-iron technology to architecture, and by the 1840s it represented an economically viable alternative to traditional timber and masonry construction. Exponents such as Joseph Paxton in England and James Bogardus in New York popularized it thereafter, bringing cast-iron construction to the forefront of contemporary building technology by midcentury. The dismal performance of structural cast-iron components under typical high-temperature fire conditions, however, discredited the material. As for wrought iron, it was still too difficult to manufacture to find widespread application in buildings, and fire effects were equally problematical. Experimenters therefore turned to developing methods of fireproofing iron by applying masonry and concrete coverings to a metal core. In the 1850s and 1860s advances in nonreinforced, mass concrete were taking place as well—particularly in the United States, where the material was finding widespread application.

English and French engineers in the 1850s were the first to break new ground in the exploration of iron and concrete used in a composite way. These pioneers sought more

William Ward House, Port Chester, New York. Built by William E. Ward between 1873 and 1876, this still-standing house was the first building in the United States to be constructed entirely of reinforced concrete. The floor slabs are made of concrete and wrought-iron I-beams. The columns are hollow cylinders, reinforced with rods bent into a series of hoops. All exterior walls and partitions are of reinforced concrete as well. Note the detailing of the concrete to emulate stonework. (Courtesy of the American Society of Civil Engineers)

fundamentally a structural formulation by which wrought iron's tensile strength and the compressive strength inherent in concrete might each be exploited in combination with the other, forming a single composite material. William Wilkerson of England received the first patent for such a material—a grid of wire rope embedded in a concrete slab—in 1854, although he was preceded in 1848 by Josef Lambot's curious iron-reinforced, poured-concrete boat. By 1867 the French engineer Josef Monier had improved on Wilkerson's idea: Monier reinforced ornamental cast-concrete garden tubs and pots against both tensile and torsional stresses by means of wire mesh. The Paris Exhibition of the following year displayed further advances by the Frenchman François Coignet, who employed iron bars for reinforcement in a concrete slab; William Ward was to adapt Coignet's ideas for his Port Chester house. In 1877 Monier received a patent for a system for reinforcing concrete columns and girders with a grid of iron rods; he promoted his system for rail and highway bridges. The German engineer G. A. Wayss soon further developed some of Monier's ideas and contributed to an understanding of the underlying structural principles present. In 1879 the French engineer François Hennebique demonstrated that he understood the relation between compressive and tensile stresses in a concrete slab by designing an iron-reinforced slab in a systematic fashion. By this time, however, Ward had completed his house, anticipating Hennebique's designs and the American Thaddeus Hyatt's similar breakthroughs by several years.

In the United States a less advanced industrial base hampered large-scale experimentation with iron-reinforced concrete. Freer experimentation, however, took place than in Europe. American engineers exploring new materials concentrated on the more humble uses to which such materials might be applied—in particular, the application of nonreinforced concrete to such unconventional uses as domestic architecture. This freedom made possible the development of a substantial body of knowledge regarding the properties of unreinforced concrete, and its subsequent adaptation to diverse needs. Thus, as early as 1835, Obadiah Parker of New York City employed unreinforced concrete for bearing walls, entablatures, cornices, and columns in a Greek revival house. Nine years later Joseph Goodrich of Milton, Wisconsin, built his home from precast-concrete blocks, termed "artificial stone," using portland cement imported from England. Engineers used mass concrete for the footings of the landmark Starrucca Viaduct of the New York and Erie Railroad in 1848, perhaps the first such adoption of concrete for this structural function (see chapter 4, Starrucca Viaduct). It was later used in many other projects as well.

William Ward of Port Chester, New York, was a charter member of the American Society of Mechanical Engineers and a successful businessman. By the 1860s his Russell, Burdsall, and Ward Bolt and Screw Company on the Byram River in Port Chester was a flourishing venture—due in large part to his business acumen and inventive genius—but Ward had become fascinated with the evolving technology of concrete, particularly the experimental work then going on in Europe. Although without formal training in either civil or chemical engineering, Ward grasped the essential difficulties confronted by Wilkerson, Monier, and Hennebique during a visit to England and the Continent in 1867. He concluded a paper read before the American Society of Mechanical Engineers in 1873 by anticipating the efforts of Hennebique in France and his countryman, Hyatt: "The utility of both iron and beton [concrete] could be greatly increased for building purposes through a properly adjusted combination of their special physical properties, and very much greater efficiency be reached through their combination than could possibly be realized by the exclusive use of either material separately, in the same or in equal quantity."

Ward sought to embody his vision of the potential of reinforced concrete in a conspicuous form—his own grand suburban residence. It must be remembered that at this time the dogma of stylistic revivalism prevailed, limiting breakthroughs in building form to the arena of industrial and exhibition structures. Ward's choice of a combined Hudson River Gothic and French Second Empire style, if in retrospect rather absurd, was very much in the vernacular of the day. Robert Mook of New York City, a fashionable (and now forgotten) architect, was hired to produce the necessary plans. Ward, however, formulated the structure of the building on his own, and acted as builder as well. Construction commenced in 1873 and proceeded for the next four years. Uncertainties regarding concrete curing in cold weather combined with Ward's devotion to meticulous supervision and ongoing experimentation to slow the building's erection. At one point, when the second-floor parlor slab had been standing for a year, Ward trial-loaded the concrete at its center point with 26 tons of dead weight for a winter's duration. To his delight, deflection in the 18-foot span measured only 0.01 inch the following spring.

If Ward's structure seems conservative by today's standards, it must be remembered that the house was erected entirely without the benefit of precedent. Ward had only local carpenters and unskilled laborers with whom to build his house; even formwork for the concrete pouring had to be explained step by step to the befuddled carpenters. The elaborately coffered ceilings of the major rooms were made of 5- to 8-inch-deep I-beams set and encased in concrete to form the ceiling ribs. A 3-inch layer of concrete was added to the tops of the beams. The layer incorporated $\frac{5}{16}$-inch-diameter iron rods, placed toward the bottom of the slab thickness, in an 8-inch-square grid. Ward's understanding of structure was evident here. The I-beam ribs were made monolithic with the floor slab, "to utilize their tensile quality for resisting strain below the neutral axis." Originally Ward intended to use inverted T-beams, in which a greater proportion of the metal would be available for resisting tensile stress below the neutral axis. At the time, however, I-beams were available as a standard rolled item, whereas T-beams would have to be specially manufactured.

The 3-inch slab was placed in two layers, the first only 1 inch thick to provide sufficient coverage beneath the reinforcement rods, upon which the upper 2 inches were placed in a fairly dry mix and carefully tamped around the metal bars. A top course, ½ inch of fine cement, was then placed, allowed to harden, and subsequently rubbed with sand and stone to simulate polished sandstone. In rooms where carpeting was to be applied, metal sockets to receive brass plugs attached to the carpet's underside were cast into the floor slab. Over 13,000 square feet of floor and roof slabs were constructed in this manner, all under the careful scrutiny of William Ward himself. Elaborate detailing of the interior finishes, particularly of the paneled ceilings, was achieved with plaster applied directly to the concrete surface. Wood furring and lathing strips were entirely absent from the house.

Innovation characterized the house throughout. Interior partition walls were cast-in-place concrete, 2½ inches thick instead of the 8-inch thickness required by conventional masonry construction. Exterior walls were cast as hollow-core units. A 6- to 10-inch spacing was created between 2½-inch-thick, cast-in-place panels, attached to each other every 2 to 3 feet. The cavity wall was incorporated into the building's heating system, the basement furnace supplying hot air through the cavity walls to all parts of the house. This forced-air system was one of the first of its kind in the United States. The spacious veranda, surrounding over half of the building's perimeter, had a sloping roof supported by hollow concrete columns. These columns, incorporating rainwater leaders within their cores, contained circular reinforcements attached to vertical bars. The columns' internal rain leaders connected with additional leaders incorporated within the exterior hollow-core walls. Thus the house's exterior remained uncluttered by gutters and downspouts. A network of piping connected the rainwater leaders with a pump and storage tank contained within the house's square tower. Ward even installed a windmill (since removed) to power the storage tank pump. Finally, still experimenting, Ward cantilevered a balcony 4 feet from the wall's edge at the building's rear, dramatically demonstrating the structural freedom implicit in reinforced concrete.

By 1876 the Ward House was essentially complete, and labeled "Ward's Folly" by a doubting newswriter. Expense was not spared in the building's furnishings or in the elaborate decoration of its exterior. Plastered outer walls with superficial detailing, particularly the redundant corner quoining, evoked the images required of the building's adopted style yet could not entirely mask the simple boldness achieved with reinforced concrete.

William Ward's achievement lay in his remarkable fusion of prevailing architectural form with a major technological innovation. With time would come experiments by others that would explore innovative forms perhaps more naturally a consequence of the characteristics of the material itself. But the Ward House remains a remarkable first exercise.

Union Station, St. Louis

Union Station in St. Louis, Missouri, is a huge Victorian building with grand spaces and a clock tower rising 232 feet. Built between April 1, 1892, and September 1, 1894, the station provided a central terminus for the many railroad lines converging on St. Louis—nine from the west, thirteen from across the Mississippi River. Its train shed was one of the largest in the world. A unique subway system, connecting the tracks to the terminal building, transported baggage back and forth. The shed made use of the longest crescent trusses used in any building in the world up to that time. Developments in reinforced-concrete flat-slab design also occurred here during an expansion of the station in 1929; at that time the station had the longest-span flat slab carrying railroad loadings in the world.

An invitational competition was held in 1891 for the design of the station on a 42-acre site. Theodore C. Link of St. Louis was the winning architect and received the commission. George H. Pegram, also of St. Louis, was the consulting engineer who designed and patented the unique crescent trusses in the train shed.

The site had been decided upon by the station's board of directors in 1890. Influencing their choice was the fact that there were to be no through trains coming into St. Louis. Rather, passengers traveling east or west beyond St. Louis transferred at this point. The decision to make the new terminal an "end station" rather than a "through station" had itself been determined by the topography of the area, which confined railroad approaches to a narrow valley. The site for the building, then, was established to accommodate this type of approach. The terminal was to include a headhouse, a great train shed, a midway, a powerhouse, and several auxiliary buildings. The track layout was designed to have end tracks in the train shed. This arrangement minimized the distance for passengers to walk along the platform and placed the coal-burning, steam-powered locomotives as far as possible from the passenger concourse and headhouse. In order to minimize interference with passengers, baggage and mail were handled in subway tunnels that rose to

Union Station, St. Louis, completed in 1894. The 232-foot clock tower dominates this recent photograph. In its heyday during the early decades of this century the station was a major hub of railroad traffic. Twenty-two lines from the east and west terminated here. (Courtesy of the American Society of Civil Engineers)

platform levels near the outer ends of the platforms. The station was one of the first buildings in St. Louis to be lighted entirely by electricity, originally using over 3,600 handmade bulbs.

The five-story headhouse contained waiting rooms (including a separate one especially for female travelers), a ticket office, dining rooms, facilities for baggage and mail, a hotel, and stores. The building was almost a city in itself. Its great hall had a barrel-vaulted ceiling reaching 65 feet above the floor level. The original dimensions of the headhouse were 80 feet by 456 feet; later additions have extended its length to 753 feet, including the Terminal Hotel.

Many unique and costly materials were used in creating this marvelous building. The interlocking floor tile in the Grand Hall and dining room came from England, the mosaic flooring in the ladies' room from Belgium, the floor tile in the basement from Germany. There are also Numidian marble from Africa, Siena and white marbles from Italy, Vert Champagne marble from France, and Alps green marble from Switzerland. Other marbles came from Georgia, Vermont, and New York. Stained glass and ornamental iron and copper work are also present.

Considerable difficulty was encountered in the construction of the foundations. To the east the ground was undermined with a series of caves and vaults, the remains of the oldest brewery in the city. To the west was Chouteau Pond; the remains of willow stumps and old log cabins were encountered—even the hulls of old boats, 20 feet below the surface. The footings were made of concrete, with a few exceptions in which steel cantilevers were used. The tower foundation was disconnected from the rest of the foundations. Slip joints were used between it and abutting walls. The lower floors of the headhouse have cast-iron columns, while the upper floors are timber with some iron and steel girders. The added Terminal Hotel is similar but less well constructed.

The train shed was immense. Originally 630 feet long and 606 feet wide, and extended in 1904 to a length of 810 feet, it was larger than the train sheds in Boston, Philadelphia, Chicago, New York, London, Frankfurt, and Cologne. Ninety-eight feet high in the center, the space covered is divided into five spans by four rows of intermediate columns—a center span of about 141 feet, two spans of about 139 feet, and two spans of about 91 feet. The columns in each of the inner rows are spaced 60 feet apart and support the roof trusses. The columns of the outer rows are spaced at 30 feet. George Pegram's crescent-shaped roof trusses, spanning the distances between these rows of columns, were the longest in the world in 1894. The patented truss system was made up of a combination of wrought iron and steel. The compression members were built up of angle members less than 4 inches by 4 inches in dimension. Construction of the shed's foundations began on April 21, 1892, and was completed on September 23 of that year. The structure was ready for use on November 25, 1893.

The space between the headhouse and the shed—called the midway—was covered by a light trussed roof of glass and iron. This passenger concourse extended the full length of the station between the headhouse and a protective barrier located a short distance from the ends of the station tracks. The midway is 50 feet wide by 606 feet long and was separated from the train shed by a high ornamental iron fence with sixteen gates.

Governing the movements of the trains was an interlocking plant. Train movements were particularly complex because of the necessity for each train to back into its assigned station track. Located in a 25-foot by 57-foot tower on top of the powerhouse, facing the track, the interlocking system controlled every movement of switch, signal, engine, or train within 1,500 feet. Train movements were directed through overhead and trackside signals. The plant itself consisted of compressors and electric storage batteries, the interlocking machine, the announcing instruments, and the switches and signals. The interlocking machine, the largest in the world at the time, consisted of levers in parallel rows and a mechanism made up of the locking shafts, the locking bars, and the locks themselves. Its frame accommodated 215 levers and was capable of providing 1,827 possible routes for train movements.

The powerhouse south of the train shed contained the massive boilers, engines, dynamos, air compressors, and heating plant

The crescent-shaped roof trusses in the Union Station train shed were the longest in the world at the time of construction. (Courtesy of E. Littmann, Photographer)

for the station. The brick structure is 67 feet by 134 feet. Fronting on the midway was a kiosk with a clock and a tower containing the station master's and telegraph offices. On the 20th Street side of the shed was the more recent baggage room building, measuring 20 feet by 300 feet. The lower floor was used entirely for baggage, while the second was used for offices, storerooms, conductors' rooms, and other functions. A 40-foot by 70-foot two-story building designed for handling the U.S. Mail was located south of the baggage room.

Train operations were at an all-time high at Union Station in 1923, when there were no less than 143 arrivals and 143 departures daily. Expansion again occurred in 1929 with the addition of tracks to the west. Since there was to be a subway under these tracks, a reinforced-concrete flat slab was designed to carry them instead of the steel girders used in the original deck. This slab was noteworthy for being the world's longest-span flat slab carrying railroad traffic at the time.

In 1976 this magnificent structure was designated a National Historic Landmark. The last train was operated in the station in 1978. The St. Louis Union Station Association then acquired the property and began a program of refurbishment. The headhouse and adjacent Terminal Hotel building now form part of a new hotel. The small rooms of the original hotel (typically 8 feet by 10 feet) have been upgraded to modern standards, while the main public rooms have been carefully restored. The great barrel-vaulted Grand Hall is the lobby of the new hotel. Most of the huge train shed has been opened to the sky and landscaped, with gardens and a small lake. An atrium rotunda has been added in the original hotel.

Engineering studies made during the renovations indicated that the design standards and procedures used for the trusses were similar to today's. The worst deterioration in the train shed was found at the south end, where the sulfur-bearing smoke from locomotives had been concentrated. Here corrosion was a significant problem. In other parts of the structure slots with bolts that had originally been installed to allow for temperature expansion and contraction of the trusses had long since rusted tight, some apparently by 1915. Earlier reports indicated that some of the trusses had consequently deflected as much as 5 inches on hot days. Truss members were found to have been made of material that was actually stronger than originally suggested.

The rehabilitation involved removing the roof in many areas, sandblasting exposed truss members, and covering them with a weatherproofing paint. The temperature expansion and contraction problem was addressed by placing new midspan joints to allow movement in the north-south direction, while roller joints made with Teflon sliding pans replaced older joints in the other direction. Exterior columns were reinforced.

Construction of the hotel additions under the shed involved putting in place a number of new piles. Post-tensioned reinforced-concrete plate floors were laid as well. In the headhouse and related old hotel, some members were found overstressed. Some of the original cast-iron columns were removed to make larger spaces, and their loads were transferred to other columns. Many of the existing timber beams were reinforced. Rehabilitation of the headhouses also involved extensive roof construction.

A new skylight has been installed in the midway area, and the old clerestory windows replaced. Whenever possible, old entries and gates have been saved. One of the main entryways, the stone-arch porte cochere on short brick pilasters, had cracked extensively because of arch thrusts, and a new stainless steel tension rod was installed.

The great building has now regained some of its former grandeur—grandeur that had slowly slipped away with the decline of the railroads.

The Peavy-Haglin Grain Elevator

Completed in 1900, the Peavy-Haglin grain elevator in St. Louis Park, Minnesota, near Minneapolis, was the first circular reinforced-concrete grain elevator constructed in North America, and possibly the world. It was the prototype for the structures now ubiquitous in the landscape of the midwest. The successful performance of the 125-foot-tall structure proved that reinforced concrete, a relatively new material at the turn of the century, could economically be used to sustain the difficult loading conditions created by fluctuating levels of granular material in tall towers.

By the late nineteenth century the Midwest was producing enormous quantities of grain. Storage of this bounty was an ever-present problem, with grain elevators proving an effective solution. From about the 1880s grain elevators were of wood-cribbed construction, which, although it had advantages over the previous balloon-frame method of building, was more expensive and still highly inflammable. The common positioning of elevators by railroad tracks, and hence their exposure to sparks from steam locomotives, increased the fire hazard, resulting in rising insurance costs. By 1900 almost all wood-cribbed "country" elevators (in contrast to the larger "terminal" elevators) were clad with protective galvanized-iron sheeting. Other approaches to fireproof construction were tried—including experiments with steel, tile, brick, granite, and concrete block—but none really took hold.

By 1899 Minneapolis had become the flour-milling center of the world and one of the nation's largest primary grain markets. Elevator storage capacity in the city reached 27 million bushels that year. Not surprisingly, the city's elevator owners were actively involved in the new construction experiments. One such individual was Frank H. Peavy, who had built up a large grain company. Born in Eastport, Maine, in 1850, Peavy had boarded a Chicago-bound train the day the Civil War ended (having sold papers for two years to pay his fare) and had ended up broke in Sioux City, Iowa, in 1866. There he went into

Peavy-Haglin grain elevator, Minneapolis. This structure, completed in 1900, was the first circular reinforced-concrete grain elevator built in North America. The form was soon adopted across America's wheat-growing heartland. (Courtesy of the Peavy-Haglin Company)

selling farm implements and soon organized a grain company, which he eventually moved to Minneapolis.

Peavy entered into discussions about a new grain elevator with a builder named Charles F. Haglin. Haglin had left his parents' home in upstate New York and set out for Minnesota at the age of twenty. Trained as an architect, he had become more interested in construction and had moved into the contracting business. It is not known precisely how the idea of building a grain elevator out of concrete was conceived. But it soon became the conviction of Haglin that such a structure was possible, and it became the desire of Peavy to get one built by backing it. They met with considerable skepticism from other contractors and owners, however, many of whom felt that a tank of solid concrete would lack "give" and either explode or crack badly. There was also uncertainty about how well the grain itself would keep in a concrete environment. But the decision was made to try it.

The elevator was designed as a single reinforced-concrete cylinder, 125 feet high, with an inside diameter of 20 feet. The walls were to be 12 inches thick at the base, tapering to 8 inches at the top. Some brass-rod reinforcement is believed to have been used. Construction, begun in the summer of 1899, was accomplished using round wood forms braced with steel hoops. Concrete was poured into the forms. As the concrete dried, the forms were pulled up and braced, and another layer was poured.

In the fall, when the elevator had reached a height of 68 feet, construction was stopped. Whether or not the hollow, monolithic concrete structure was a practical design was still in doubt. It was decided to test the structure by filling it. Grain was brought in railroad cars, shoveled into a bucket elevator, carried to the top of the concrete structure, and dumped in—to remain until the spring, when the condition of the grain would be tested, as well as that of the structure when the grain was drawn off.

During the winter Haglin and Frank T. Heffelfinger, Peavy's son-in-law, traveled to Europe to survey grain elevator building practices there—for it was rumored that concrete elevators were already in use in various locations. Another purpose of the trip was to ascertain if it might be possible for Peavy to build and operate grain elevators in Russia. Visits were made to Hamburg, Brunswick, Copenhagen, Budapest, Braila, Galatz, Vienna, Paris, several towns in Russia, Amsterdam, Rotterdam, and Antwerp. At Brunswick, for example, they met with one G. Luther, the designer of an elevator constructed of "Hennebique" concrete and steel. Elevators of this type were square and had been in use in Braila and Galatz in Romania for five or six years. Haglin and Heffelfinger eventually traveled to those cities, also stopping in Copenhagen to see the "bins of Monier." (François Hennebique and Josef Monier were accomplished European experimenters with reinforced concrete; see the description of the William Ward House, earlier in this chapter.) They examined many grain storage structures but found none that was both round and made of reinforced concrete. Some of Hennebique's structures in Braila—composed of hexagonal bins with rounded corners, fitted together like the cells in a honeycomb—impressed Heffelfinger by their ingenious design; nonetheless, he was "fully convinced that our construction is all right and even better than this." In Russia Heffelfinger had an interview with the finance minister, who informed him that his government would allow Americans to build grain elevators but not to run them. Heffelfinger concluded that this would not be profitable.

The travelers returned to Minneapolis in the spring, in time for the drawing off of the grain loaded into the 68-foot concrete elevator the previous fall. Onlookers drew back, putting about a block between them and the structure. Many thought that the great tube would burst upon emptying. Haglin, however, stood at the base of the elevator. As the lever was pulled, grain flowed smoothly through the ramp at the base of the tank into a pit 8 feet below ground. The monolithic tube remained intact, and the grain was found to be in perfect condition.

For further experiment the elevator was built up to its present height of 125 feet in the spring of 1900. Curiously, it was never used commercially to store grain after these early experiments. Shortly afterward Peavy's company undertook to build a large concrete and steel mesh elevator at Duluth, which was designed by Haglin. The use of reinforced concrete for making grain elevators was well on its way.

The circular concrete form explored by Peavy and Haglin soon prevailed throughout the Midwest. These structures were pronounced by the architect Le Corbusier to be "the magnificent First Fruits of the new age." Peavy and Haglin's prototypical structure still stands at the southeast corner of the intersection of Highways 7 and 100 outside Minneapolis, near the tracks of the Chicago, Milwaukee, St. Paul, and Pacific Railroad. In 1969–1970 engineers employed by the present owner investigated the base of the elevator and found timber pilings that were deteriorating. The foundation was replaced with steel-reinforced concrete to a depth of 10 feet to prevent sinking. The structure seems destined to tower above the landscape for some time to come. It was placed on the National Register of Historic Places in 1982.

The Ingalls Building

The world's first reinforced-concrete frame skyscraper was built in Cincinnati, Ohio, in 1903. The sixteen-story, 210-foot-high Ingalls Building (today renamed the ACI Building) more than doubled the height previously achieved in reinforced-concrete structures. The art of high-rise construction, pioneered in the American Midwest during the last quarter of the nineteenth century, had heretofore been closely linked with the development of steel framing technologies. Reinforced concrete, however, was chosen for the Ingalls Building by its architects, Elzner and Anderson of Cincinnati, for two primary reasons: its cost savings compared with steel framing and its fireproofing advantages. The building's construction proceeded in 1902–1903 amid speculation about its imminent collapse. Many people could not believe a structure this tall made of material poured into temporary wooden molds would ever stand. A number of engineers predicted failure occasioned by concrete shrinkage or wind loading. One newspaper editor is rumored to have spent an entire night watching the partially completed building, hoping to scoop the story of its collapse. Nevertheless, after eight uneventful months of construction the building was successfully completed. It has served admirably to this day, long since quieting even the most doubting critics.

Brought together in the design of this structure was a body of diverse theories developed independently in Europe and the United States regarding the properties of metal-reinforced concrete. The Ingalls Building first synthesized in one project developments in reinforcement theory upon which the success of high-rise concrete frame structures was, and remains, wholly dependent. Together with the landmark Ward House in New York, erected in 1873–1876, the Ingalls Building of Cincinnati brought the United States to the forefront in the design of reinforced-concrete building systems.

Whereas modern concrete reinforcement perhaps began in 1848 with Josef Lambot's iron-reinforced, poured-concrete boat, the scientific principles underlying the utilization of steel for tensile reinforcement were not to

For the reinforced-concrete columns Hooper displayed considerable design skill as well. Column lines appeared in the floor plan at three locations, within the building's two longitudinal walls and dividing the interior space into unequal bays. The rectangular columns, 30 inches by 34 inches for the first ten floors and 12 inches square thereafter, were tied to the floor girders by means of concrete haunches. These haunches contained steel bars, bent to a 45-degree angle extending outward and tied into each girder's reinforcement. The haunches resisted the diagonal punching shear stress induced at the slab-column connection. By this monolithic construction Hooper effected a rigid joint between the horizontal spanning and vertical support systems, increasing the building's resistance to lateral forces via frame action.

Within the columns themselves Hooper placed two vertical rows of five square twisted bars, to resist bending stresses transferred to the columns by the diaphragm action of the floor slabs. Following Considère's innovation of several years earlier, the parallel rows of tensile reinforcement were bound together with hoops, thus resisting the outward force induced by the buckling actions of columns under compressive load. Additional compressive reinforcement, in the form of four smooth rounded bars, bound together by hoops and diagonal ties to resist buckling, was placed in the first ten floors of columns. Hooper's detailed knowledge of the latest advances in reinforced-concrete theory was evident throughout. He here used smooth rounded bars rather than Ransome's square twisted shape. Above the tenth floor Hooper figured the decreased column loadings could be carried entirely by the concrete, and thus eliminated the compression reinforcement.

For the building's exterior Elzner and Anderson conformed to the stylistic dictates of classic revivalism, in its heyday in turn-of-the-century America. The lower three floors were clad in Vermont marble, the upper twelve in enameled gray brick. The attic floor and cornice, the latter hung from the roof slab cantilevered 5 feet beyond the building's walls, were faced with hollow-tile terra cotta. The facing materials were attached to the concrete frame with wrought-iron anchors that were cast into the concrete, bent upward after the formwork was removed, and grouted into the cladding material. Additionally, at each floor level the slab was poured with a 3-inch lip extending beyond the frame; the marble, brick, or terra cotta was then laid directly upon this ledge and secured to the wall by the metal anchors. One-inch face tiles were then secured to the concrete lip edge to create a continuous facade surface.

Construction of the Ingalls Building, excluding the outfitting of rented office space, was accomplished in an eight-month period that ended in the spring of 1903. With the exception of concrete-mixing and hoisting machinery, designed by the Ransome Concrete Machinery Company of San Francisco, the structure was built essentially by hand. A labor force of sixty men was involved, pouring each monolithic floor slab as a continuous operation in order that joints be avoided except where planned. Columns were poured a level at a time, although the steel reinforcement was carried above each floor level, in order that rigid joints between floor slab and column be created. Reinforcing bars within the columns were spliced together by wrapping lengths of smaller bars around each splice or by means of pipe sleeves, into which each bar end was grouted. The cost of the building, again excluding interior office finishes, came to $400,000, quite likely less than a steel-frame structure would have run.

Cincinnati's Ingalls Building introduced high-rise reinforced-concrete frame construction into the United States. Standing at the end of an epochal period of theoretical development, it brought together the diverse strands of concrete-reinforcing technology in one project that revolutionized concrete building construction around the world. The techniques and principles pioneered by Elzner and Anderson and Henry Hooper were so effective that many of the basic methods of reinforced-concrete frame construction practiced today remain little changed from 1903.

The Goodyear Airdock

The Goodyear Airdock in Akron, Ohio, was built in 1929 to house U.S. Navy dirigibles. It is among the largest buildings ever constructed, containing over 55,000,000 cubic feet of interior space and more than 364,000 square feet of uninterrupted floor space—for many years more than in any other building in the world. Its design involved some of the first studies in the aerodynamics of buildings.

The airdock is 1,175 feet long, 325 feet wide, and 211 feet high. It is as tall as an eleven-story building and shaped approximately like a semiparaboloid in section, thus resembling a huge quonset hut with rounded ends. The building was specifically intended by the Goodyear-Zeppelin Corporation to house the USS *Akron* and the USS *Macon,* two 785-foot dirigibles, but was designed to hold airships considerably larger than these.

The aerodynamic shape of the structure was extremely important in its design for several reasons. First, the pressure and suction forces associated with wind blowing against and around it are extremely large in a building of this size. In addition, the design had to ensure that the launching and docking operations of the big airships were not complicated by unusual air currents created by the presence of the building itself or by the opening and closing of its great doors. Observations at a hangar at Lakehurst, New Jersey, for example, indicated that the currents around the sliding doors might attain a velocity twice that of the prevailing winds. This often prevented the launching or docking of an airship.

Before starting the design of the full-scale building, Dr. Karl Arnstein, director of engineering of the Goodyear-Zeppelin Corporation, conducted extensive tests on a model in a wind tunnel at the Guggenheim School of Aeronautics of New York University. These tests demonstrated the superiority of the shape he had devised, both with regard to the magnitude and distribution of the forces created by the action of the wind on the structure and with regard to minimizing the interference of the building with normal air currents around its ends.

The general design of the airdock was carried out by the firm of Wilbur Watson and Associates, along the lines proposed by the Stress Analysis Department of the Goodyear-Zeppelin Corporation. The structure consists of eleven parabolic arches spaced at 80 feet on center and connected by a system of horizontal and vertical bracing trusses placed between the upper and lower chords of the arches. These trusses also carry light trussed rafters spaced 10 feet apart, on which are placed Z-bar purlins spaced 8 feet apart. At each end of the shell are placed two diagonal arches that meet the end arches. Only the center arches have fixed shoes; the balance are supported on rockers, allowing the structure to expand and contract from the center to the ends. The doors are built up of similar arched and braced ribs.

The substructure consists of concrete footings for the arches, carried on vertical and inclined concrete piles driven to rock, and concrete ties that were laid across the building on sand and clay after the muck was removed. These heavily reinforced ties carry the horizontal components of the thrusts from the great arches. Concrete door circles for supporting the rails carrying the doors were also placed on concrete piles. About 1,300 McArthur-type piles were used.

The engineers of the erection department of the American Bridge Company developed the erection plan for the steelwork. The technique consisted of erecting the lower sections of an arch—about 100 feet in height—upon temporary bents and then assembling the center portion of the arch on the ground in a cradle, afterward lifting it into position by means of counterweights carried on the side sections already erected. This process involved the use of eight lines of railroad tracks, six inside the building and two outside. Seven locomotive cranes were used on these tracks, two with maximum reaches of 125 feet.

The erection bents for supporting the side sections consisted of two steel columns, each supported by a 300-ton jack at its base. The function of the jack was to spread the tops of the lower sections enough for the center section to pass by them and then to effect the

Goodyear Airdock, Akron, Ohio. This interior view shows three U.S. Navy dirigibles housed inside the structure, which was completed in 1929. For many years this building was the world's largest in terms of uninterrupted floor space, with over 364,000 square feet. The design of the structure involved pioneering studies in the aerodynamics of buildings, as well as many unusual engineering innovations. (Courtesy of the American Society of Civil Engineers)

closure between the side and center sections. At the tops of these columns were sheaves for carrying the counterweight boxes.

Erection of steel started on April 29, 1929. The first sets of arch sections were raised into position on May 22. The hoisting and connecting of each arch took about six hours. All the arches were in place by November 25.

The doors were marvels in themselves. Looking something like the peel of half an orange section, each door was held with a pin at its pointed top and rested on rollers at its base. Each was 202 feet high and 214 feet wide at the base. A geared, rack-drive system, connected to a 125-horsepower motor, was installed for the specific purpose of moving them. The doors could be opened and closed in five minutes.

Today, the building is still an imposing presence. The last dirigible constructed there was a U.S. Navy ZPG-3W in 1960. The airdock has since been used for various research, production, and development purposes.

Chapter Ten

Urban Planning

Introduction

Increasingly, the American city has come to depend on technology to accomplish its vital functions. That the civil engineering profession has influenced the shaping of American cities is without question, but precisely identifying the multiple and often indirect means by which this influence has been exerted is not easy. In some instances, civil engineers have been engaged in direct planning activities (such as the establishment of street patterns and other urban features) that have fundamentally and decisively characterized urban environments. Sometimes these direct planning activities were visionary; on other occasions they were pragmatic responses to immediate problems. In many other instances, analyses of urban problems based on function and economics have determined more indirectly the form of a city and its infrastructure systems. Specific engineering activities involving urban transit, maritime and hydraulic engineering, power, public health, and other areas have all contributed to the shaping of our urban environments. Technology, through engineering, made it possible for our cities to grow, and the development of urban society, in turn, encouraged the advance of technology.

Direct planning activities have a long history in this country. Planned new communities date back as far as the building of the town of Sudbury, Massachusetts, in 1638. In the same year New Haven was founded on a nine-square plan of enormous blocks, with the central square set aside as a common. The orthogonal grid towns of Philadelphia, laid out in 1683 by Thomas Holme, and Savannah, designed in 1733 by James Oglethorpe, clearly convey some of the technical concerns of their designers. Both made extensive use of squares. The Philadelphia plan—characterized by a uniform street plan, uniform setbacks and building placements, and uniform open spaces—shows the influence of those of many European cities; Holme himself was a frequent European visitor. Oglethorpe's remarkable plan of Savannah, however, may well have had its origins in layouts for military encampments used earlier in Europe. Through its European lineage, the profession of civil engineering itself in America has many of its antecedents in military engineering.

Not all early cities adopted the grid approach. Annapolis's plan was based on a square and two large circles with radiating diagonal streets. It was developed in 1695 by Francis Nicholson, who produced Williamsburg's plan four years later. The plan for Washington, D.C., also reflects a carefully considered departure from the grid approach to street layout. Pierre Charles L'Enfant, the French military engineer who has been credited as its author—with help from the famed surveyor Andrew Ellicott—was a prominent early practitioner of urban planning who had received formal professional training. Many others, often surveyors, had had little in the way of education.

Despite interesting variations, the general planning attitude based on the grid and square prevailed in urban areas. With time, however, increasing land values in some cities caused a premium to be placed on intensive land use, leading to the demise of many existing squares once reserved for open space and to the elimination of such spaces in new plans. The plan adopted by New York City in 1811, for example, paid little attention to topography, except in the upper part of the island, and included few open spaces. The grid approach has remained a staple of American city planning and was particularly reinforced by land-grant and related surveying practices in the development of the Midwest and West.

The forms our cities have taken also owe much to the work of the engineers who developed systems for water supply and control, urban transportation, and power. The planning of these systems not only responded to the city fabrics already present but influenced future developments as well, in either anticipated or unanticipated ways. Examples of public works that have stimulated urban growth may be found wherever one looks. The rapid expansion of New York City in the nineteenth century is inconceivable without the building of the Croton water supply system—which was a case of foresighted urban planning. Many urban features now considered amenities also resulted from such public works accomplishments—such as Philadelphia's Fairmount Park, established to preserve the purity of the Schuylkill River as part of the water supply system developed by Benjamin Latrobe. The impacts of urban transit systems—trolley cars, subways, elevated railroads—have been profound. The location and layout of many of Boston's now prominent suburbs, for example, cannot be fully explained except in connection with the routes of nineteenth-century streetcar lines. Many times, as in the case of the development of tenement areas in upper New York City as a result of the subway routes selected, the consequences for urban form were either simply unanticipated or different from what was anticipated.

Most of the civil engineers of the nineteenth century who influenced urban environments did not think of themselves as acting in the specific context of city planning, as we employ the term today; and they certainly did not cast themselves as practitioners of the professional discipline now given that name. The "engineer as city planner" is a fairly recent phenomenon—indeed, the profession of city planning as it is now conceived did not really emerge until the beginning of the twentieth century. The concept of city planning, of course, is quite old, but it was only in the twentieth century that professional organizations of planners arose and universities and colleges began offering formal training in the field. Influential in the development of the profession were the early-twentieth-century municipal reform movements, especially those that dealt directly with the physical problems associated with urban growth. The planning and zoning ordinances that grew out of these movements also stimulated the growth of the planning profession.

Engineers and architects were attracted to the fledgling city planning field and became influential groups within it. Many civil engineers began specifically identifying themselves as "municipal engineers" during this period. Nelson P. Lewis, a municipal engineer involved in city planning in New York City during the progressive era, stated in his presidential address to the Municipal Engineers of New York in 1904 that "municipal engineering in its broad sense does not only apply to the construction work incidental to sewers, pavements, docks, water supply, bridges, lighting, transit, parks, etc., but it should take account of all the structures which go to make up the modern city and the physical characteristics which make it beautiful or ugly, healthy or unsanitary, dignified or trivial, stimulating to our civic pride or prompting apology to our visitors." Lewis himself had been appointed secretary to a commission established by the New York Board of Aldermen in 1903 to report on a comprehensive plan for the city; he went on to become a superb example of the civil engineer as city planner.

In the years just before and after the turn of the century, most of the direct impetus for city planning came from advocates of the City Beautiful movement, which had arisen out of the classically designed Columbian Exposition in Chicago (1893). The exposition came to be called the White City, because it stood so much apart from the dirty and cramped cities of the day. It ushered in a new interest in planning and designing cities on the grand scale—with broad avenues arranged on axes, monumental civic buildings, great plazas, and a meticulous attention to landscaping. Frederick Law Olmsted, Daniel Burnham, John Nolen, and others from the fields of architecture and landscape architecture were leaders of the movement. Engineers such as Nelson

Lewis, although they supported many of the City Beautiful aims and were often involved in its causes, took issue with the basic approaches used and with what they perceived to be the limited scope of the movement. They continued to view the solution of technical problems as fundamental to planning activities and consequently assigned priority to transportation, lighting, air quality, and so forth, over what they considered the purely aesthetic concerns of the City Beautiful movement—which, they argued, left too many urban problems untouched. George Swain, a professor of civil engineering at Harvard University, declared at the 1912 National Conference on City Planning that the task of the urban planner "seems to have been considered, in its inception, primarily as an architectural and sociological one" but that "as its problems become more concretely defined, it will be found to be fundamentally more and more an engineering problem." Members of other professions concerned with urban problems—social workers, housing officials, and so on—also attacked the movement as unresponsive to planning needs as they perceived them.

By 1910 the City Beautiful movement had almost run its course, to be replaced by an emphasis on the City Efficient. Practicality and reliance on technical experts were hallmarks of this new movement, and the civil engineer as city planner came into his own. Most engineers involved in planning at the time stressed economy. Many turned to city administration, for them a heretofore largely unexplored role. Engineers became the backbone of most planning commissions.

The City Efficient movement proved to have shortcomings of its own, however. Many urban issues, particularly social problems, could not be addressed by technical solutions alone. In other areas well-meant technical solutions proved counterproductive in the long run. The advent of the automobile, for example, with consequent traffic snarls and many other previously unencountered problems, was a prime city planning issue of the day. The City Efficient movement's emphasis on economy led to planning responses that involved the widening and extension of streets, often at the expense of neighborhoods and of such amenities as parks. Further, this preoccupation with responding to the automobile tended to mean neglect of other forms of urban transportation.

The 1920s and 1930s saw the maturing of the city planning profession and the establishment of professional planning schools. The development of the discipline brought with it a turning away from the extremes of the City Efficient movement toward a more balanced approach to planning. Since that period civil engineers have remained active participants in most of the country's city planning activities.

The City Plan of Savannah

In 1885 a visitor to Savannah, Georgia, wrote: "There are far vaster and wealthier cities with much more commerce and culture than this city, but for architectural simplicity and natural beauty, for the indescribable charm about its streets and buildings, its parks and squares . . . there is but one Savannah. Without a rival, without an equal, it stands unique." The two square miles of old Savannah retain their appeal for the twentieth-century visitor. It is a town of elegant eighteenth- and nineteenth-century buildings, set on high basements, ornamented with ironwork, and approached by steep stairs that border the streets. Other eastern cities possess similar architectural treasures, but what gives Savannah its special character is its series of interconnected garden squares, heavily planted with live oak and magnolia. This graceful solution to the problems of urban life must be credited to the plan for the city that was made by its founder, James Oglethorpe, in 1733. For nearly 120 years his unique Savannah plan was conscientiously followed as the city expanded from 114 colonists and six squares to more than 14,000 residents when the last garden squares were laid out in the 1850s. Because of the plan and the adherence to it, Savannah holds a distinguished place in the history of city planning.

Savannah, Georgia. Peter Gordon's drawing of 1734 shows the early stages of execution of James Oglethorpe's 1733 plan, which has produced a city internationally known for the beauty of its neighborhood squares. (Library of Congress)

Of his siting for what was to be the southernmost English colony in the New World, Oglethorpe wrote in 1733:

> Went myself to view the Savannah River. I fixed upon a healthy situation in about ten miles from the sea. The river here forms a half-moon, along the south side of which the banks are about forty foot high, and on the top flat, which they call a bluff. The plane high ground extends into the country five or six miles, and along the river-side about a mile. Ships that draw twelve foot water can ride within ten yards of the bank. Upon the riverside in the center of the plain I have laid out the town.

Oglethorpe thus also claimed credit for the city plan.

Some writers have effectively argued that in locating Savannah Oglethorpe conformed to suggestions for siting cities in coastal areas contained in a manuscript expounding Vitruvius's rules for the location of cities. His possible familiarity with these rules, however, does not account for the unique street plan that was developed and implemented. Peter Gordon's famous 1734 drawing of the city shows both the plan used and the repetitive wooden houses built to conform to Oglethorpe's stipulations. Some eighty buildings are shown, including houses, a public store, and a mill.

The basic planning unit, the ward, was 675 feet square and was initially repeated six times. In each of the four corners of a ward were two lines of five town lots. Each corner set of ten lots constituted a tithing. At the center of each ward lay its square. In each direction streets linked the squares at their midpoints. Another street separated the city into north and south wards. The two lines of town lots in each corner of a ward were separated by a narrow lane—arguably the earliest colonial example of alleys, which were later added to Philadelphia's plan and eventually adopted throughout the country.

The central settlement was but one piece of the entire plan. In addition to his town lot, every male emigrant—each of whom was approved by the trustees and sent to the colony at public expense—was given a 5-acre garden plot beyond the commons belt that wrapped the town, and a 44.9-acre farm farther out. In return for the land and his transport to the colony, each emigrant was expected to stay a minimum of three years and to plant 100 mulberry trees on each of the 44.9 acres. These trees were to be used to nourish the silkworms for the industry that was intended to support the colony. Beyond the town's land were a number of villages. Each was—according to Oglethorpe's friend Francis Moore, writing in 1736—associated with one tithing in the town, so that "in case a war should happen the villages without may have places in the town, to bring their cattle and families into for refuge, and to that purpose there is a square left in every ward, big enough for that outwards to encamp in."

Moore's account suggests a possible utilitarian reason for the existence of the squares. His comments may have been made, however, to calm would-be colonists' fears about the safety of the colony, situated so precariously in a largely unsettled territory. More also points out that "houses are built at a pretty large distance from one another for fear of fire: the streets are very wide, and there are great squares left at proper distances for markets and other conveniences."

The original settlement contained only six squares. For a time during the middle of the eighteenth century Savannah was a palisaded city. The well-known W. G. deBrahm plan of 1757 shows it surrounded by a wall with six gates. Growth was slow, and few visitors had kind words about Savannah or could even perceive its symmetrical design because of the few buildings present.

The city was incorporated in 1789 and began prospering. The close of the eighteenth century coincided with the beginning of a new era for Savannah. Nearby, a young New Englander, Eli Whitney, produced a working model of the cotton gin, foreshadowing the reign of King Cotton. It was during these years that the city began to concern itself with civic amenities. Trees were planted and the squares defined by barrier chains. Following the War of 1812 prosperity came to Savannah in a clear way with the era of the "merchant princes" and their handsome residences. A Rhode Islander, Isaiah Davenport, built many finely crafted buildings. The young Englishman William Jay imparted a sense of architectural elegance to the city, by now becoming a place of brick and stone.

The antebellum years were good to Savannah. Handsome churches and residences were built. It was in the 1850s that the last of its 24 squares were laid out and the final extension of Oglethorpe's original plan completed. The city was largely spared the ravages of Sherman's conquering army during the Civil War but endured some hard economic times afterward. Today, although it has continued to expand, Savannah retains its particularly humane character.

The fact that Savannah's expansion into the 1850s followed Oglethorpe's 1733 plan is remarkable. Given the success of the plan, questions naturally arise about its origin. Was it Oglethorpe's own inspiration and accomplishment, or did other individuals or precedents strongly influence him? While it seems most likely that this particular plan was indeed the work of Oglethorpe and not of a contemporary, it is also clear that the under-

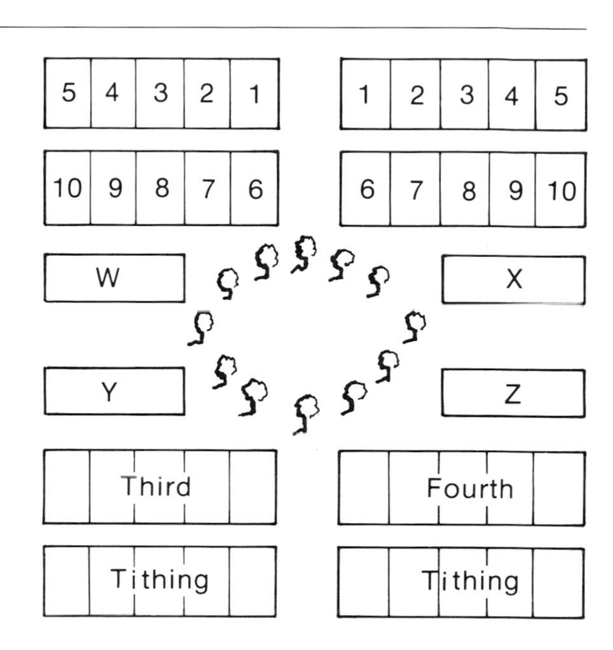

A typical Savannah ward.

lying principles derive from earlier sources.

By 1733 there were several American examples of towns laid out around one or more public squares. In 1638 New Haven was founded on a nine-square plan of enormous blocks, with the central square set aside as a common. Since there is no exact precedent for this form or its dimensions, it is generally considered an American innovation. Philadelphia, which was laid out in 1683 by Thomas Holme, William Penn's surveyor general, may not have originally been planned to include its five squares. Penn, who intended the city to become a great one, is himself credited with the addition of the squares to the gridiron pattern. He may have been influenced in this by the various plans for the rebuilding of London, particularly Richard Newcourt's, which the Philadelphia plan closely resembles. Penn, who had lived in London during the plague of 1665 and the fire of 1666, wished to plan a city that would be subject to neither. He was also perhaps influenced by the plans of the Irish new towns, with which he was familiar.

But although the planning of towns around public squares was not unknown in the colonies, and although Oglethorpe was probably acquainted with Philadelphia's layout, there seems to have been no direct link between the older colonial plans and Oglethorpe's design for Savannah. Certainly the important and unique characteristic of the Savannah plan was its regular series of carefully repeated squares. One South Carolina town, Childsbury, was laid out in 1707 with more than one square; but this plan was not widely copied, nor is it clear whether Oglethorpe even knew of it.

William Harden speculated in 1885 that the source of Oglethorpe's plan was a 1723 volume entitled *Villas of the Ancients Illustrated,* published by the architect Robert Castell, a friend of Oglethorpe's. Another suggestion is that Oglethorpe derived the plan from the ideas of other men experienced in laying out cities. One such was William Bull of South Carolina, who had replotted Charleston in 1722 based on the original plan of the 1670s and who later aided Oglethorpe in selecting and staking out the site of Savannah. There are also some similarities in the land allotment scheme suggested by William Purry for the colony of Purrysburgh, also on the Savannah River, which he was organizing. There proved to be a marked contrast, however, between the actual layouts of the two towns. Close analysis leads one away from assigning the source of Oglethorpe's plan to either Castell, Bull, or Purry—although Bull unquestionably made contributions to the development of the city, and some tenuous links can be made to Purry.

Other suggestions point to earlier precedents, ranging from the rigid military pattern of the old Roman castra to the plan of the new city of Peking. It is not unlikely that town planning experiences in Europe proved influential. The association with the military fortress towns or bastides developed in the Middle Ages in France, Spain, and Wales, for example, is certainly an inviting one in view of the similarities present—the grid plan, the walled perimeter, the restricted size, and the need to establish secure footholds. Yet there are no known direct links between the Savannah plan and these towns. Other inviting but unproven antecedents include the grid and radial plans for ideal cities developed during the Renaissance, which contained public open spaces. Although these plans have a strong visual appeal to us today, their utilitarian intent cannot be overlooked. The open spaces were ideal for the mustering of troops, and the straight and broad streets perfect for moving troops rapidly across town, giving the authorities better control over insurgencies than was possible in the cramped and crooked medieval street network.

England generally stood apart from such town planning activities. Some small new towns were established, however, in Ireland in 1611, and these were apparently known by, and may have been of significance to, some early American colonial leaders. Londonderry, in particular, bears the marks of European ideal city designs, containing a central square intersected by streets at the midpoints of each of its sides.

Besides Ireland's new towns, English town planning theory was given another impetus by the great London fire of 1666. The plans submitted for the rebuilding of the city reveal the extent of the effect of European planning—they largely used orderly grid or radial schemes. A noteworthy example is Richard Newcourt's plan, which featured five open squares—the largest in the center, with one smaller square centered in each quadrant, each quadrant comprising its own parish. Traces of Newcourt's scheme can be found in several early American town plans, including that of Philadelphia, mentioned earlier. The attractions for English planners of such ordered city plans were chiefly public convenience and civic monumentality, not military control. London is still famous for its residential squares. These were extremely popular and, coming as they did at the beginning of the colonial period, undoubtedly had some influence on American town planning.

There is some evidence that Oglethorpe was aware of some of the plans proposed for the rebuilding of London, but in spite of the striking similarities between his plan and Newcourt's there is no certainty that he had specific knowledge of the latter. Surely Oglethorpe's incorporation of residential squares is not surprising, in light of the fashionableness of squares in London when he lived there in the early part of the eighteenth century. But, as persuasively argued by Bannister in 1966, Oglethorpe's views were seemingly shaped less by fashion than by his military background and the colony's defense needs. He was most likely familiar with the numerous military manuals on fortification design of his day and may well have visited some fortress towns of continental Europe. Although Oglethorpe's personal military experience was limited, his father and other family members were connected with the military. Oglethorpe was, therefore, certainly aware of the ancient practice of castrametation—the laying out of encampments for troops in the field in an undeviating, orderly arrangement that permits the army, no matter where it is located or what unexpected circumstances arise, to assemble and function efficiently. During the Renaissance, military theorists adapted the rules for castrametation to their own purposes. One late-sixteenth-century manual by the Neapolitan Cesare de'Evoli, entitled *Delle Ordinanza et Battaglie,* presents an encampment scheme for a site in open country on the bank of a stream. In this scheme an orthogonal pattern of major and minor roads defines blocks, several of which each contain a central parade and commissary square that is served by streets intersecting it at the midpoints of its sides. The four L-shaped pieces of each block bear a startling similarity to Oglethorpe's Savannah blocks. Such theoretical designs could have reached Oglethorpe via an English military manual by Robert Barret, *The Theorike and Practike of Modern Warres,* published in London in 1598. In it Barret gives a plan and instructions for an encampment sited on a river; the plan clearly resembles Oglethorpe's.

It is probable that the military theory that was an integral part of Oglethorpe's background came nicely, though perhaps unconsciously, to bear when he was given responsibility for establishing an English outpost in the New World. Although Oglethorpe altered the encampment scheme to suit his particular needs, the defensive capability of the town plan is undeniable. For example, each ward of four tithings had its constable, and each of the tithings was required to supply ten men ready to bear arms at all times. Each of the ten stood guard every fourth night. The plan served the settlement's defensive needs well, although it probably did not contribute much to appearances in the early days, since the squares were unplanted, open, muddy plots.

Fortunately, in later years when attacks by the Spanish were no longer imminent, the residents of Savannah made an enlightened decision concerning the use of the squares. In 1810 the first decoration of the squares began. They were later spotted with patriotic monuments, the trees grew luxuriant, and new squares were laid out. The inhabitants of this city thus preserved its unique configuration at first, perhaps, by neglect but later because of pride in their treasure. Only one square has been destroyed.

Savannah's plan was not widely copied, which is probably to be regretted. The traits of orderliness, scale, intimacy, and openness contained in Oglethorpe's design served both military requirements and those of urban civilization, because they were fitted to fundamental human needs. Today more than ever, the plan bears witness to Oglethorpe's genius in achieving an appropriate ordering of spaces and human relationships, which, though it served his immediate purpose, was not tied to that particular purpose for its success.

The Kansas City Park and Boulevard System

The system of parks and boulevards proposed for Kansas City, Missouri, in 1893 and substantially completed by 1915 represents a pioneering effort both to regulate city growth and to improve the present and future urban environment. The plan provided a means whereby the residential aspects of urban life could be improved within a larger framework of dividing and structuring city districts.

By the 1850s Kansas City represented an important link in the route westward from St. Louis and the Mississippi River, and the last quarter of the nineteenth century brought a commercial and industrial boom to the area. Perhaps the most visible manifestation of the growing city's prosperity was the construction of the first railroad bridge across the Missouri River, a feat completed in 1867. With success, however, came urban sprawl, resulting in unpaved streets of alternating dust and mud, an inadequate sewer system, and haphazard housing for middle- and lower-income groups. In addition, the steep bluffs on either side of the Missouri aggravated the poor housing and sanitary conditions by concentrating growth in low-lying areas.

The park and boulevard plan that was finally evolved to regulate growth and enhance city life had its roots largely in the suggestions of Kersey Coates, a Quaker originally from Pennsylvania. Coates envisioned a boulevard encircling the city as early as 1856. For many years he advocated his ideas without much support, but in 1881 William Rockhill Nelson, publisher of the *Kansas City Star,* joined the cause. Coates and Nelson succeeded in gaining additional support from various prominent citizens of Kansas City, including August R. Meyer, a German-born naturalist who had made a fortune in mines and smelting and who became president of the Park Board on its establishment in 1892.

The board hired George E. Kessler, a landscape architect born and trained in Germany, to design a park system for the metropolitan area. Kessler, educated in civil engineering at the University of Jena and in landscape architecture at the Weimar Institute, possessed a design attitude that represented a break from current European practices based on

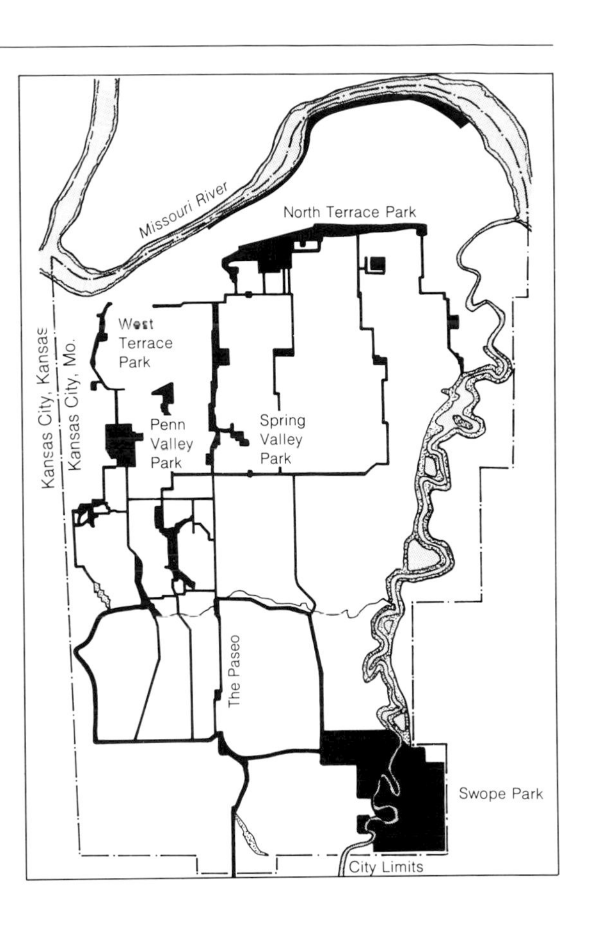

Park and boulevard system, Kansas City, Missouri. Designed by George Kessler in 1893, with major additions in 1896, this system represents a superb planning effort aimed at regulating city growth and improving the urban environment.

the axial plans and formal gardens long advocated by the Ecole Des Beaux Arts in Paris. Instead Kessler believed in integrating park development with existing land forms and structuring urban development in sympathy with the existing landscape.

The Kessler plan, submitted in 1893, contained a system of parks on the river bluffs, linked to each other and to the city's commercial and industrial sectors by a series of scenic boulevards. Three public parks—North Terrace, atop the river bluffs at the city's northern edge; West Terrace, similarly sited at the western edge; and Penn Valley, to the south—were the basic elements of the scheme. A network of boulevards, each conceived as a parkway defining residential neighborhoods, linked the three parks. Smaller neighborhood parks, interspersed so as to give access by foot for all residents, extended the park system. Finally, a formal development consisting of nine block-long parks linked to each other was sited on the principal boulevard, the Paseo, as a means of identifying the principal residential districts.

A significant addition to the original plan came in 1896, with the gift by Thomas H. Swope of 1,334 acres of undeveloped land four miles southeast of the city proper. Kessler incorporated the land into his park system through a series of additional boulevards, which served as a framework for future planned residential development to the city's south while providing direct access to the city's business districts.

The virtue of Kessler's plan lay in its successful integration of a rational street system within the existing topography. The design simultaneously defined city sectors, according to appropriate use of the land forms, and established a system of roads through and between these sectors that were both scenic and direct. The effect was preservation of the unspoiled high lands, whose own features as well as the outlooks they provided were striking, for residential development. The bluffs themselves, affording the most spectacular views, were preserved as parkland for access by all. Commercial, industrial, and transportation development was contained within the low-lying areas, adjacent to water and rail links and sited on land appropriate for this use. The boulevard system further subdivided

West Terrace Park, one of the major features of the Kansas City park and boulevard system. The photograph dates from 1908. (Courtesy of the American Society of Civil Engineers)

residential areas into neighborhood units, accentuated by small parks, and provided a system of scenic links to the city's other sectors.

In the absence of any zoning regulations, Kessler's plan served to discourage the development of new congested residential districts, a phenomenon Kansas City and all major American cities suffered during the nineteenth century in periods of rapid industrial expansion and growth. Furthermore, the division of the metropolitan area into "suggested" sectors enhanced the attractiveness of established commercial, industrial, and transportation zones to prospective businesses and prevented their siting in residential districts. The consequences of Kessler's 1893 plan are evident today, as it has been and continues to be the principal urban design tool regulating Kansas City's growth.

Specific engineering advances were also achieved in the implementation of Kessler's plan. Some of the first specifications for pavements, curbs, gutters, and walks were produced in the course of construction. Innovative practices were employed at the Penn Valley Park site in the design and building of earth dams, subsurface drains, and an impoundment lake to control erosion. The steep slopes of the river bluffs at the West Terrace and North Terrace parks required unusual designs for both retaining walls and drainage provisions. With the advent of the automobile the boulevard system was repaved in bituminous macadam pavement, at an early stage in the development of paving systems. Moreover, the legal and engineering provisions made for future boulevard widening anticipated the auto age, ensuring the plan's continued use into the present day.

The Kansas City plan had a nationwide impact. Its success, pronounced by both the National Conference on City Planning in 1914 and the American Institute of Architects in 1917, also ensured George Kessler's future. In the role of landscape architect and city planner he served on planning commissions in Dallas, Wichita Falls, Oklahoma City, Memphis, Cincinnati, Indianapolis, St. Louis, and Mexico City. Kessler's masterwork, however, is still the pioneer project at Kansas City, which helped to shape the character of the western American city.

The Charles River Basin

Boston is among the most beautiful of American cities—a fact attributable in large part to the presence of the Charles River Basin, which separates Boston and Cambridge. The broad fresh-water basin provides not only a focus for the city but a recreational resource as well. Sailboats and crew boats use its waters; Boston's nineteenth-century fabric of brick buildings, softened by Storrow Park, flanks one side, and the campus of MIT overlooks the other. The stately Longfellow Bridge, with its arches and towers, crosses the basin and defines it. This incalculable urban asset is the result of the damming of the Charles River, completed in 1910, to flood its tidal flats.

Prior to 1903 salt-water tidal inflows from Boston Harbor filled the estuary of the Charles River, alternating with outflows that left broad, unsightly, and often malodorous mudflats along each shoreline. By the late 1800s most of the marshy territory adjacent to the river had been filled in and built upon, bringing ever-increasing portions of the populace face to face with twice-daily displays of estuarial muck and leading to growing public pressure for a dam across the Charles that would form a basin of sufficient depth to continuously cover the flats. In fact, a dam had been proposed as early as the beginning of the nineteenth century. Various more recent proposals favored half-tide dams, full-height dams, a fresh-water basin, a salt-water basin, and several possible dam locations.

In 1891, at his inaugural, Boston's Mayor Matthews spoke of "the opportunity for making the finest water park in any city in this country . . . an imitation of the plan adopted by the city of Hamburg. . . . We should dam up the stream . . . and lay out a series of boulevards along the basin thus created." A group named the Committee on the Charles River Dam became the primary public force favoring a dam. Its opponent was the Beacon Street Committee, which represented the shoreline landowners, primarily along Beacon Street, and local wharf interests.

The issue came to a head in 1893–1894. The Massachusetts legislature had funded an investigation of the condition of and the means for improving "the beds, shores, and waters of the Charles River between the Charles River Bridge and the Waltham Line." The conclusion of this study was in favor of a dam and a fresh-water basin. Howls of protest arose from the Beacon Street Committee, which, although motivated chiefly by commercial concerns, cited the prediction of some experts that the dam would create a malodorous lake, with huge, floating masses of excrement, unaffected any longer by daily tidal cleansing. This appalling vision took sufficient hold of the public imagination to prevent any action on the mudflat problem until 1901, when once again a background study was funded, this time under the guidance of an extremely dedicated and effective civil engineer, John Ripley Freeman.

Freeman had already had twenty-five years of civil engineering experience when asked to advise the Committee on the Charles River Dam. He had spent ten of them working on hydraulics projects in New England. He also knew many of the people involved in the basin controversy. The chairman of the dam committee was Henry Pritchett, president of MIT, Freeman's alma mater. Freeman himself was a member of the MIT Corporation. Freeman engaged many other experts in his work, including Charles T. Main, an MIT classmate, and the eminent civil engineer Hiram Mills.

Assisted by specialists in a number of fields, Freeman conducted his researches throughout 1902, assembling enough data to discount all the pseudoscientific proclamations that had fueled the opposition to the dam. The work began with new surveys of the basin area, in order to reflect recent dredgings, but eventually involved the following: the continual monitoring of air and water temperatures at selected points; the charting of the paths and velocities of currents; the mapping of channels; the measure of flow rates of sewer discharges, under normal conditions and during storms; chemical and bacterial analyses of fresh- and salt-water samples from various points and under varying conditions of water motion, oxygenation, and temperature; the inspection of harbor bottom borings with reference to potential

Charles River Basin, Boston. Completed in 1910, the damming of the river was a masterpiece of environmental engineering that converted a huge expanse of malodorous estuarial muck into a unique urban asset. In this aerial photograph from 1925, the dam and locks are barely visible beyond the Longfellow Bridge. The Back Bay area is in the foreground. (Courtesy of the Prints and Photographs Department, Boston Public Library)

structural integrity, as well as to thickness of silt deposits subject to tidal scouring; and the collection of other relevant information.

Freeman's work was remarkable for its thoroughness and for its attention to what would today be termed the probable "environmental impacts" of the proposed dam. He found, for example, that fresh water in the area was less likely to precipitate sludge and give off odors than salt water, and that the rate of pollution abatement in oxygenated water was not reduced by lack of water movement. He also found that the existing tidal flows mainly moved the same water back and forth and thus slowed the dilution of contaminants. He even found that the elimination of tidal inflows would not seriously affect air temperatures in the area, thus responding to opponents' fears that there would be a loss of cooling influence on hot days associated with influxes of sea water.

In sum, Freeman demonstrated that the dam and basin would in actuality function positively against pollution, malaria, and storm flooding and would have no substantial effect on summer temperatures or the harbor channels. Freeman was thus able to raise the quality and broaden the scope of the investigation to such an extent that his eventual support for the dam was based on rock-solid scientific refutation of the objections raised by various dam opponents. His report was made available to the committee and thence to the legislature in early 1903.

Freeman's final recommendations included many specific suggestions. He recommended that the dam be built so as to create a basin in the lower Charles River and that it be built at the site of the aging Craigie Bridge, so that the dam and the much-needed new bridge, by being virtually one and the same structure, could render substantial benefits at reduced costs. His original plans also included a park area. Further, he recommended that the dam be full rather than half height to take advantage of fresh water's greater antipollution effect, as well as to maximize recreational boating safety; and that a constant basin level be established to protect adjoining properties from high-tide storm sewage overflows and subsequent cellar floodings and to expedite navigation in the basin area irrespective of tide. Other recommendations were that marginal conduits approximately 16 feet in diameter be placed near the shorelines, terminating via tide gates below the dam (in order to further protect the basin from any excessive discharges from contiguous waterways—the Fens basin, the Broad Canal, the Lechmere Canal, and various sewer lines); and finally that various other ancillary measures be undertaken to help the basin exert positive effects on the entire surrounding system of waterways.

The projected costs of the dam and basin were only slightly higher than the costs of providing similar benefits by any other means, such as heavy dredging of the estuary. By filling it with water, such dredging could be limited to the amount of material required to build shore embankments. Because fresh water had been found to be preferable for pollution control, expensive tidal sluices were unnecessary. Most important, because the dam would incorporate a replacement for Craigie Bridge and could be built for about the same cost as simply replacing the bridge—which had to be done in any case—the basin represented a substantial bargain.

A large lock to facilitate commercial shipping was included in the plan. Given that the railroads were taking over most of the loads, this appears to have been done at least partly to mollify the Beacon Street Committee.

Enabling legislation having been passed in 1903, construction began under the able direction of Frederic P. Sterns. The project was completed in 1910. Its chief component, the Charles River Dam, included a roadway, a large lock for the passage of boats, sluice gates, and overflow conduits. A large recreational area was part of the dam complex design.

Studies were made for dams of various widths, from one a little greater than 100 feet wide to one having a park of several acres on the upstream side. It was decided to continue the 488-foot width required by the lock nearly across the river. This made an area to be filled in the river of about 6.9 acres, of which 5.7 acres would be used for park purposes and 1.2 acres for the roadway.

The first contract for construction was signed on January 14, 1905, and included, among other things, the lock, sluices, harbor and basin walls, and the earth filling between. The award was made to the Holbrook, Cabot, and Rollins Corporation of Boston. Later contracts were let for the dam, pile driving in the Broad and Lechmere canals, sections of the embankment, and other work. In order to construct the lock, a cofferdam was built enclosing about 3.5 acres. The water was then pumped out and piles driven to receive the concrete. By January 1, 1907, a total of 9,969 foundation piles had been driven.

A critical part of the construction was the creation of a shutoff dam capable of acting quickly to change the water level in the basin. This device was necessary because the swift flow of the tide and of the river would otherwise prevent earth fill from being deposited in such a way that it would remain in place. The shutoff dam was composed of six rows of round piles, cross-braced, running across the river. Six-inch yellow pine sheeting was driven between the middle rows and cut off slightly below mean low water in Boston Harbor. Earth was eventually filled in on both sides of the sheeting and surrounded by riprap to prevent scour. The driving of the piling was difficult, because of the swift flow of the river, but it was accomplished successfully.

The method used for shutting off the river involved a series of gates. There were 82 of them, running in grooves built on heavy uprights. The gates were 6-inch by 8-inch timber, braced lengthwise and diagonally, forming solid pieces, 10 feet wide by 15 feet high. Across the top of the uprights were placed timbers, and the gates were held to these by ropes. On October 20, 1908, the ropes were cut, and in 7 seconds all of the gates were dropped into place. Rubber hose had been nailed to the bottom edge of each gate so that the joint between the gate and the sheet piling might be as close as possible. As soon as the river was closed off, the uprights projecting above the round piling were cut off and removed, and large shovels and dredges began at once to pile earth against the shutoff. This filling was continued until the structure became an earth dam with a wooden core wall. The finished dam consisted of two concrete retaining walls (faced with granite) with earth fill between.

The discharge of the river was provided for by eight sluices, each 7.5 feet wide by 10 feet high and controlled by a sluice gate operated by electricity. The roofs of the sluices and the lock for small boats and launches were made of concrete reinforced by steel beams.

The lock consisted of a reinforced-concrete structure with expansion joints, placed on piles and including the foundations for a Scherer rolling lift bridge at the roadway. Also included were recesses into which the great steel lock gates are drawn when the lock is opened, bollards, and electrically operated capstans to aid vessels passing through the lock. The lock gates were made of structural steel beams, plates, and girders and were supported on two 4-wheel trucks running on heavy steel rails at right angles to the center line of the lock. As the gates were to be operated in winter, they were heated by steam pipes, where necessary, to keep ice from forming against their sides and between the bearing surfaces.

The Charles River Basin was an immediate success. Some repairs and alterations have been required over the years. In 1955 a hurricane caused flood flows exceeding those designed for, leading to construction of a new dam that was dedicated on May 24, 1978. The new dam is located a half-mile below the original one. Three smaller locks, instead of one large one, brought increased passage efficiency through the new dam, but craft still have to clear the bottleneck created by the old dam's lock. A major improvement consisted in the new dam's contribution to water quality in the basin, which had suffered increasing salinity due to sea water intrusion through the lock. Because this salt water sinks to the bottom, it tends to stay in the basin. The dam included a pumping station drawing from 20 feet below the surface, thus reversing most of the saline intrusion at the locks. In addition, bubblers were installed upstream to promote the flushing of deep saline pockets and to destratify the basin water, with consequent water quality improvement.

The Charles River Dam was the chief component in the construction of the Charles River Basin. This photograph, taken by H. Shattuck in 1908, shows one of several pieces of floating construction equipment used during the making of the dam. The view is from the upstream side on the Boston end. (Courtesy of the Prints and Photographs Department, Boston Public Library)

This photograph from 1906 provides a general view of the construction of the lock, showing the pile and concrete foundations used. In order to construct the lock, a cofferdam was built enclosing about 3.5 acres. The Charlesbank recreation grounds and the old river wall are on the left. Cambridge is in the distance. (*Harvard Engineering Journal,* 1907)

Increased storm sewage overflows were addressed by a new chlorination and detention plant between the two dams.

It is a measure of Freeman's achievement that water quality in the basin has only recently been considered below desirable standards, following a half-century of increasing pollutive strain. The quality and depth of his pioneering investigations swept objections firmly aside and helped provide Boston with an urban jewel comparable to Frederick Law Olmsted's "Emerald Necklace," which rings the city with some of the loveliest waterways and parklands in the country. Though less informed by the older Olmsted's nostalgic sense of the landscape, which was brought to bear on the design of Boston's Fenway, than by scientific and commercial concerns, the basin nonetheless contributes to Boston's environment at every level, from pollution control to the provision of a beautiful river park, a recreational resource of irreplaceable value, and a source of civic pride.

Chapter Eleven

Power Systems

Introduction

Although early Spanish explorers and settlers in the Southwest introduced hydropower techniques directed at utilizing the area's sparse water resources for irrigation, it was the colonists of New England who first made extensive use of water power—the continent's most significant and obvious energy source. The rivers and streams of New England offered settlers a vast supply of cheap power. Simple waterwheels on both vertical and horizontal axles, familiar to these settlers from their homelands, were constructed to power sawmills, gristmills, and powder mills, soon after settlements were established. As the pattern of towns and villages spread westward over the following two hundred years, mills—both for agricultural and for industrial functions—played a vital role in these settlements.

The nineteenth century brought industrialization to America, spurring major advances in the application of hydropower, principally in the northeastern states. The complex at Paterson, New Jersey, started in 1793, is a fine example of the use of hydropower. Another is the Merrimack Manufacturing Company of Lowell, Massachusetts, where engineers and inventors achieved a number of technological breakthroughs, including the introduction of breast-back and pitch-back wheels of cast-iron construction, 10 to 20 feet in diameter, to replace wooden predecessors. These were in turn replaced by even more efficient turbines. Equally important, the Merrimack Company, in the vanguard of nineteenth-century hydropower technology, established an "engineer of the corporations at Lowell," thereby acknowledging the need for continual technological development. This post was soon filled by James Francis, whose innovations, achieved in collaboration with Uriah Boyden, also of Lowell, advanced the technology of hydropower deployment to a level of precision beyond that of any other industry in America at the time. Their development of the turbine to replace the great breast wheels was a significant step forward. Francis's seminal publication of 1855, *Lowell Hydraulic Experiments,* is regarded today as a classic in the field of hydraulic science. At a time when skilled labor was in short supply, in a young and largely undeveloped country, and with virtually no tradition in engineering science to follow, Francis and Boyden achieved unprecedented success in exacting scientific research and technical precision.

The discovery of gold at Sutters Mill in 1848 and the California gold rush of the following year stimulated innovative developments in hydropower technology as well. The achievements of Francis and Boyden were of little use to gold miners in the Sierra Nevada. The requirements for hydraulic mining were in marked contrast to the needs of the great mills of the East. The power of the fast-running mountain streams of the West had to be made to serve the extraction techniques of fluming, digging shafts and drifts, and panning, which required base camps incorporating sawmills, forges, grinding machines, and stamp mills for crushing the mined ore. Innovations included the invention of a pressure nozzle in 1857; the creation of the hurdy-gurdy wheel, which was predominant during the 1860s and 1870s; and the introduction of the even more efficient Pelton impulse wheel in 1880.

The establishment in 1882, in Appleton, Wisconsin, of one of the world's first central-station hydroelectric plants, serving a network of private and commercial customers, marked a breakthrough in the use of the timeless energy resource of hydropower. Electricity almost immediately came into widespread use for a variety of applications, including lighting. Before the advent of electricity, lighting for buildings and streets had come first from various forms of oil lamps and then from illuminating gas, the latter developed in the late eighteenth century by heating coal and drawing off the volatiles. Gas lighting was installed in Newport, Rhode Island, in 1812. Baltimore followed in 1817, and soon gas lighting appeared in many other cities—New York (1823), Boston (1828), and New Orleans (1832) among the earliest. The use of whale oil increased as well, and after 1859 kerosene became a widely used fuel for rural lighting. Natural gas was first used in Fredonia, New York, in 1821. But it was electricity that really altered the face of society. Twelve carbon arc lamps were erected in 1877 in Cleveland; San Francisco adopted arc lights in 1880; New York installed them on Broadway in the same year; and Philadelphia used them in 1881. Edison's development of the first practical incandescent light in 1878 was, of course, a turning point in the evolution of electric lighting, and George Westinghouse's development of alternating-current transmission was an event of comparable importance to the distribution and utilization of the new form of power. New York City placed a steam-powered central generating plant in service in late 1882. Soon there were many electric companies in New York and other cities, providing power for manifold uses. Major advances in the transmission of electrical power were achieved in 1889, with the activation of the network from the falls at Oregon City to Portland, Oregon, and in 1892, when the network developed by the Folsom Water Power Company, which supplied Sacramento, California, with power over a 22-mile line, began operation.

Large hydroelectric installations were rapidly developed as a consequence of the growing demand for electricity. A complex at Niagara Falls, for example, started operation in 1895. At the turn of the century the federal government finally became involved in the production of hydroelectric energy, which until then had been controlled primarily by private industry. With the funds made available under the 1902 Reclamation Act, stored waters behind several western dams became significant producers of electricity. The need for regulation and coordination in the use of the country's water power reserves soon became apparent, as conflicts developed between private and public interests. Legislation in 1905 and 1906 provided regulatory milestones. Theodore Roosevelt urged further regulation in the public interest in his message to Congress of 1908. A major battle between legislators supporting private power interests and conservationists occurred in 1912. Eastern versus western needs for water and water power also gave rise to conflict. A 1914 conference of nine western states, for example, emphasized state claims to water resources and power within their boundaries. After further contention and compromise the Federal Water Power Act of 1920 was finally signed into law. Among other things, this act created the Federal Power Commission and established as a principle a federal role in the development of the nation's water power resources. Federal sponsorship of hydroelectric projects, such as the enormous developments on the Columbia River and the Tennessee River basin, continued, amidst controversy. These were major civil engineering endeavors, of course, and among the largest public works projects ever undertaken by this nation.

The Great Falls Raceway and Power System

Great Falls raceway and power system, Paterson, New Jersey. Construction of this system began in 1793. America's first major planned industrial complex, it has been in constant use since its inception. The Great Falls of the Passaic River are shown. (Courtesy of the American Society of Civil Engineers)

The complex of raceways, dams, and buildings erected between 1793 and 1912 at the Great Falls of the Passaic River, in Paterson, New Jersey, represents the first site of planned industrial development in the United States. In December 1791 Alexander Hamilton (then U.S. secretary of the treasury), with the help of a New York friend named William Duer, founded the Society for Establishing Useful Manufactures. Hamilton's goal for the corporation was to demonstrate that American manufacturing was not only desirable but profitable—an idea he had previously expounded in detail. He advocated the development of a manufacturing community on a scale large enough to make it competitive with those of Britain. Realizing that such a venture would require an abundant source of power, he began a search for an appropriate site.

New Jersey had been decided upon as the general location for the society's activities, for political and financial reasons. Hamilton and Duer hoped to attract money from neighboring New York and Pennsylvania in order to stimulate an underdeveloped New Jersey economy. In August 1791, even before the corporation was officially in existence, Hamilton appointed William Hall and Joseph Mort to travel throughout the state in search of possible sites for the manufacturing center. Hamilton directed their attention to one location that he had visited as a Revolutionary War officer, the Great Falls of the Passaic River. The falls, 77 feet high and 280 feet wide, are formed where the Passaic River breaks through the 800-foot ridge of the Watchung Mountains. Hall and Mort reported back to Hamilton in September, and although their report lacked details concerning the cost of construction on the site, they proclaimed Great Falls to be "one of the finest situations in the world." Hamilton also had in his employ an Englishman named William Marshall, who proposed that the society make what would today be called a hydrographic survey of the area. Marshall's concerns were not only with the quantity of water available for use as a power source but also with its velocity, the amount of head, and the nature of the underlying rock. His awareness of the need for this information was uncommon at the time, and his suggestions were ignored by the society, which wanted to get construction under way as soon as possible. Duer did send Marshall to visit Great Falls with a Frenchman named Allon, but the two experienced communication difficulties. Moreover, they were under strict instructions not to question the locals about the area for fear of driving up land prices. Their expedition, therefore, hardly provided the society with an adequate survey.

The proposal that came out of the journey seems to have been drawn up by Duer and Allon alone, without Marshall's help. It had two parts: (1) a transportation–hydraulic power canal from Great Falls all the way to the head of navigation on the Passaic at Vreeland's Point, where locks would permit the passage of boats from the canal to the river; and (2) the creation of a manufacturing town at the east end of the canal, near modern Passaic, where mill sites would be located along tiers of raceways following the contours of the land.

The society, wishing to examine other possible sites, advertised in December 1791 for interested localities to furnish information on water power, the availability of building supplies, transportation, and population in their areas. At the monthly meeting in January 1792 three rivers were under consideration: the Delaware, the Raritan, and the Passaic. Duer, convinced of the superiority of his Vreeland Point plan—for which Allon had produced a mistakenly low estimate—went ahead with it. The consequences were tragic. The financial panic of 1792 destroyed Duer, and he spent the last years of his life in debtor's prison. The panic nearly ruined the society as well; but even worse, it drove off eager new investors, and the optimistic attitude of the populace became one of caution.

In May of that year the society made its decision in favor of the Passaic location and appointed a three-man committee to purchase land. The committee visited the Great Falls area, accompanied by General Philip Schuyler, Hamilton's father-in-law and a man with technical experience. Schuyler had been the first American to make use of steam-driven water pumps, which he had installed in his New Jersey mines. He was also involved with the Western Inland Lock Navigation Company, one of the predecessors of the Erie Canal. Schuyler was in all probability responsible for most of the decisions concerning the location cf the power canal. He may also have helped locate the path of the raceway. Another collaborator in the early planning stages may have been Christopher Colles, author of the 1774 plan for New York City's first waterworks. For financial rather than engineering reasons the plans for a canal were temporarily abandoned. In all, the committee purchased 700 acres for the society's operations at Great Falls.

The topography of the Passaic area presented serious engineering problems, for the river at this point cut down between two stone ridges. Bringing water from the river at a point above the falls required cutting through the rock embankment. On the other side of the rock the land fell off into a large ravine, which had to be crossed before reaching ground in which a raceway could be dug. Standing between the ravine and the buildable ground was another section of the main ridge of rock, which blocked the path of the canal to the east. If these three obstacles could be overcome, however, the work would then consist of common cut-and-fill construction.

Various methods were suggested for bringing the water across the ravine, the most practical of which seemed to be the use of wooden troughs supported by an earthen wall. In an effort to generate other possibilities, the society advertised for bids on the project, allowing bidders leeway to invent their own construction solutions. The bids that came back presented the society with a rude shock. Not only were there no new solutions, the bids were for partial work only and the estimates were very high. Dismayed, the society decided to construct the raceways itself and asked Hamilton to recommend a trained engineer who might serve as superintendent. Hamilton suggested Pierre Charles L'Enfant, a Frenchman who had received some architectural and engineering training in Europe and who had just finished a design for the streets of the new capital city of Washington.

The society appointed L'Enfant, whose first action was to dismiss the plan for a wooden aqueduct. Instead he proposed to carry the water across the ravine on a stone arched aqueduct, its canal large enough for barges and flanked by towpaths and a carriage way. The enormous structure would have been about 55 feet wide, a rival even for the Roman aqueducts. L'Enfant also proposed building a 100-foot by 10-foot reservoir at the end of the canal across the ravine, to even out the flow of the river. This last was a sensible idea, since the minimum low-water flow of the Passaic is one-fourth its normal flow and one-hundredth its maximum. From the reservoir two main raceways would carry water to the mills, one of which would be built immediately, the others later as business grew.

Although L'Enfant's plan was accepted by the society initially, it soon came to be regarded as unrealistic. History did bear out the validity of the scheme, however: a structure following it in many details was eventually constructed. L'Enfant's plan was indeed impractical for America at the end of the eighteenth century. A grand, attractive, and permanent masonry edifice was exactly what would have been built in Europe; but the Society for Establishing Useful Manufactures was in too much of a hurry, lacked too many of the skills and workers, and was without the necessary financial backing for such a project. L'Enfant, however, persisted in his views; and this attitude, combined with other manifestations of an unaccommodating personality, caused enormous difficulties. The construction site lacked overall supervision, because L'Enfant, used to the European method, where the engineer supervised only the initial stages of the project and left the construction to skilled workmen, was by temperament and training unable to provide the hour-by-hour, task-by-task direction that unskilled American workers needed. Nor did he get along well with Hall and Mort, whose proposals for the raceways had been rejected in favor of his own.

This uncomfortable situation continued until February 1, 1793, when Peter Colt of Hartford was appointed to take over the position of superintendent. Colt had financial and business experience but knew nothing of textile manufacturing or engineering. When he arrived at Great Falls, he found progress to be poor and the workmen dissatisfied. L'Enfant had left for the winter, assuming that no work could be done when the ground was frozen. Although L'Enfant was probably right, his action was regarded with displeasure by Colt and the society, since his communications with them were scanty. The society was also unhappy over L'Enfant's preoccupation with plans for grand avenues for the new town. On his return in the spring they urged him to limit his concerns to the aqueduct itself, giving him supervision of that alone and putting Colt in charge of all other parts of the project. L'Enfant and the aqueduct were now clearly on trial. L'Enfant assured the society that he could bring water across the valley to the mill during the working year, but on June 9, when he had been missing for some time and it had become clear that his goal would not be achieved, the directors dismissed him and decided to stop work on the aqueduct. They deemed the latter a necessary decision because progress had been slow and funding was inadequate. L'Enfant departed with the drawings for the raceway and the city, and it was only after some hesitation that the society paid him the remainder of his wages.

With the failure of L'Enfant's grand scheme the responsibility for a solution fell to Colt, who solved the problem in a simple and economical manner. Rather than carry water across it, Colt made use of the ravine as a reservoir; from the reservoir the water passed through a gap in the rocks into a single raceway, which fed the mill wheels. Although the use of the ravine as a reservoir cost the society 20 feet of water head that would have been available had the water been carried across the top on an aqueduct, the masonry project became a simple earth-moving task. By July 1794 the water power system was complete. A wooden dam across the Passaic, above the falls, diverted the river water into the reservoir; it then passed through the raceway into a flume, on to the waterwheel in the mill, and back into the Passaic through a drainage channel. The channel was built along the hillside, probably by cutting into the hill and then using the resulting fill to

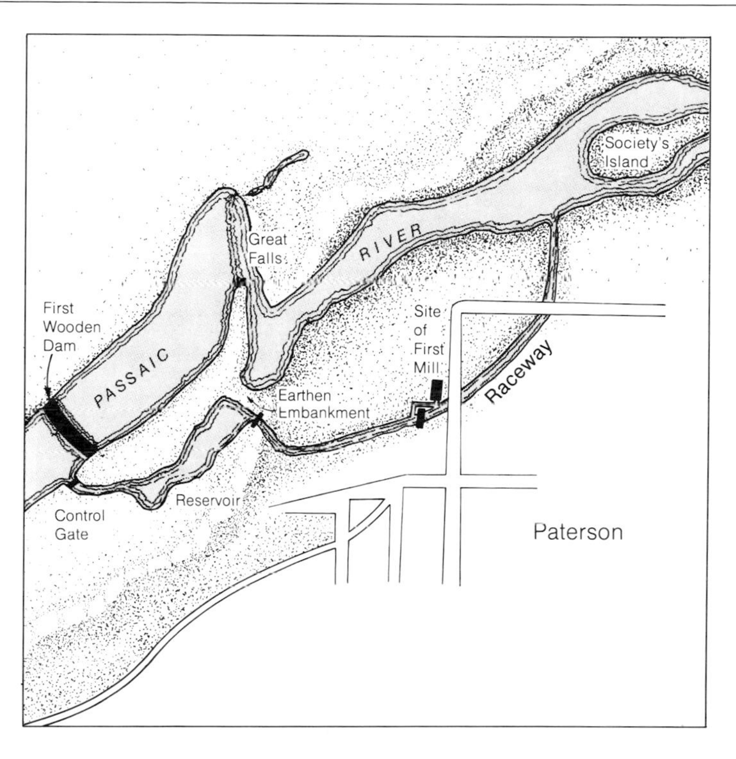

This map shows the components of the Great Falls project as of 1797. (Adapted from HAER drawing, Library of Congress)

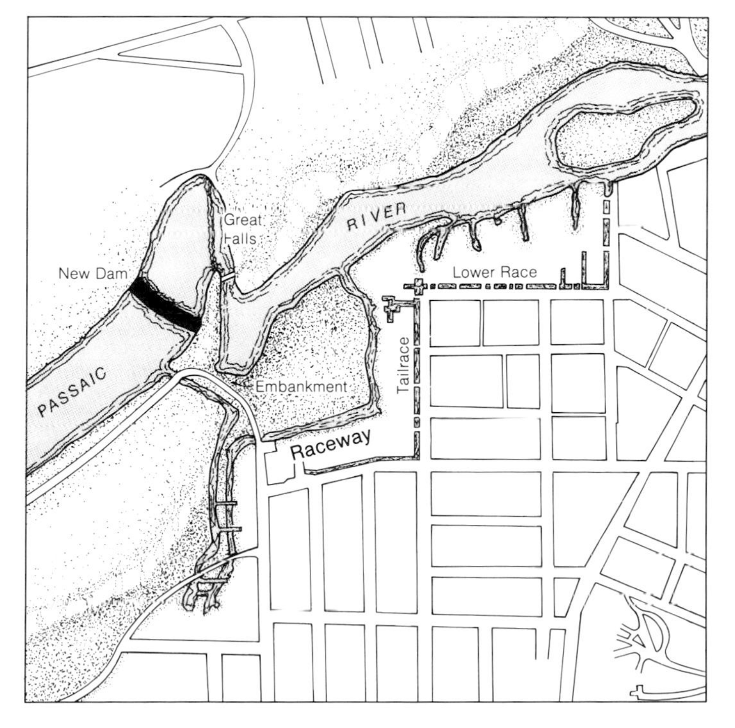

With time the Great Falls site developed into an extensive complex. (Adapted from HAER drawing, Library of Congress)

build up a 20-foot-wide embankment on the downslope side. It is possible that the cut-sandstone inner wall of the channel was placed at this time; but it was more likely added during a later expansion. The outer embankment wall was constructed of rough fieldstone, braced occasionally along its length.

Once water power had been established, the next step was to develop a spinning business, and in this the Society for Establishing Useful Manufactures was less successful than it had been with the raceway. In America at this time there was no extensive knowledge of the technical end of the textile manufacturing business, although there were enterprising men willing enough to claim expertise. After a small and unsuccessfull venture with its own mill, the society's funds ran low and it suspended manufacturing operations in January 1796. Despite this failure the society still controlled a large power project, capable of running several mills. Its members began to debate the questions—new to America—whether to sell the power, whether to let other companies develop it, whether to rent it. From 1800 on, this corporation functioned as what we would now call a real estate developer, abandoning Hamilton's dream of a model manufacturing enterprise. At Colt's suggestion, lots were sold with water power rights, the price based on the cost of the channel and dam.

Colt also planned the extension of the canal and the addition of a tailrace that would drive another set of mill wheels; construction of these improvements occurred in 1807. In the late 1820s, under the engineering direction of his son John, a major realignment of the hydraulic system occurred, which permitted the construction of a new upper tier of mill seats. The last major modification of the system came in 1838, when leakage through the earthen dam—which had been predicted by L'Enfant—forced the society to fill the dam and cut a new channel around the reservoir into the rocky river edge. A mortar- and iron-reinforced masonry dam, 8 to 12 feet high and 210 feet long, was built downstream to divert river water into the new channel. The total length of the races, then numbering three, was about one mile.

The Great Falls raceway and power system has since been in continuous use. Its abundant power supply attracted engineers and industry to the area, leading to an improved textile industry, the development of new textile machinery, and the appearance of several noteworthy inventions, including the Colt revolver and the Rogers and other steam locomotives. Great Falls has served as a center for the manufacture of silk, flax, and jute products, early Wright aircraft engines, and wearing apparel; more recently it has been used for the generation of hydroelectric power. Over its nearly two-hundred-year history the site has proved abundantly Alexander Hamilton's claim that American manufacturing capability was both desirable and profitable.

The Lowell Water Power System

The water power system at Lowell, Massachusetts, developed largely between the years 1823 and 1880, helped create one of the country's first great industrialized centers—a city that has come to symbolize the Industrial Revolution in America. The complex of canals, gatehouses, dams, and hydraulic turbines helped initiate changes at Lowell that forever altered the course of American industrial practices. The water power experiments carried out here by James Francis and Uriah Boyden during the 1840s and 1850s won widespread acclaim as important contributions to the evolving field of hydraulic engineering. The early social experiments at Lowell in creating a beneficial work environment in the context of an industrial complex remain as intriguing as the later exploitations of immigrant workers is appalling. Lowell was the quintessential nineteenth-century American industrial city, a place both good and bad.

The presence of the Pawtucket Falls on the Merrimack River led to the eventual development of an industrial complex at what is now Lowell. Early in the history of Massachusetts the falls presented a barrier to the flow of goods from New Hampshire down the Merrimack to Boston. Leaders from Newburyport, at the mouth of the Merrimack, incorporated as the Proprietors of Locks and Canals on the Merrimack River in 1792, with the intention of building a canal to bypass the falls. By 1796 the resulting Pawtucket Canal was completed at East Chelmsford. This stimulated the construction of a few small mills in the area. The small transportation canal soon fell into disuse, however, because of the generally waning fortunes of Newburyport and the completion of the Middlesex Canal, which connected the upper Merrimack with Boston directly. By 1821 the Pawtucket Canal was in poor physical condition.

During this same period other merchants in Massachusetts were beginning to explore the potential of water power in connection with the introduction of industrial practices developed in European cities. A Boston merchant, Francis Cabot Lowell, visited textile factories

Lowell water power system, Lowell, Massachusetts. A symbol of the Industrial Revolution in America and of the societal changes it wrought, Lowell was one of the country's first great manufacturing centers. C. H. Vogt's 1872 lithograph shows the city in its prime. The modified view depicts the power canal system as it was developed at that time, and indicates which of the buildings shown in the 1872 view are still extant (albeit usually in modified forms). New mill and manufacturing buildings continued to be built, and old ones altered, into the twentieth century. With time most were torn down. But enough remain to give a visitor a glimpse of Lowell as it once was. Some important Lowell buildings built after 1872 are also shown on the modified view. (The 1872 view and permission for modification are courtesy of the Museum of American Textile History, North Andover, Massachusetts)

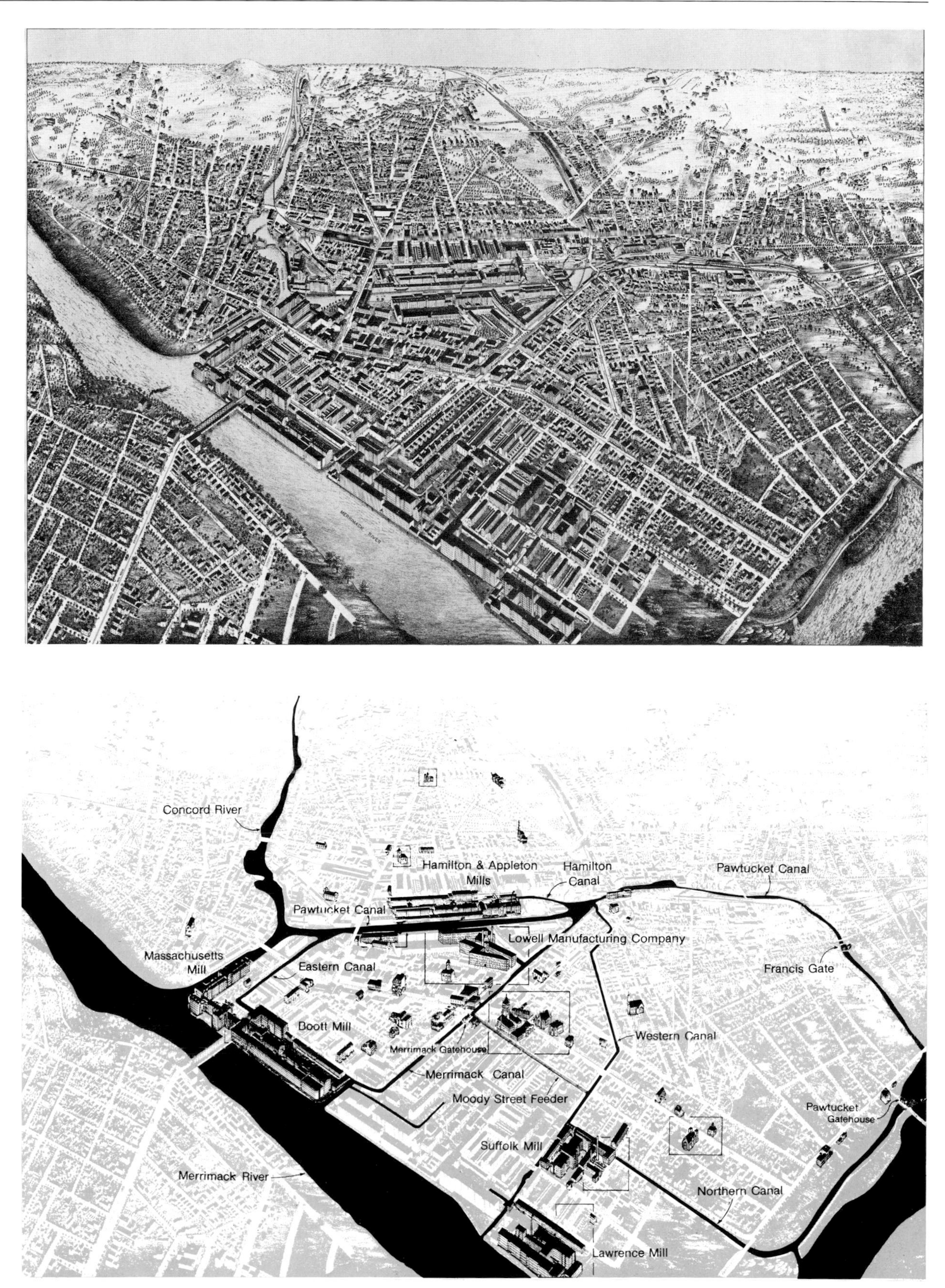

in England and Scotland in 1811 and developed both an appreciation for the technological advances he observed there and an abhorrence of the exploitation of labor. He soon headed a group of Boston merchants, called at first the Boston Associates and later the Boston Manufacturing Company, which started a textile mill at Waltham, Massachusetts, in 1813. Lowell's observations of English power looms led to the development in 1813 of a similar loom by the highly regarded mechanic Paul Moody, whose machine shop was supported by the company. The power loom formed the basis for the textile revolution that was shortly to occur in America.

The mill at Waltham was successful but limited by the small power from the Charles River. Lowell himself died in 1817, but the Boston Manufacturing Company continued and began to seek alternatives to the Waltham site, soon fixing on Pawtucket Falls. In 1821 the group, now under the direction of Nathan Appleton, acquired control of the Proprietors of the Locks and Canals corporation, together with some 400 acres of land. In 1822 they formed the Merrimack Manufacturing Company. A cluster of textile mills grew up on the site, the old Pawtucket Canal now serving as the primary water power feeder. Kirk Boott became the company agent. The village of East Chelmsford was renamed Lowell, grew rapidly, and in 1836 incorporated into a city.

Leaders of the manufacturing corporations that set up mills in the area were from the outset more benign and concerned about the quality of work and life in the city that they were helping create than many of their English counterparts. Attention was paid to making of the buildings and their surrounding landscapes something more than their industrial purposes would require. Much of the labor for the mills was supplied by young female workers, recruited from the cities, villages, and farms of New England. In an era when there were few alternatives for women other than to enter into the traditional marriage role, and few ways for poor women to improve their lives by working independently, the Lowell mills offered a remarkable opportunity. Company agents hired young women, who then lived in specially established boardinghouses under carefully controlled conditions. Many women sent their wages back to their families to help see them through hard times, some created dowries, others saved up until they could set out on their own. The life was hard, with twelve-hour days, but a unique social environment developed, complete with lectures, literary magazines, and other cultural activities.

By 1836 Kirk Boott, with the help of Paul Moody, had developed a two-level canal system that provided water power for the many mills in the area. Water from the tailraces of the upper-level system was fed into a complex that powered additional mills at the lower level. A wooden dam was built at Pawtucket Falls to increase the flow into the canal system. It was replaced by a masonry structure in 1830–1833.

The year 1837 marked a turning point in the development of the Lowell water power system, with the appointment of James B. Francis as engineer to the Locks and Canals Company machine shop. Francis was born at Southleigh, Oxfordshire, England, in 1815. His father was superintendent on a Welsh railway, and he soon was set to work on associated construction projects. Francis eventually left a job on the Great Western Canal to emigrate to New York, in hopes of finding engineering work. He soon found it under the direction of the important American engineer George Whistler, on the Stonington Railroad. Francis followed Whistler to Lowell when the latter was appointed chief engineer to the Locks and Canals Company machine shop; there he worked as a draftsman. Following Whistler's resignation in 1837 Francis, then only twenty-two years old, was appointed engineer. He immediately turned his attention to meeting the ever-increasing power requirements of the expanding mill complex.

By the 1840s power shortages were common. Francis proposed the development of a second major feeder canal to supply additional water to the canal system and reduce current levels in existing canals. The plan also required that water be ponded behind the dam overnight to store a sufficient quantity for the following day. He further suggested controlling the outlets of feeders into the Merrimack River itself as a way of providing reservoirs for controlled use. Francis's studies of the mill complex's water power requirements, on which these proposals were based, had been carefully reviewed and approved by a commission consisting of Whistler, Charles Storrow, and James Baldwin, created especially for the purpose.

Francis selected a route for the new feeder canal—named the Northern Canal—largely parallel to the river and cutting through higher ground than did the Pawtucket Canal, so that water could be brought in at a higher head pressure than before. The 4,373-foot canal averaged about 100 feet in width and 15 to 21 feet in depth. A wall had to be constructed between the river and the canal. The first and shortest section consisted of a battered, dry-laid rubble wall. The next, longer section was built on an island formed by the canal excavations and consisted of rubble laid in cement. The final section, the Great River Wall, about 1,000 feet long and built in the bed of the river, rises at points to as much as 36 feet above the rapids and is made of roughly squared granite blocks, with quarried, coursed ashlar on the exterior.

Construction of the canal also involved rebuilding the Pawtucket Dam. One section was built in 1847 as part of the Northern Canal project; a second section was built in 1875. The whole dam is 1,093.5 feet long and follows the natural ledge outline of the original falls. The core is rubble set in cement. Quarried granite blocks form the exterior. A row of wooden flashboards, set along the entire length of the structure, raised the ponding level even more. A complex series of gates was also constructed as part of the Northern Canal project. This new canal initially connected with the Western Canal, but Francis eventually connected the Northern to the Merrimack Canal through three brick-vaulted tunnels. The Boott Penstock transferred some of the Merrimack Canal water to the Eastern Canal.

One part of the new works, now referred to as the Francis Gate, particularly illustrates its designer's careful and farsighted approach to the system. While designing the new dam, Francis searched exhaustively through records of past floods along the Merrimack River. He discovered a 1785 flood higher than

Line shaft for the gate-hoisting mechanisms in the Pawtucket Gatehouse. (HAER Collection, Library of Congress)

any known since, and beyond the memory of local inhabitants. Francis estimated that a repeat of this flood might, after the erection of the dam, rise over the tops of the guard gates and flood the city. He therefore designed a massive gate, 27 feet wide by 25 feet high, to be suspended over the canal lock. The gate was constructed in 1850 and soon dubbed "Francis's Folly" by locals who could not conceive its necessity. Only two years later the same flood stage as that of 1785 was reached during the spring, and the water rose to the tops of the regular barriers. The great iron link that held the gate suspended was cut, and the gate dropped in place. The water continued to rise behind it for another 22 hours. The gate prevented what would surely have been a major disaster. It was used again in 1936.

Having completed these construction projects, Francis next implemented a careful management system for allocating water power to the many mills in the complex. The development of the Northern Canal eventually led to a 50-percent gain in available power. He also persuaded the Locks and Canals Company to join forces with the Essex Company of Lawrence, another developing industrial complex on the Merrimack, in gaining control of many square miles of feeder lakes in New Hampshire.

A significant feature of the complex as completed in 1847 was the introduction of hydraulic turbines to replace the more common breast wheels then in use. These wheels were often huge, up to 30 feet in diameter, but were not highly efficient. In view of this problem, Francis had been attracted to the work of Uriah Boyden. Born in Foxboro, Massachusetts, in 1804, Boyden had worked as a farmer, a blacksmith, an assistant to his inventor brother, a surveyor for the Boston and Providence Railroad, and an assistant to Loammi Baldwin in the construction of the dry dock at the Charlestown navy yard. He had superintended the construction of the Nashua and Lowell Railroad from 1836 to 1838 and had then become engineer for the expanding Amoskeag Manufacturing Company at Manchester, New Hampshire. It was while holding this last post that Boyden devised a special outward-flow water turbine.

The Guard Locks Gatehouse contains the great gate, designed by James Francis, that saved the city from flooding in 1852 and 1936. The gatehouse is now part of the Lowell State and National Historic Park. (HAER Collection, Library of Congress)

His turbines were installed in 1846 in the Appleton cotton mills at Lowell. Their design was based on that of a Frenchman, Fourneyron, but was much more efficient. Boyden's improvements included a special scroll penstock, a suspended top bearing, and a special diffuser.

Francis soon designed a horizontal turbine of this type on which he conducted numerous tests. Under Francis's direction an extensive program of conversion to hydraulic turbines took place throughout Lowell's mills following completion of the Northern Canal.

A remarkable aspect of Francis's work, particularly in view of his limited formal education, was his insistence on a scientific basis for his designs. In 1846 he conducted tests on Boyden's turbines before he advocated switching from breast wheels. It was the series of tests he conducted in 1851 on an installation of his own design, however, that provided his greatest contributions to the field of hydraulic engineering. Francis carried out some 92 experiments on water flows, using the Hook gauge invented by Uriah Boyden and various weirs as measuring devices. He developed several improved methods for measuring and controlling water flows in channels. These experiments produced practical information that was immediately applied toward the more efficient utilization of water power resources at Lowell. The leasing of water power on an objective basis became possible, as did controlling its distribution based on defined power requirements. As knowledge of how to improve water power use increased, additional measurement systems, gates, and other control devices were installed in the Lowell system, making it far more sophisticated than comparable systems elsewhere.

Francis published his results in 1855, in a study now known as the *Lowell Hydraulic Experiments*. The full title of the work, however, perhaps gives a better indication of the character and interests of its author: *Lowell Hydraulic Experiments, being a Selection from Experiments on Hydraulic Motors, on the Flow of Water over Weirs, in Open Canals of Uniform Rectangular Section and through Submerged Orifices and Diverging Tubes, Made at Lowell, Massachusetts* (second edition, 1868). The book is a classic in its field

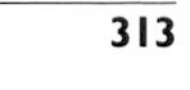

Lowell was the site of a famous series of water power experiments conducted by James Francis and Uriah Boyden. The publication of the results of these experiments in 1855 contributed greatly to the field of hydraulic engineering. The tradition of experimentation and the careful monitoring of flows in the canal system went on long after Francis's seminal experiments. Two assistants are shown measuring levels in 1899. (Courtesy of the University of Lowell)

and helped usher in a new era in water power utilization and engineering. Francis became well known for this and other work, publishing extensively in the new *Transactions of the American Society of Civil Engineers.*

Lowell in its early days possessed the power and vigor of an expanding city with an industrial strength that was not only competitive but on the forefront. The social experiments noted earlier and the attention to specific amenities such as landscaping (some of the latter designed by Francis himself, who had also undertaken a study of botany) were made possible by this strength. But by the end of the nineteenth century the city was undergoing some dramatic changes, and not for the best.

James Francis continued to serve as engineer for the system for many years after his formative hydraulic studies, although by the late 1870s he had done as much with the system as he could. (As for Uriah Boyden, he too continued his interests in scientific matters, leaving a small fortune at his death for establishing observatories on mountain tops. During his last years the whiskered inventor lived quietly and inexpensively in a Boston hotel.) Lowell, by the latter part of the century a bustling industrial center with not only textile mills but also many support industries, as well as companies unrelated to textiles that had located there because of the available pool of skilled workers, was nonetheless already threatened by competition. By 1875, for example, Fall River to the south was producing more cotton products than Lowell. Other industrial centers existed as well. Moreover, Lowell began losing its leadership role in technological innovation. The increasing competition from other mills led new managers to reduce workers' wages, and amenities all but disappeared. The spirit of the place became one of mercenary drive and exploitation of the cheap labor afforded by waves of immigrants. Labor movements developed, and bitter disputes between labor and management marked the turn of the century. Lowell underwent a precipitous decline as the twentieth century wore on. The general collapse of the textile industry in New England hit this city particularly hard. Many

Members of the Society of Civil Engineers on the Merrimack Canal, 1875. (Courtesy of the University of Lowell)

A late-nineteenth-century view of one of the repair shops of the Proprietors of Locks and Canals. (Courtesy of the University of Lowell)

Maintenance of the canal system was a constant need. This 1901 photograph shows construction workers on the Pawtucket Canal. (Courtesy of the University of Lowell)

mills closed, and buildings were demolished. Lowell became one of the most severely depressed cities in the country. The misguided urban renewal program of the 1950s threatened what was left of the great complex.

Model cities programs attempted to revitalize Lowell in the late 1960s. In 1972 the city council sought to focus new development around an urban park consisting of the remains of the once-great industrial complex. President Carter eventually signed a law establishing the Lowell National Historic Park and the Lowell Historic Preservation District, to be administered by the National Park Service in cooperation with the state of Massachusetts, the city of Lowell, and several other smaller organizations. An extensive rehabilitation program was developed with the aim of preserving the remains of the industrial complex and its water power system and making its technology understandable to the visiting public. Today, tour boats travel down the great canals, and guides explain the workings of the gatehouses and other features to vacationers and schoolchildren.

A turn-of-the-century view down Dutton Street. (Courtesy of the University of Lowell)

The Pelton Impulse Waterwheel

The waterwheel invented and patented by Lester A. Pelton in 1878–1880 at Camptonville, California, represents a significant development in the history of water power. Within twenty years of its patent the Pelton wheel was to be found on five continents, in diverse sizes and applications.

Nineteenth-century industrialization sparked major breakthroughs in the application of hydropower, principally in the New England states. The textile industry, and in particular the manufacturing centers in the river valleys of Massachusetts and Rhode Island, developed large and sophisticated machinery. The Merrimack Manufacturing Company of Lowell, Massachusetts, for example, sponsored many technological innovations, such as the introduction of turbines, and introduced careful water-power management systems. These advances, largely the result of the collaboration of James Francis and Uriah Boyden, brought hydropower technology to new levels of sophistication.

The California gold rush at midcentury stimulated innovation in hydropower technology as well. Both the mountainous terrain where the search for gold took place and the extraction techniques that were devised, however, called for an application of hydrotechnology far different from that at Lowell. The precision and scrupulous refinements of Francis and Boyden's carefully installed, fixed-place, relatively low water-head systems were of no use to California gold miners. In the first place, the latter required portable equipment. In the absence of scientific methods for systematizing the location and exploration of hidden gold and silver veins, miners' successes were governed by chance and insight. Equipment had to be simple and easily assembled. Furthermore, the lack of railroads through the rugged country restricted transport to mule and horseback. The relatively high water-head pressures of mountain streams, in contrast to those of the geologically older and broader rivers of New England, also required a unique response in their application to mining operations. As a result, the development of hydropower equipment to meet the miners' needs focused on a

The Pelton impulse waterwheel shown here was constructed around 1895 by the Pelton Water-Wheel Company of San Francisco for the North Star Mining Company, Grass Valley, California. (*Transactions of the American Society of Civil Engineers* 36, no. 788 [1896])

set of criteria that were highly restrictive and quite distinct from those successfully confronted by Francis and Boyden.

The fast-running mountain streams promised an abundant supply of power, vital to hydraulic mining operations. The extraction techniques of fluming, digging shafts and drifts, and panning required base camps incorporating sawmills, forges, grinding machines, and stamp mills for crushing the mined ore. In addition, water nozzles were used to break down earth banks and expose the ore. All of these facilities required large amounts of cheap power, produced by portable plants. Gas engines, developed only to a primitive and unreliable level, were unfeasible; steam-powered machinery was bulky and dependent on skilled mechanics. Thus it was natural for the California miners to turn to water for their power requirements.

The first waterwheels employed in the mining country were the cumbersome overshot type fabricated in wood, their design derived from early New England wheels. As textile manufacturers in New England had done before them, the first miners discovered the inadequacies and inefficiencies of these wheels. Alternatives were constantly sought. Indeed, the gold rush was marked in the U.S. Patent Office by hundreds of patents for largely useless inventions intended to improve the inefficiencies of water-powered mining equipment. A significant advance was achieved in 1857 with the invention of the pressure nozzle, which directed small-diameter high-velocity water jets at high head pressures against earth banks in fluming operations.

Another significant, and anonymous, innovation brought about during the gold rush was the hurdy-gurdy waterwheel. This was a natural outgrowth of the environment—fashioned out of the materials at hand, easily transportable, and simple to construct. The hurdy-gurdy was fabricated of triangular blocks of wood about 4 inches thick, sandwiched between wooden shrouds of circular shape and attached to a central axle by four or six spokes. The wheel was of the undershot type, powered by a water spray directed tangentially against its underside. The spray was controlled in turn by a pressure nozzle, thus allowing for limited variation in wheel speed by adjustment of the jet speed and direction. In this form the hurdy-gurdy supplied the power demands of gold and silver fields in both California and Nevada during the 1850s and 1860s.

Progress in the mines, however, led eventually to the mining of impoverished ores and consequently to the use of larger stamp mills requiring more power to extract the precious metals. It also became increasingly obvious that the hurdy gurdy's efficiency could and should be improved. Curved buckets of cast-iron construction would more effectively transmit the water force directed against them to axial rotation; and cast iron would be far more durable against the pressurized sprays of silt-laden water. In 1866 the Pacific Iron Works of San Francisco produced its first cast-iron hurdy-gurdy with curved buckets, rendering obsolete the established wooden models. Further improvements were needed, however, since power demands for stamp milling in particular were still in excess of that provided by the cast-iron hurdy-gurdy.

Lester A. Pelton crossed the continent from his native Ohio in 1850, joining thousands of his countrymen in the quest for wealth in the California gold fields. Like many others, he failed to realize his dream in the prospecting camps of the Sierra Nevada. It was rather through his skill in carpentry and metal working that Pelton successfully established himself by 1864 in Camptonville, California, serving the mining camps along the Yuma River. He constructed homes, a schoolhouse, and stamp mills, and assisted in the disassembly and erection of mining camps as they pursued the strata of gold through the hills and embankments of the Sierra Nevada. His enthusiasm, however, was devoted to the improvement of the waterwheel, and he was forever tinkering with modifications to the basic hurdy-gurdy design.

By the 1870s the principles underlying the modern pressureless turbine were well known—the tangential approach of water to bucket surface, reversal of the water flow within the bucket area, and ejection of the spent water without interference in the wheel's rotation. Pelton's invention in 1878 of the wheel that bears his name came about through the synthesis of these three principles. In achieving this he modernized the mining industry, making possible an efficiency of power production previously unattainable.

Pelton concentrated his efforts toward the last of the stated objectives—ejecting the spent water without impeding the wheel's rotation. He isolated the problem by observing how pressurized water, ricocheting from the bucket's base and walls, struck the succeeding bucket, thereby resisting the wheel's induced forward motion. The answer, and the key to Pelton's breakthrough, came quite by accident. Experimenting with a small 2-foot-diameter waterwheel, Pelton accidentally dislodged the shaft key securing the wheel to its axle; the subsequent skewed rotation of the wheel resulted in the water jet striking primarily against the bucket's side and ejecting away from the succeeding bucket. Despite the increased friction at the axle, the wheel increased speed, now unimpeded by backsplash resistance at the wheel edge. After further exploration of this phenomenon, Pelton devised a split-bucket design, with a center splitter, curved bottom, and flared sides. Constructed of cast iron, buckets of this new form were riveted to cast-iron wheels made with iron spokes and sealed bearing axles.

This simple but radical improvement provided horsepower ranges and efficiency ratings of ample magnitude to meet the mining and hydroelectric requirements of the western states well into the twentieth century. By 1880, having secured the patents to his invention, Lester Pelton was on the path to the fortune he had sought thirty years earlier. The Pelton impulse wheel was an immediate success. Minor modifications to reduce water turbulence and eddying within the bucket, which had caused excessive wear from the sand- and silt-laden waters of the mining camps, were carried out by William A. Doble in the 1890s. By 1900, having incorporated his operations with a San Francisco foundry, Pelton saw his impulse wheels installed in

North and South America, Europe, Africa, and Australia. Moreover, the simultaneous development of hydroelectric power generation, when combined with Pelton's wheel, provided the necessary machinery for electrifying urban areas across the country. By the early twentieth century the Pelton wheel was available in sizes from 20 feet to 4 inches in diameter, for applications ranging from the world's largest mining operations to precision dental drilling.

The Vulcan Street Plant

The Vulcan Street power plant, on the banks of the Fox River in Appleton, Wisconsin, represents a pioneering effort in hydroelectric power generation in nineteenth-century America. Utilizing dynamos designed and fabricated by the Edison Electric Company of New York, the Vulcan Street plant, which went into operation in September 1882, was the first central-station plant in the world to generate electrical power from impounded water for a network of private and commercial customers. An earlier hydroelectric installation at Minneapolis, Minnesota, completed seven months before, was more simply powered by the force of a natural drop of water at St. Anthony's Falls. The Appleton installation, by contrast, featured a designed network of flumes, wooden penstocks, and head gate controls, to regulate the water pressure powering the plant's turbine and dynamos. Thus, the Vulcan Street plant's significance is to be found in the innovative synthesis of diverse nineteenth-century engineering developments. By wedding the advances achieved in waterwheel design to the ongoing development of Edison electrical generating systems, this installation first demonstrated the successful conversion of cheap water power into electrical power. Beginning on the night of September 30, 1882, an Edison Company K-type dynamo, installed in an existing paper mill and powered by the mill's water turbine, supplied electricity for lighting systems in nearby mills, a pulp processing plant, and the mill owner's home. Seven weeks later a second dynamo, installed in its own shed and linked to its own water turbine and feeder network, duplicated this achievement of hydroelectric power. Within four years of these pioneering demonstrations, Appleton possessed the West's first regularly scheduled and continuously maintained electric trolley car system.

The successes at Appleton in the early 1880s were due to the acumen of four of the town's business leaders, and to geography. Incorporated as a town at midcentury, Appleton, in central Wisconsin, was a thriving industrial community by 1880. The Fox River, on the banks of which Appleton had been founded and developed, exhibited a 38-foot level drop within the town's boundaries. The location was therefore advantageous for the establishment of industry dependent on water power. Further, Appleton's proximity to Green Bay, 29 miles to the east, provided an important link to Great Lakes shipping and the burgeoning iron ore industry. Rail lines from Chicago tied Appleton to the transportation nexus of the rapidly expanding West. Within thirty years of its incorporation, the town displayed diverse industrial development: sawmills, paper and pulp mills, flour and woolen mills, blast furnaces, wagon wheel factories, and machine shops lined the banks of the Fox River and two man-made canals as well.

The water power that drove Appleton's midcentury growth held equal promise for hydroelectricity. Thomas Edison's accomplishments with electrical generation, and the development of a marketable lightbulb, made electric lighting a realistic possibility in American cities by 1880. On September 4, 1882, twenty-six days before the first Appleton success, the Edison Company's Pearl Street plant in New York City began producing electricity from steam power. Wiring installations to light New York's First District, powered by the Pearl Street plant, were already under way. In Chicago the Western Edison Electric Light Company, incorporated in May 1882, began licensing Edison plants, as yet unbuilt, to light cities in Illinois, Iowa, and Wisconsin. The *Edison Company Bulletin,* published in New York but widely distributed in American cities, hailed the advent of the electrical age and ended each issue with a list of fatalities and injuries caused by accidents involving gas lighting.

The possibility of electric lighting, powered by water, attracted the attention of Appleton's H. J. Rogers, president of both the Appleton Paper and Pulp Company and the Appleton Gas Light Company. Foreseeing the ascendancy of hydroelectricity and its adverse effect on his gas light business, Rogers secured from Western Edison the exclusive right to light Appleton with the Edison electric lamp. He obtained financial backing for his enterprise, the Appleton

Vulcan Street plant, Appleton, Wisconsin. The first central-station plant in the world to generate electrical power from impounded water, this installation began serving a network of private and commercial customers in 1882. (Courtesy of the American Society of Civil Engineers)

Edison Electric Light Company, from fellow businessmen A. L. Smith, H. D. Smith, and Charles Beveridge. The grand house overlooking the river that Rogers was at the time engaged in building was to be equipped with this latest technology.

By early 1882 Rogers had ordered two K-type dynamos from the Edison Company in New York, and Edward T. Ames, an engineer with Western Edison, had been hired to supervise the first dynamo installation in the beater room of Rogers's pulp mill. Wiring installations in this mill, in Rogers's paper mill next door, and in his new home were carried out during this period. The wiring and hardware components that were used reflect the state of the electrical art at the time. Distribution wiring between the mill and Rogers's home was of uninsulated copper without fuse protection; in the event of a storm or fallen wire, the entire system would have to be turned off before repairs could be made. The wires in Rogers's home were covered with thin cotton insulation and attached to wall surfaces with wooden cleats. When a wire passed through wall partitions, cloth tape was wrapped around it to provide additional insulation. Sockets and switch handles were constructed of wood. The lightbulbs, purchased from the Edison Company, contained carbonized bamboo filaments in an airless glass casing. Finally, voltage regulators to control the dynamo's output were entirely absent.

By late September 1882 the first installation was complete. On the night of the 27th, with considerable fanfare—and amid much public skepticism—the dynamo was linked to the pulp mill's water turbine, without result. Edward Ames, the supervising engineer from Chicago, was hastily recalled and set about varying the wiring patterns at the dynamo, at first without success. Finally, on the night of September 30, the correct configuration was achieved and electrical current flowed to Rogers's new home for the first time.

The pulp beaters in the mill that housed this dynamo imposed a varying load on the mill's water turbines; the latter's consequent irregular motion in turn affected the dynamo. Without voltage regulators, the current thus produced varied in intensity, causing lightbulbs to dim and glow in cadence with the pulp beaters. The second dynamo, however, was already being installed in the Vulcan Street plant, which contained its own waterwheel and head-gate control—designed by Edward O'Keefe, a local engineer—and thus promised a more regulated current flow. The Vulcan Street plant commenced successful operations on November 25, 1882, initially lighting the homes of electric company backers A. L. and H. D. Smith. By December the two installations were powering electric lights in five Appleton mills and six homes, the first demonstration of central-station hydroelectric power from a controlled head source in the world. On January 16, 1883, Appleton's Waverly Hotel was electrically illuminated, the first such hotel in the American West.

Progress was rapid from this point, but not without early difficulties. With its penstock and head-gate system, the Vulcan Street output could be regulated only through maintenance of steady water pressure upon the waterwheel buckets. Without voltage regulators, however, varying service loads affected the entire system. The plant supervisor, gauging the dynamo's output by eying a lightbulb in the plant, could not make minor adjustments with the control means at hand. The practice of burning light bulbs from dusk until dawn, which served to maintain constant loads, was encouraged by the establishment of set monthly fees for electrical service by the Appleton Edison Electric Light Company. Not incidentally, the electric company was the town's supplier for Edison lightbulbs as well.

By 1886 Rogers had purchased two additional, larger generators, placing them in a new plant near the Vulcan Street installation. This new plant was the precursor of the Wisconsin Michigan Power Company. The original K-type dynamos were subsequently relocated to the enlarged central facility, and Appleton's first three wire distribution systems, providing far safer and more consistent service, went into operation. The electrification of Appleton's trolley car system took place in the same year, providing one of America's first regularly operated and electrically powered trolley car networks.

The Vulcan Street plant, left to ruin after the removal of its dynamo in 1886, was reconstructed in exact replica in 1932 as part of a celebration marking the fiftieth anniversary of the world's first central-station hydroelectric plant. The H. L. Rogers home, the first residence to be lighted from a central hydroelectric plant, has also been restored and contains much of the original wiring installation.

The Folsom Hydroelectric Power System

The Folsom power system earned a place in engineering history by achieving the first significant long-distance transmission of electric current. On July 13, 1895, an 11,000-volt, three-phase alternating current was transmitted from the hydroelectric power station at Folsom, California, to the city of Sacramento, 22 miles away. In Sacramento the power was used not only to light shops and office buildings but also to run factories and the streetcar system. Thus the event also marked the first long-distance transmission for large-scale, multipurpose use. The powerhouse served as the prototype for the generation and transmission systems in use throughout the country today.

The development of the power system began in 1886, when the Natoma Water and Mining Company began the construction of a granite block dam across the American River just above Folsom. The purpose of the project was to create a holding pond for logs floating downstream from timberlands, as well as to provide power for a sawmill that was to be located at Folsom. An outflow canal was also planned in order to float the logs to the proposed sawmill. After building the foundation and the first 30 feet of the granite and concrete dam, the Natoma Company ran into financial difficulties and stopped work. An agreement was then reached with the state of California whereby the company would cede to the state ground for a new prison as well as certain water power rights in return for a specified amount of prison labor to help complete the dam.

When, after a long delay, work on the dam resumed, the scope of the project had changed. The state now envisioned a New England–style industrial town at Folsom, where riverside industries would be fed by the power generated at the dam. The president of the Natoma Company, Horatio Livermore, had a different vision, however. His idea was that instead of developing industry along the river, electric power should be conveyed to Sacramento, where it could be used by people and industry already there and waiting for a cheap source of power. Accordingly, Livermore formed the Folsom Water

Folsom powerhouse, Folsom, California. This hydroelectric plant was part of the first system to provide long-distance, high-voltage, three-phase transmission for significant municipal and industrial multipurpose use. Completed in 1895, the system supplied power to Sacramento, 22 miles to the southwest. (Courtesy of the American Society of Civil Engineering)

Power Company with the purpose of taking over the dam and property rights belonging to the Natoma Company. He also arranged to control the sale and distribution of electric power in Sacramento. In 1888, after a dispute over the amount of prison labor the company would receive, a new contract was entered into with the state, whereby the project was changed from the creation of a millpond with a water power adjunct to the establishment of a complete water power and electrical system. The dam and canal were to be enlarged, and a canal was to be dug from the prison powerhouse to a new powerhouse to be built by the power company 1.5 miles away.

Construction recommenced on the dam and canal, under the direction of Chief Engineer P. A. Humbert. By January 1889 the first water was turned through the prison powerhouse. All the power generated in the prison, however, was solely for prison use. In 1893 the dam was finally completed. The Folsom Water Power Company next successfully negotiated a contract with the General Electric Company for the materials and plant for a power station that would transmit electricity from Folsom to Sacramento. This contract was considered somewhat risky for General Electric, because it had not yet been conclusively proved that long-distance transmission of commercial quantities of electricity was feasible.

Work on the powerhouse started in October 1894. The construction of the 22-mile line to Sacramento took ten months to complete. The first trials, in early July 1895, were unsuccessful; but at last all the difficulties were cleared up, and on July 13 current from Folsom reached the Sacramento substation. To celebrate the occasion, a grand Electric Carnival was held on September 9. Some 30,000 residents and visitors attended.

The Folsom powerhouse and transmission system were conceived and built at a time when knowledge about such matters was in an experimental, infant stage. The first commercial electric power had been generated earlier, but industries and engineers were faced with the practical problem of its distribution to and use by parties not directly attached to the power station. A long-distance transmission had occurred in 1889, when a small amount of current at 4,000 volts was sent some 14 miles to Portland, Oregon. Other small transmissions were attained in 1891 in Colorado over a distance of 2.6 miles, to power a 100-horsepower motor; in 1892 in Bodie, California, where a single-phase, 3,500-volt, 13-mile transmission was made to operate a mill; and in 1893 in Redlands, California, where the first three-phase transmission of electric power took place, over a distance of 8 miles. Each of these experiments, however, had merely provided power for a single mine or mill, or had been used entirely for lighting. From the start the power generated at Folsom was used to run the Sacramento streetcar system, factories, and machine shops, as well as to provide municipal and commercial buildings with electricity for lighting and other purposes.

The physical plant of the power generating system at Folsom consisted of three basic parts—the dam, the canal, and, most important, the powerhouse. The dam, 89 feet high and 650 feet long, had a bottom width of 87 feet and a top width of 24 feet. It contained 78,500 cubic yards of granite block masonry, sealed with 20,000 barrels of cement. The impounded lake that the dam created was 3 miles long. The canal from the dam to the powerhouse was given a bottom width of 35 feet, a top width of 50 feet, and a depth of 8 feet; much of it was excavated into solid granite. At the canal's entrance were three enormous head gates, operated by hydraulic rams. Sand gates were also provided, to catch sand that may have come through the head gates. The canal ended in a large granite forebay, 150 feet long, 100 feet wide, and 12 feet deep, divided into two sections by a granite wall. The forebay served to lower the velocity of the water before it reached the turbines. Because of the extensive hydraulic mining for gold that had occurred on the American River, large quantities of sand and gravel were being carried downstream. The halving of the forebay permitted one section to be cleaned of accumulated silt while the powerhouse took on water from the other. From the forebay the water fell into the four hydraulic turbines through four penstock pipes, 8 feet in diameter. Each of the turbines consisted of two opposed McCormick runners mounted on a horizontal shaft connected to the generators inside the powerhouse. The turbines operated under a head of 55 feet of water and achieved 1,265 horsepower at 300 revolutions per minute.

In the powerhouse, located at the end of the forebay, each of the four turbines was directly connected to a 750-kilowatt three-phase alternator, manufactured by General Electric. These were reported to be the largest such generators constructed up to that time. The generators delivered a 60-cycle current at 800 volts. They were remarkable also in that they were among the last to be constructed with a fixed field and a revolving armature. From the generators the current fed through a switchboard to step-up transformers, and from the transformers 11,000 volts was fed to the lines going to Sacramento, terminating at the substation of the Sacramento Power and Light Company. The 22-mile connection consisted of twelve wires on two lines of poles. The lines were divided so that they could be connected, cut off, or cross-connected in the event of trouble.

A supplementary lower power station was added to the complex in 1897 in order to take advantage of the water's 26-foot fall after it leaves the powerhouse. This was equipped with larger insulators, and the potential carried was raised from 11,000 volts to 60,000 volts.

The Folsom powerhouse was given to the state of California in 1952 to become a part of the state park system. Until that time it had remained in continuous operation. The original equipment is now on display, and tours are given of the complex.

The Sault Ste. Marie Hydroelectric Power Complex

The hydroelectric complex completed in 1902 at Sault Ste. Marie, Michigan, is still the largest low-head facility in the United States. When it was designed, the plant was the largest in the world in terms of the volume of water passing through its penstocks. The canal had the largest water-carrying section in the United States, carrying 30,000 cubic feet of water per second. It now generates electricity for 50 percent of the Upper Peninsula of Michigan.

The falls at Sault Ste. Marie presented both a barrier to shipping traffic and a potential for water power to nineteenth-century inhabitants of the area. The enormous water power potential led to a proposal by the Sault Ste. Marie Falls Company in 1844 to build mills on four islands, and another by Samuel Whitney in the same year for a power canal to bypass the rapids at the site. These attempts failed, apparently because of financial difficulties. A ship canal (which had been authorized by the state of Michigan in 1837) was completed in 1855. Increases in traffic in the ship canal led to the construction of another ship lock in 1871, which revived interest in water power development. In the late 1870s H. W. Seymour laid initial plans for a power canal following the same route proposed by Whitney earlier. The local company formed by Seymour was sold out to a group of LaCrosse, Wisconsin, businessmen who hired engineers to develop plans and cost estimates. Construction began in 1888. A decision to widen the canal led to new expenditures, and by 1891 work was dormant as the company sought new investors. Meanwhile, on the Canadian side, construction of a power canal was started in 1889 by the Ontario Sault Ste. Marie Water, Light, and Power Company. This project stopped in 1894 when the walls of the canal collapsed. In this same year the enterprising Francis Clergue, a lawyer with a degree from the University of Maine, visited the area and became interested in the hydroelectric power possibilities of the unfinished canals. He returned east, found financial backing, and soon

Sault Ste. Marie hydroelectric complex, Michigan. Completed in 1902, it remains the largest low-head facility in the United States. (Courtesy of the Edison Sault Electric Company)

bought out the Canadian canal. In 1895 he lined up sufficient capital to acquire the canal on the Michigan side.

At that time calcium carbide, valuable as an illuminating gas, was produced by the Lake Superior Carbide Company. It and two other manufacturers combined to form the Union Carbide Company in 1898. The new company had an interest in expanding, a move that would require large quantities of cheap electrical power. This coincided well with Clergue's desire to have a major power consumer lined up as a customer prior to the construction of a hydroelectric plant. Clergue signed a contract with Union Carbide for the supply of power.

Under Clergue's leadership the project began to take shape. Hans A. E. von Schon, a German immigrant who had served with the U.S. Army Corps of Engineers, was employed as chief engineer. He remained with the project from conception through construction to completion. A graduate of the Royal Prussian Military Academy in 1869, he had already worked on a wide variety of engineering projects, which had provided him with experience in surveying, excavation techniques, and hydraulics, as well as structural, mechanical, and electrical work. He proved an excellent manager.

The team of engineers working under von Schon's direction was a particularly capable one. Mortimer Barnes, who acted as chief assistant engineer in 1896, was involved in the design of the head gates, penstock bulkheads, forebay, turbine installation, and cofferdams. He also did the hydrographic survey of the St. Marys River. Later he went on to assist Joseph Ripley in designing the locks on the Panama Canal. Albert Crane, an 1891 civil engineering graduate of Cornell, replaced Barnes when he left. Crane later became a major figure in the design of dams, hydroelectric stations, and irrigation projects. Alfred Boller, a Rensselaer and University of Pennsylvania graduate who was an expert on structures and foundations, was retained as a general consultant. Boller had previously been involved with the construction of the foundation for the Statue of Liberty. One of the nation's leading hydraulic and civil engineers, Alfred Noble, was also retained as a consultant on the project. Noble became involved in the works at the outlet of Lake Superior, the canal head gates, and the powerhouse design and construction. Ralph Modjeski, later to design the Oakland Bay bridge, was engaged on Noble's recommendation to design the head gates and inspect the steel works. Other engineers included John Bogart, John Kennedy, George Wisner, Leonard Davis, and Gardner Williams. Williams was widely known for his experimental work at the hydraulic laboratory at Cornell University, one of the largest facilities of its kind in the country.

Excavation of the canal began in 1898 and required massive amounts of dirt removal. Construction of the forebay and powerhouse was also started in 1898 and required leveling the area in front of the powerhouse, forming embankments, and constructing entrances to the penstocks. Work was eventually completed in 1902. The length of the canal from the powerhouse to the intake is 11,850 feet, its width varies from 200 to 220 feet, and it is approximately 24 feet deep. Four steel head gates control the canal entrance. The powerhouse is a remarkable 1,340 feet long and 80 feet wide. Initially 33-inch horizontal turbines were installed in pairs on the eighty penstocks. The turbine units in turn drove generators that developed electric power.

The laying of the cornerstone—a premolded concrete block—of the powerhouse on the concrete tailrace base, September 10, 1900. (Library of Congress)

The rear of the powerhouse is shown in this photograph, taken July 26, 1901. Beneath the partially completed wall of the generator room are the completed tailraces. The cofferdam is to the right. (Library of Congress)

Chapter Twelve

Surveying and Mapping

Introduction

The practice of surveying and mapping has been, throughout America's history, an important element in the development of the civil engineering profession, as well as in the growth of the nation. From the pioneer mapmakers of colonial times—who can be considered as among the first civil engineers in this country—to the developers of today's interstate highways, surveyors have literally laid the path for America's expansion.

By the time of the first European settlements in America, maps of the New World had long been available. Based primarily on early sea explorations, these were almost all limited to coastal regions. Thomas Harriot, an English mathematician, came with Richard Grenville around 1585 to Virginia, where he made some surveys. John Smith produced maps of the Chesapeake Bay area in 1612 and parts of New England in 1616. Others developed navigational maps during this period. The first settlements on the eastern seaboard necessitated some form of land division and distribution, and as the population of the colonies grew, further land acquisition was needed. Early colonial surveyors in the seventeenth century include Nathaniel Foote, who helped lay out Wethersfield, Connecticut; Nathaniel Woodward and Solomon Saffrey, who surveyed the Massachusetts–Rhode Island border in 1542; Andries Hudde, who worked in New Amsterdam; John Bonner and Thomas Holme, who made maps of Boston and Philadelphia; Daniel Leeds, who surveyed in New Jersey; and Edward Pennington, who worked in Pennsylvania. Joshua Fry was a later surveyor of note. He surveyed part of the Carolina border, published *Maps of the Inhabited Parts of Virginia* in 1751, and was a professor of mathematics at William and Mary College in Virginia. He also approved young George Washington's registration as a surveyor in Culpeper County. Practitioners of the day, whose chief instruments were simple open-vaned compasses and chains, nonetheless produced some excellent surveys.

Colonial surveyors were of course concerned with creating accurate maps for the sake of proper land distribution among the settlers in already inhabited areas, and they also were instrumental in the development of new settlements farther inland. In addition, surveyors were responsible for resolving many boundary disputes that threatened the stability and cohesiveness of the new colonies. One of the most famous of the early surveys was initiated to settle a property dispute between two families but established what later became one of the most important political, social, and economic demarcation lines of all time. From 1763 to 1767 Charles Mason and Jeremiah Dixon conducted a careful survey aimed at ending a conflict over land holdings between the Penn and Calvert families. It was a technical feat of the first order. The line laid out by Mason and Dixon was later extended farther west.

By the latter part of the eighteenth century the science and practice of surveying and mapmaking had made notable advances. The efforts of the Rittenhouse brothers contributed a large share of these improvements. David Rittenhouse, who also did several significant surveys in his own right, was led by his interests in mathematics and astronomy into instrument making. He and his brother Benjamin devised a number of high-quality instruments, used by the notable surveyor Andrew Ellicott, among others.

During the postrevolutionary period a number of skilled surveyors were at work, including Thomas Marshall, Andrew Porter, and Robert Erskine. This period also saw perhaps the most influential of all events in the development of land division and use in the United States—the passage of the famous Land Ordinance of 1785. This act not only established the boundaries of the new western states, by a series of specific longitude and latitude lines, but also provided the foundation for the public land system in the United States. The ordinance was originally meant only for the Ohio Territory, but eventually almost all the nation's western land would follow the same system of rectangular survey. The ordinance provided for a survey of new lands, to be organized in townships of one square mile each. Each township was to be oriented along lines of longitude and latitude and was to be referenced back to principal meridians and base lines. The work was to be done by private surveyors under the supervision of a newly created government agency, the General Land Office. This ordinance, and supplementary acts passed later, served to shape the nation's land use patterns by virtue of the way the land was subdivided.

Thomas Hutchins, born in New Jersey in 1730, became responsible for most of the surveys immediately resulting from the Land Ordinance of 1785. His duties included a survey of the whole Northwest Territory. Hutchins came to bear the title Geographer to the United States.

Other surveying tasks of importance to the nation were undertaken during this era as well. One of the commissioners involved in the later extension of the line established by Mason and Dixon was the Pennsylvanian Andrew Ellicott, one of the country's foremost early surveyors. Noteworthy is his location of the boundary between the United States and the Spanish territory that is now Florida, accomplished between 1796 and 1800. Ellicott was also involved in the development of the plan for Washington, D.C. Once the Constitution had been ratified and George Washington had taken office as president, plans were begun for the capital of the United States. Pierre L'Enfant produced a plan consisting of a regular grid intersected by long diagonal avenues between monuments. Ellicott worked with L'Enfant for about a year, in 1792. L'Enfant was later dismissed by President Jefferson, after numerous conflicts with the commissioners in charge, and when he left he took all his plans and studies of Washington with him. Ellicott was given the task of drawing up a new plan, working from his memory of what he and L'Enfant had developed; his proposal was eventually adopted. Ellicott was aided by his brothers and by Benjamin Banneker, son of an African slave, who had a remarkable talent for surveying and mapping. Banneker later produced an astronomical almanac that received international attention.

A most significant event for the field occurred a few months after the opening of the War of 1812, when the U.S. Army established the Corps of Topographical Engineers. Later temporarily disbanded, but reinstated in 1816, these engineers engaged in surveys in border or frontier areas and for fortifications. Stephen Long, a major in the corps, conducted a number of expeditions in the Great Plains and the Rocky Mountains, during which he compiled extensive information on their topography, geology, and ethnology, and drew many maps of this part of the West. (Long's Peak in Colorado was named after him.) Isaac Roberdean, another leading figure in the corps, was appointed head of the U.S. Topographic Bureau on that agency's creation. At this time the influential Board of Internal Improvements, which sponsored a significant amount of surveying for road and canal expansions, was also established.

Private surveyors were at work as well, charting paths for the young country's rapidly expanding transportation networks. Many were employed by the turnpike companies then building roads westward and between existing towns. The construction of canal systems also required surveyors, who could plan the most economical routes between existing rivers and lakes. In 1762 the first survey of a canal route in America was done for the Union Canal from Reading to Middletown, Pennsylvania. Virtually all canal projects had need of experienced surveyors, as witnessed by Loammi Baldwin's aggressive attempts to obtain the services of William Weston for the Middlesex Canal in Massachusetts. (Baldwin ended up doing much of the surveying himself, but with an instrument borrowed from Weston.) Most of the builders of the Erie Canal, such as James Geddes and Benjamin Wright, had no initial canal engineering background but were rather trained as surveyors or as road and canal contractors. With the knowledge and experience gained during the course of the Erie's construction, these men went on to become the engineers of later canals and of railroads as well. (For accounts of the building of the Middlesex and Erie canals, see chapter 1.)

As the nineteenth century progressed, the railroad was fast becoming the quickest and most economical form of transportation in the country. Of fundamental importance to its development was the General Survey Act of 1824, under which the Army Corps of Engineers completed 61 railroad surveys, the first in 1826 for a railroad connecting the James and Roanoke rivers. The experienced Stephen Long, for example, conducted surveys for the Baltimore and Ohio Railroad and published a manual on location surveys (he later became a successful designer of covered truss bridges: see chapter 4, Carrollton Viaduct). Army engineers continued to aid the railroad industry until 1838, when the practice was discontinued because of abuses to the system.

Railroad surveyors had a great deal more to consider in their task than did the early road planners. Because the trains could not make sharp turns and could not travel on tracks steeper than a certain angle, location was extremely important. As the railroads pushed west, surveying parties were always one step ahead of the construction crews. No sooner had the surveyors planted their stakes than the builders took them up and laid down the tracks. The survey of lands adjacent to railroad routes was also extremely important, particularly during the building of the transcontinental railroads. Attitudes taken at the time still influence current land use patterns.

Often an obstacle such as a river or mountain could not be avoided during the construction of a railroad, and a surveyor was called in to work directly with the construction crew on a bridge or tunnel site. Such was the case with the Hoosac Tunnel in western Massachusetts (see chapter 5). The tunnel, which took twenty-four years to finish, was planned to create an easy railroad link between the Hudson River at Albany and the seaport of Boston. Because of the extreme length of the proposed tunnel, elaborate new surveying techniques were employed to ensure that the two parts of the tunnel would eventually line up. It was also necessary to work from a central shaft, and owing to the accuracy of the survey the alignment of this central section with those dug from the east and west portals was off by only $\frac{7}{16}$ inch.

In these and other projects the surveyor held broad and primary responsibilities. The more recent evolution of this field, however, has been accompanied by a sharpening of distinctions among the types of individuals engaged in surveying and by a narrowing of their roles, culminating in the emergence of a class of land surveyors distinct from the traditional civil engineer. Nonetheless, surveying remains an important aspect of the civil engineering profession, with its own vital contributions to make.

The Mason-Dixon Line

Named for its surveyors, Charles Mason and Jeremiah Dixon, the Mason-Dixon Line has been regarded as the dividing line between the North and the South ever since its adoption, before the Civil War, as one of the boundaries between slave states and free states. The line was originally plotted between 1763 and 1767 to settle a boundary dispute between the Penn and Calvert families. A highly significant event in the history of geodetic measurement, it is also a classic example of the use of the best scientific and technological minds to solve an urgent practical problem.

In 1681 William Penn obtained from Charles II a land grant west of Delaware and north of Maryland, whose southern boundary was to begin as a "12-mile circle" from New Castle, Delaware, and to continue as a line running westward. The grant apparently overlapped with that of Charles Calvert, the third Baron Baltimore. The Calvert family had been granted Maryland in 1632 by Charles I. The ambiguity of the boundary's location led to a protracted dispute over the right of title and the payment of taxes, with skirmishes and riots occasionally occurring along the border.

At last in 1732 the families executed an agreement, recognized by England's High Court of Chancery, fixing the boundary. Unfortunately, however, the language of the 1732 agreement was not precise enough to settle all questions—such as whether the 12-mile circle meant a circle of 12-mile radius, diameter, or circumference, whether an east-west line meant a line 90 degrees to a meridian or a parallel of latitude, and how distances were to be measured over the surface of the earth. Litigation continued, and several attempts were made to locate the prescribed line on the ground, but none progressed very far.

In the 1750s advice was sought from many scientists and academicians as to how to resolve these issues. In 1760 a final deed was drawn at New Castle in an attempt to close the dispute. Distances were to be measured horizontally, the 12-mile circle was to be taken as a 12-mile radius, and a parallel of latitude was to be used. More specifically, the agreement stated that there was to be an

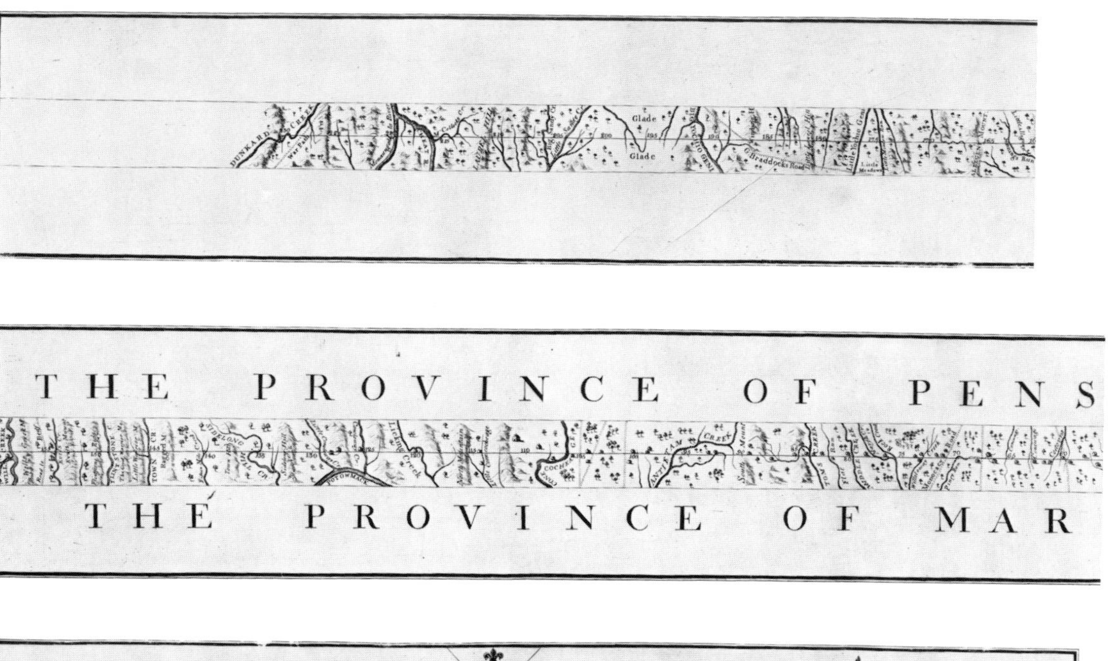

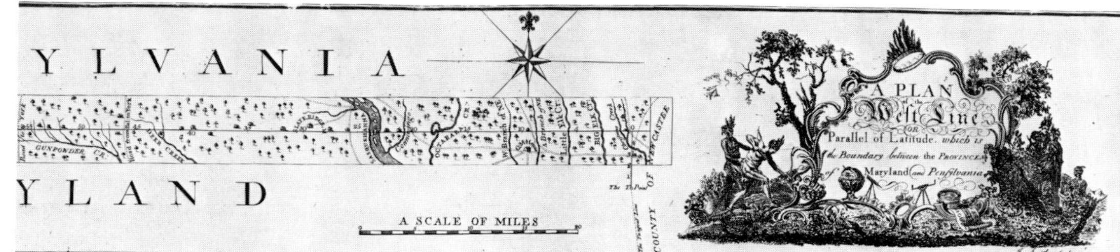

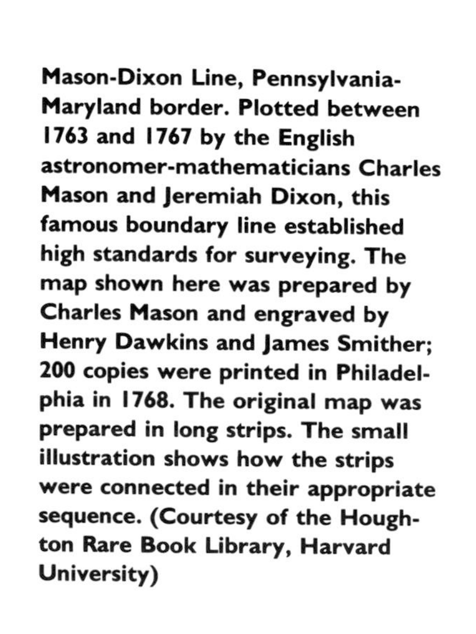

Mason-Dixon Line, Pennsylvania-Maryland border. Plotted between 1763 and 1767 by the English astronomer-mathematicians Charles Mason and Jeremiah Dixon, this famous boundary line established high standards for surveying. The map shown here was prepared by Charles Mason and engraved by Henry Dawkins and James Smither; 200 copies were printed in Philadelphia in 1768. The original map was prepared in long strips. The small illustration shows how the strips were connected in their appropriate sequence. (Courtesy of the Houghton Rare Book Library, Harvard University)

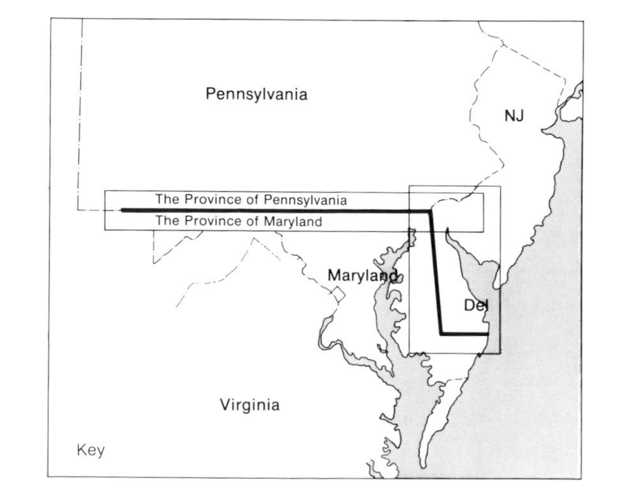

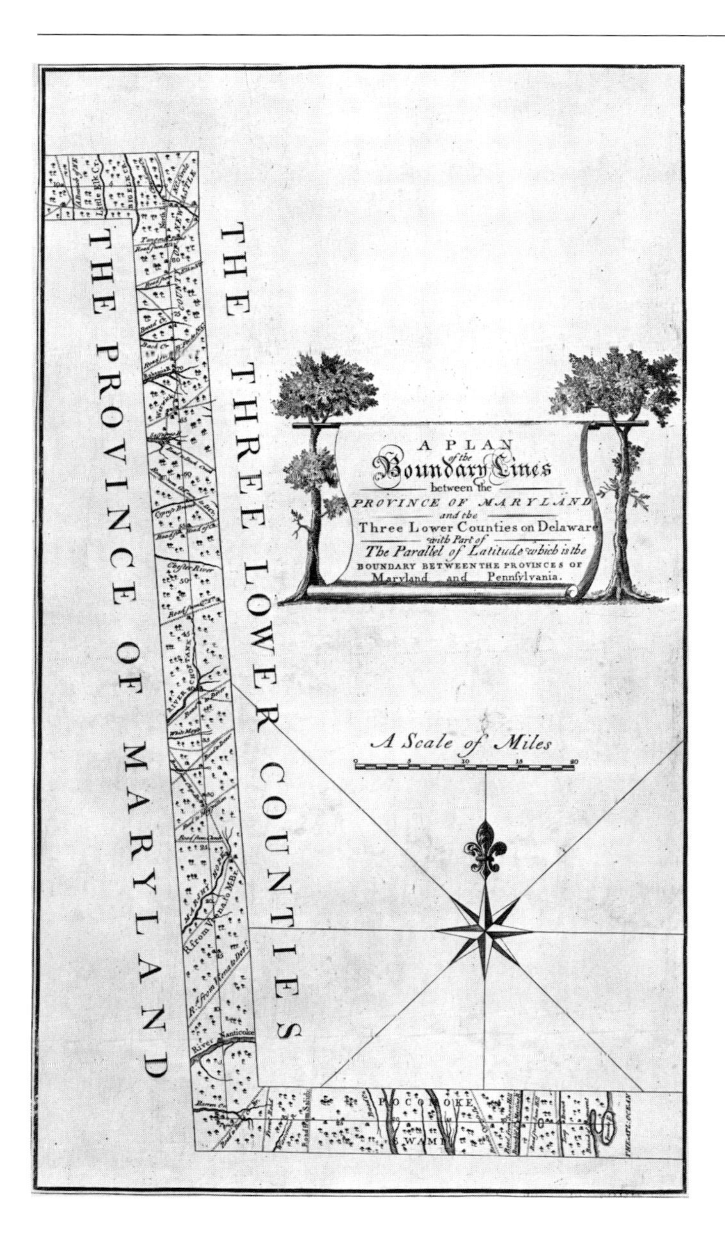

equal division of the Delmarva Peninsula from Cape Henlopen northward; that the northern boundary of Lord Baltimore's grant was to be 15 miles south of the city of Philadelphia (later the north wall of a house on the south side of Cedar Street was selected as the latitude of the southernmost point in Philadelphia); that the northern boundary of present Delaware (then the "Three Lower Counties" of Pennsylvania, granted to the Penns in 1685) was to be defined by a circle of 12-mile radius around the center of New Castle (that center having been established by the survey commissioners in 1750 as the belfry of the county courthouse); that a line was to be run from the midpoint of the southern border of Penn's domain to a point tangent to the circle of 12-mile radius, then to follow the circumference of the circle until a point was reached where the line could be run due north from the tangent point to the northern boundary of Lord Baltimore's domain; and, finally, that the western limit of Penn's domain was to be 5 degrees of longitude west of Delaware Bay. The line was to be marked in a particular manner: at the end of approximately every fifth mile the surveyors would plant a stone graven with the arms of the Penns on the north side and the arms of the Calverts on the south. Intermediate miles were to be marked with smaller stones bearing *P* on one side and *M* on the other.

In 1763 two English astronomers and mathematicians, Charles Mason (1728–1786) and Jeremiah Dixon (1753–1779), were selected by the interested parties to conduct the survey. The choice of these men and of the instruments they were to use was made on the advice of the same eminent scientists who had been consulted on the terms of the agreement. Both Mason and Dixon were assistants to the famous astronomer Dr. James Bradley at the Royal Observatory at Greenwich, where they performed measurements of the transit of Venus for determining the earth's distance from the sun.

Dr. Bradley himself had laid the groundwork for an accurate measurement of latitude in the survey by his discoveries of precession and mutation of the earth's axis in 1747. His colleague Dr. John Bevis suggested using an equatorial telescope, which was evaluated and found not sufficiently accurate by Mason and Dixon as a tool for measuring latitudes. Another colleague, John Robertson, the director of the Royal Naval Academy at Portsmouth and an advisor to the Penns, recommended that a zenith sector be used to take measurements for latitude, feeling that it was the most accurate instrument then available for measuring the zenith, or overhead passage, of selected stars, from which latitudes could be calculated. (Robertson also suggested triangulation principles for running the north-south line between Delaware and Maryland.) The acceptance of Robertson's suggestion concerning the zenith sector led in turn to a consideration of Mason and Dixon as the surveyors, since they were acquainted with its use. The actual instrument they eventually employed was made by John Bird, a cloth weaver who had become an expert in dividing clock dials and who later developed other instruments for Dr. Bradley.

Other scientists contributed their advice. Daniel Harris, master of the Mathematical School of Christ's Hospital and also an advisor to the Penns, suggested a method for running the east-west line. Dr. John Blair, Fellow of the Royal Society, approved the methodology developed by Robertson and stated his belief that the zenith sector was the most appropriate instrument for use. Nevil Maskelyre, a colleague of Mason and Dixon, discovered a fault in the design of previous zenith sectors (due to the manner in which the plumb line was suspended) and suggested the secant method used by Mason and Dixon for establishing a parallel of latitude.

The secant method consisted of running arcs of great circles that intersected the desired parallel of latitude. The arcs were allowed to extend beyond the points of intersection. The arc of a great circle then formed a secant, and offsets from the arc were used to establish the location of the parallel. Mason and Dixon selected arcs of 10-minute length, or approximately 11.5151 statute miles. Latitude checks were made at the ends of the arcs and corrections applied. Selected stars were used. Corrections were made for mutation, precession, annual aberration, and refraction. Corrections for proper motions and parallax were not made, either because of their very minor effects or because of a lack of precise knowledge of those effects. Principles of spherical trigonometry were employed to determine the bearing angles for running the arcs. Although the zenith sector was the principal instrument used, Mason and Dixon also made use of a direction transit, a Hadley navigator's quadrant, 66-foot chains, an astronomical clock, wooden rods 16.5 feet long with spirit levels, and other rods. Of extreme significance were an up-to-date star catalog by Dr. Bradley, tables for astronomical corrections, and seven-place logarithmic tables for making calculations involving sines and cosines.

In the journal kept by Charles Mason and Jeremiah Dixon it is noted that from November 16 to December 18, 1763, they were occupied in setting up an observatory and waiting for clear nights. They began their observations for the latitude of the southern point of the city of Philadelphia on December 19, 1763, and continued until January 4, 1764, using John Bird's zenith sector provided with a micrometer. Finding this point of departure impractical, they moved to the vicinity of the forks of Brandywine Creek and again observed for latitude, at last opening a line through the forest from this location. On June 1, 1765, they were at the tangent point of the New Castle circle and found a direction for running a north line from that point by observing the transit of the polestar and four other stars. At the point of intersection of the prescribed latitude and meridian lines they "placed a post." Afterward they set a line toward the Susquehanna and extended their parallel of latitude toward the Delaware River.

On August 11, 1767, the surveyors reached the east bank of the Youghiogheny River, and on September 19 they were near the Monongahela. As they progressed farther into the wilderness, their difficulties increased. On September 29, "twenty-six of our men left us. They would not pass the river for fear of the Sharnes and Deleware Indians. But we prevailed upon 15 men to proceed with us." More men joined them from Fort Cumberland. On October 9 they crossed Drunkard Creek. Indian troubles continued. They finally set up a post some 233 miles from the beginning of the west line (the northeast corner of Maryland). This point was on Brown's Hill, near present-day Morgantown, West Virginia. The distance surveyed was 233 miles, 13 chains, 68 links from the beginning and 230 miles, 18 chains, 21 links from the northeast corner of Maryland. The surveyors had not quite made their goal of "5 degrees of longitude west" because of the Indian troubles, but they had come within 21 miles, 769.1 feet of it. On October 20, 1767, Mason and Dixon began to open up the line back eastward, setting markers as they went. On November 29, they discharged most their hands and returned to Philadelphia.

Dixon died in England in 1779. Mason too had returned to England, to work on astronomical tables, but had eventually returned to Philadelphia, where he died in 1786.

In 1784 the western end of the line was established as a result of a land controversy between Pennsylvania and Virginia. Of the surveyors involved, Andrew Ellicott was the most noted. Ellicott went on to make several other important surveys, especially that of the boundary between the United States and Florida in 1796–1800 (see the next section, on Ellicott's Stone).

In 1850 a fallen stone marking the northeast corner of Maryland caused Lieutenant Colonel James D. Graham, an officer of the U.S. Topographical Engineers, to review the work of Mason and Dixon. His work corroborated that of his predecessors in "all important particulars," although some errors in tracing the curve between the tangent and intersection points were discovered and corrected. Part of the line was resurveyed a second time in 1883 by C. H. Sinclair, a geodesist with the U.S. Coast and Geodetic Survey. His line passed within 1.5 inches of Mason and Dixon's station on Brown's Hill. Between 1901 and 1903 the northern part of the line was again resurveyed under the direction of a commission jointly appointed by Pennsylvania and Maryland to settle local property disputes.

Ellicott's Stone

The only known remaining marker of a boundary line of great significance in the early history of the United States, Ellicott's Stone is located a few miles north of Mobile, Alabama. It is named after the surveyor and mathematician Andrew Ellicott, who between 1796 and 1800 conducted a survey establishing the boundary between what was then the United States' Mississippi Territory and Spain's West Florida possession.

Under the terms of the Pinckney Treaty of 1795 the two nations had agreed on 31 degrees north latitude as the boundary between their territories. But neither the Spanish nor the Americans knew exactly where this line lay. In order to forestall disputes, President George Washington personally appointed Major Andrew Ellicott to work with Spanish counterparts in carrying out a survey.

Born a Quaker in 1754 in Bucks County, Pennsylvania, Ellicott was known in his day for his expertise in many fields, among them surveying, mapping, astronomy, mathematics, and city planning. As a young man he studied under the eminent mathematician Robert Patterson at the University of Pennsylvania. He earned the rank of major in the Revolutionary War. In 1782 he published the *United States Almanac.* One of Ellicott's most important early undertakings as a surveyor was his participation in the westward extension of the Mason-Dixon Line in 1784. Over the next ten years he taught mathematics in Baltimore, conducted boundary surveys in western Pennsylvania and New York, served a term in the Maryland legislature, made the first topographic study of the Niagara River, and collaborated on the planning of the new capital, Washington, with the French engineer Pierre L'Enfant. After L'Enfant's conflict with the city commissioners and his subsequent resignation, Ellicott carried out the surveying and planning of the new city himself, with the help of his remarkable self-trained assistant, Benjamin Banneker. Ellicott's other projects prior to the Florida survey included laying out the towns of Erie, Franklin, and Warren, Pennsylvania, between the years 1794 and 1796. His brother, Joseph, was also involved in surveying during this period and made maps of western New York that were soon to become invaluable in the building of the Erie Canal. It was Joseph Ellicott too who was instrumental in getting Benjamin Wright and James Geddes selected as engineers for the construction of the Erie.

The United States–Spanish West Florida boundary survey required all of Andrew Ellicott's technical and diplomatic skills. Much of what we know about the project comes from his journal, in which were recorded his adventures through the territory and many climatic observations. Ellicott started near the Mississippi River on April 11, 1798, and worked eastward to the Chattahoochee, encountering many difficulties along the way, foremost among them the problem of making accurate measurements in terrain composed of thick forests and swamps. He arrived at the Mobile River on March 18, 1799, and completed a course of observations that resulted in the placement of the marker now called Ellicott's Stone.

The stone is located on a rise near the bank of the Mobile River, between what are now the Cold Creek and Chastang stations on the Mobile and Birmingham Railroad line. An irregularly shaped piece of sandstone about 2 feet long, 8 inches thick, and 3 feet high, the marker bears the carved legend "U.S. Lat. 31, 1799" on its north side and "Dominio de S. M. Carlos IV, Lat. 31, 1799" on its south side. It is now identified by a metal marker affixed to a high fence that gives it protection. Despite the hindrances of treacherous terrain and relatively crude instruments, the survey—if Ellicott's Stone is a fair representative—was remarkably accurate.

The results of the survey brought disappointment to Spain: it was found that both Natchez and the then-thriving town of St. Stephens, Alabama's early capital, were on the United States side of the 31st parallel.

Ellicott himself went on to perform more surveys and become secretary to the Pennsylvania Land Office. He got involved in the mapping of the Georgia-Carolina border (Georgia was recalcitrant in the matter of payments to Ellicott, because the surveys had added territory to the Carolinas). His last major survey, in 1817, was a consequence of the

Ellicott's Stone, Mobile County, Alabama. This sandstone marker, located a few miles north of the city of Mobile, is the only visible remainder of the survey conducted between 1796 and 1800 to establish the 31st parallel as the border between the United States and Spanish West Florida. Andrew Ellicott was appointed by President George Washington for the task. (Courtesy of the American Society of Civil Engineers)

Treaty of Ghent, which ended the War of 1812. This project established the location of the 45th parallel on the New York–Canada border. Ellicott had in 1813 accepted a professorship at West Point; there he returned after the 1817 survey and remained—enjoying a close friendship with the academy's superintendent, the influential engineer and educator Sylvanus Thayer—until his death in 1820.

The Massachusetts Base Line

The base line established by Simeon Borden in 1831 as the basis for a survey of the state of Massachusetts was an outstanding achievement in precision measurement. The 39,000-foot line, between points in Hatfield and South Deerfield, in western Massachusetts, brought American skill in geodetic engineering to international notice.

In March 1831 the Massachusetts legislature resolved that a survey of the state be conducted on trigonometric principles. The objective was to attain the superior accuracy of trigonometric surveying for an area of a size for which astronomical surveys only had formerly been used. The governor appointed Robert Paine as principal engineer; his assistant for topographical survey was Colonel James Stevens. Paine assumed the role of astronomical surveyor. After Stevens's resignation in 1834, Borden carried his part of the project to completion, publishing the map of Massachusetts in 1844.

Borden, a native of Fall River, Massachusetts, had acquired only a rudimentary education in country schools when he left at thirteen to help his mother manage her farm. He continued to study applied mathematics at home, however, and eventually became superintendent of a machine shop—which led to his being hired to devise and employ the measuring instruments used in the base line survey. The establishment of an accurate base line was the key to the success of the undertaking.

Borden created a measuring device consisting of two 50-foot rods, one of steel, the other of brass. The use of two rods enabled him to compensate for thermal expansion of the metal, so that distances measured by the device would remain essentially constant over a wide range of temperatures. The rods were enclosed in a "strongly and firmly soldered tin tube between 7 and 8 inches in diameter," as Borden noted in a paper presented before the prestigious American Philosophical Society in Philadelphia in 1841. To improve accuracy, compound microscopes with cross hairs were focused on the end of the measure,

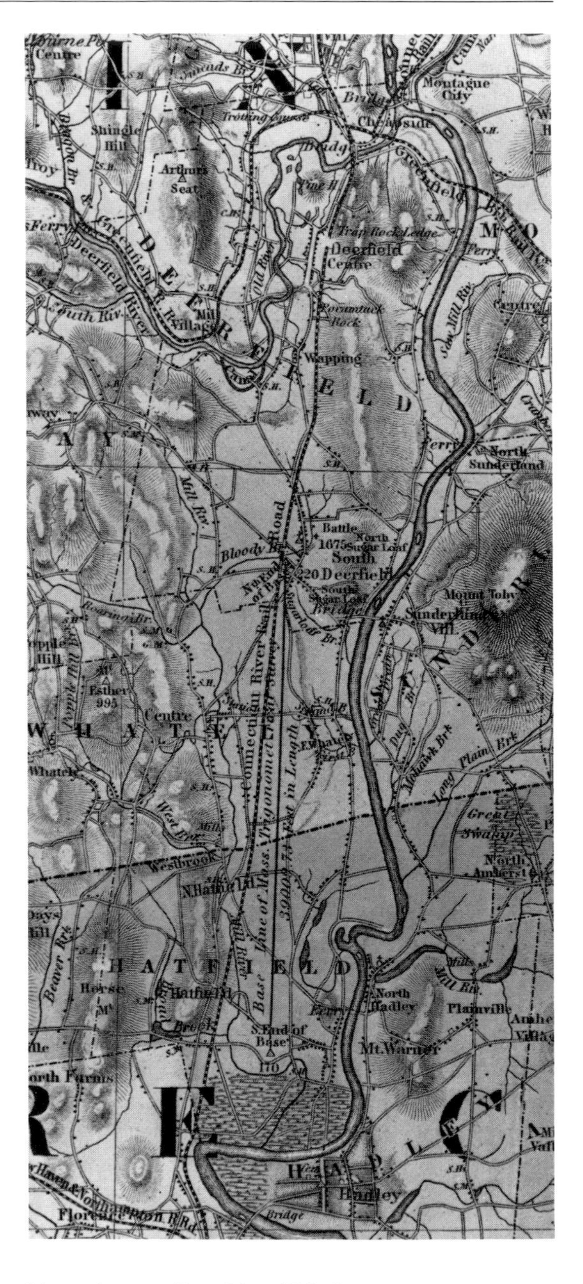

Massachusetts Base Line. This feat of precision measurement in 1831 helped make possible an accurate survey of the state of Massachusetts. The 39,009.73-foot line ran from a point in Hatfield, Hampshire County, to another in South Deerfield, Franklin County. (Courtesy of the Northampton Public Library, Northampton, Massachusetts)

marked by a simple cross upon a plate of silver. Borden developed this remarkable device from scratch, yet it was the most accurate of its kind in this country at the time.

Using this instrument, Borden began his survey in 1831. Two sets of measurements were taken, one in each direction. The base line ran from a point in Hatfield, Hampshire County, to a point in South Deerfield, Franklin County. The southerly terminus marker, near Bridge Street in Hatfield, was discovered by Arvo A. Solander, a civil engineer and land surveyor from nearby South Hadley, Massachusetts, in 1973. It is a boulder roughly 3 feet in diameter, into which is inserted a copper bolt. It was covered by 10 inches of soil, as compared to the 18 inches reported in 1831. The line is 39,009.73 feet long.

International Boundary Marker Number I

With the Treaty of Guadalupe Hidalgo, signed in the Mexican village of that name in 1848, the Mexican War ended and Mexico ceded to the United States the territory that now comprises California, Nevada, Utah, and parts of Arizona, New Mexico, Colorado, and Wyoming. The boundary between the two nations having been established in principle, a survey was ordered to locate it on the ground. International Boundary Marker Number 1, located on the west bank of the Rio Grande in New Mexico between El Paso, Texas, and Juarez, Mexico, marks the easternmost end of the land boundary (the Rio Grande forming the rest of the border) and stands as a monument to the professional skills of the Emory-Salazar Commission, the American surveying team that determined its location in 1855. Built that same year, the marker is situated with its south half in Mexico and its north half in the United States; its exact location is latitude 31°47′01.627″ and longitude 106°31′45.108″. The monument is of stone, 12 feet high and 5 feet wide at its base. In 1966 it was refaced with white marbleized concrete, and a concrete base was installed.

The U.S.-Mexico border was surveyed by a team headed by William Emory, an 1831 graduate of the United States Military Academy at West Point. Emory was working as principal assistant on the Northeastern Boundary Survey, establishing the U.S.-Canada border, when the Mexican War began in 1846. He was then assigned as chief engineer officer to General Stephen Kearny's Army of the West, which campaigned in New Mexico and California. Raised to the position of acting assistant adjutant general, Emory was later assigned as chief astronomer in charge of running the boundary line between the United States and Mexico. His task lasted from 1848 until 1853. In 1854 Emory received a presidential appointment to survey the boundary of the additional land acquired from Mexico through the Gadsden Purchase of 1853. After completing his work in 1857, he served with distinction in the Civil War.

During his later service in the Topographical Engineers, Emory made many contributions to understanding the geography of the West. Kearny's route had been across largely unknown territories. Emory undertook a survey of a military, scientific, and economic nature that would give the government an idea of the regions traversed. He subsequently prepared his well-known *Notes of a Military Reconnaissance,* which contained large-scale maps and significant geographical measurements, including records of latitudes and longitudes, altitudes, the dimensions of streams, gradients, and other similar data—all important in documenting the nature of the Southwest. The War Department thought so highly of his *Notes* that it was printed in a 10,000-copy edition. This work undoubtedly influenced considerations concerning the admittance of western territories as states.

Chapter Thirteen

Coastal Facilities

Introduction

The fine natural harbors found by the colonists of the New World at first required little of the artificial improving that had been practiced for centuries in the Old. In ancient India, Greece, and Egypt, as well as in modern Europe, the need to provide safe moorage, safe passage along coasts, and facilities for building and loading ships had given rise to superb civil engineering accomplishments. Soon, however, the intensifying use of the North American coast demanded that the settlers draw on this engineering heritage to develop marine support facilities ranging from wharves, breakwaters, and dry docks to ever-needed lighthouses.

The first lighthouse in America was built on Brewster Island in Boston Harbor in 1716, with a number of others following shortly afterward. In 1789 Congress passed the first federal public works law, which provided for federal control of navigational aids. The Cape Henry lighthouse, completed in 1792 at the entrance to the Chesapeake Bay, was the first public work built by the United States. Later periods saw extensive building programs for lighthouses as the maritime power of the new nation began to emerge. Often lighthouses were built and operated by private contractors for the government. Eventually, in 1852, Congress appointed a Lighthouse Board, consisting of naval officers, army engineers, and civilians, to oversee the lighthouse system. An army engineer was soon after appointed to each of twelve districts to supervise construction and repair.

Under the guidance of the Lighthouse Board, improved navigational aids began to be developed, as well as improved construction methods. The harsh sea environment necessitated a high quality of design and construction—as evidenced by the disastrous failure of the first lighthouse at Minot's Ledge off the Massachusetts coast, a poorly conceived and executed structure that collapsed during an intense storm in 1851, carrying two keepers to their deaths. Its replacement, superbly designed and crafted by General Joseph Totten in 1860, still stands. The Lighthouse Board was disbanded in 1910 and replaced by the Bureau of Lighthouses, later renamed the Lighthouse Service. This was in turn abolished in 1939 and its activities transferred to the U.S. Coast Guard.

Military needs also led to the construction of coastal works, such as naval stations and shipyards. Throughout the colonial and revolutionary eras the latter were all privately owned. The Algerian pirate crises of 1794, however, prompted the new nation's secretary of war, Henry Knox, to authorize the construction of six frigates and the leasing of shipyard facilities in which to build them. Eventually the threat subsided, and only the *Constitution, Constellation,* and *United States* were launched by 1797. But trouble with France soon developed, and additional vessels were authorized. In 1798 the Navy Department was created and Benjamin Stoddert designated as its first secretary. Realizing the need for a system of construction and support facilities, Stoddert moved to create several shipyards. Their locations, of course, became the subject of intense political maneuvering and pressure; the towns of Portsmouth, Washington, Boston, Norfolk, Philadelphia, and Brooklyn were eventually selected as sites for the yards.

The War of 1812 demonstrated the need for more extensive facilities and, in particular, for dry docks. Stoddert had recognized this need earlier but could never secure adequate funding. In 1825 Secretary Benjamin Southward recommended the construction of two dry docks, one at the Boston yard and the other at Norfolk. These were built under the direction of Loammi Baldwin, Jr., and put into service in 1833.

Their careful craftsmanship also attests to the perception of early engineers of the need to build well in marine environments. From this time on, shipyard developments continued at a rapid pace.

The improvement of coastal waterways, river entrances, and harbors for navigational purposes was another area in which notable civil engineering feats were accomplished. The federal government has been the chief author of waterway improvement since 1824, when a landmark case gave it control over navigation within the limits of every state of the union. Based on this decision, two authorizations were soon passed in Congress that enabled federal support for surveys and other engineering work on roads and canals, and for the improvement of navigation on the Ohio and Mississippi rivers. The latter authorization formed the basis for subsequent Rivers and Harbors acts, which provided for extensive federal involvement in the construction of waterways and coastal facilities.

Major river improvement projects included the remarkable work at the mouth of the Mississippi conducted by that most versatile of engineers, James B. Eads. Eads, who had engineered the first major bridge across the Mississippi at St. Louis, undertook in 1875 to open the mouth of the Mississippi to deep-water navigation by building a series of jetties in such a way that the river itself would scour a channel and prevent silt buildup. This project, completed in 1879, and other works turned the Mississippi River into an even more valuable national asset than it had been; today the Mississippi is linked to an extensive system of coastal waterways still actively in use along the Gulf Coast and the eastern seaboard.

Of many remarkable harbor improvement projects, two deserve particular mention. In 1852, in a major construction undertaking, New York's port facilities were enormously improved by the removal of the rocks at Hell Gate in the East River. In the West, improvements to the San Francisco harbor also date to the 1850s, with the dramatic removal by blasting of Blossom Rock, which sat only five feet below the surface in the main ship channel east of Alcatraz Island. Other noteworthy coastal improvements were the Houston Ship Channel, constructed between 1912 and 1914, and the Intracoastal Waterway, begun in 1919, which provides sheltered passage for commercial and pleasure boats along the Atlantic and Gulf coasts. The Army Corps of Engineers was instrumental in most of these improvements.

The Naval Dry Docks at Boston and Norfolk

The dry-docking facilities built at the U.S. Navy's Boston and Norfolk shipyards during the period 1827–1834 appreciably increased the effectiveness of the young nation's naval forces by greatly reducing the time that vessels were out of service for underbody cleaning and repairs. The significance of the dock facilities to the civil engineering profession was notable as well. Produced from the same set of plans, both were the work of Loammi Baldwin, Jr., of Boston. Without a precedent in this country, and with only local materials and the limited skilled labor and mechanical equipment then available in America, Baldwin designed and constructed two superb facilities whose functional success established a prototype on which many of the navy's future dry dock installations were based.

The early military experience of the United States, in particular the American Revolution and the War of 1812, vividly demonstrated the need for a coastal defense system centered on a strong navy. Important to the formation of such a naval force were land-based support facilities in strategic locations. As early as 1798, America's first secretary of the navy, Benjamin Stoddert of Maryland, recognized the value of dry-docking facilities for such a force. The common practices of "heaving down" and "careening" a vessel in order to clean and repair its hull were time-consuming and dangerous processes. The ship was also entirely vulnerable to attack, and such necessary maintenance depleted the ranks of an already underequipped navy. Consequently, Stoddert advocated the erection of dry docks at the entrance to Chesapeake Bay in the south and at Boston, Portsmouth, or New York to the north. Congress apparently only half-heartedly agreed, appropriating the inadequate sum of $50,000 in 1799 for the dry docks. Because of the insufficient allocation these installations were not realized at the time.

The War of 1812 and the burning of Washington, the new capital, finally convinced Congress of the woeful inadequacy of American naval resources. Secretary of the Navy Benjamin Southward reviewed the subject of

Dry docks at the Charlestown Navy Yard, Boston, Massachusetts (*top*), and the Gosport Navy Yard, Norfolk, Virginia (*bottom*). These structures, among the earliest of their type in the Western Hemisphere, were designed and built by Loammi Baldwin, Jr., between 1827 and 1834. Both served the United States Navy for well over a century. The Charlestown Navy Yard, now a national park, is also the home of the USS *Constitution*. (Courtesy of the American Society of Civil Engineers)

dry docks in 1825, after strongly worded reports from the Naval Commissioners in 1815 and 1821 argued the desperate need for such facilities. Southward presented proposals for dry docks at Norfolk and Boston to Congress, and in 1827 Loammi Baldwin, Jr., was commissioned to prepare plans for the undertaking.

For Baldwin, whose father was well known for the design of Massachusetts's Middlesex Canal at the turn of the century (see chapter 1), the project culminated a career of many accomplishments. Having started out as a lawyer on his graduation from Harvard in 1800, Baldwin found only frustration in this field. He achieved early success, however, in designing one of America's first steam-powered fire engine pumpers, used in Groton, Connecticut, for almost seventy years. By 1814 he was in residence as an engineer in Charlestown, Massachusetts, designing and supervising the construction of fortifications against the British at Fort Strong in Boston Harbor. As engineer in charge of public improvements for the city of Boston in 1819, Baldwin extended Beacon Street beyond the Boston Common into the Charles River marshes. Intended to create a milldam that would provide water power, the scheme proved unsuccessful; but it formed the cornerstone for the later project creating Back Bay on land reclaimed from the river. On his return from a trip to Europe in 1825, Baldwin supervised parts of the design and erection of Boston's Bunker Hill Monument, collaborating with Alexander Parris for the first time on the 220-foot granite tower. During this period Baldwin also made proposals for a canal from Boston across Massachusetts to New York State, connecting with the Erie Canal. The proposals, never carried out, included a daring plan for a tunnel through the Hoosac Mountain, an idea that came to fruition fifty years later as the largest railroad tunnel in the Western Hemisphere (see chapter 5). In duties outside New England, Baldwin oversaw numerous public works projects in the state of Virginia between 1817 and 1820, and designed Pennsylvania's Union Canal, an outstanding 79-mile system of interconnected waterways and locks between Reading and Middletown. He also acted as a consulting engineer in the review of the Charlestown shipyard facilities prior to commencing the dry dock project.

Baldwin received the commission for the two dry docks and undertook the task with extreme rapidity, planning from the outset an identical configuration for both sites. Having observed examples of functioning drydocks in England and Antwerp during visits in 1807 and 1824, he resolved to duplicate successfully these achievements without the advantages available in industrialized Europe. Work commenced on the Boston dock, in the Charlestown Navy Yard, in June 1827. The Norfolk installation, in the Gosport Navy Yard, was started in November of the same year. Baldwin appointed Alexander Parris, the noted Boston architect known for his use of granite, as his principal assistant at Charlestown, and Captain William P. S. Sanger of the navy to a similar position at Gosport. During the following seven-year period Baldwin divided his time between the two sites, overseeing construction in Boston during the summer months and residing at Norfolk during the winter. In each case his appointed assistant was in charge during his absence.

At each site construction began with the erection of a cofferdam to protect the excavation from the waters of the harbor and tidal forces. Double rows of sheet piling, spaced 8 to 13 feet apart, were driven into the harbor mud, and the space between was filled with loose stone and earth in order to resist the considerable lateral thrusts of tidal forces. By the spring of 1828 excavation was under way at both sites, almost entirely carried out by hand because of the limited availability of steam power at the time. Wooden bearing piles, spaced 3 feet on center both ways, were driven at both sites in the summer of 1828. Incredibly enough, the bearing-pile machinery was also man powered, utilizing human treadmills to raise the gravity-operated pile hammers. Although some resistance to this mode of operation was encountered, particularly in Boston, the primitive process went steadily forward. The wooden base timbers, 12 inches square and of treated white oak, were laid in the early months of 1829. The spaces between the timbers were filled in with broken stone. Two layers of 3-inch pine planking, laid diagonally, rested upon the timbers, supporting in turn a layer of oak timbers, 16 inches square. The space between the timbers was filled with brick laid in waterproof cement, thus presenting a level surface upon which the granite stone flooring could be laid.

Baldwin designed the dry docks for cut-stone flooring and side walls, using granite from the Quincy, Massachusetts, quarries for both installations. A relatively high level of craftsmanship in stone masonry existed in New England at the time, mainly because of the existence of the Quincy and Vermont stone quarries. Baldwin took full advantage of this resource, planning for both dry docks to incorporate face-hammered granite stonework on all floor and interior wall surfaces. Waterproof cement, composed of fine sand and hydraulic lime, was used in the setting beds throughout. Granite from the Quincy quarries was hauled to the Neponset River terminus on the Granite Railway, another notable civil engineering achievement of the time, whereupon it was barged to the Charlestown site or shipped by sea to the Norfolk installation. The masonry cornerstones for both projects were laid in the summer of 1829, and for the next four years stonework proceeded, weather permitting.

The masonry floors varied in thickness from 5.5 feet at the dock head to 4 feet at the base. Side walls were set to contain a series of offsets, so that the 35-foot thickness of the wall at the dock floor reduced to 7 feet at the wall's top. The thickened bottom walls were required to counteract the greater water pressure at those times when ships were floated into the dock for servicing. The offsets allowed for a series of work platforms from which the ships could be serviced, as well as convenient stops against which the shoring used to stabilize the dry-docked ship could be anchored. Six flights of stairs leading to the dock floor were incorporated into each facility. Finally, brick culverts of an oval arch design were built within the dock's stone walls, leading to wells from which steam-powered lift pumps drained the water into the sea. When completed, each dock contained 500,000 cubic feet of Quincy granite.

Baldwin's design for the gates was equally innovative. Each dry dock had an inner gate, consisting of two wooden doors that turned outward, and an outer, floating gate. The massive wooden doors of the former moved on friction rollers resting upon cast-iron plates set into the stone flooring. A series of chain-driven, steam-powered pumps operated the turning doors. This gate was designed with a convex curvature against the harbor's waters, using the arched configuration to provide additional bracing against tidal forces. This concept was later employed in the development of the modern concrete arch dam. Beyond the turning gate was a floating gate of massive wood construction, designed to be floated into location and set into notches in the stone walls. The floating gate served as an additional buffer against harbor tides, which in Boston varied to a vertical distance of 11 feet. Moreover, Baldwin foresaw the building of larger ships by the navy and thus provided for an extra length of approximately 50 feet. Both types of gates at Boston and Norfolk were sheathed in copper, to prevent attack on the wood by sea worms.

By the late spring of 1833, despite a severe New England winter in 1830–1831 that had greatly restricted stone quarrying operations at Quincy, both installations were substantially complete. Few structures of such immense dimensions had been built in the United States at this time. The docks measured 314 feet in length, 100 feet in width, and 32 feet in depth at Charlestown, slightly less at Norfolk due to a shallower excavation. The adaptation of the arch principle to his gate design illustrates Baldwin's ingenious modification of the European turning gate from iron construction to wood. In the stone floor and wall construction Baldwin applied the arch principle once again, laying the stone in each course as an inverted arch from side wall to side wall, thus resisting the upward hydraulic pressure exerted from beneath the dock floor. Finally, by utilizing the oval arch throughout the network of culverts within the side walls, Baldwin again achieved greatest strength at points of maximum pressure, allowing an economy of means with only minimal skilled forces. The assistance of

The USS *Delaware* was the first ship dry-docked in the Gosport Navy Yard at Norfolk. She entered the dock on June 17, 1833. (Drawn on stone by G. Lehman from a sketch by J. Bruff; Naval Historical Center photograph)

Totten's stone lighthouse replaced an iron openwork tower constructed 1847–1849. In April 1851 a violent and prolonged gale lashed the Massachusetts coast, causing the supporting pilings to begin giving way. The two keepers threw out a bottle with a message about their plight and frantically rang the tower's bell, but they could not be reached by rescue boats. The tower finally collapsed into the sea, carrying the hapless men to their deaths. (Courtesy of the Prints Department, Boston Public Library)

The impossibility of rescuing the keepers during the 1851 gale is easily understood when one views this engraving, which dates from before the iron tower's collapse. It shows how keepers sometimes got to the tower in rough seas. (Courtesy of the Prints Department, Boston Public Library)

under water. Iron shafts 20 feet high were set into eight of the piling holes that remained from the previous structure; the center hole was left open. Thirty-five hundred tons of Quincy granite was shipped uncut to Government Island, near the ledge, where it was fashioned into 1,079 stones averaging 2 tons each—a total of 2,300 tons. In order to ensure a perfect fit, the blocks were prefitted on the island before being shipped piecemeal to the ledge.

After two years of foundation work—during which at one point the entire work was washed out—the first stone of the tower was laid on July 9, 1857. Seven enormous blocks were supplied to form the base, which was concreted to the rock ledge. Temporary cofferdams made from sandbags protected the mortar from wave action, while iron straps laid between the courses kept the stones apart until the concrete hardened. After the first courses had been laid and work could be performed above water level, construction speed increased, but the task was still slow and arduous. By the end of 1859 only the thirty-second course, 62 feet above low water, had been reached; and it was not until June 29, 1860, only one day short of the day construction work had begun five years before, that the last stone was set in place. The keeper's house at the top was ready in August, and on the 22nd of that month the first light shone from the new Minot's Ledge Light.

The $330,000 that was required to build this superb structure seems well spent today. The Minot's Ledge Light has withstood the fiercest gales of this century and the last. Only the slightest vibration occurs in the tower during strong winds and high seas, which have swept over the top of the 97-foot shaft. These great waves have done no more than break windows. The lighthouse remains in excellent condition and is structurally sound. In 1894 it was given a new flashing lantern whose rhythm of one, four, three—construed by romantics to spell out "I love you"—has inspired the epithet "Lovers' Light." Since 1947 the keeper's house has been empty; the present electric light is visible for 15 miles. Together with England's famous Eddystone Light, Minot's Ledge stands as one of the two foremost wave-swept lighthouses in the world.

The South Pass Navigation Works

In 1879 James B. Eads completed a feat of hydraulic engineering that gave the Mississippi River basin direct deep-draught access to the oceans of the world. This internationally acclaimed achievement is known as the South Pass Navigation Works.

The Mississippi River joins the Gulf of Mexico by a series of small mouths and passages. As water velocity decreases through these many channels, the river's load of silt precipitates and is formed into bars that block navigation. During the nineteenth century New Orleans and other river ports were frequently rendered inaccessible to large ships because of these blockages in the delta. Many attempts were made to loosen the bars or provide passage through them. In 1726 the French dragged an iron harrow through Southwest Pass in an effort to loosen it. A bucket-drag dredge was used in 1837; harrowing was again tried in 1852; timber jetties were tried in 1857; and another harrowing was tried in 1860. None of these attempts yielded lasting results. Bars simply reformed.

After the Civil War there was intense pressure on the U.S. Army Corps of Engineers from various commercial and farming interests to make the Mississippi navigable for deep-water ships. Congress soon authorized funds to improve the mouth of the river. A unique steam-powered dredge was tried in 1868, but the attempt again failed. Captain Charles W. Howell of the corps was subsequently charged with making surveys and estimates for a ship canal that would bypass the river mouth. The completion of the Suez Canal in 1869 renewed interest in canal construction. A board of engineers was convened and evaluated alternative methods of opening up the river to navigation. Studies were made of similar projects in Europe. Attempts, based on dredging and the use of dikes and jetties, had been made to improve the mouths of the Rhone, Vistula, and Danube rivers. The results of these European works were ambiguous at best. The board's report recommended a canal only if all other techniques proved failures. Howell argued

South Pass Navigation Works, Mississippi. This project by James B. Eads opened the Mississippi River to ocean-going ships in 1879. The photograph, from 1949, shows South Pass, with the head of the pass just visible in the distance. The dike system is evident in the foreground. (Courtesy of the U.S. Army Corps of Engineers)

that this was indeed the case and recommended a canal at Fort St. Philip. He was supported by General Pierre Beauregard and General Andrew Humphreys, both extremely influential engineers of the time. But not all engineers in the corps agreed. Another review board met in June 1873 and approved the canal—although the president of the board, Colonel John Bernard, submitted a minority report strongly advocating the use of jetties at South Pass (a short but narrow pass that had been largely ignored previously). Given the high projected cost of the canal, controversy over the canal proposal continued.

Meanwhile, James B. Eads had urged a jetty plan to a group of visiting congressmen in May 1873. By the end of the same year he had submitted a formal proposal to open Southwest Pass. Eads was already famous for his magnificent bridge across the Mississippi at St. Louis (completed in 1874; see chapter 4, Eads Bridge). His plan was to create a channel 28 feet deep and to build jetties for scouring action to maintain the channel. The issue went to Congress, which was faced with Howell's canal proposal and Eads's jetty proposal. The House approved an appropriation for a canal, but the Senate rejected the bill. A new review board was established that was composed of both army and civilian engineers. In its January 1875 report to Congress, a jetty system was recommended for South Pass—a recommendation supported by two of the three corps engineers. South Pass was chosen apparently because if the experiment failed no permanent damage would be done: with its 8-foot depth, the pass was worthless as it stood. After long and heated debate, Congress approved Eads's proposal. But Eads would only be paid in increments depending on the widths and depths of the channels obtained and kept clear. Not only did he have to create a deep-draft channel, but he had to maintain it for twenty years before he would be fully paid.

On June 14, 1875, Eads began work at South Pass. Dredging was soon proceeding at the rate of 200 feet per day. The channel had been dredged to a depth of 20 feet and a width of 200 feet by late December 1876. By July 8, 1879, it had reached a depth of 30 feet. Eads succeeded in maintaining this depth until the termination of his contract twenty years later.

Parallel jetties, wing dikes, and T-dams were also built. The east jetty ranged 11,700 feet by 1877, and the west, 10,125. At the head of South Pass a T-dam was built to deepen the channel at that point. Wing dikes perpendicular to the direction of the current were built to force the current into a narrow cross section and thus to increase scouring action. Still dams were built across Southwest Pass and Pass à Loutre to send even more water down South Pass.

While construction was under way, Eads constantly struggled for money and against General Andrew Humphreys. By this time Humphreys was Chief of Engineers. Congressional approval of Eads's proposal had strongly angered Humphreys, who felt that the approval was an attack on the Engineer Corps. An accomplished hydraulic engineer in his own right, he had also written extensively about how a jetty system could never succeed at the mouth of the river. Howell also fought against the work and commissioned detailed studies to demonstrate how the mouth was indeed shoaling up again as predicted by Humphreys. A series of remarkable and detailed studies on flows and the rate of new bar formation was conducted by Captain Micah Brown—who found that a channel was indeed forming, and growing wider and deeper as the scouring action provided by the jetties did its work. Sediment was being carried down the continental slope into extremely deep water. These surveys were sent up through official channels and finally to Humphreys. Eads was not allowed to see them. In spite of the reports, Howell continued his public opposition and predicted that shoaling would shortly occur. With Humphreys and Howell against him, Eads found it difficult to get funding for ongoing work. He bought the support of General Beauregard for the jetties and eventually prevailed upon the secretary of war to order the Engineer Corps to provide access to the surveys. The remarkable nature of Eads's accomplishment became known as the survey results came to light.

The project was an immediate and lasting success, proving that the construction of parallel jetties and other structures at the mouth of an alluvial river could indeed serve to scour a navigable deep-water channel. The benefits to the port of New Orleans were soon evident. The city's decline was reversed, as deep-draft vessels found safe, easy access to its facilities. To the civil engineering profession, too, this project brought benefits in the form of advances in hydrological science and in the understanding of the mechanics of bar formation.

The South Pass jetty system, 1949. (Courtesy of the U.S. Army Corps of Engineers)

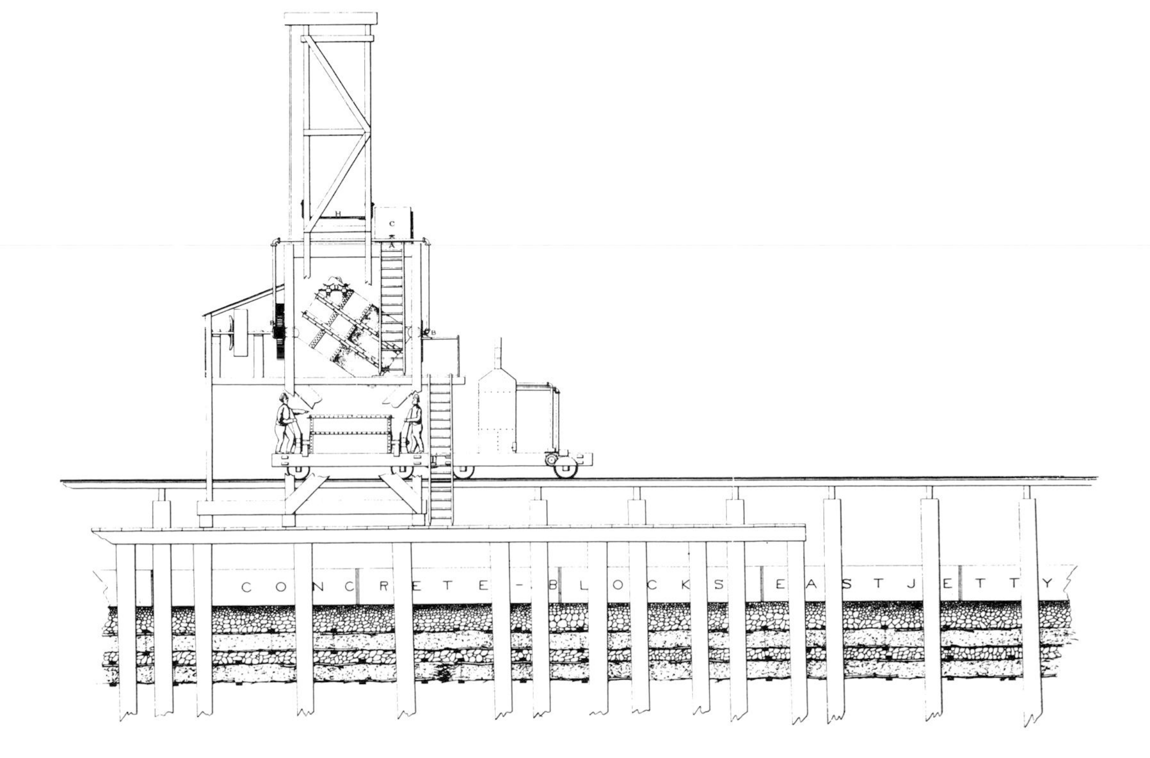

Jetty construction. First, a line of pilings was driven. Willow mattresses were then sunk with broken stones and attached to the pilings. Alternate courses of willow mattresses and broken stone were next added. A railroad line was then constructed to carry concrete mixers and dump cars. A final concrete cover completed the structure. (*Transactions of the American Society of Civil Engineers* 8, no. 183 [1884])

Chapter Fourteen

Airports

Introduction

In modern American society, characterized as it is by rapid movement—the instant dispersal of information, the quick and widespread distribution of goods, and ease of travel for people—air transportation has assumed a dynamic role. Aviation as a mode of transit and interchange is of equal importance with the automobile, the telephone, or television. One of the factors leading to this result has been the development of the American airport, whose early history was marked by innovative developments in landing systems, air traffic control, and airfield design.

From the time of the Wright brothers' first flight through the early 1920s, relatively simple takeoff, landing, and support facilities such as those at Hammondsport, New York, served aircraft for the military, private enterprise (ex–war flyers and barnstormers), and the mail service. It was not until 1926, when the Air Commerce Act was passed, that the importance of aviation for America's future was recognized through the establishment of a mechanism for both regulating and providing services for commercial air flight. With this act the United States Department of Commerce became responsible for the registration and licensing of pilots and aircraft and for the establishment of air traffic laws. The federal government did not assume responsibility for air traffic control, however, leaving this task instead to state and local governments, airport operators, or even the private airline operators themselves. The basis of the modern airport was thus established.

Commercial air travel increased significantly in the years 1926–1929. During this time such aspects as soil conditions, drainage, wind velocity, orientation, and proximity to ground transportation became familiar factors in the design of airports, and the civil engineer became crucially involved in runway siting, design, and construction issues. The airports constructed at Newark and Cleveland in this period were prototypes for today's facilities.

During the Depression there was a constant improvement of airport facilities through the WPA and other national relief programs. Gradually the needs of both airplanes and the people who used them became better understood, leading engineers to define more clearly the design requirements for airports. Uniform conventions for marking approaches and obstructions and for delineating boundaries to airfields were established in coordination with the Department of Commerce. In 1933 the Civil Aeronautics Act was passed, providing government subsidy for new facilities within the guidelines set by the Air Commerce Act of 1926.

The Second World War brought about the biggest surge in aviation in the century. As a result, under Defense of Landing Areas and Defense of Civil Areas programs, many more airports were established and improved. In 1946 the significant increase in the volume of air traffic stimulated passage of the Federal Airport Act, which established new guidelines for classifying airports and provided some funding for development. Air traffic control became the major issue of the time and continues to be a central aspect of airport design today, utilizing advances in radio, radar, and computer applications to guide and land planes.

Drawing on the precedents set by the engineers of the 1920s, airports have reached an advanced level of technology and efficiency based on a coordination of complex use with public and private needs. The development of airport design has helped make commercial aviation, in the span of little more than half a century, an absolutely necessary part of our nation's industry, business, and life style.

Cleveland-Hopkins International Airport

The Cleveland-Hopkins International Airport, still in its original location ten miles from the center of Cleveland, Ohio, has from the outset produced landmark developments in aviation engineering. Early advances in airfield lighting systems, pioneered at Cleveland, allowed for night flying on the first transcontinental air mail service. Cross-country overnight mail service helped demonstrate to Americans in the 1920s the tremendous potential of air transport for rapid goods and information exchange. As the permanent home of the National Air Races, first staged in 1929, Cleveland Airport attracted widespread publicity to the early aviation industry and popularized the exploits of leading flyers of the day, such as Jimmy Doolittle. Perhaps Cleveland's greatest contribution to aviation development, however, was its control tower of 1929. As the first such facility in the United States, Cleveland's tower marked the advent of ground control for air traffic. In addition, radio communication from the control tower advised pilots of wind, air traffic, and field conditions both en route and at landing fields. This greatly increased the safety of air transportation before the introduction of radar.

The founding of the airport at Cleveland in 1925 was due largely to that city's leadership as a freight shipping terminus in the nineteenth and early twentieth centuries. Prior to the railroad era, canals and other waterways had carried most of America's goods and had brought vitality and growth to Cleveland as a major port of interchange between Great Lakes industries and the inland river transport network. Shipping and shipbuilding industries flourished in nineteenth-century Cleveland, dominating the city's economic structure into the twentieth century. With the introduction of commercial air passenger and freight transport in the 1920s, Cleveland city officials sought to establish a major airfield facility, hoping to maintain Cleveland's importance in a marketplace being revolutionized by the new technologies of aviation.

Cleveland-Hopkins International Airport, Cleveland, Ohio. Completed in 1925, this was the first American airport to provide an integrated, engineered system of runways, floodlighting for runways, and a terminal complex consisting of both operational buildings and hangars. This 1937 view is of the boarding area. (Courtesy of the HAER Collection, Library of Congress)

Until 1925 limited air mail and shipping services used Cleveland's Glenn Martin Field, a small facility crowded within the city limits. The introduction of night flying by the U.S. postal service, however, rendered this field obsolete, as the lack of clear air approaches and lighting made night landings unsafe. William R. Hopkins, the Cleveland city manager, and John "Jack" Berry, an engineering officer for the postal service, lobbied for a new airfield, to be located on 1,040 acres of semi-improved flat farmland, ten miles southwest of downtown Cleveland. Although labeled by the press "Major Berry's Folly" because of its outlying location, the site was a fortunate one: the surrounding farmland allowed for clear air approaches from all sides, a streetcar link to the center of town terminated at the site's edge, and tracks from Cleveland's principal railroads passed within a quarter-mile. By an act of the city council the 1,040-acre plot was purchased in early 1925; airfield construction commenced shortly thereafter.

With Jack Berry supervising, an initial area of 510 acres was readied in 1925 for the airport. A main 48-inch drainage line, over a mile in length, was placed beneath the landing area, connected to over 13,000 feet of 12-inch lateral piping. Sixty thousand feet of French drains, spaced on 20-foot centers, were constructed and subsequently tied into the lateral piping. Incorporating a network of manholes and catch basins, this drainage system provided an all-weather landing field, efficiently draining runoff and capillary water buildup in the clay subsoil. A 3,600-foot by 4,200-foot sodded landing field was then prepared, in accordance with accepted layouts of the day. Sod provided the most economical landing surface for the light aircraft of the 1920s. An almost square landing field design was adopted by Berry in order that planes might always take off and land into the wind without delay from other aircraft. Concrete taxiways of 100-foot widths were subsequently placed along the north, east, and south perimeters of the airfield, in order that aircraft ground movements would not interfere with takeoffs and landings.

A pioneering night lighting system was incorporated into the layout in 1930, ensuring Cleveland's position as a terminal along the first scheduled transcontinental air mail route. Flush-mounted boundary lights, spaced at 200-foot intervals, outlined the landing area. Four 1.5-million-candlepower beacon lights illuminated the night sky in intersecting arcs, locating the field for night pilots; a 0.5-billion-candlepower arc floodlight, mounted in a modified lighthouse tower, illuminated the landing area. This pioneer light is now at the Smithsonian Institution in Washington, D.C.

The airport's formal dedication, highlighted by the arrival of the first nighttime transcontinental air mail flight, took place on July 1, 1925, with an estimated 250,000 spectators in attendance. The first scheduled passenger service (Thompson Aero Service, now part of TRW), between Cleveland and Dearborn, was also offered in 1925. An average of 300 aircraft per month used the new airport. By 1928, when the major commercial lines had established themselves at Cleveland, 1,100 planes embarked per month. The first hangar facilities at the airport were built by the private carriers on land leased to them by the city. The Ford Motor Company built its own hangar and used the famous Ford Trimotors to carry critical components from Detroit to Cleveland for the local assembly plant.

By 1929 a central passenger terminal and administration building was complete. Topping the structure stood the country's first control tower, a glass-enclosed octagonal room containing radio communication equipment for centralized air traffic control in the terminal's vicinity. In the same year the National Air Races were brought to Cleveland, the main event being the Labor Day Thompson Trophy Race, featuring low-level pylon racing. Bleachers to accommodate over 100,000 spectators were constructed along the airfield's western edge; the airport's size permitted commercial aviation to proceed unimpeded during the air races. The airport became the home of not only the Thompson Trophy Race, which developed land-plane speed records, but also the Grieve Trophy Race for closed-course racing and the Bendix Trophy Race for cross-country flight from Los Angeles.

Through the 1930s Cleveland Airport grew steadily in size and commercial capacity, operating as an important terminus for passenger, freight, and air mail traffic. More than 200,000 travelers passed through its portals annually, and a new landing field, 1,800 feet by 1,300 feet, was laid to accommodate both increased aircraft weight and increased traffic. With the onset of World War II, the National Advisory Committee on Aeronautics, predecessor to the National Aeronautics and Space Administration, established an engine propulsion laboratory at the airfield in 1940, the first such facility in the United States. Throughout the war, however, commercial air transport continued at Cleveland Airport.

Cleveland constructed a new $8-million terminal complex in 1952–1956, adding "Hopkins" to the facility's name in honor of the field's first promoter. A new 9,000-foot jet runway was added in 1962, and a $6.5-million addition to the terminal complex was completed in 1968. During this latter improvement Cleveland's rapid transit system was extended to link directly to the upgraded passenger facilities.

The greatest advances in aviation technology achieved at Cleveland Airport have been in air traffic control. The founding of the country's first control tower facility at Cleveland in 1929 significantly advanced the safety and efficiency with which the increasingly crowded airways, particularly around airports, could be used. Shortly after the First World War the U.S. government had put its surplus military aircraft on the market at bargain prices, effecting a sharply increased risk of midair collisions. In recognition of this problem the International Commission for Air Navigation had been established in 1919 to make available advice and information useful in the planning and execution of all types of air flight. During the 1920s, as commercial air transport companies mushroomed across the country, terminal traffic control was accomplished by flagmen on the runway edges, guiding and coordinating takeoffs and landings through visual contact with pilots. Mishaps were not infrequent, particularly during

The boarding gates of the Cleveland-Hopkins Airport as they were in 1937. (Courtesy of the HAER Collection, Library of Congress)

inclement weather and at night. As a consequence of the Air Commerce Act of 1926, the Department of Commerce established air traffic rules but left the actual task of control in local hands. Cleveland Airport took a major step forward with the erection of its control tower in 1929.

Located atop the central administration building, the glass-enclosed tower permitted operators an unobstructed view of the entire landing field and surrounding air traffic. In 1929 Berry had visited airports in England and had taken note of the methods used there for controlling air traffic by radio. Claude King, later the city's airport commissioner, was at the same time experimenting with radios from lake boats and had already installed radio equipment in the existing administration building. When Berry returned, the two compared notes and set up a system. Early in 1930 ground-to-air radio transmitters and receivers were installed in the tower, linking the tower operators to the pilots of all incoming and departing flights. In this fashion, control of the airfield and surrounding airspace was completely centralized; communication between operators and pilots permitted safe separation between airborne craft and allowed for orderly and safe takeoff and landing operations. The first tower operators were employees of the city of Cleveland who had earned certification by passing an examination developed by the Department of Commerce. Within five years of Cleveland's pioneering installation, twenty major American airports had installed radio-equipped control towers, establishing the mechanism by which all air traffic is still handled.

Together with the Newark and Chicago airports, Cleveland Airport was also the site of one of America's first en-route traffic control centers, providing constant ground monitoring of postal air flights across the route's eastern sector. Installed in Cleveland's new control tower, the long-range radio equipment supplied air mail pilots with constant weather and traffic information across the middle range of the mail route; Newark and Chicago operators provided information at either end of the journey. Within a few years the entire transcontinental mail route was similarly equipped, based upon the successful Newark-Cleveland-Chicago prototype. Comprehensive monitoring of private commercial air transport was slower in developing, however; not until 1943, amid military concern for emergency preparedness, did a federal take-over of control tower and control center operations introduce a national air traffic control network. The measures at Cleveland Airport carried out by the city government in 1929–1930 can be said to have provided the basis upon which today's complex air traffic control system was implemented.

Newark International Airport

Newark International Airport, New Jersey, where flight operations commenced on October 1, 1928, is one of the nation's oldest commercial airports, a pioneering facility at which significant innovation in air transport and landing operations technology has continuously taken place. Beginning with its asphalt-topped runway of 1928, one of the first hard-surfaced landing strips at any commercial airfield in the United States, Newark Airport has sponsored breakthroughs in both navigational and lighting systems that revolutionized commercial air flight both in this country and around the world. Additionally, its strategic location within the New York–New Jersey metropolitan area has played a vital role in the development of this region as a major commercial, manufacturing, and industrial center.

As early as 1914 the city of Newark began formulating plans for a major airfield, to be the eastern terminus for air travel in the United States. The city commissioners and Mayor Thomas Raymond envisioned the airfield, in fact, as but one component of a great shipping terminus, adjacent to New York's financial center, that would incorporate road, rail, ship, and air transport. Newark, and in particular its southeastern section of swamps and marshland, presented an ideal location for such a facility. The lowlands were owned by the city, the result of a 1907 state legislative enactment permitting their purchase for reclamation and development. Rail lines from the six major eastern railway companies already traversed the swamps, and Newark Harbor bounded the area on one side. Moreover, the harbor and swampland offered clean air approaches from the south and west, allowing for an airport contained within the urban area. Fog and air pollution—the latter already a problem in Manhattan by the 1920s—were not found over the Newark swamplands, further enhancing the site's attractiveness. Finally, a major endorsement for the Newark airport and shipping terminal complex came in 1927 from Secretary of Commerce Hoover's committee on commercial air transportation, which expressed its enthusiastic support for both the terminal concept and the location of a landing field on the reclaimed swampland.

Even given all these positive factors, the Newark airport would certainly not so soon have become a reality had it not been for Charles Lindbergh's epochal transatlantic flight in May 1927. Public interest in air flight mushroomed with Lindbergh's success. Air service companies and landing strips appeared almost overnight across the country. Plans for Newark's airfield proceeded apace.

The infill and construction of an airfield upon reclaimed swampland was itself a major undertaking. Under the supervision of James Costello, chief engineer of the city's Department of Public Works, construction began in April 1928 with the leveling and infilling of the marsh. A deep-sea shipping channel, to be terminated with concrete docking facilities within 1,000 feet of the landing field, was already under way; mud from the channel dredging operations provided fill through hydraulic sluicing methods for the field. Three existing creeks, running across the site, were diverted under the fill into 60-by-65-inch precast-concrete pipe, partly supported by piles driven through the muck. A 12-inch cast-iron water supply line and a sewer within a sheeted trench were also installed before dry fill could be placed and compacted. Throughout the airfield construction, time was an essential factor: Newark officials wanted to secure air mail contracts from the United States Post Office, which were due for assignment in August 1928. In the interest of expediency, curious items found their way into the wet fill, including seven thousand discarded Christmas trees and two hundred old safes donated by a Newark junk dealer.

By the time dry fill operations began, the swampland level had been raised 6 feet; it included 6 miles of subsurface main and lateral drain and sewer piping. Altogether, 4 miles of creek waters had been diverted into the concrete piping. Dry fill, consisting principally of sand and industrial waste fly ash, was then put down by a fleet of almost ninety trucks, operating continuously despite extremely soft and uneven sub-base conditions. Bulldozers on crawler tractors spread and compacted the dumped material, totaling 1,500,000 cubic yards. At the airfield's northwestern edge New Jersey State Highway 25, concurrently under construction, was linked by access road to the field. The new highway terminated at the recently completed Holland

Newark International Airport, Newark, New Jersey. This pioneering commercial airport was opened on October 1, 1928. Its 1,600-foot runway was one of the first hard-surfaced runways to be constructed at an American airport and served as a prototype for those of today. Newark Airport was the site of many innovations in navigational systems. (Courtesy of the Port Authority of New York and New Jersey)

Tunnel, and thus the Newark airfield was just 9 miles by road from Times Square in the heart of Manhattan. A 120-foot by 120-foot hangar of steel-frame construction, incorporating prefabricated steel trusses, was under way at the airfield as well. Its column-free interior provided shelter for up to twenty-five planes in the early days of airport operations.

By August 1928 grading operations for the first 68-acre tract were complete; construction to complete the proposed 400-acre landing field continued for years, incorporating 420 acres of reclaimed land by 1940. A 1,600-foot-long, asphalt-topped runway climaxed the construction operations. Through careful planning, and despite the haste characterizing the operations, all electric wiring and telephone cabling was placed underground to prevent obstruction in landing and takeoff exercises; all vertical obstructions, such as smokestacks and chimneys, in the airfield vicinity were illuminated with floodlights and had red lights mounted on them as well. On October 1, 1928, just seven months after construction began, Newark Airport was formally opened.

Scheduled airline service did not begin until 1929; nevertheless, the airfield was an immediate success. Short flights over New York City, costing five dollars, attracted huge crowds, and many came simply to observe the rapidly developing air transport industry. The U.S. Post Office transferred its eastern terminus to Newark from Hadley Field in nearby New Brunswick during 1929, thus centering its operation in the New York metropolitan area. By 1930 the four largest passenger carriers, Colonial Airways (now American Airlines), Eastern Air Transport (now Eastern Airlines), National Air Transport (now United Airlines), and Transcontinental and Western Airways (TWA), maintained scheduled airline service from Newark to as far away as Los Angeles, Miami, and Montreal. Transcontinental air mail service extended to Los Angeles, with Chicago, Cleveland, and Newark as the principal termini. By Labor Day, 1930, Newark was the busiest commercial airport in the world, with only Floyd Bennet Field in Brooklyn providing competition for the New York market. Famous aviators of the 1930s, including Lindbergh, Amelia Earhart, and Howard

Hughes, frequented Newark in their numerous attempts to establish transcontinental time records.

Newark's prominence in commercial aviation during the 1930s spawned technological innovation, particularly with regard to navigational safety. Night flying began at the field during 1929, with the aid of a battery of floodlights, mounted on a wooden platform and trained on the landing area. A 500,000-candlepower revolving beacon served to locate the airfield from above, and lamps mounted on vertical obstacles warned of surrounding dangers. Many of America's early airports, notably that at Cleveland, used similar lighting systems. During the early 1930s many improvements were carried out at Newark, particularly the installation of flush-mounted landing lights down the runway's centerline and the placing of white cobblestones along the runway paths. Coupled with the advent of landing lights on the aircraft itself, these innovations, first tried at Newark, improved night landing operations considerably. Experimentation continued during 1932 with the installation of signaling wires placed at right angles to, and in advance of, the runway's path; a radio signal emitted from the wire registered a click in the pilot's headset, thus guiding the aircraft onto the current approach path to the runway. This now primitive-sounding method was vitally important to the development of instrument landing systems and was first explored at Newark's airport.

By 1939, however, Newark's leadership in commercial air transport was being successfully challenged by New York's newly completed municipal facility, La Guardia Airport, built principally with WPA funds. Located in the borough of Queens on Flushing Bay, La Guardia featured runway and passenger facilities for both land craft and the day's most popular air carrier, the seaplane. A modern control tower and capacious hangars rendered Newark's existing facilities, haphazardly developed during the 1930s, obsolete. In addition, La Guardia had four fully paved runways, the longest of which stretched 6,000 feet, designed for loads up to 25 tons per square foot. Newark, by contrast, had only a 3,700-foot main runway, mostly of an oiled-cinder surface. Such a surface was incapable of absorbing the loadings to be imposed by the four-engine DC-4 aircraft that was then on the drawing boards of McDonnell-Douglas. Moreover, four existing hangar structures, built by the early commercial carriers in 1929–1930, now impeded the use of a centralized passenger terminal constructed in 1935.

By May 30, 1940, all commercial and air mail operations had been transferred to La Guardia. Newark's control tower was officially closed by the Civil Aeronautics Authority.

The entrance of the United States into World War II breathed new life into Newark Airport. The Army Air Force took control of the facility in June 1942 and over the next four years constructed three fully paved runways, each 4,000 feet long. Accommodations for fighter planes and four-engine bombers, flown to Newark for processing prior to overseas shipment, were also built, including concrete taxiways, staging aprons, and hangar facilities. By 1946 commercial and shipping services had returned to Newark.

On March 22, 1948, Newark Airport operations were transferred to the regionally based Port Authority of New York and New Jersey. Newark's importance as an air mail terminus and cargo shipping center had become evident during the facility's dormancy; much of the industrial and manufacturing base of northern New Jersey had developed with the airport as its principal air shipping facility. Recognizing its role in the region's economy, the Port Authority undertook a large-scale modernization of Newark Airport during the early 1950s. As part of the Authority's regional strategy, Idlewild Airport (now Kennedy International) was developed as the eastern terminus for international commercial air traffic; Newark, closely tied to Manhattan by road and rapid transit links, would primarily serve national and intracontinental passenger traffic. Reinstatement to its former prominence as a cargo and mail shipping center for the region was planned as well. A new passenger terminal, containing more than five times the floor space of the 1935 facility, was erected at a cost of $8.5 million in 1953.

Innovative navigational systems were incorporated into a new instrument-landing runway at Newark in 1952. The $9-million pavement, 7,000 feet long and 200 feet wide, was made fully bidirectional in 1958, the first such facility in the country. Paralleling the runway's full length, a 100-foot-wide taxiway was constructed, with connecting taxiways to the runway every 1,200 feet. This feature allowed a greater number of landings and takeoffs, since the runway surface was quickly vacated by incoming planes. Together with the bidirectional instrument approach systems at the runway's ends, sophisticated lighting systems permitted landings in almost any weather condition or emergency, giving Newark the most advanced guidance capability for aircraft landings of its day. Centerline approach lights, pioneered at Newark, were extended to 3,000 feet beyond both runway ends: the condenser-discharge, or strobe, high-intensity lights flashed in rapid sequence, providing a "moving" light to guide aircraft to the runway centerline. Such installations have since become standard at airports around the world. Precision approach radar (PAR), newly available to commercial aviation, was also installed with the new runway. Finally, state-of-the-art transmissometer and ceilometer equipment, automatically measuring and transmitting to the control tower ceiling and visibility indices at the runway's approach ends, was incorporated into the 7,000-foot runway—another first in the nation. Newark's modernization was completed in 1959–1960 with the erection of a 150-foot-high control tower and a large air cargo center.

By the arrival of the jet age in the 1960s and 1970s, Newark Airport and the Port Authority had regained a leading role in the aviation industry. New facilities were constructed. Port Authority engineers pioneered the development of low-cost runway pavement design and construction methods, by utilizing compacted sections of sand, fly ash, lime, and portland cement to construct pavements capable of withstanding enormous loads at significantly lower costs than more conventional concrete pavements—thus continuing Newark's tradition of innovation in airport technology.

Appendix One

Additional Landmark Projects

In addition to those described in the main body of the text, other civil engineering achievements have been officially nominated or already designated as landmark projects by the American Society of Civil Engineers. Some were added too recently to permit the preparation of full accounts; about others there exists too little reliable information. Such projects are listed here, in alphabetical order. (A supplementary list, containing the newest landmarks, appears on page 366.)

Atlantic City Municipal Convention Hall, Atlantic City, New Jersey. This was the world's largest auditorium and had the greatest permanent three-hinged arch roof span when completed in 1929.

Bailey Island Bridge, Harpswell, Maine. This structure, completed in 1928, crosses a 1,150-foot stretch of high tidal water and swift currents, and is subject to heavy ice floes in the winter. A special split-stone crib construction, with a concrete deck and a concrete span over the navigation channel, was devised to meet these difficult design conditions.

Bailey Island Bridge, Harpswell, Maine. (HAER Collection, Library of Congress)

Stone crib construction, Bailey Island Bridge. (HAER Collection, Library of Congress)

Bayonne Bridge, New Jersey and New York. Designed by Othmar Ammann, this 1,675-foot steel arch bridge was the longest-span bridge of its type when completed in 1931 and remained so until the completion of the New River Gorge Bridge in West Virginia in 1977.

Cape Cod Canal, Barnstable County, Massachusetts. This was America's first major sea-level canal when opened in 1914. It spared thousands of sea-going vessels the hazardous passage around Cape Cod. When enlarged in 1940 to 34 feet deep and 480 feet wide, it became the widest in the world. William Barclay Parsons was its chief engineer.

Chesapeake and Delaware Canal, Delaware City, Delaware, and Chesapeake City, Maryland. Begun by Benjamin Latrobe, Jr., this canal was completed between 1825 and 1829 by Benjamin Wright, Canvass White, and Nathan Roberts, all of whom had formerly worked on the Erie Canal. The Chesapeake and Delaware went out of service in 1921, was converted into a sea-level canal in 1927, and remains in service today.

Bayonne Bridge, New Jersey and New York, under construction. (Courtesy of the Port Authority of New York and New Jersey)

Cape Cod Canal, Massachusetts, under construction. (Courtesy of the American Society of Civil Engineers)

Columbia-Wrightsville Bridge, Columbia and Wrightsville, Pennsylvania. Completed in 1930, this one-mile bridge was the longest multiple-arch reinforced-concrete bridge in the world at the time.

Davis Island Lock and Dam, Ohio River, Allegheny County, Pennsylvania. This lock facility had the world's first rolling gate when it was constructed between 1878 and 1885. It served as a prototype for many similar facilities. The project also included the largest movable dam built during the nineteenth century.

First concrete pavement, Bellefontaine, Ohio. Laid in 1893, this pavement represented the first use of portland cement concrete in public road construction. It was the forerunner of many thousands of miles of such construction across America.

Going-to-the-Sun Road, Glacier National Park, Montana. Completed in 1933, this road was the first major transmountain scenic highway built in the United States. In addition to overcoming enormous engineering difficulties, construction of the road included precedents for environmental sensitivity that were followed in the construction of later roads in national parks.

Golden Gate Bridge, San Francisco. A widely recognized civil engineering accomplishment, the bridge has a 4,200-foot main span and 746-foot-high towers. Construction was started in January 1933 and completed in May 1937.

Golden Gate Bridge, San Francisco. (Courtesy of the Golden Gate Bridge, Highway, and Transportation District)

Construction of the piers of the Golden Gate. (Courtesy of the Golden Gate Bridge, Highway, and Transportation District)

Golden Gate: construction of the road deck. (Courtesy of the Golden Gate Bridge, Highway, and Transportation District)

Golden Gate: working on the deck-suspending cables. (Courtesy of the Golden Gate Bridge, Highway, and Transportation District)

Granite Railway, Quincy, Massachusetts. This railroad, completed in 1826 before the widespread use of steam locomotives, demonstrated the advantages of rail transport and introduced many technical features, such as switches, the turntable, and double-truck railway cars.

High Bridge over the Kentucky River, Jessamine and Mercer counties, Kentucky. The first major cantilever bridge in the United States, the High Bridge was built between 1876 and 1877 by Charles Shaler Smith. It had three principal spans of 375 feet each. The structure used the foundations of an earlier uncompleted bridge by John A. Roebling. Because of increases in train loads it was replaced in 1911 by a bridge of similar construction by Gustav Lindenthal.

Mormon Tabernacle, Salt Lake City. Originally constructed in 1867, the 150-foot-long lattice timber arches were a remarkable achievement for engineers of the young West.

Old Columbia Scenic Highway (U.S. Highway 30), Oregon. Built between 1913 and 1922, this project involved 74 miles of roadways, tunnels, bridges, viaducts, and overlooks.

Public Land System point of beginning, Columbiana County, Ohio. The U.S. Public Land System was established in 1785 as a result of the famous land ordinance of that year, providing a framework for the growth of the country into the western territories. Initial surveys were done under the direction of Thomas Hutchins, United States Geographer.

Granite Railway, Quincy, Massachusetts. (Courtesy of the American Society of Civil Engineers)

High Bridge over the Kentucky River, Jessamine and Mercer counties, Kentucky. (Photograph by Eddie B. Smith, courtesy of the American Society of Civil Engineers)

Mormon Tabernacle, Salt Lake City. An old view of the great arches. (HAER Collection, Library of Congress)

A recent view of the arches of the Mormon Tabernacle. (HAER Collection, Library of Congress)

Rockville Stone Arch Bridge, Rockville, Pennsylvania. This bridge is considered by many to be the zenith of American stone arch construction. It was opened in 1902 as one of the longest and widest multiple stone arch railway bridges in the world.

Rocky River pumped-storage hydroelectric plant, New Milford, Connecticut. This facility was the first major pumped-storage hydroelectric power project in the United States. It was completed in 1925.

Second Street Bridge, Allegan, Michigan. Erected in 1886 by the famous King Iron Bridge and Manufacturing Company, this structure, with its 225-foot double-intersection through truss, represented the culmination of an era in which wrought iron was used as the construction material of choice in bridges. Shortly afterward most bridges were constructed of steel.

Snoqualmie Falls Cavity Generating Station, Washington. The concept of an underground hydroelectric station was first successfully realized at the Snoqualmie Falls site in 1899.

Statue of Liberty, New York Harbor. While primarily a symbol of Franco-American friendship, the construction of the Statue of Liberty also involved many unique and difficult engineering problems.

Watertown Arsenal, Watertown, Massachusetts. The first major engineering testing laboratory in the United States was located at this site. The laboratory conducted tests on the properties of materials and developed testing procedures and machines from 1859 through the early part of the twentieth century, when the National Bureau of Standards began similar operations.

Rockville Stone Arch Bridge, Rockville, Pennsylvania. (Courtesy of the American Society of Civil Engineers)

Snoqualmie Falls Cavity Generating Station, Washington. (Courtesy of the American Society of Civil Engineers)

Statue of Liberty, under construction. (Library of Congress)

Watertown Arsenal, Watertown, Massachusetts. A rail-mounted 12-inch Howitzer is shown. (Courtesy of the U.S. Army Materials Technology Laboratory, Watertown, Massachusetts)

Snoqualmie Falls
Cavity Generating Station
Going-to-the-Sun Road
Mullan Road
Old Columbia Scenic Highway
Rogue River Bridge
Buffalo Bill Dam
Pelton Impulse Water Wheel
Central Pacific Railway
Promontory Point
Bidwell Bar
Suspension Bridge
Marlette Lake Water System
Folsom Hydroelectric Power System
Mormon
Tabernacle
Union Pacific Railway
Golden Gate
Alvord Lake Bridge
Bridgeport Covered Bridge
Moffat Tunnel
Cheesman Dam
Gunnison Tunnel
Durango-Silverton Railroad
Cumbres and Toltec
Railroad
First Owens River-
Los Angeles Aqueduct
Embudo Stream-Gauging Station
Salt River Project
Theodore Roosevelt Dam
Elephant Butte Dam
International Boundary Marker No. 1
Acequias
Stone Arch Bridge
Peavy-Haglin Grain Elevator
Milwaukee Metropolitan Sewage Treatment Plant
Second Street Bridge
Cortland Street Bridge
Chicago Water Supply System
Reversal of the Chicago River
Park and Boulevard System
Eads Bridge
Union Station
Chain of Rocks
Sault Ste. Marie Hydroelectric Complex
Vulcan Street Plant
Ohio State
Canal System
National Road
See Inset
Louisville Waterworks
High Bridge
Naval Drydocks
Fink Deck-Truss Bridge
Pattison Tunnel
Charleston-Hamburg Railroad
City Plan of Savannah
King's Road
Castillo de San Marcos
Ellicott's Stone
Eads' South Pass
Navigation Works

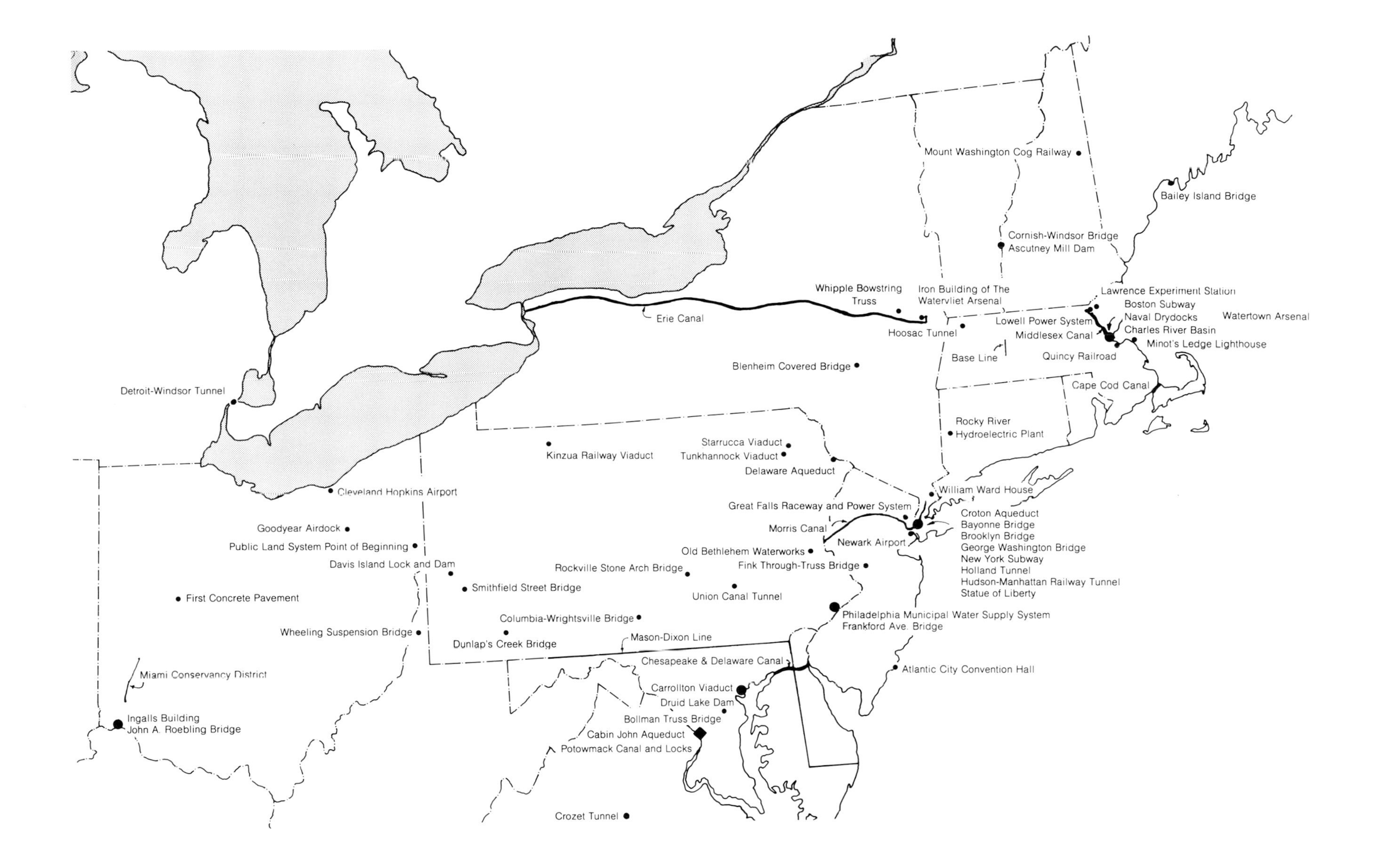
Mount Washington Cog Railway
Bailey Island Bridge
Cornish-Windsor Bridge
Ascutney Mill Dam
Whipple Bowstring Truss
Iron Building of The Watervliet Arsenal
Lawrence Experiment Station
Boston Subway
Naval Drydocks
Watertown Arsenal
Charles River Basin
Minot's Ledge Lighthouse
Erie Canal
Lowell Power System
Hoosac Tunnel
Middlesex Canal
Base Line
Quincy Railroad
Blenheim Covered Bridge
Cape Cod Canal
Detroit-Windsor Tunnel
Rocky River Hydroelectric Plant
Starrucca Viaduct
Kinzua Railway Viaduct
Tunkhannock Viaduct
Delaware Aqueduct
William Ward House
Cleveland Hopkins Airport
Great Falls Raceway and Power System
Croton Aqueduct
Bayonne Bridge
Brooklyn Bridge
George Washington Bridge
New York Subway
Holland Tunnel
Hudson-Manhattan Railway Tunnel
Statue of Liberty
Goodyear Airdock
Morris Canal
Newark Airport
Public Land System Point of Beginning
Old Bethlehem Waterworks
Davis Island Lock and Dam
Rockville Stone Arch Bridge
Fink Through-Truss Bridge
Smithfield Street Bridge
First Concrete Pavement
Union Canal Tunnel
Philadelphia Municipal Water Supply System
Frankford Ave. Bridge
Columbia-Wrightsville Bridge
Wheeling Suspension Bridge
Mason-Dixon Line
Dunlap's Creek Bridge
Chesapeake & Delaware Canal
Atlantic City Convention Hall
Miami Conservancy District
Carrollton Viaduct
Druid Lake Dam
Ingalls Building
John A. Roebling Bridge
Bollman Truss Bridge
Cabin John Aqueduct
Potowmack Canal and Locks
Crozet Tunnel

Bibliography

General

Baker, M. N. *The Quest for Pure Water.* New York: American Water Works Association, 1949.

Billington, D. P. *Tower and Bridge.* New York: Basic Books, 1983.

A Biographical Dictionary of American Civil Engineers. New York: American Society of Civil Engineers, 1972.

Condit, Carl. *American Building Art—The Nineteenth Century.* New York: Oxford University Press, 1960.

Condit, Carl. *American Building Art—The Twentieth Century.* New York: Oxford University Press, 1960.

FitzSimons, Neal. "Benchmarks and Who Am I." Series, *Civil Engineering,* 1965–1969.

Graydon, Lawrence P. "A Brief History of Engineering Education in the United States." *Engineering Education,* December 1977, 246–263.

History of Public Works in America. Chicago: American Public Works Association, 1976.

Hunter, Louis C. *A History of Industrial Power in the United States, 1780–1930.* Charlottesville, Virginia: The University Press of Virginia, 1979.

Jacobs, David, and Anthony E. Neville. *Bridges, Canals and Tunnels: The Engineering Conquest of America.* New York: Van Nostrand, 1968.

McGivern, James. *The First Hundred Years of Engineering Education in the United States.* Spokane, Washington: Gonzaga University Press, 1960.

Newlon, Howard, Jr., ed. *A Selection of Historic Papers on Concrete.* Detroit: American Concrete Institute, Publication SP-52.

Maas, A. *Muddy Waters: The Army Engineers and the Nation's Rivers.* Cambridge, Massachusetts: Harvard University Press, 1951.

Raafat, Aly A. *Reinforced Concrete in Architecture.* New York: Reinhold, 1958.

Reier, Sharon. *The Bridges of New York.* New York: Quadrant, 1977.

Robbins, Frederick W. *The Story of Water Supply.* London: Oxford University Press, 1946.

Straub, Hans. *A History of Civil Engineering.* Cambridge, Massachusetts: MIT Press, 1960.

Walker, M. E. M. *Pioneers of Public Health.* London: Oliver and Boyd, 1930.

Chapter One: Canals

Potowmack Canal

Bacon-Foster, C. *The Potomac Canal Route to the West.* Washington, D.C.: Columbia Historical Society, 1912.

Brown, Alexander Crosby. "America's Greatest 18th-Century Engineering Achievement." *Virginia Cavalcade,* Spring 1963, 40–46.

Durham, C. J. S. *Washington's Potowmack Canal Project at Great Falls.* Fairfax, Virginia: Fairfax County Park Authority, 1957.

"Great Falls Canal and Locks: Civil Engineering Landmark." *Civil Engineering,* November 1972, 53–56.

Harlow, Alvin F. *Old Towpaths.* Port Washington, New York: Kennikat Press, 1954.

Middlesex Canal

Clarke, Mary Stetson. *The Old Middlesex Canal.* Melrose, Massachusetts: Hilltop Press, 1974.

Dame, Lorin L. "The Middlesex Canal." *The Old Residents' Historical Association* 3, no. 1, September 1884.

Dickinson, Samuel N. *Caleb Eddy, Agent of the Corporation—Historical Sketch of the Middlesex Canal with Remarks for the Consideration of the Proprietors.* Boston, 1843.

Harlow, Alvin F. *Old Towpaths.* Port Washington, New York: Kennikat Press, 1954.

Mann, Moses W. *The Middlesex Canal, an 18th-Century Enterprise.* Reprinted in The Boston Society Publications, no. 6. Boston: Old State House Press, 1910.

Roberts, Christopher. *The Middlesex Canal, 1793–1860.* Cambridge, Massachusetts: Harvard University Press, 1938.

Erie Canal

Jensen, Albert C. "Engineering Clinton's Ditch." *Civil Engineering* 33 (September 1963): 48–50.

Whitford, Noble E. *History of the Canal System of the State of New York.* Albany, New York: State Engineer and Surveyor's Office, 1906.

Morris Canal

"Inclined Planes of the Morris Canal." *American Canals* 37 (May 1981): 6.

Stevenson, David. *Sketch of Civil Engineering in America.* London: John Weale, 1838.

Van Kirk, Eileen M. "In Search of the Morris Canal." *New Jersey Outdoors* 5 (September–October 1978).

Ohio State Canal System

Klein, Benjamin. *The Ohio River Collection.* Cincinnati: Young and Klein, 1969.

Ohio State Archeological and Historical Society. "History of the Ohio Canals." 1905.

Scheiber, H. N. *The Ohio Canal Era.* Athens, Ohio: Ohio University Press, 1969.

State of Ohio Board of Canal Commissioners. *Annual Reports,* 1822–1835.

Wilcox, F. *The Ohio Canals.* Kent, Ohio: Kent State University Press, 1969.

Chapter Two: Roads

King's Road

Abbey, Kathryn Trimmer. *Florida, Land of Change.* Chapel Hill: University of North Carolina Press, 1941.

Boyd, Mark. "The First American Road in Florida." *Florida Historical Quarterly* 14 (October 1935): 77.

Bullen, Ripley P. "Fort Tonyn and the Campaign of 1778." *Florida Historical Quarterly* 28, no. 1 (July 1949).

Loomes, Charles. "The King's Road of British East Florida." Unpublished paper, St. Augustine Historical Society, Florida.

Mowat, Charles Loch. *East Florida as a British Province, 1763–1784.* Berkeley and Los Angeles: University of California Press, 1943.

Schaub, James H. "The King's Road." Unpublished paper, Department of Civil Engineering, University of Florida, Gainesville, Florida.

National Road

Hulbert, Archer Butler. *The Cumberland Road,* vol. 10 of *Historic Highways of America.* Cleveland: Arthur H. Clark Company, 1904.

Jordan, Philip D. *The National Road.* New York: Bobbs-Merrill, 1948.

Searight, Thomas B. *The Old Pike, a History of the National Road.* Uniontown, Pennsylvania: published by the author, 1894.

Mullan Road

Coleman, Louis C., and Leo Rieman (B. C. Payette, compiler). *Captain John Mullan: His Life Building the Mullan Road: As It Is Today and Interesting Tales of Occurrences along the Road.* Montreal: printed privately for Payette Radio, 1968.

McGregor, Alexander C. "The Economic Impact of the Mullan Road on Walla Walla, 1860–1883." *Pacific Northwest Quarterly,* July 1974, 118–129.

Chapter Three: Railroads

Charleston-Hamburg Railroad

Derrick, Samuel M. *Centennial History of the South Carolina Railroad.* Reprint. Spartanburg, South Carolina: The Reprint Company, 1975.

LeCato, John M. "Ezra Miller: Builder of the 'Best Friend.' " Unpublished paper. Charleston, South Carolina: Charleston Historical Society.

Transcontinental Railroad

The American Heritage History of Railroads in America. New York: American Heritage, 1975.

Brettell, Richard R. *Historic Denver—The Architects and Architecture,* 8–19. Denver: Historic Denver, 1973.

Dodge, Grenville M. *How We Built the Union Pacific Railway.* Denver: Sage Books, 1965. Reprinted from a private edition issued c. 1911–1914.

Excursion of the Directors and Stockholders of the Central Branch, Union Pacific Railroad, November, 1866. New York: New York Printing Co., 1867.

Knell, K. A., and P. L. Trinder. "Indians and Buffalos Permitting." In Mayer, Lynne Rhodes, and Kenneth E. Vose, *Makin' Tracks.* New York: Praeger, 1975.

Ream, John. *Building the Pacific Railroad.* Brigham City, Utah: City Chamber of Commerce, 1966.

White, Henry Kirke. *History of the Union Pacific Railway.* Chicago: University of Chicago Press, 1895.

Mount Washington Cog Railway

Anderson, Leon W. "Mountain Climbing Wonder Honored." Concord, New Hampshire: Village Press, 1978.

Cumbres and Toltec Scenic Railroad

"History Runs on a Three-Foot Gauge—Cumbres & Toltec Scenic Railroad." *Arizona Highways,* July 1973.

Wilson, Spencer, and Vernon Glover. *The Cumbres and Toltec Scenic Railroad: The Historic Preservation Study.* Albuquerque, New Mexico: University of New Mexico Press, c. 1980.

Durango-Silverton Railroad

Beehe, Lucius. "The Narrow Gage Railroads of Colorado." In *Bulletin No. 67A* of the Railway and Locomotive Historical Society, Boston, August 1946.

Harrington, E. R. "The Little Giant Takes the Count." *Professional Engineering,* November 1968.

"Two Railroads Designated Civil Engineering Landmarks." *Civil Engineering,* April 1963.

Workers of the Writers Program (compilers). *Colorado—A Guide to the Highest State,* 48, 70–77, 343, 417. New York: Hastings House, 1951.

Boston Subway

Boston Transit Commission. *First Annual Report,* August 15, 1895.

Boston Transit Commission. *Second Annual Report,* August 15, 1896.

Boston Transit Commission. *Third Annual Report,* August 15, 1897.

Boston Transit Commission. *Fourth Annual Report,* August 15, 1898.

Scientific American, September 5, 1896, 204ff.

Scientific American, September 18, 1897, 184ff.

New York Subway

"Beach's Broadway Subway." In *Paper 41: Tunnel Engineering in the U.S.,* 227–231. Washington, D.C.: *National Museum Bulletin 240,* 1966.

Hastings, Paul. *Railroads: An International History.* London: Ernest Benn, 1972.

Holbrook, Stewart H. *The Story of American Railroads.* New York: Crown, 1947.

Honey, Lewis H. *A Congressional History of Railways in the United States.* Madison: University of Wisconsin Press, 1908. Reprinted by Augustus M. Kelley, New York, 1968.

Lankton, Larry D. "Movers and Shakers." In *Civil Engineering,* American Society of Civil Engineers, November 1979, 73–75.

Rudy, William H. "The Subway Story." New York: New York Post Corporation, 1970. Reprinted from the *New York Post Magazine,* February 23–28, 1970.

White, Edward and Muriel. *Famous Subway Tunnels of the World,* 73–85. New York: Random House, 1953.

Chapter Four: Bridges

Frankford Avenue Bridge

Blakiston, Mary. *Lower Dublin Township.* Philadelphia: City Historical Society, December 1906.

Eve, Sarah. "Extract from the Journal of Miss Sarah Eve." *Pennsylvania Magazine* 5, May 4, 1773.

Philadelphia County. *Second Court Record,* 1683.

Willits, J. Pearson. "The Pennypack in Lower Dublin Township." Philadelphia: City Historical Society, October 1911.

Carrollton Viaduct

Hill, Forest G. *Roads, Rails and Waterways: The Army Engineers and Early Transportation,* 100–107. Norman, Oklahoma: University of Oklahoma Press, 1957.

Hungerford, Edward. *The Story of the Baltimore and Ohio Railroad, 1827–1927,* 62–65. New York: G. P. Putnam's Sons, 1928.

Dunlap's Creek Bridge

Colburn, Z. "American Iron Bridges." *Institution of Civil Engineers* 22, Session 1862–63.

Condit, Carl. *American Building Art—The Nineteenth Century,* New York: Oxford University Press, 1960.

Ellis, Franklin. "History of Fayette County, Pennsylvania." Philadelphia: L. H. Everts, 1882.

Hill, Forest G. *Roads, Rails and Waterways: The Army Engineers and Early Transportation.* Norman, Oklahoma: University of Oklahoma Press, 1957.

Hungerford, Edward. *The Story of the Baltimore and Ohio Railroad, 1827–1927.* New York: G. P. Putnam's Sons, 1928.

Starrucca Viaduct

Jacobs, David, and Anthony E. Neville. *Bridges, Canals and Tunnels: The Engineering Conquest of America.* New York: Van Nostrand, 1968.

Tyrrell, H. G. *History of Bridge Engineering.* Chicago, 1911.

Young, William S. "The Starrucca Viaduct." *Railroading* 48 (Fourth Quarter 1973): 3–32.

Delaware Aqueduct

Roebling, J. "Suspension Aqueduct on the Delaware and Hudson Canal." *American Railroad Journal* 22 (1849): 21ff.

Vogel, Robert M. "Roebling's Delaware and Hudson Canal Aqueducts." *Smithsonian Studies in History and Technology,* no. 10. Washington, D.C.: Smithsonian Institution Press, 1971.

Wheeling Suspension Bridge

American Society of Civil Engineers Newsletter, West Virginia Section 9, August 1969.

Kemp, Emory L. "Ellet's Contribution to the Development of Suspension Bridges." *Engineering Issues,* American Society of Civil Engineers, July 1973, 331ff.

Kemp, Emory L. "Links in a Chain: The Development of Suspension Bridges 1801–70." *The Structural Engineer* 57A, no. 8 (August 1979): 50f.

Lewis, G. D. *Charles Ellet, Jr.* Urbana, Illinois: University of Illinois Press, 1968.

Steinman, D. B. "Dedication of the Wheeling Suspension Bridge as a National Monument." *The West Virginia Engineer,* June 1956, 8–10.

Tyrrell, H. G. *History of Bridge Engineering.* Chicago, 1911.

Cornish-Windsor, Blenheim, and Bridgeport Covered Bridges

Allen, Richard. *Covered Bridges of the Middle Atlantic States.* Brattleboro, Vermont: Stephen Greene Press, 1959.

Allen, Richard. *Covered Bridges of the Northeast.* Brattleboro, Vermont: Stephen Greene Press, 1957.

Battison, Edwin A. "The Cornish-Windsor Covered Bridge and the Ascutney Mill Dam." Unpublished paper.

Condit, Carl N. *American Building: Materials and Techniques from the Beginning of the Colonial Settlements to the Present.* Chicago: University of Chicago Press, 1968.

Morley, S. Griswold. *The Covered Bridges of California.* Berkeley: University of California, 1938.

Whipple Bowstring Truss

Condit, Carl. *American Building Art—The Nineteenth Century,* 113–118. New York: Oxford University Press, 1960.

Vogel, Robert, ed. *A Report of the Mohawk-Hudson Area Survey,* 137–150. Washington, D.C.: Smithsonian Institution, 1973.

Bollman Truss Bridge

Condit, Carl. *American Building Art—The Nineteenth Century,* 118–124. New York: Oxford University Press, 1960.

Vogel, Robert M. "The Engineering Contributions of Wendel Bollman." Paper 36, *United States National Museum Bulletin 240,* 77–104. Washington, D.C.: Smithsonian Institution, 1964.

Fink Through-Truss and Deck-Truss Bridges

"Memoir of Albert Fink." *Transactions of the American Society of Civil Engineers* 41 (1898): 626–638.

Newlon, Howard. "Report on Virginia's Fink Deck Truss." Prepared for the Virginia Section of the American Society of Civil Engineers. Undated.

Bidwell Bar Suspension Bridge

Butte Record, May 26, 1855, p. 2.

"Cables of 90-Year-Old Bridge Still in Sound Condition." *Engineering News Record* 137, no. 16 (October 17, 1946): 106.

North Californian, January 9, 1857, p. 2.

Sacramento Daily Union, August 9, 1856, p. 2.

Cabin John Aqueduct and Bridge

Myer, Donald Beekman. "Bridges and the City of Washington." *National Geographic* 12 (December 1897).

John A. Roebling Bridge

Bouscaren, G. "Restoration of the Cable Ends of the Covington and Cincinnati Suspension Bridge." *Transactions of the American Society of Civil Engineers* 20 (1893): 47–56, 358–371.

Engineering Record 7 (1893): 434–435.

Engineering Record 38 (1898): 314–316, 554–555.

"Largest Bridge in the World." *The Engineer* 20 (London, 1865): 354.

Roebling, J. A. "The Cincinnati Suspension Bridge." *Engineering* 4 (London, 1867): 22ff.

Scientific American 40 (1879): 335, 337.

Steinman, D. B. *Builders of the Bridge.* New York: Harcourt, Brace, 1945.

Stern, Joseph S., Jr. *The Great Suspension Bridge at Cincinnati, Ohio.* Cincinnati: Centennial Year Publication, 1966.

Eads Bridge

Cooper, Theodore. "The Erection of the Illinois and St. Louis Bridge." *Transactions of the American Society of Civil Engineers,* 1875, 239–254.

Kouwenhoven, J. A. "The Designing of the Eads Bridge." *Technology and Culture* 23 (1982): 535–568.

Miller, Howard S., and Quinta Scott. *The Eads Bridge.* Columbia, Missouri: University of Missouri Press, 1979.

Woodwards, Calvin C. *A History of the St. Louis Bridge.* St. Louis, Missouri: G. I. Jones, 1881.

Kinzua Railway Viaduct

Engineering News 24 (July 5, 1890): 7ff.

Engineering Record, 42, no. 22 (December 1, 1900): 508ff.

Railroad Gazette, November 30, 1900, 787ff.

Smithfield Street Bridge

Black, Archibald. *The Story of Bridges,* 211. London: Whittlesey House, 1936.

Lindenthal, Gustav. "The Monongahela Bridge." *Engineering News* 10 (1883): 314–315.

Lindenthal, Gustav. "Rebuilding the Monongahela Bridge." *Transactions of the American Society of Civil Engineers* 12 (May 1883): 353ff.

Brooklyn Bridge

"Cable Fastening of the East River Bridge." *Scientific American* 37, no. 6 (August 11, 1877): 79f.

"Cable Making of the East River Bridge." *Scientific American,* 37, no. 5 (August 4, 1877): 63f.

Collingwood, Francis. "Notes on the Masonry of the East River Bridge." *Transactions of the American Society of Civil Engineers* 6 (1877): 7–27.

Collingwood, Francis. "Progress of Work at the East River Bridge." *Transactions of the American Society of Civil Engineers* 9 (1880): 162–172.

"The Great Suspension Bridge between New York and Brooklyn." *Scientific American* 23, no. 20 (November 12, 1870): 307f.

"Launching of the Great Caisson for the Brooklyn Terminus of the East River Bridge." *Scientific American* 23, no. 2 (July 9, 1870): 15–16.

McCullough, D. *The Great Bridge.* New York: Simon and Schuster, 1972.

Reier, Sharon. *The Bridges of New York,* 8–26. New York: Quadrant, 1974.

Steinman, D. B. *Builders of the Bridge.* New York: Harcourt, Brace, 1945.

Trachtenberg, A. *Brooklyn Bridge: Fact and Symbol.* New York: Oxford University Press, 1965.

"The Wire for the East River Bridge." *Scientific American* 36, no. 9 (March 3, 1877): 127f.

Stone Arch Bridge at Minneapolis

"Ancient Masonry Bridge Saved by Double Jeopardy Rescue Operation." *Engineering News-Record.* New York: McGraw-Hill, 1965.

"Northstar News Heading." *Northstar News,* Minneapolis, December 1971.

Alvord Lake Bridge

"Reinforced Concrete." Chicago: Concrete Reinforcing Steel Institute, 1928.

"The Story of Cement, Concrete, and Reinforced Concrete." *Civil Engineering,* November 1977, 63ff.

Cortland Street Drawbridge

Becker, Donald N. "Development of the Chicago Type Bascule Bridge." *Proceedings of the American Society of Civil Engineers,* February 1943.

Tunkhannock Viaduct

The Bridge Was Built. Nicholson Area Library, Pennsylvania, 1976.

Simpson, C. W. "Construction Methods on the Tunkhannock and Martin's Creek Viaducts, Lackawanna Railroad." *Proceedings of the American Concrete Institute* 2 (1916): 100ff.

Rogue River Bridge

McCullough, C. B., and Albin L. Gemeny. "Application of the Freyssinet Method of Concrete Construction to the Rogue River Bridge in Oregon." Salem, Oregon: Oregon State Highway Commission, 1933.

McCullough, C. B., and E. S. Thayer. *Elastic Arch Bridges.* New York: Wiley, 1931.

Oregon State Highway Commission. *Ninth Biennial Report,* 1930.

Oregon State Highway Commission. *Twelfth Biennial Report,* 1936.

George Washington Bridge

Ammann, O. H. "The George Washington Bridge: General Conceptions." *Transactions of the American Society of Civil Engineers* 97 (1933): 1–65.

Black, Archibald. *The Story of Bridges.* New York: McGraw-Hill, 1936.

Steinman, D. P., and S. R. Watson. *Bridges and Their Builders.* New York: Dover, 1941.

Chapter Five: Tunnels

Pattison Tunnel

Corlew, Robert E. *A History of Dickson County,* 24–25ff. Tennessee Historical Commission, 1956.

Union Canal Tunnel

Kirby, Richard S., and Philip G. Laurson. *The Early Years of Modern Engineering.* New Haven, Connecticut: Yale University Press, 1932.

Sandstrom, Gosta E. *Tunnels.* New York: Holt, Rinehart, and Winston, 1963.

Shank, William H. *The Amazing Pennsylvania Canals.* York, Pennsylvania: Historical Society of York County, 1965.

Stevenson, David. *A Sketch of Civil Engineering in America.* London: John Weale Architectural Library, 1838.

Tanner, H. S. *Description of the Canals and Railroads of the United States.* New York: T. R. Tanner and J. Disturnell, 1840.

Waggoner, Madeline Sadler. *The Long Haul West: The Great Canal Era, 1817–1850.* New York: G. P. Putnam's Sons, 1958.

Crozet Tunnel

Board of Public Works, Virginia. *34th Annual Report,* 1849, 328–338.

Couper, William. "Claudius Crozet: Soldier, Educator, Engineer." *Southern Sketches,* first series, no. 8. Charlottesville, Virginia: Historical Publishing Co., 1936.

Nelson, James Poyntz. *Four Tunnels in the Blue Ridge Region of Virginia.* Richmond, Virginia: Mitchell and Hotchkiss, 1917.

Warden, William E., Jr. "Claudius Crozet: Napoleon's Captain versus the Blue Ridge." *Lexington Gazette,* October 22, 1857, 1.

Hoosac Tunnel

Bond, E. A. *History of the Hoosac Tunnel.* Boston: George W. Armstrong Co., 1880.

Byron, Carl. *A Pinprick of Light: The Troy & Greenfield Railroad and Its Hoosac Tunnel.* Brattleboro, Vermont: Stephen Greene Press, 1974.

Dalrymple, O. *History of the Hoosac Tunnel.* North Adams, Massachusetts, 1880.

Troy and Greenfield Railroad and Hoosac Tunnel Commissioners. *Annual Reports,* 1867–1885.

"Tunnel Engineering—A Museum Treatment." *Contributions from the Museum of History and Technology,* bulletin 240, paper 41 (1966): 208–215. Washington, D.C.: Smithsonian Institution.

Hudson-Manhattan Railroad Tunnel

"Progress of the Great Railway Tunnels under the Hudson River." *Scientific American* 63, no. 18 (November 1, 1890): 279–280.

Spielman, A., and C. Brush. "The Hudson River Tunnel." *Transactions of the American Society of Civil Engineers* 9 (July 1880): 259ff.

"The Tunnel under the Hudson River." *Scientific American* 42, no. 19 (May 8, 1880): 290f.

Gunnison Tunnel

Golze, Alfred Rudolph. *Reclamation in the United States.* New York: McGraw-Hill, 1952.

Huffman, Roy E. *Irrigation Development and Public Water Policy.* New York: Ronald Press, 1953.

Rockwell, Wilson. *New Frontier.* Denver, Colorado: World Press, 1938.

U.S. Department of the Interior, Bureau of Reclamation. *Federal Reclamation Projects: Water and Land Resource Accomplishments, 1902–1910.* Washington, D.C.: U.S. Government Printing Office, 1904–1911.

Holland Tunnel

Boardman, F. W. *Tunnels.* New York: Henry Z. Walch, 1960.

Gray, C., and H. Hagen. "The Eighth Wonder: The Holland Vehicular Tunnel." *Annual Report,* Smithsonian Institution, 1930.

Singstad, O. "A Year's Operating Experience with the Holland Vehicular Tunnel." *Engineering News Record,* December 27, 1928.

Moffat Tunnel

Beaver, Patrick. *A History of Tunnels.* Secaucus, New Jersey: Citadel Press, 1972.

Betts, C. A. "Completion of Moffat Tunnel of Colorado." *Transactions of the American Society of Civil Engineers,* paper no. 1771 (1928): 333.

Denver Municipal Facts, 5–7, 8–9, August–September 1923, September–October 1925.

Keays, R. H. "Construction Methods on the Moffat Tunnel." *Transactions of the American Society of Civil Engineers* 92 (1928): 63.

Detroit-Windsor Tunnel

"Constructing the Detroit-Windsor Tunnel." *Civil Engineering,* April 17, 1929.

"The Detroit-Canada Vehicular Tunnel." *Engineering News Record,* October 1931.

Chapter Six: Water Supply and Control

Acequias of San Antonio

Chabot, Frederick. *Indians and Missions,* 23ff. San Antonio, Texas: Naylor Printing Company, 1930.

Hutchins, Wells A. "The Community Acequia: Its Origins and Development." *The Southwestern Historical Quarterly* 31 (1927–1928): 201.

Yoakum, Henderson K. *History of Texas, from Its First Settlement in 1685 to Its Annexation to the United States in 1846,* 49ff., 122ff. New York: Redfield Company, 1855.

Bethlehem Waterworks

Baker, M. N., ed. *The Manual of American Water Works.* New York: Engineering News, 1888, 1889–1890, 1890–1891, 1897.

Blake, Nelson Manfred. *Water for the Cities.* Syracuse, New York: Syracuse University Press, 1956.

Levering, Joseph M. *A History of Bethlehem, Pennsylvania, 1741–1892.* Bethlehem, Pennsylvania: Times Publishing, 1903.

Waterman, Earle Lytton. *Elements for Water Supply Engineering.* New York: Wiley, 1934.

Philadelphia Municipal Water Supply System

Blake, Nelson Manfred. *Water for the Cities.* Syracuse, New York: Syracuse University Press, 1956.

Eberlein, H. D. "The Fairmount Park Waterworks—Philadelphia." *Architectural Record* 62 (1927): 57–67.

Fitz-Gibbon, C. "Latrobe and the Centre Square Pump House." *Architectural Record* 62 (1927): 18–22.

Hamlin, T. *Benjamin Henry Latrobe.* New York: Oxford University Press, 1955.

Keyser, C. S. *Fairmount Park.* Philadelphia: Claxton, Remsen, and Haffelfinger, 1872.

U.S. Department of the Interior, Heritage Conservation and Recreation Service, Historic American Engineering Record. *Rehabilitation: Fairmount Waterworks 1978—Conservation and Recreation in a National Historic Landmark.* Washington, D.C.: U.S. Government Printing Office, 1979.

Croton Aqueduct

Blake, Nelson Manfred. *Water for the Cities.* Syracuse, New York: Syracuse University Press, 1956.

Scientific American, July 19, 1890, 40–41.

Chicago Water Supply System

Anon. *The Great Chicago Lake Tunnel.* Chicago: Jack Wing, 1867.

Baker, M. N., ed. *Manual of American Water Works, 1888.* New York: Engineering News, 1889.

Christensen, Daphne, ed. *Chicago Public Works: A History.* Chicago: Department of Public Works, 1975.

DeBerard, W. W. *A Century of Water Service for Chicago—1852–1952.* Paper presented to the Construction Division of the American Society of Civil Engineers, Centennial Convention, Chicago, September 4, 1952.

Ericson, John. *The Water Supply System of Chicago.* Chicago: Department of Public Works, 1924.

McAlpine, William J. *Report Made to the Water Commissioners of the City of Chicago, September 26, 1851, on Supplying the City with Water.* Chicago: Seaton and Peck, 1851.

Siegel, Arthur, ed. *Chicago's Famous Buildings.* Chicago: University of Chicago Press, 1965.

"Water Supply Tunnels of Chicago." *Engineering News* 44, no. 16 (October 18, 1900).

Wilson, James G., and John Fiske, eds. *Appleton's Cyclopaedia of American Biography* 1. New York: Appleton, 1887.

Wilson, James G., and John Fiske, eds. *The National Cyclopaedia of American Biography* 9. New York: White, 1907.

Embudo Stream-Gauging Station

Frazier, Arthur H., and Wilbur Heckler. *Embudo, New Mexico, Birthplace of Systematic Stream Gaging.* Geological Survey Professional Paper 778, U.S. Department of the Interior. Washington, D.C.: U.S. Government Printing Office, 1972.

Marlette Lake Water System

Shamberger, Hugh A. *The Story of the Water Supply for the Comstock.* Geological Survey Professional Paper 779, Nevada Department of Conservation and Resources and U.S. Geological Survey. Carson City, Nevada: Nevada Historical Press, 1969.

First Owens River–Los Angeles Aqueduct

Annual Reports of the Bureau of the Los Angeles Aqueduct. Los Angeles: Department of Public Works, 1910–1912.

Complete Report on Construction of the Los Angeles Aqueduct. Los Angeles: Department of Public Service, 1916.

Heinly, B. A. "Carrying Water through a Desert: The Story of the Los Angeles Aqueduct." *National Geographic* 21, no. 7 (July 1910).

Wood, Richard. *The Owens Valley and the Los Angeles Water Controversy: Owens Valley as I Knew It.* Stockton, California: Pacific Center for Western Historical Studies, University of the Pacific, 1973.

Miami Conservancy District

Construction Plant, Methods, and Costs: Technical Report Part X. Dayton, Ohio: State of Ohio, Miami Conservancy District, 1925.

The Miami Conservancy Bulletin. Dayton, Ohio: State of Ohio, Miami Conservancy District, August 1918–December 1923.

Morgan, Arthur E. *The Miami Conservancy District.* New York: McGraw-Hill, 1951.

Chapter Seven: Environmental Engineering

Louisville Waterworks

Baker, M. N. *The Quest for Pure Water.* New York: American Water Works Association, 1949.

Fuller, G. W. "Water Works." *Transactions of the American Society of Civil Engineers* 92 (1928): 1209–1224.

Chain of Rocks

Baker, M. N. *The Quest for Pure Water.* New York: American Water Works Association, 1949.

Blake, N. M. *Water for the Cities.* Syracuse, New York: Syracuse University Press, 1956.

Fuller, G. W. "Water Works." *Transactions of the American Society of Civil Engineers* 92 (1928): 1209–1224.

Lawrence Experiment Station

Proud Heritage: A Review of Lawrence Experiment Station, Past, Present, Future. Commonwealth of Massachusetts, September 25, 1953.

Sanitas: Quarterly Publication of the Division of Sanitary Engineering 10. Boston: Massachusetts Department of Public Health, 1962.

Reversal of the Chicago River

Barrett, George F. *The Waterway from the Great Lakes to the Gulf of Mexico.* Chicago: Sanitary District of Chicago, 1926.

Chicago Public Works: Chicago, A History. Chicago: Department of Public Works, 1973.

Mayer, Harold M., and Richard C. Wade. *Chicago: Growth of a Metropolis.* Chicago: University of Chicago Press, 1969.

Williams, C. Arch. *The Sanitary District of Chicago: History of Its Growth and Development.* Chicago: Sanitary District of Chicago, 1919.

Milwaukee Metropolitan Sewage Treatment Plant

Description of Milwaukee's Activated Sludge Sewage Disposal Project. Sewerage Commission of the City of Milwaukee, 1923.

Leary, R. D., and P. Werner. *Development of a Waste Water Treatment System for the Milwaukee Metropolitan Sewerage District.* Sewerage Commission of the City of Milwaukee, Metropolitan Sewerage Commission of the County of Milwaukee, 1973.

Chapter Eight: Dams

Ascutney Mill Dam

Allen, Zachariah. *The Science of Mechanics.* Providence, Rhode Island: Hutchens and Corey, 1829.

"ASCE News." *Civil Engineering* 40, no. 7 (July 1970): 45.

Battison, Edward A. "Ascutney Gravity-Arch Mill Dam, Windsor, Vermont, 1834." *Journal of the Society for Industrial Archeology* 1, no. 1 (Summer 1975): 53–58.

Battison, Edward A. "The Windsor-Cornish Covered Bridge and the Ascutney Mill Dam." Unpublished article.

Condit, Carl. *American Building Art—The Nineteenth Century.* New York: Oxford University Press, 1960.

Druid Lake Dam

Bassell, Burr. *Earth Dams: A Study.* New York: Engineering News, 1907.

Blake, Nelson M. *Water for the Cities: A History of the Urban Water Supply Problem in the United States.* Vol. 3, Maxwell School Series. Syracuse, New York: Syracuse University Press, 1956.

Martin, Robert K. "Report of the Assistant Engineer." *Report of the Water Department, City of Baltimore.* Baltimore, January 1, 1868.

Quick, Alfred M. "The High Earth Dam Forming Druid Lake, Baltimore Water Works." *Engineering News* 47, no. 8 (February 20, 1902): 158–159.

Wegmann, Edward. *The Design and Construction of Dams.* Eighth ed. New York: Wiley, 1927.

Cheesman Dam

The Cheesman Dam. Denver Board of Water Commissioners, May 1962.

Cox, James L. *Metropolitan Water Supply: The Denver Experience.* Boulder, Colorado: Bureau of Governmental Research, University of Colorado, 1967.

Harrison, Charles. "Cheesman Dam." *Proceedings of the American Society of Civil Engineers,* May 4, 1904.

Buffalo Bill Dam

Bureau of Reclamation, U.S. Department of the Interior. *Reclamation Project Data.* Washington, D.C.: U.S. Government Printing Office, 1961.

Smith, Norman. *Man and Water: A History of Hydro-technology.* London: Peter Davies, 1975.

Warne, William E. *The Bureau of Reclamation.* New York: Praeger (Praeger Library of U.S. Government Departments and Agencies), 1973.

Theodore Roosevelt Dam and Salt River Project

Bureau of Reclamation, U.S. Department of the Interior. *Reclamation Project Data.* Washington, D.C.: U.S. Government Printing Office, 1961.

Smith, Norman. *Man and Water: A History of Hydro-technology.* London: Peter Davies, 1975.

Warne, William E. *The Bureau of Reclamation.* New York: Praeger (Praeger Library of U.S. Government Departments and Agencies), 1973.

Elephant Butte Dam

Bureau of Reclamation, U.S. Department of the Interior. *Federal Reclamation Projects: Water and Land Resource Accomplishments, 1902–1910.* Washington, D.C.: U.S. Government Printing Office, 1904–1911.

Bureau of Reclamation, U.S. Department of the Interior. *Reclamation Project Data.* Washington, D.C.: U.S. Government Printing Office, 1961.

Golze, Alfred Rudolph. *Reclamation in the United States.* New York: McGraw-Hill, 1952.

Huffman, Roy E. *Irrigation Development and Public Water Policy.* New York: Ronald Press, 1953.

Rockwell, Wilson. *New Frontier.* Denver, Colorado: World Press, 1938.

Smith, Norman. *Man and Water: A History of Hydro-technology.* London: Peter Davies, 1975.

Warne, William E. *The Bureau of Reclamation.* New York: Praeger (Praeger Library of U.S. Government Departments and Agencies), 1973.

Chapter Nine: Buildings

Castillo de San Marcos

"Fort Marion, St. Augustine, Florida." *American Architect* 146, no. 2631 (March 1935): 33–40.

Manucy, Albert C. *The Building of Castillo de San Marcos.* Washington, D.C.: U.S. Government Printing Office, 1942.

"Refortification." *Historic Preservation* 10, no. 2 (1958): 50–53.

Iron Building of the Watervliet Arsenal

Condit, Carl W. *American Building Art—The Nineteenth Century.* New York: Oxford University Press, 1960.

Thomas, Selma. "Cast-Iron Storehouse, 1859." In Robert M. Vogel (ed.), *A Report of the Mohawk-Hudson Area Survey.* Washington, D.C.: Smithsonian Institution.

William Ward House

Boorstin, Daniel J., ed. *The Chicago History of American Civilization.* Chicago: University of Chicago Press, 1968.

Condit, Carl. *American Building Art—The Nineteenth Century.* New York: Oxford University Press, 1960.

Kramer, Ellen W., and Aly A. Raafat. "The Ward House: A Pioneer Structure of Reinforced Concrete." *Journal of the Society of Architectural Historians* 20, no. 1 (March 1961): 34–37.

Raafat, Aly Ahmed. *Reinforced Concrete in Architecture.* New York: Reinhold, 1958.

Ward, William E. "Beton in Combination with Iron as a Building Material." *American Society of Mechanical Engineers Transactions* 4, 1883.

Wight, Peter B. "Concrete as a Building Material—A Remarkable House at Port Chester." *American Architect and Building News* 2, no. 86 (August 18, 1877): 266–267.

Wight, Peter B. "The Pioneer Concrete Residence of America." *Architectural Record* 25, no. 5 (May 1909): 359–363.

Union Station, St. Louis

"Romanesque Reused." *Civil Engineering,* September 1985, 41ff.

Peavy-Haglin Grain Elevator

Heffelfinger, Ruth. "A Pioneer Venture in Grain Storage." *Minnesota History* 37, no. 1 (March 1960): 14.

Riley, Robert B. "Grain Elevators: Symbols of Time, Place, and Honest Building." *AIA Journal* 66, no. 12 (1977): 50–55.

Ingalls Building

Boorstin, Daniel J., ed. *The Chicago History of American Civilization.* Chicago: University of Chicago Press, 1968.

Condit, Carl. *American Building Art—The Nineteenth Century.* New York: Oxford University Press, 1960.

Condit, Carl. "The First Reinforced-Concrete Skyscraper." *Technology and Culture* 9, no. 1 (January 1968): 1–33.

Diehl, John A. "Landmarks in Reinforced Concrete." *Civil Engineering* 45, no. 11 (November 1975): 87–89.

Elzner, A. O. "The First Concrete Skyscraper." *Architectural Record* 15, no. 6 (June 1904): 531–544.

"The Tallest Ferro-Concrete Building in the World." *Engineering News,* July 30, 1903.

Goodyear Airdock

"Saving the Largest Landmark." *Preservation News,* March 1974.

Watson, Wilbur. "Building the World's Largest Airship Factory and Dock." Wilbur Watson and Associates, Cleveland, Ohio, 1929.

Chapter Ten: Urban Planning

City Plan of Savannah

Bannister, Tupin C. "Oglethorpe's Sources for the Savannah Plan." *Journal of the Society of Architectural Historians* 20 (May 1961): 47–62.

Federal Writers Program. *Georgia.* Athens, Georgia: University of Georgia Press, 1940.

Lane, Mills. *Savannah Revisited: A Pictorial History.* Savannah, Georgia: Beehive Press, 1973.

Stevenson, Frederick, and Carl Feiss. "Charleston and Savannah." *Journal of the Society of Architectural Historians* 10 (December 1951): 3–9.

Kansas City Park and Boulevard System

Board of Park Commissioners of Kansas City. *The Park and Boulevard System of Kansas City.* Kansas City, Missouri, 1914.

Proceedings of the Ninth National Conference on City Planning, Kansas City, Missouri, May 7–9, 1917, pp. 106–116. New York, 1917.

Wilson, William M. *The City Beautiful Movement in Kansas City.* Columbia, Missouri: University of Missouri Press, 1964.

Charles River Basin

Barrett, R. E. "A Resume of the Charles River Basin Project." *Harvard Engineering Journal* 5, no. 4 (January 1907).

Holmes, A. "Some Piledriving Experiments in Connection with the Construction of the Charles River Dam." *Harvard Engineering Journal* 6, no. 3 (November 1907).

Rogers, M. T. "Closing the Charles River Dam." *Harvard Engineering Journal* 7, no. 3 (November 1908).

Chapter Eleven: Power Systems

Great Falls Raceway and Power System

Fries, Russel I. "European vs. American Engineering: Pierre Charles L'Enfant and the Water Power System of Paterson, N.J." *Northeast Historical Archaeology* 4, nos. 1 and 2 (Spring 1975): 68–96.

Herrick, C. M. *A History of Industrial Paterson.* Paterson, New Jersey, 1882.

Lowell Water Power System

Francis, James B. *Lowell Hydraulics Experiments, Being a Selection from Experiments on Hydraulic Motors, on the Flow of Water over Weirs, in Open Canals of Uniform Rectangular Cross Section and Through Submerged Orifices and Diverging Tubes, Made at Lowell, Massachusetts.* 2d ed. New York: Van Nostrand, 1868.

Fitz-Gerald, Desmond, Joseph P. Davis, and John R. Freeman. "James Bicheno Francis, A Memoir." *Journal of the Association of Engineering Societies* 13, no. 1 (January 1894): 1–9.

Hunter, Louis. *Waterpower: A History of Industrial Power in the United States, 1780–1930.* Charlottesville, Virginia: University Press of Virginia, 1979.

Malone, Patrick M. *Canals and Industry Engineering at Lowell, 1821–1880.* Lowell, Massachusetts: Lowell Museum, 1983.

Pelton Impulse Waterwheel

American Society of Civil Engineers. *Transactions* 36, no. 788 (1896).

Durand, W. F. "The Pelton Water Wheel." *Mechanical Engineering,* June–July 1939.

Gravander, Axel. "Pelton Wheel Draws Notice in Steel Magazine." *Grass Valley Union,* Nevada City, California, August 13, 1955.

Journal of the Franklin Institute 140, no. 3 (September 1895).

Mining and Scientific Press. New York: Dewey and Co., Publishers, and U.S. and Foreign Patent Agenda, November 6, 1880, and April 23, 1881.

National Cyclopedia of American Biography, vol. 13, p. 602. New York: James T. White and Co., 1906.

Smith, Norman. *Man and Water: A History of Hydro-Technology.* London: Peter Davies, 1975.

Vulcan Street Plant

Condit, Carl. *American Building Art—The Nineteenth Century.* New York: Oxford University Press, 1960.

Edison Electric Company Bulletin 13 (August 28, 1882), 15 (December 20, 1882), 17 (April 6, 1883). New York: Edison Electric Company.

Reid, A. J. *Illustrated Annual Review of the Appleton Post, Devoted to the City of Appleton, Wisconsin: Its Water Power and Industries.* Appleton: Post Publishing Co., 1879.

Stover, Frances. "First Hydroelectric Plant." *Milwaukee Journal,* May 23, 1953.

Wisconsin Michigan Power Co. *Vulcan Street Plant.* Commemorative booklet. Appleton, Wisconsin, September 15, 1977.

Folsom Hydroelectric Power System

Coleman, Charles M. *PG&E of California,* 116–127. New York: McGraw-Hill, 1952.

Low, George. "Folsom-Sacramento Electric Power Transmission." *Journal of Electricity* 1, no. 3 (1896): 1–14.

Sault Ste. Marie Hydroelectric Power Complex

Reynolds, T., and R. Wilson. "Michigan–Lake Superior Power Company: Hydroelectric Plant and Power Canal." *Historic American Engineering Record,* MI-1, 1978.

Chapter Twelve: Surveying and Mapping

Mason-Dixon Line

"Astronomy and Geology as Applied By Mason and Dixon." *Virginia Journal of Science* 19 (1968): 172.

Clarke, H. W. "Report of the Regents Boundary Commission upon the New York and Pennsylvania Boundary." Albany, New York: Weed, Parsons and Company, 1886.

Cope, Thomas D. "The Apprentice Years of Mason and Dixon." *Pennsylvania History,* July 1944.

Cope, Thomas D. "Westward Five Degrees Longitude." *Proceedings of the Pennsylvania Academy of Science,* 1948.

Cope, Thomas D. "Mason and Dixon: English Men of Science." *Delaware Notes,* 1949.

Cummings, H. M. *The Mason-Dixon Line—Story for a Bicentenary, 1763–1963.* Harrisburg: Commonwealth of Pennsylvania, Department of Internal Affairs, 1962.

Encyclopedia Britannica, 11th ed., s.v. "Mason and Dixon Line."

Johnston, William F. *Pennsylvania Boundaries: Reports of the Joint Commissioners and of Colonel Graham, U.S. Engineers.* Harrisburg, Pennsylvania: J. M. G. Lecure, 1886.

Latrobe, John H. B. "An Annual Discourse Delivered before the Pennsylvania Historical Society." Philadelphia: Carey, Lea and Carey, 1828.

Ellicott's Stone

A Biographical Dictionary of American Civil Engineers. New York: American Society of Civil Engineers, 1972.

Hamilton, Peter J. *Colonial Mobile.* Boston: Houghton Mifflin, 1897.

Massachusetts Base Line

Borden, Simeon. "Account of a Trigonometrical Survey of Massachusetts." *Transactions of the American Philosophical Society at Philadelphia* 9 (1846): 33–91.

"National Historic Civil Engineering Landmark Nomination Form." History and Heritage Committee, Boston Society of Civil Engineers, 1981.

Solander, Arvo. "Massachusetts Base Line." Boston: Massachusetts Association of Land Surveyors and Civil Engineers, September 1970.

International Boundary Marker No. 1

Encyclopedia Americana, 11th ed., s.v. "Emory, William Hemsley."

"Notes on U.S.-Mexico Relations." St. James Rectory, Clovis, New Mexico, August 1950.

Chapter Thirteen: Coastal Facilities

Naval Dry Docks at Boston and Norfolk

Armstrong, Ellis L., ed. *History of Public Works in the United States, 1776–1976.* Chicago: American Public Works Association, 1976.

"Dry Docks at Charlestown." *American Magazine of Useful and Entertaining Knowledge* 1 (Boston 1835): 298–300.

Lull, Edward P. *History of the United States Navy Yard at Gosport, Virginia.* Washington, D.C.: United States Navy, 1874.

Stuart, Charles B. *The Naval Dry Docks of the United States.* New York: Van Nostrand, 1870.

Vose, George L. *A Sketch of the Life and Works of Loammi Baldwin.* Boston: Geo. H. Ellis, 1885.

Minot's Ledge Lighthouse

Holland, Francis R. *America's Lighthouses.* Brattleboro, Vermont: Steven Greene Press, 1981.

South Pass Navigation Works

Corthell, E. L. "The South Pass Jetties." *Transactions of the American Society of Civil Engineers* 7 (June 1878): 130–158.

Corthell, E. L. "The South Pass Jetties—Ten Years' Practical Teaching in River and Harbor Hydraulics." *Transactions of the American Society of Civil Engineers* 8 (October 1884): 313–330.

Schmidt, Max E. "The South Pass Jetties." *Transactions of the American Society of Civil Engineers* 8 (August 1879): 189–226.

Chapter Fourteen: Airports

Cleveland-Hopkins International Airport

Armstrong, Ellis L., ed. *History of Public Works in the United States, 1776–1976.* Chicago: American Public Works Association, 1976.

"The Development and Operation of Cleveland Airport." *Airways Age,* August 1928, pp. 14–17.

Division of Information and Complaints, City of Cleveland. "Cleveland's Airport." November 1, 1929.

"Hopkins-International Airport, Fiftieth Anniversary History." Cleveland: Cleveland-Hopkins International Airport, 1985.

Wood, John W. *Airports: Some Elements of Design and Future Development.* New York: Coward-McCann, 1940.

Newark International Airport

"Airports in Pictures." *Airports Magazine,* October 1928, p. 25. Washington, New Jersey.

"Airports in Pictures—120 Ft. Trusses for Newark Hangar." *Airports Magazine,* November 1928, pp. 26–27. Washington, New Jersey.

Armstrong, A. H. "Newark Airport History." New York: Port Authority of New York, April 1, 1955.

Armstrong, Ellis L., ed. *History of Public Works in the United States, 1776–1976.* Chicago: American Public Works Association, 1976.

"Newark's New Airport for the Metropolitan Area." *New York Times,* February 19, 1928, Aviation Section, p. 16.

Port of New York Authority. "History of Newark Airport." 1959.

Port of New York Authority. "Fact Sheets: Newark Airport; New Control Tower, Newark Airport; Instrument Runway (4-22), Newark Airport." October 1960.

"Variety Features Airport Construction at Newark, N.J." *Construction Methods,* November 1928, pp. 26–27.

Wood, John W. *Airports: Some Elements of Design and Future Development.* New York: Coward-McCann, 1940.

Index